Advances in Measurement Systems

Advances in Measurement Systems

Editor:

Harinirina Randrianarisoa, Ph.D.

ArclerPress

www.arclerpress.com

Advances in Measurement Systems

Editor: Harinirina Randrianarisoa, Ph.D.

708 3rd Avenue, 6th Floor
New York
NY 10017
United States of America
www.arclerpress.com

ISBN: 978-1-68094-366-5

Library of Congress Control Number: 2016950943

About the Editor

Harinirina Randrianarisoa, Ph.D.

Hari Randrianarisoa obtained is Ph.D from Institute of Electronics, Microelectronics and Nanotechnology (IEMN) of the University of Science and Technology – Lille (France). His thesis focused for designing and manufacturing of thermal micro sensors. His interests span from on renewable energy especially solar energy. He has a strong knowledge in technology transfer field. He is currently working as a Business advisor for economic development organization in Québec City and as a consultant Scientist advisor.

Contents

List of Contributors

Diogo Ribeiro
Polytechnic of Porto, School of Engineering, Porto, Portugal

Rui Calçada
University of Porto, Faculty of Engineering, Porto, Portugal

José Ferreira
Enermeter - Sistemas de Medição, Braga, Portugal

Teresa Martins
Enermeter - Sistemas de Medição, Braga, Portugal

William P. Fisher Jr.
LivingCapitalMetrics.com, Sausalito, CA, USA
BEAR Center, Graduate School of Education, University of California, Berkeley, CA, USA

A. Jackson Stenner
MetaMetrics, Inc., Durham, NC, USA
School of Education, University of North Carolina, Chapel Hill, NC, USA

Ken Krechmer
757 Greer Road, Palo Alto, CA 94303, USA

Roberto Di Marco
Department of Mechanical and Aerospace Engineering, 'Sapienza' University of Rome, via Eudossiana, 18, Roma 00184, Italy
Department of Mechanical Engineering, The University of Sheffield, Sir Frederick Mappin Building, Mappin Street, Sheffield S1 3JD, UK

Stefano Rossi
Department of Economics and Management, Industrial Engineering (DEIM), University of Tuscia, Via del Paradiso, 47, 01100 Viterbo, Italy

Enrico Castelli
Laboratory of Movement Analysis and Robotics (MARlab), Children Hospital 'Bambino Gesù', via della torre di Palidoro, 00050 Passoscuro, Rome, Italy

Fabrizio Patanè
School of Mechanical Engineering, 'Niccolò Cusano' University, Via Don Carlo Gnocchi, 3, 00166 Rome, Italy

Claudia Mazzà
Department of Mechanical Engineering, The University of Sheffield, Sir Frederick Mappin Building, Mappin Street, Sheffield S1 3JD, UK
INSIGNEO Institute for In Silico Medicine, The University of Sheffield, The Pam Liversidge Building, Sir Frederick Mappin Building, Mappin Street, Sheffield S13JD, UK

Paolo Cappa
Department of Mechanical and Aerospace Engineering, 'Sapienza' University of Rome, via Eudossiana, 18, Roma 00184, Italy
Laboratory of Movement Analysis and Robotics (MARlab), Children Hospital 'Bambino Gesù', via della torre di Palidoro, 00050 Passoscuro, Rome, Italy

Barouch Giechaskiel
Sustainable Transport Unit, Joint Research Centre, Institute for Energy and Transport, European Commission, Varese, Italy

Francesco Riccobono
Sustainable Transport Unit, Joint Research Centre, Institute for Energy and Transport, European Commission, Varese, Italy

Theodoros Vlachos
Sustainable Transport Unit, Joint Research Centre, Institute for Energy and Transport, European Commission, Varese, Italy

Pablo Mendoza-Villafuerte
Sustainable Transport Unit, Joint Research Centre, Institute for Energy and Transport, European Commission, Varese, Italy

Ricardo Suarez-Bertoa
Sustainable Transport Unit, Joint Research Centre, Institute for Energy and Transport, European Commission, Varese, Italy

Georgios Fontaras
Sustainable Transport Unit, Joint Research Centre, Institute for Energy and Transport, European Commission, Varese, Italy

Pierre Bonnel
Sustainable Transport Unit, Joint Research Centre, Institute for Energy and Transport, European Commission, Varese, Italy

Martin Weiss
Sustainable Transport Unit, Joint Research Centre, Institute for Energy and Transport, European Commission, Varese, Italy

Georgios Charalampous
Department of Mechanical Engineering, Imperial College London, London SW7 2AZ, UK

Constantinos Hadjiyiannis
Department of Mechanical Engineering, Imperial College London, London SW7 2AZ, UK

Yannis Hardalupas
Department of Mechanical Engineering, Imperial College London, London SW7 2AZ, UK

Anthony Simm
School of Engineering and Computing Sciences, Durham University, United Kingdom

Qing Wang
School of Engineering and Computing Sciences, Durham University, United Kingdom

Songling Huang
State Key Laboratory of Power Systems, Department of Electrical Engineering, Tsinghua University, Beijing, China

Wei Zhao
State Key Laboratory of Power Systems, Department of Electrical Engineering, Tsinghua University, Beijing, China

Martinus de Kanter
Chair for Laser Technology, RWTH Aachen University, Aachen, Germany

Julian Meyer-Kirschner
Aachener Verfahrenstechnik – Process Systems Engineering, RWTH Aachen University, Aachen, Germany

Jörn Viell
Aachener Verfahrenstechnik – Process Systems Engineering, RWTH Aachen University, Aachen, Germany

Alexander Mitsos
Aachener Verfahrenstechnik – Process Systems Engineering, RWTH Aachen University, Aachen, Germany

Michael Kather
Institute of Technical and Macromolecular Chemistry, RWTH Aachen University and DWI Leibniz Institute for Interactive Materials e.V., Aachen, Germany

Andrij Pich
Institute of Technical and Macromolecular Chemistry, RWTH Aachen University and DWI Leibniz Institute for Interactive Materials e.V., Aachen, Germany

Christoph Janzen
Fraunhofer Institute for Laser Technology (ILT), Aachen, Germany

Johan Nohlert
Department of Signals and Systems, Chalmers University of Technology, SE-41296 Göteborg, Sweden

Livia Cerullo
Department of Signals and Systems, Chalmers University of Technology, SE-41296 Göteborg, Sweden

Johan Winges
Department of Signals and Systems, Chalmers University of Technology, SE-41296 Göteborg, Sweden

Thomas Rylander
Department of Signals and Systems, Chalmers University of Technology, SE-41296 Göteborg, Sweden

Tomas McKelvey
Department of Signals and Systems, Chalmers University of Technology, SE-41296 Göteborg, Sweden

Anders Holmgren
AstraZeneca R&D, Pharmaceutical Development, SE-431 83 Mölndal, Sweden

Lubomir Gradinarsky b,
AstraZeneca R&D, Pharmaceutical Development, SE-431 83 Mölndal, Sweden

Staffan Folestad
AstraZeneca R&D, Pharmaceutical Development, SE-431 83 Mölndal, Sweden

Mats Viberg
Department of Signals and Systems, Chalmers University of Technology, SE-41296 Göteborg, Sweden

Anders Rasmuson
Department of Chemical and Biological Engineering, Chalmers University of Technology, SE-41296 Göteborg, Sweden

Jeremy Talbot
School of Engineering and Computing Sciences, Durham University, DH1 3LE, UK

Qing Wang
School of Engineering and Computing Sciences, Durham University, DH1 3LE, UK

Neil Brady
Nikon Metrology UK Ltd, Staffordshire B77 5ES, UK

Roger Holden
Nikon Metrology UK Ltd, Staffordshire B77 5ES, UK

Pablo Cotera
Center for Engineering and Industrial Development - CIDESI, Querétaro, Querétaro, Mexico

Miguel Velazquez
Center for Engineering and Industrial Development - CIDESI, Querétaro, Querétaro, Mexico

David Cruz
Center for Engineering and Industrial Development - CIDESI, Querétaro, Querétaro, Mexico

Luis Medina
Center for Engineering and Industrial Development - CIDESI, Querétaro, Querétaro, Mexico

Manuel Bandala
Center for Engineering and Industrial Development - CIDESI, Querétaro, Querétaro, Mexico

Muhammad Asif Arain
Mobile Robotics & Olfaction Lab, Center of Applied Autonomous Sensor Systems (AASS), School of Science & Technology, Örebro University, Örebro SE-70182, Sweden

Marco Trincavelli
Mobile Robotics & Olfaction Lab, Center of Applied Autonomous Sensor Systems (AASS), School of Science & Technology, Örebro University, Örebro SE-70182, Sweden

Marcello Cirillo
Mobile Robotics & Olfaction Lab, Center of Applied Autonomous Sensor Systems (AASS), School of Science & Technology, Örebro University, Örebro SE-70182, Sweden

Erik Schaffernicht
Mobile Robotics & Olfaction Lab, Center of Applied Autonomous Sensor Systems (AASS), School of Science & Technology, Örebro University, Örebro SE-70182, Sweden

Achim J. Lilienthal
Mobile Robotics & Olfaction Lab, Center of Applied Autonomous Sensor Systems (AASS), School of Science & Technology, Örebro University, Örebro SE-70182, Sweden

Hiroshi Murakami
Department of Mechanical Systems Engineering, Faculty of Environmental Engineering, The University of Kitakyushu, 1-1 Hibikino, Wakamatsu-ku, Kitakyushu, Fukuoka 808-0135, Japan

Akio Katsuki
Department of Mechanical Engineering, Faculty of Engineering, Kyushu University, 744, Motooka, Nishi-ku, Fukuoka 819-0395, Japan

Takao Sajima
Department of Mechanical Engineering, Faculty of Engineering, Kyushu University, 744, Motooka, Nishi-ku, Fukuoka 819-0395, Japan

Mitsuyoshi Fukuda
Department of Mechanical Systems Engineering, Faculty of Environmental Engineering, The University of Kitakyushu, 1-1 Hibikino, Wakamatsu-ku, Kitakyushu, Fukuoka 808-0135, Japan

Libo Zhao
State Key Laboratory for Manufacturing Systems Engineering, Collaborative Innovation Center of Suzhou Nano

Yingjie Hu
State Key Laboratory for Manufacturing Systems Engineering, Collaborative Innovation Center of Suzhou Nano

Tongdong Wang
State Key Laboratory for Manufacturing Systems Engineering, Collaborative Innovation Center of Suzhou Nano

Jianjun Ding
State Key Laboratory for Manufacturing Systems Engineering, Collaborative Innovation Center of Suzhou Nano

Xixiang Liu
State Key Laboratory for Manufacturing Systems Engineering, Collaborative Innovation Center of Suzhou Nano

Yulong Zhao
State Key Laboratory for Manufacturing Systems Engineering, Collaborative Innovation Center of Suzhou Nano

Zhuangde Jiang
State Key Laboratory for Manufacturing Systems Engineering, Collaborative Innovation Center of Suzhou Nano

Xiyuan Chen
School of Instrument Science and Engineering, Southeast University, Nanjing 210096, China
Key Laboratory of Micro-Inertial Instrument and Advanced Navigation Technology Ministry of Education, Nanjing 210096, China

Hong Zhang
School of Instrument Science and Engineering, Southeast University, Nanjing 210096, China
Key Laboratory of Micro-Inertial Instrument and Advanced Navigation Technology Ministry of Education, Nanjing 210096, China

Sheng Zou
School of Instrument Science and Engineering, Southeast University, Nanjing 210096, China
Key Laboratory of Micro-Inertial Instrument and Advanced Navigation Technology Ministry of Education, Nanjing 210096, China

Francisco J. Azcona
Centre for Sensors, Instruments and Systems Development (CD6)-UPC BarcelonaTech, Rambla St. Nebridi 10, Terrassa E-08222, Spain

Ajit Jha
Centre for Sensors, Instruments and Systems Development (CD6)-UPC BarcelonaTech, Rambla St. Nebridi 10, Terrassa E-08222, Spain

Carlos Yáñez
Centre for Sensors, Instruments and Systems Development (CD6)-UPC BarcelonaTech, Rambla St. Nebridi 10, Terrassa E-08222, Spain

Reza Atashkhooei
Centre for Sensors, Instruments and Systems Development (CD6)-UPC BarcelonaTech, Rambla St. Nebridi 10, Terrassa E-08222, Spain

Santiago Royo
Centre for Sensors, Instruments and Systems Development (CD6)-UPC BarcelonaTech, Rambla St. Nebridi 10, Terrassa E-08222, Spain

Stefan Tenbohlen
Institute of Power Transmission and High Voltage Technology, University of StutFtgart, Stuttgart 70569, Germany

Sebastian Coenen
Institute of Power Transmission and High Voltage Technology, University of StutFtgart, Stuttgart 70569, Germany

Mohammad Djamali
Institute of Power Transmission and High Voltage Technology, University of StutFtgart, Stuttgart 70569, Germany

Andreas Müller
TÜV Süd Energietechnik, Filderstadt 70794, Germany

Mohammad Hamed Samimi
Institute of Power Transmission and High Voltage Technology, University of StutFtgart, Stuttgart 70569, Germany

Martin Siegel
Institute of Power Transmission and High Voltage Technology, University of StutFtgart, Stuttgart 70569, Germany

José Manuel Andújar Márquez
Escuela Técnica Superior de Ingeniería, Universidad de Huelva, Ctra. Palos de la Ftra.-Huelva s/n, Huelva 21819, Spain

Miguel Ángel Martínez Bohórquez
Escuela Técnica Superior de Ingeniería, Universidad de Huelva, Ctra. Palos de la Ftra.-Huelva s/n, Huelva 21819, Spain

Sergio Gómez Melgar
Escuela Técnica Superior de Ingeniería, Universidad de Huelva, Ctra. Palos de la Ftra.-Huelva s/n, Huelva 21819, Spain

Elvira Baygildina
Department of Electrical Engineering, Lappeenranta University of Technology, P.O. Box 20, Lappeenranta FI-53851, Finland

Liudmila Smirnova
Department of Electrical Engineering, Lappeenranta University of Technology, P.O. Box 20, Lappeenranta FI-53851, Finland

Kirill Murashko
Department of Electrical Engineering, Lappeenranta University of Technology, P.O. Box 20, Lappeenranta FI-53851, Finland

Raimo Juntunen
Department of Electrical Engineering, Lappeenranta University of Technology, P.O. Box 20, Lappeenranta FI-53851, Finland

Andrey Mityakov
Department of Electrical Engineering, Lappeenranta University of Technology, P.O. Box 20, Lappeenranta FI-53851, Finland

Mikko Kuisma
Department of Electrical Engineering, Lappeenranta University of Technology, P.O. Box 20, Lappeenranta FI-53851, Finland

Olli Pyrhönen
Department of Electrical Engineering, Lappeenranta University of Technology, P.O. Box 20, Lappeenranta FI-53851, Finland

Pasi Peltoniemi
Department of Electrical Engineering, Lappeenranta University of Technology, P.O. Box 20, Lappeenranta FI-53851, Finland

Katja Hynynen
Department of Electrical Engineering, Lappeenranta University of Technology, P.O. Box 20, Lappeenranta FI-53851, Finland

Vladimir Mityakov
Department of Thermodynamics and Heat Transfer, Saint-Petersburg State Polytechnical University, Polytechnicheskaya 29, Saint-Petersburg 195251, Russia

Sergey Sapozhnikov
Department of Thermodynamics and Heat Transfer, Saint-Petersburg State Polytechnical University, Polytechnicheskaya 29, Saint-Petersburg 195251, Russia

Huajun Li
State Key Laboratory of Industrial Control Technology, College of Control Science and Engineering, Zhejiang University, Hangzhou 310027, China

Haifeng Ji
State Key Laboratory of Industrial Control Technology, College of Control Science and Engineering, Zhejiang University, Hangzhou 310027, China

Zhiyao Huang
State Key Laboratory of Industrial Control Technology, College of Control Science and Engineering, Zhejiang University, Hangzhou 310027, China

Baoliang Wang
State Key Laboratory of Industrial Control Technology, College of Control Science and Engineering, Zhejiang University, Hangzhou 310027, China

Haiqing Li
State Key Laboratory of Industrial Control Technology, College of Control Science and Engineering, Zhejiang University, Hangzhou 310027, China

Guohua Wu
Key Laboratory of Complex Systems Modeling and Simulation of Ministry of Education, Hangzhou Dianzi University, Hangzhou 310018, China

Preface

Measurement systems include all components in a chain of hardware and software that leads from the measured variable to process data. New measurement techniques and technologies will stimulate innovation in products, processes and services, and also help to develop new skills in the workforce. With regards to measurement systems research and development, the advances discussed in this book provide new research ideas and new techniques for measurement systems in various fields as in nanotechnology, in robotic, in electricity, in optical systems and others. This book described the design and the concept of the monitoring, the control of processes and operations. Many different aspects of this field are explored in a vast number of different ways. The experimental engineering analysis is also discussed to check behaviour of the systems and to validate theoretically predict.

Editor
Harinirina Randrianarisoa, Ph.D.

CHAPTER 1

MEASUREMENT OF THE DYNAMIC DISPLACEMENTS OF RAILWAY BRIDGES USING VIDEO TECHNOLOGY

Diogo Ribeiro[1], Rui Calçada[2], José Ferreira[3] and Teresa Martins[3]

[1] Polytechnic of Porto, School of Engineering, Porto, Portugal

[2] University of Porto, Faculty of Engineering, Porto, Portugal

[3] Enermeter - Sistemas de Medição, Braga, Portugal

ABSTRACT

This article describes the development of a non-contact dynamic displacement measurement system for railway bridges based on video technology. The system, consisting of a high speed video camera, an optical lens, lighting lamps and a precision target, can perform measurements with high precision for distances from the camera to the target up to 25 m, with acquisition frame rates ranging from 64 fps to 500 fps, and be articulated with other measurement systems, which promotes its integration in structural health monitoring systems. The system's performance was evaluated based on two tests, one in the laboratory and other on the field. The laboratory test evaluated the performance of the system in measuring the displacement of a steel beam, subjected to a point load applied dynamically, for distances from the camera to the target between 3 m and 15 m. The field test allowed evaluating the system's performance in the dynamic measurement of the displacement of a point on the deck of a railway bridge, induced by passing trains at speeds between 160 km/h and 180 km/h, for distances from the camera to the target up to 25 m. The results of both tests show a very good agreement between the displacement measurement obtained with the video system and with a LVDT.

INTRODUCTION

The monitoring of the structural integrity of bridges is a current and relevant topic of research in structural engineering, motivated by the need of the owners and managers of infrastructures to reduce the costs of inspection and maintenance, while ensuring that the structural behavior and the security level are appropriate throughout the bridge's useful life [1-4]. This monitoring is based on the continuous or periodical acquisition of an appropriate set of relevant measures of the structural response based on a network of transducers.

For railway bridges, one of the most relevant measures of the structural response is the dynamic displacement of the structure due to the passage of trains [5-6]. The displacement measurement systems can be classified as contact and non-contact measurement systems [2-3, 5].

The contact systems generally encompass high precision displacement transducers, such as LVDT, cable type, etc. The main disadvantage of these systems is their dependence on a

Citation: Diogo Ribeiro, Rui Calçada, José Ferreira and Teresa Martins, "Measurement of the Dynamic Displacements of Railway Bridges Using Video Technology," DOI: http://dx.doi.org/10.1051/matecconf/20152402007.

reference because they are based on the relative displacement measurement between a point of the structure and a close point that can be considered fixed. This fixed point can, in many instances, be difficult, or even impossible, to materialize as, for example, when the structure is located over a watercourse, a road or in a rugged terrain, etc.

The non-contact measurement systems rely on laser technology [7], radar technology [8], GPS technology [9] and on the processing and analysis of images obtained by video technology [3-4, 6, 10] or digital photography [1, 11].

The application of video technology based systems in the measurement of dynamic displacement of bridges due to high speed rail traffic is only briefly reported in the literature. The study carried out by Lee et al. [5] on the measurements of a railway bridge of the Korean high speed railway network was the only one identified.

Systems based on image processing have the advantage of establishing a good compromise between the acquisition frame rate and the resolution. These systems usually include a video camera, an optical lens, lighting lamps and a precision target fixed to the structure. Typical resolutions obtained in displacement measurements of bridges are within 0.1 mm, for distances of 10 m between the video camera and target, and 1.0 mm, for a distance of 100 m [9]. These systems have been used in static and quasi-static measurements for the performance of load tests on bridges [1-2, 4, 11-12]. More recently Bell et al. [1] and Malesa et al. [6] applied a technique, often referred as Digital Image Correlation (DIC), that doesn't require a target because is based on the tracking of a predefined pattern or texture of the structure.

This article describes the development of a noncontact dynamic displacement measurement system based on video technology. The measurement system consists of a high speed video camera, an optical lens, lighting lamps and a precision target. The system has the following main advantages: versatility, as it can easily be articulated with other measuring systems, it can be programmed for acquisition frame rates from 64 fps to 500 fps, and it can perform dynamic displacement measurements for distances from the camera to the target up to 25 m with high precision.

The system's performance was evaluated based on two tests, one in the laboratory and other on the field. The laboratory test evaluated the performance of the system in measuring the displacement of a steel beam, subjected to a point load applied dynamically, for distances from the camera to the target between 3 m and 15 m. The field test allowed evaluating the system's performance in the dynamic measurement of the displacement of a point on the deck of a railway bridge, induced by passing trains at speeds between 160 km/h and 180 km/h, for distances from the camera to the target up to 25 m.

MEASUREMENT SYSTEM BASED ON VIDEO TECHNOLOGY

Architecture

The measurement system based on video technology comprises the components presented in Figure 1.

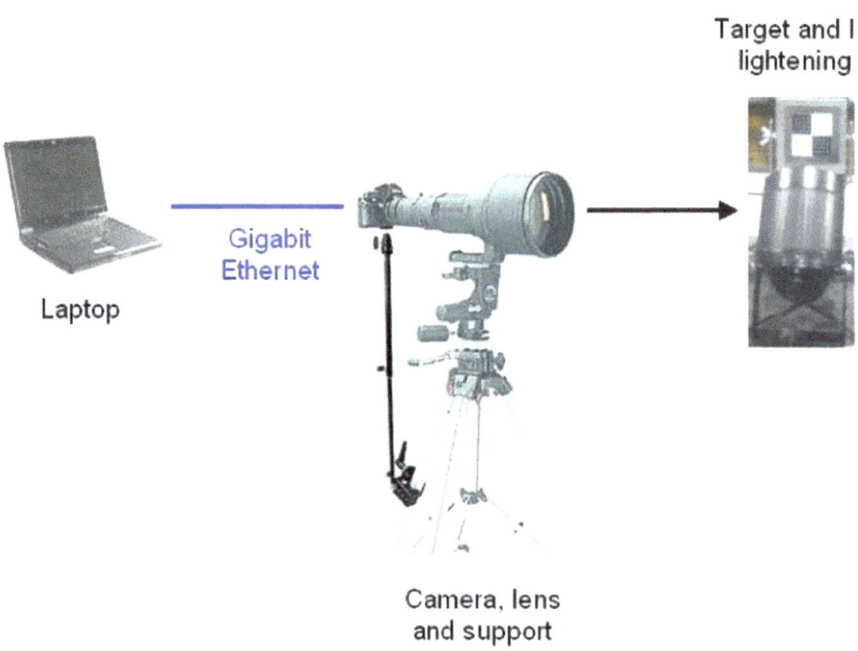

Figure 1. Measurement system based on video technology.

Table 1 shows the characteristics of the different components of the system.

The main component of the system encompasses the high speed video camera, the lens and the magnification elements. The GENIE video camera, model H1400, has a maximum resolution of 1400×1024 pixels and an adjustable acquisition frame rate up to 500 fps. This camera presents the highest sensitivity in wavelengths between 600 nm and 700 nm, located in the area of the near-infrared. It is equipped with a Nikon lens model 80- 400 VR F/4.5-5.6D ED. The magnification elements (duplicator and quadruplicator) enable to increase the depth of the field of view, and are useful in situations involving high distances between the video camera and the object.

Table 1. Technical characteristics of the systems' components

Component	Brand / Model	Technical characteristics
Laptop computer	Clevo / D901C 	Operating system: Windows Pro X64 CPU: Intel Core 2 Quad RAM: 4 Gb DDR3 Discs: 2 × SATA 320 Gb 7200 rpm 16 Mb Ethernet Board: Intel Pro 1000
Video camera	Genie / HM1400 	Maximum resolution: 1400 × 1024 Frequency at maximum resolution: 64 fps Pixel size: 7.4μm Sensor: CMOS Color: Monochrome Output: Gigabit Ethernet Lens mount: Type C + F-Mount adaptator Dimensions: 44×29×67 mm Mass: 115 g Sensitivity to visible and near IR (400 a 900 nm)
Lens	Nikon / 80-400 VR F4.5-5.6D ED 	Zoom: 80 to 400 mm Luminosity: F 4.5-5.6 Manual zoom and luminosity control Minimum focus distance: 2.3 m Dimensions: 91 × 171 mm^2 Weight: 1.36 kg
Magnification elements	Duplicator 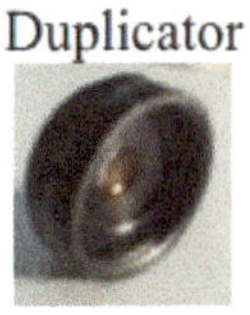	Magnification: 2.0x extender Mount: CS-Mount Mounting threads: (2) M3 Position: between lens and video camera Outer diameter: 28 mm Length: 12 mm
	Quadruplicator 	Magnification: 4.0x extender Mount: CS-Mount Mounting threads: (2) M3 Position: between lens and video camera Outer diameter: 28 mm Length: 33 mm
Precision target		Distance between reference points: 5 (± 0.01) mm Dimensions of the sticker: 50×50 mm Dimensions of the base: 70×70×5 mm Material of the sticker: acrylic Material of the base: stainless steel

Tripod and accessories	Manfrotto / 058B 	Height minimum / maximum: 44 / 217 cm Weight: 6,15 kg Maximum load: 12 kg Bubble leveling Accessories: rubber footings, hydrostatic ball head and arm
Lamps	Derwent / MF100 	Infrared - Incandescent Bulb life: average 8000 h Aperture angles: 10°, 30°or 60° IR filters: 730 or 830 nm

The camera stand is composed of a tripod, a hydrostatic ball head and a support arm. The video camera is indirectly supported on the structure of the tripod with the assistance of an extensible support arm. The height of the tripod is generally limited to 1 m, for stability reasons. The lens is attached to the tripod through a plate, which, in turn, is coupled to a hydrostatic ball head that enables small or large amplitude movements in all directions.

A laptop computer is used for storing and processing images. The images are directly stored in two hard drives running on RAID 0 system. The connection between the video camera and the laptop computer is performed by a Gigabit Ethernet interface that enables the data transference at a speed of approximately 1 Gb/s, for distances up to 100 m, without resorting to repeaters.

In order to measure the displacement of a point on a given object, a precision target should be applied to that point. The precision target consists of an acrylic sticker glued to a metallic plate, consisting of four black and white squares ensuring a good contrast between the different colored areas and also a good reflectivity to the infrared (IR). Each square has imprinted a grid of reference points with a regular spacing equal to 5 mm (tolerance of ± 0.01 mm) used to scale and calculate the image resolution.

The illumination of the target is performed by IR incandescent lamps and generally it is necessary to use two of them, in order to minimize the influence of natural light.

The displacement measurement system based on video technology can easily be articulated with other measurement systems. Figure 2 shows the articulation with a National Instruments' cDAQ-9172 data acquisition system, used in the practical applications of this paper, later described in Sections 3 and 4.

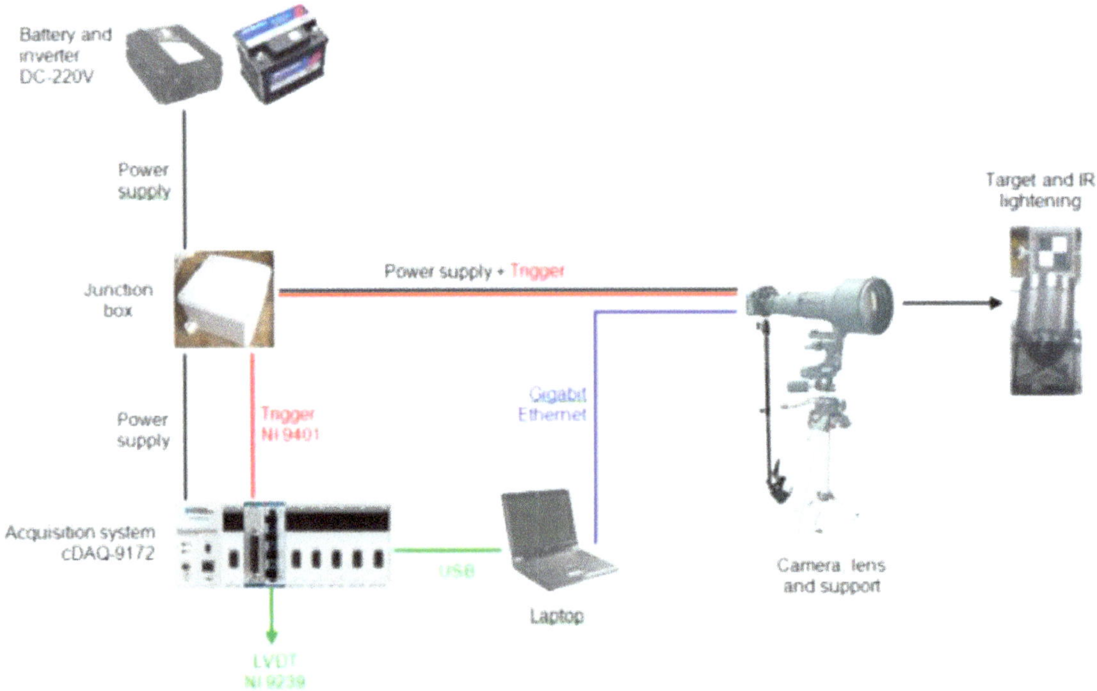

Figure 2. Articulation between the measurement system based on video technology and the National Instruments' cDAQ-9172 measurement system.

The cDAQ-9172 measurement system allows the data acquisition from different sensors, namely LVDTs, accelerometers, strain gages, etc., through adequate analogue modules. The data synchronization is ensured through an analogue-digital module (NI 9401) by a trigger signal. The power supply of both systems is guaranteed by a set of batteries.

Requirements and Performance

The system's basic requirement is to enable the performance of non-contact dynamic displacements measurements in distances to the objects up to 25 m, with an acquisition frame rate up to 500 fps.

The theoretical resolution of an image (R) can be calculated by the application of the following expression [13]:

$$R = \frac{D_s dN}{n_{px}.f} \quad (1)$$

where D_s is the dimension of the camera's sensor, equal to 3.6 mm for the Genie camera that was used, d is the distance from the camera to the object, N is the Nyquist factor, equal to 2, n_{px} is the number of active pixels of the camera on a given direction and f represents the focal length.

Table 2 presents the number of active pixels of the video camera's sensor according to two directions (1 and 2) for various acquisition rates. These directions generally correspond to the vertical and horizontal directions, or vice versa, depending on the position of the camera in relation to the object. Scenario A considers the maximum number of active pixels in direction 1 while scenario B considers the maximum number of active pixels in both directions.

Table 2. Number of active pixels for various image acquisition rates

Acquisition rate [fps]	Number of active pixels			
	Scenario A		Scenario B	
	Dir 1	Dir 2	Dir 1	Dir 2
64	1400	1024	1400	1024
100	1400	658	960	960
250	1400	260	880	413
500	1400	131	880	203

The increase of the image acquisition rate leads to a reduction in the number of active pixels of the camera. In scenario A, for the higher acquisition rates, such as 250 fps and 500 fps, the camera can provide the maximum number of active pixels in one direction while, in the other direction, the number of active pixels is significantly reduced. This feature is useful for the evaluation of displacements in a preferential direction. In scenario B, due to the architecture of the camera, above 250 fps it is impossible to reduce the number of active pixels in direction 1.

Figure 3 presents the values of the theoretical resolution of the image associated to direction 1 of the camera's sensor, according to the distance to the object for various focal lengths. The results consider the maximum number of active pixels of the camera's sensor and are valid for any value of the image acquisition rate. The focal length of the lens, equal to 400 mm, can be increased to 800 mm or to 3200 mm by using a duplicator, or a duplicator associated with a quadruplicator, respectively.

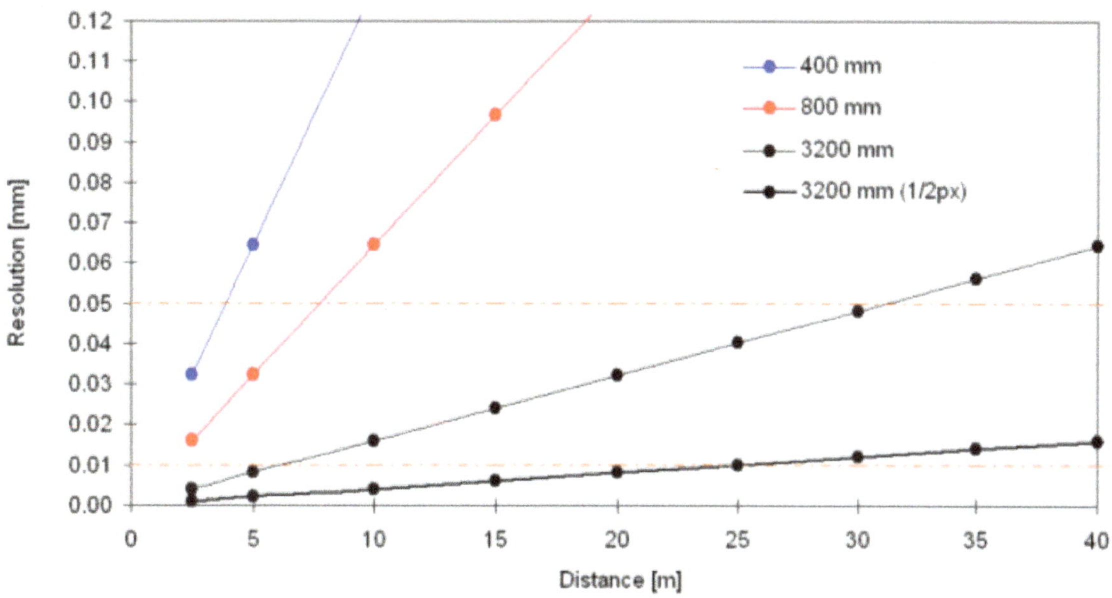

Figure 3. Theoretical resolution of the images according to the distance to the target and the focal length.

For distances to the object in the order of 5 m it is possible to obtain an image resolution of approximately 0.01 mm by selecting a focal length of 3200 mm. For distances in the order of 25 m, to obtain a resolution of 0.01 mm, it is necessary to use a focal length of 3200 mm and apply an image processing technique at subpixel level, in this case, at half pixel level [15]. Under these conditions the illumination of the target should be a critical aspect, since the use of the lens and magnification elements significantly reduces the light entrance into the camera's sensor.

Software

The image acquisition and processing software was developed based on the SAPERA [14] vision software and on a set of functions specifically developed for this work and compiled in C++ environment. The graphical interface of the developed software, presented in Figure 4, includes several functionalities, namely, the definition of the regions of interest, the definition of the camera's parameters (acquisition frame rate, gain, shutter speed), the selection of the type of trigger, the monitoring of the image quality parameters (brightness and contrast) and the automatic processing of acquired images [15].

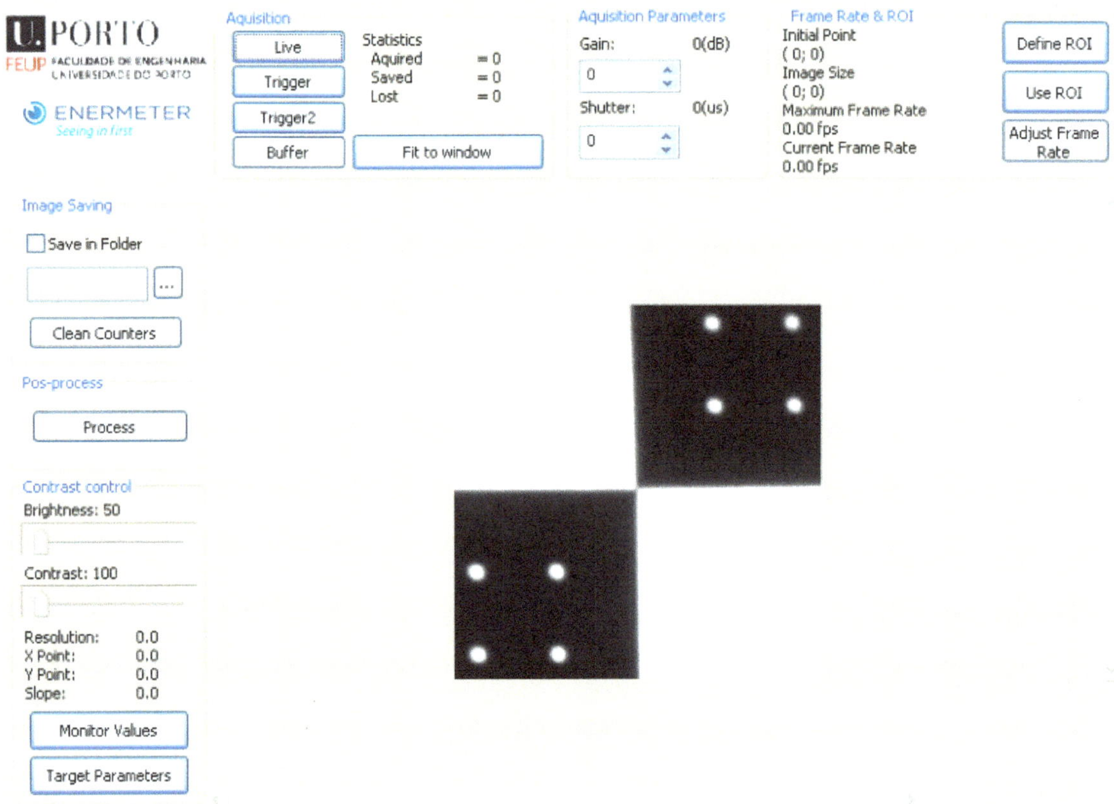

Figure 4. Graphical interface of the software for the acquisition and processing of images.

The definition of a region of interest of a field of view is performed graphically by selecting a portion of the image using a rectangular window. This option enables to reduce the size of the images and, thereby, increase the acquisition frame rate, without altering the resolution of the images. The gain option enables to digitally increase the intensity of the gray levels of the image in situations where the field of view presents low luminosity. The trigger can be internally or externally activated [15]. The brightness and contrast are indicators of image quality monitored at the level of the region of interest. The optimal value of image brightness is equal to 50% while the optimal value of contrast is close to 100% [13].

LABORATORY TEST

The laboratory test aims at evaluating the performance of the video technology based system through the confrontation of displacement measurements obtained with this system and with a conventional displacement sensor (LVDT).

Description

The laboratory test consisted in the measurement of the vertical displacement of a simply supported IPE100 metallic beam with a span of 0.90 m, obtained by the application of a point load. The point load was applied dynamically to the midspan by means of a servo hydraulic actuator assembled on a reaction frame, according to the details presented in Figure 5 a).

a)

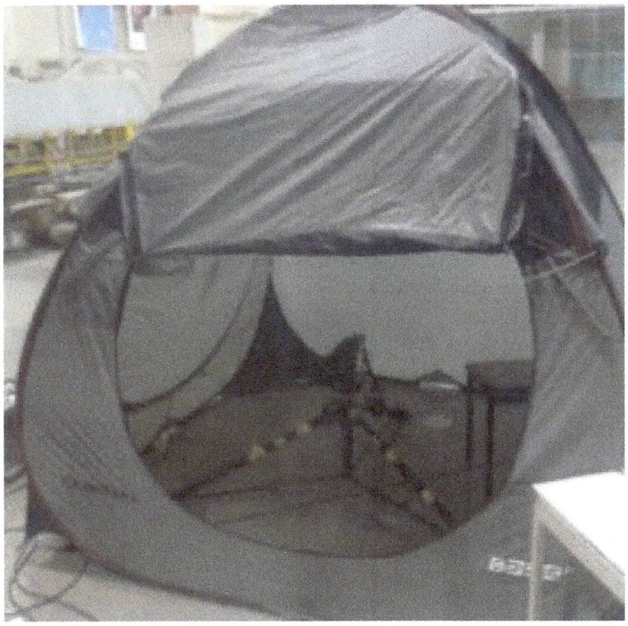

b)

c)

Figure 5. Laboratory test: a) overview; b) camera's protection stall; c) target and lighting lamp.

The camera was positioned at 3 m, 5 m, 10 m and 15 m from the target and placed inside a stall to minimize the movements of the stand that could be caused by the action of air flows (Figure 5 b)). The illumination was ensured by one or two IR incandescent lamps (Figure 5 a) and c)). The target was set on a metallic segment bonded to the lower flange of the beam as shown in Figure 5 c). The metallic segment enables the rotation of the target to ensure that its position is perpendicular to the camera. This operation was conducted with the assistance of a digital inclinometer. The LVDT was placed in the same section of the beam through a magnetic base attached to the reaction frame (Figure 5 c)). The LVDT is a RDP, model DCTH100AG, with a measuring range equal to ± 2.5 mm and a precision in the order of 0.001 mm.

The conditions of the measurements performed by the video system are summarized in Table 3. This table contains information about the distance to the target, the resolution of the images and the optical accessories used.

Table 3. Laboratory test: distance from the camera to the target, image resolution and optical accessories

Distance to target [m]	Resolution [ppm]	Optical accessories		
		Lens	2x	4x
3	20.9	×		
5	24.4	×	×	
10	22.4	×		×
15	14.7	×		×

The second IR incandescent lighting was used for the distances of 10 m and 15 m, having a decisive role in the maintenance of the image quality in terms of brightness and contrast since, for these distances, the duplicator was replaced by the quadruplicator, which led to a significant brightness reduction in the field of view.

For distances of 3 m, 5 m and 10 m from the target the image resolution ranged from 20 ppm to 25 ppm. For the distance of 15 m the images were acquired at a lower resolution and close to 15 ppm. This option was somewhat deliberate since the maintenance of the images' resolution would involve the combined use of the duplicator and the quadruplicator. However, the combined use of these magnification elements is not viable with the available means of lighting, since the reduction of light in the sensor of the camera is very significant.

Results

Figure 6 presents the results of the vertical displacement measurement at the midspan of the beam, by the video system and the LVDT, for distances from the camera to the target equal to 3 m, 5 m, 10 m and 15 m.

The analysis of the graphs supports the conclusion that there is a good agreement between the displacement records measured by both systems, especially in the records obtained for distances of 3 m and 5 m from the camera to the target.

The absolute errors of the displacement measurement performed with the video system, taking as exact the measurement performed with the LVDT, are equal to 0.018 mm (3 m distance to the target), 0.019 mm (5 m distance to the target), 0.031 mm (10 m distance to the target) and 0.042 mm (15 m distance to the target). The errors were obtained based on the maximum difference registered between the displacements measured by the video system and by the LVDT. It is possible to conclude that the errors increase as the distance from the camera to the target increases.

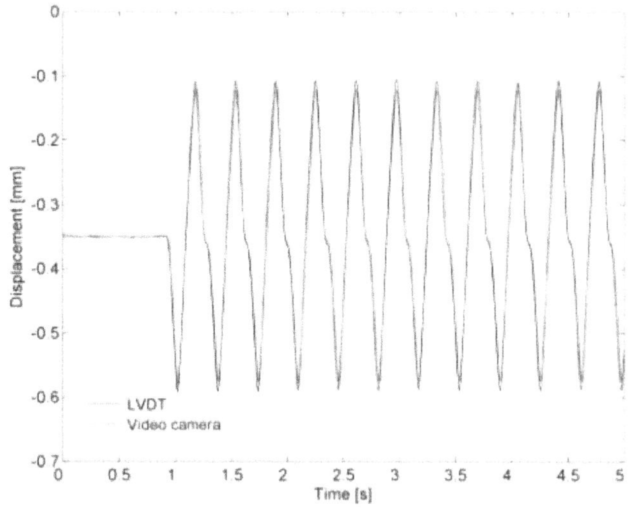

a)

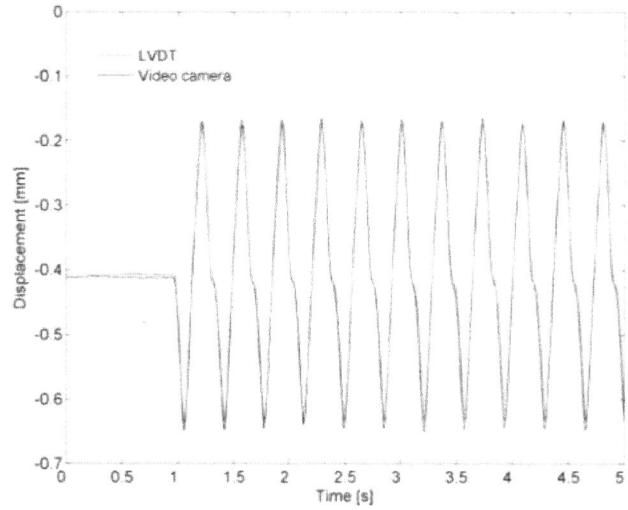

b)

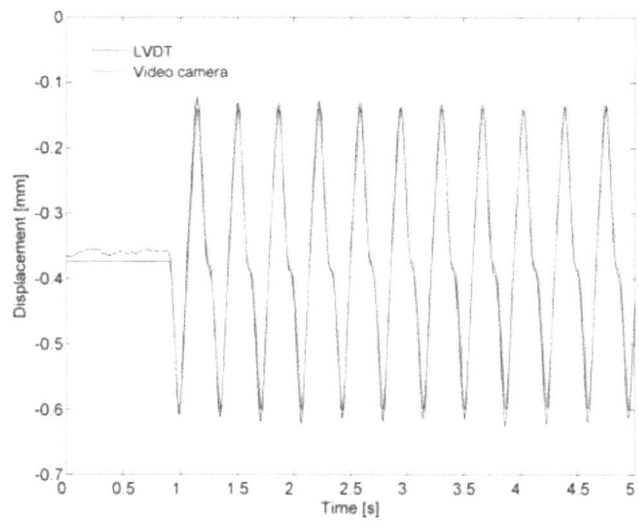

c)

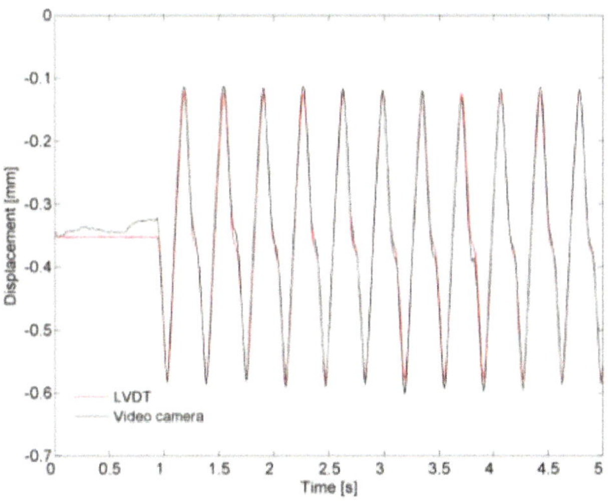

d)

Figure 6. Comparison of the vertical displacement of the beam measured with the video system and with the LVDT, for static and dynamic loadings, for camera to target distances of: a) 3 m; b) 5 m; c) 10 m; d) 15 m.

Figure 7 compares the real and theoretical precision of the video measurement system. The theoretical precision (pT) is obtained by the application of the following expression:

$$p_T = \frac{1}{R} N \qquad (1)$$

where R is the resolution of the image expressed in pixels/mm and N is the Nyquist factor associated to the camera's sensor CMOS, which is equal to 2 pixels.

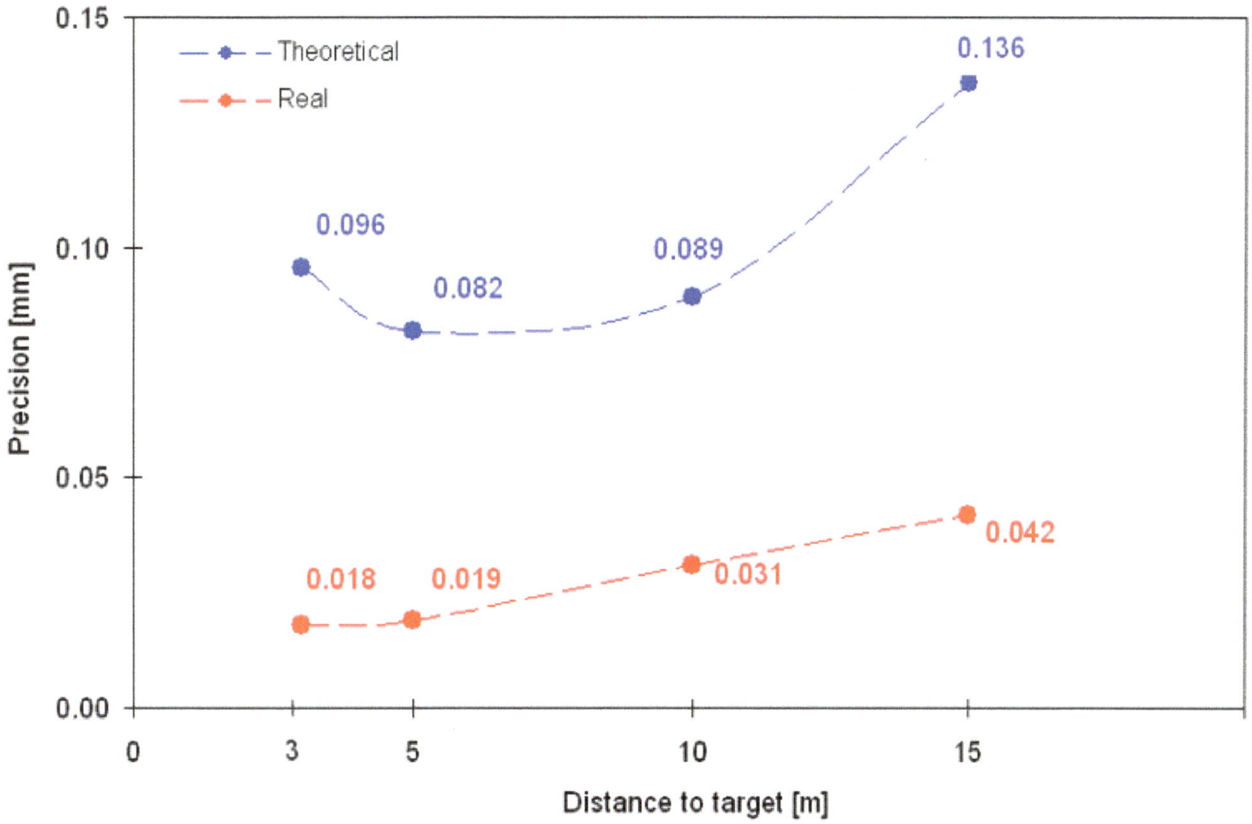

Figure 7. Comparison between the real and theoretical precisions in laboratory testing, according to the distance to the target

The analysis of the figure shows that the real precisions can be significantly higher than the theoretical precisions for all tested distances, demonstrating the efficiency of the image processing technique at subpixel level [15].

FIELD TEST

The field test involved the dynamic measurement of displacements of the São Lourenço bridge and aims evaluating the performance of the video system in an external environment. This bridge, located at the Northern line of the Portuguese railways', is a bowstring type bridge, consisting of two half decks with a 42 m span, formed by a prestressed concrete lower slab laterally suspended by metallic arches. The slab is suspended by means of metallic hangers and diagonals near the beginning of the arches [16]. Figure 8 presents a lateral view of the São Lourenço railway bridge.

Figure 8. Lateral view of São Lourenço railway bridge with the identification of the measurement point.

Description

The measurement of the vertical displacement of point A of the deck, identified in Figure 8, was performed for distances from the camera to the deck equal to 5 m, 10 m, 15 m and 25 m. The camera was installed in an adjacent terrain next to the bridge, inside a protective stall in order to reduce the influence of wind and possible air flow resulting from the passage of trains at high speed. The target was attached to a metallic plate of the hanger located at the lower level of the main beam.

a) b)

Figure 9. Images of the LVDT and target placed on point A: a) overview and b) detail.

The LVDT was positioned with the help of a magnetic base, which is supported by a metallic structure fixed to the ground. The LVDT is a RDP, model ACT1000A, with a measuring range equal to ± 25 mm and a precision in the order of 0.01 mm. The lighting system, comprising two IR incandescent lamps supported by a metallic tripod, was installed at approximately 0.30 m from the target. The illumination lamps could also be supported by other type of accessories which do not require contact with the ground. The lamps were switched on just before the passing of the train to prevent overheating. Figure 9 illustrates an overview and a detail of the measurement site.

The measurement conditions of the video based system are summarized in Table 4, including the distance to the target, the image resolution and the optical accessories used.

Table 4. Test conditions with the video measurement system

Distance to the target [m]	Resolution [ppm]	Optical accessories		
		Lens	2x	4x
5	21.0	×	×	
10	12.5	×	×	
15	14.5	×		×
25	20.0	×	×	×

It is possible to conclude that the resolution of the system ranged between 12.5 ppm and 21 ppm in all measurements. The existing natural light in the field allowed the camera's sensor to have an available amount of light superior to that of the laboratory test. However, the luminosity showed significant variability during measurement due to the alteration of weather conditions. This fact led to the necessity of introducing a digital gain in the images in order to stabilize the values of the brightness and contrast parameters.

Results

Figure 10 shows the vertical displacement records from point A of the deck, obtained through the LVDT and through the video system, with the camera located at 5 m, 10 m, 15 m and 25 m from the target, for the passage alfa pendular trains, at speeds near 180 km/h, and intercity trains at speeds near 160 km/h. The load schemes of these trains are presented in [15].

The observation of the graphs demonstrates that, for all considered distances, there is a very good agreement between the displacement records measured with the video system and with the LVDT.

The absolute errors of displacement measurements performed with the video system, taking as exact the measurements performed with the LVDT, are equal to 0.067 mm (5 m distance to the target), 0.103 mm (10 m distance to the target), 0.107 mm (15 m distance to the target) and 0.274 mm (25 m distance to the target). The errors refer to the largest registered difference between the displacement measured by the video system and by the LVDT. Just as in the laboratory test there is a tendency for error increase as the distance to the target increases.

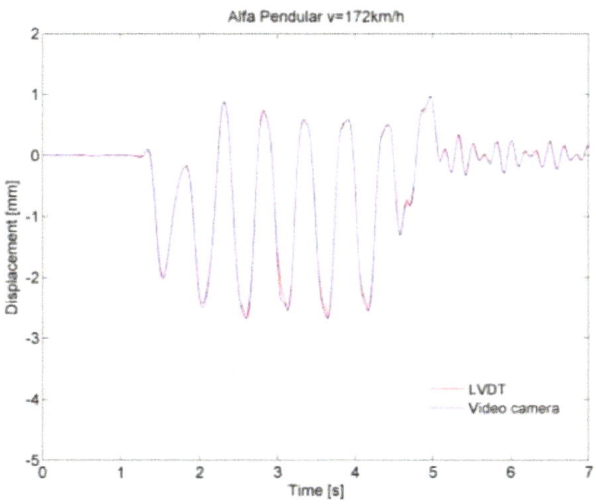

a)

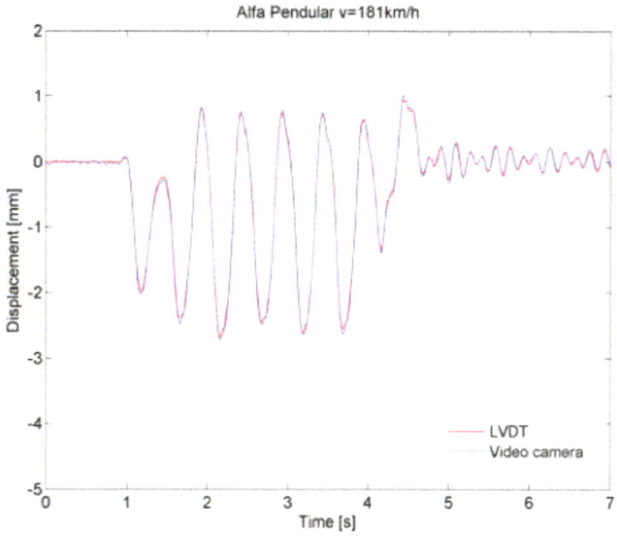

b)

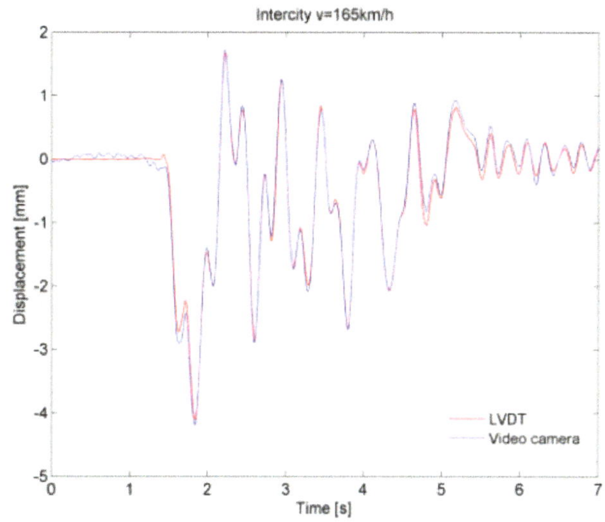

c)

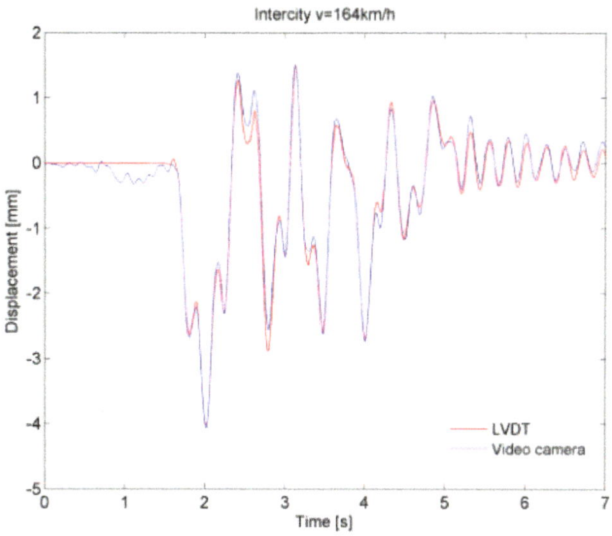

d)

Figure 10. Comparison of the measurements of vertical displacement of the deck with the video system and with the LVDT, for the passage of alfa pendular and intercity trains, for camera to target distances of: a) 5 m; b) 10 m; c) 15 m; d) 25m

Figure 11 presents a comparison of the real and theoretical precisions of the video measurement system.

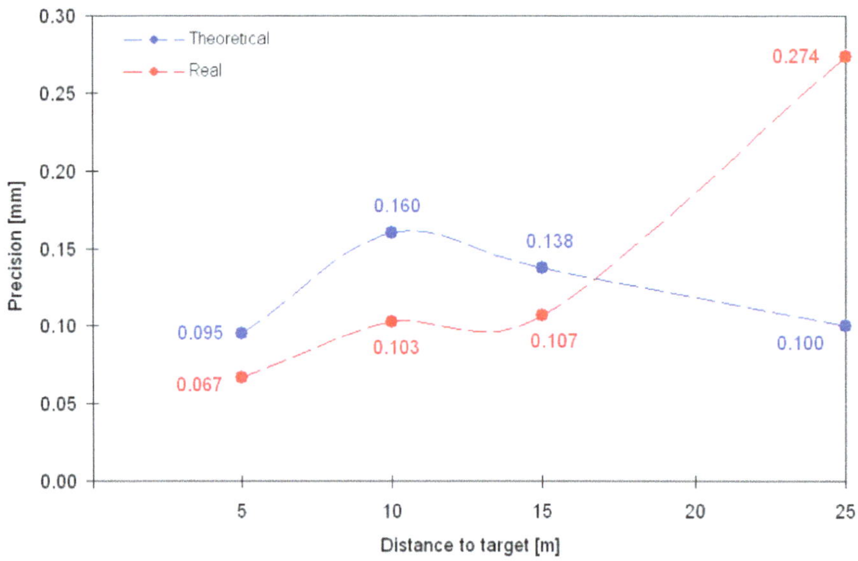

Figure 11. Theoretical and real precision of the measurements performed in the field according to the distance from the camera to the target.

The analysis of the figure shows that the real precision values are lower than the theoretical precision values, except for the distance of 25 m. It should also be noted that the precision gain resulting from the application of the image processing technique at subpixel level was significantly lower than that observed in the laboratory test. The reasons for this reduction are possibly related to the fact that the values of the brightness and contrast parameters are not equal to the optimal values, and to the existence of noise in the images due to the inclusion of digital gain. The unusual behaviour of the images' resolution for the distance of 25 m can be justified by the inclusion of a very high value of digital gain, which aimed to compensate for the reduced light input into the camera's sensor, due to the combined use of the duplicator and quadruplicator.

CONCLUSIONS

This article describes the development of a non-contact dynamic displacement measurement system for railway bridges based on video technology and its application in laboratory and field testing.

The laboratory test of a beam subjected to static and dynamic loading resulted in a good agreement between the measurements obtained with the video system and with the LVDT. The obtained precision was below 0.05 mm for distances from the camera to the target up to 15 m.

The field test of the São Lourenço railway bridge presented a good agreement between the displacement measurements obtained with the video system and with the LVDT, for the passage of trains at speeds between 160 km/h and 180 km/h. The obtained precision was below 0.1 mm for distances from the camera to the target up to 15 m, and in the order of 0.25 mm for a distance of 25 m.

In future works, the authors intend to introduce improvements in the measurement system based on video technology, in order to: i) apply it to distances from the structure to the camera of over 25 m, which may involve the use of a wide-ranging telescopic lens; ii) allow the simultaneous displacement measurement of various points of the structure, developing the software in order to process multiple regions of interest in the images associated with several targets, and iii) discard the placement of targets in the structure, through the implementation and application of DIC techniques.

ACKNOWLEDGMENTS

The present work has been funded by the Portuguese Foundation for Science and Technology (FCT), in the context of Research Project FCOMP-01-0124-FEDER- 007195. The authors also would like to thank REFER, the company responsible for the management of the Portuguese railway network, for the support on the realization of the field tests, and to Eng. Henry Simões, from Infaimon, for the cooperation and information provided about the video camera, lens and lighting system.

REFERENCES

1. Bell E, Peddle J, Goudreau A. Bridge condition assessment using digital image correlation and structural modeling. IABMAS'12 - Bridge Maintenance, safety, management, resilience and sustainability. Dubrovnik, Croatia.2012. p. 330-7.
2. Lee J, Cho S, Shinozuka M, Yun C, Lee C, Lee W. Evaluation of bridge load carrying capacity based on dynamic displacement measurement using real-time image processing techniques. Steel Structures. 2006;6:377-85.
3. Lee J, Fukuda Y, Shinozuka M, Cho S, Yun C. Development and application of a vision-based displacement measurement system for structural health monitoring of civil structures. Smart Structures and Systems. 2007;3:373-84.
4. Lee J, Shinozuka M. A vision-based system for remote sensing of bridge displacement. NDT&E International. 2006;39:425-31.
5. Lee J, Oh J, Park M, Kwon S, Kwark J. Bridge displacement measurement system using image processing. IABMAS'06 - Bridge Maintenance, safety,

management, resilience and sustainability. Porto, Portugal. 2006.

6. Malesa M, Szczepanek D, Kujawinska M, Swiercz A, Kolakowski P. Monitoring of civil engineering structures using digital image correlation technique. ICEM 14 – 14th International Conference on Experimental Mechanics. Poitiers, France. 2010.
7. Nassif H, Gindy M, Davis J. Comparison of laser Doppler vibrometer with contact sensors for monitoring bridge deflection and vibration. NDT&E International. 2005;38:213-8.
8. Gentile C, Bernardini G. An interferometric radar for non-contact measurement of deflections on civil engineering structures: laboratory and full-scale tests. Structure and Infrastructure Engineering. 2010;6:521– 34.
9. Stephen G, Brownjohn J, Taylor C. Visual monitoring of the Humber bridge. Engineering Structures. 1993;15:197-208.
10. Chang C, Ji Y. Flexible videogrammetric technique for three-dimensional structural vibration measurement. Journal of Engineering Mechanics. 2007;133:656-64.
11. Jáuregui D, White K, Woodward C, Leitch K. Noncontact photogrammetric measurement of vertical bridge deflection. Journal of Bridge Engineering. 2003;8:212-22.
12. Kim S, Kim N. Multi-point displacement response measurement of civil infrastructures using digital image processing. Procedia Engineering. 2011;14:195-203.
13. Forsyth D, Ponce J. Computer vision: a modern approach. 2nd ed: Prentice Hall; 2011.
14. Sapera. Sapera++ LT™ 6.20 - Programmer's Manual. St-Laurent, Quebec: DALSA Corporation; 2006.
15. Ribeiro D, Calçada R, Ferreira J, Martins T. Noncontact measurement of railway bridges dynamic displacement using an advanced video-based system. Engineering Structures. 2014;75:164-80.
16. Ribeiro D, Calçada R, Delgado R, Brehm M, Zabel V. Finite element model updating of a bowstring-arch railway bridge based on experimental modal parameters. Engineering Structures. 2012;40:413-35.

CHAPTER 2

THEORY-BASED METROLOGICAL TRACEABILITY IN EDUCATION: A READING MEASUREMENT NETWORK

William P. Fisher Jr. [a,b], A. Jackson Stenner [c,d]

[a] LivingCapitalMetrics.com, Sausalito, CA, USA

[b] BEAR Center, Graduate School of Education, University of California, Berkeley, CA, USA

[c] MetaMetrics, Inc., Durham, NC, USA

[d] School of Education, University of North Carolina, Chapel Hill, NC, USA

ABSTRACT

Huge resources are invested in metrology and standards in the natural sciences, engineering, and across a wide range of commercial technologies. Significant positive returns of human, social, environmental, and economic value on these investments have been sustained for decades. Proven methods for calibrating test and survey instruments in linear units are readily available, as are data- and theory-based methods for equating those instruments to a shared unit. Using these methods, metrological traceability is obtained in a variety of commercially available elementary and secondary English and Spanish language reading education programs in the U.S., Canada, Mexico, and Australia. Given established historical patterns, widespread routine reproduction of predicted text-based and instructional effects expressed in a common language and shared frame of reference may lead to significant developments in theory and practice. Opportunities for systematic implementations of teacher-driven lean thinking and continuous quality improvement methods may be of particular interest and value.

INTRODUCTION

Metrology connects measurement applications across industrial, scientific, and practical tasks separated by space and time. Significant fractions of many nations' economic productivity are invested in ensuring traceability to standards for various units of measurement. The human, social, environmental, and economic value of the returns on these investments depends on the transparency of the measures and their integration into a wide range of decision processes at multiple organizational levels. Huge resources are required to create and maintain technologically produced effects, such as volts, seconds, or meters, with the primary return on those resources being the illusion that the effects seem to be products of nothing but completely natural processes occurring with no human intervention.

New insights into how cognitive, social and technological resources aid in creating shared cultural frames of reference have emerged from close critical study of historical and contemporary scientific modelling and metrological practices. From this perspective, science is not qualitatively different from everyday ways of thinking and relating, except in more

Citation: William P. Fisher Jr., A. Jackson Stenner, Theory-based metrological traceability in education: A reading measurement network, Measurement, Volume 92, October 2016, Pages 489-496, ISSN 0263-2241, http://dx.doi.org/10.1016/j.measurement.2016.06.036.

deliberately extending laboratory processes into the world as distributed cognitive systems supporting a range of associated problem-solving methods [1], [2], [3],[4] and [5]. Of particular interest here is the linking of specific ways in which organizations align and coordinate their processes and relationships relative to technical developments and expectations. A positive result of adopting this point of view is recognition of the value of previously obscured accomplishments in, and opportunities for, advancing the quality of research and practice in psychology and the social sciences. An illustrative example is found in the scientific modelling and metrological practices informing integrated reading assessment and instruction in education.

Transparent Instruments, Invisible Production

By definition, metrologists are doing their jobs best when no one knows they are there. Experimental scientists, for instance, may take little notice of their instrumentation until it breaks down or does not conform to expected standards. The general public and researchers in psychology and the social sciences are, then, also largely unaware of the resource-intensive work involved in establishing uniform unit standards and traceability to them [3], [4], [6] and [7].

The uniformity of the various phenomena described by natural laws allows scientists the convenient efficiency of not needing to specify scale units in statements of laws. Force equals mass times acceleration in kilograms, Newtons, and meters just as well as in pounds, poundals, and feet. The ability to skip over uniform details supports a division of labor in science that separates theoretical work from the calibration of instruments and both of these from the use of theory and instruments in experiments [8].

The convenience of separating theoretical, experimental and instrumental concerns has its drawbacks, too. Not knowing when or how reference standard units are established reinforces unexamined metaphysical assumptions—such as the idea that the universe or nature is inherently and innately numerical, quantitative, or mathematical—that rarely become explicit objects of attention.

The effect of these presuppositions is significant. Huge social, industrial, and economic efficiencies are gained by universal consensus on the facts of complex phenomena like electricity, temperature, distance, mass, and time. Though the dynamics of that consensus are complex and sometimes counterintuitive [8], making quantities seem natural is a cultural achievement of the highest order.

The advancement of science is put at risk when the historic and historical mathematical understanding of scientific objects is reified as unquestioned and unquestionable. Two questions emerge here: (1) how did the natural sciences succeed in making quantities seem so thoroughly natural [3], [4], [9], [10], [11] and [12], and (2) how might the social sciences learn from those successes? Recent advances in reading measurement embody important lessons in this regard for the social sciences.

Shortsightedly Focusing Attention on the Local Measurement Outcome

The technical processes of measurement were historically cut out of the picture of science by the positivist focus on empirical observation, as well as by the later anti-positivist focus on theoretical constraints on observation [8]. Sometimes this omission was literal and deliberate, as when a woodcut of a laboratory scene printed in its entirety in one place is trimmed in a later publication to exclude the means by which a technical effect was produced [12]. Other times the omission was metaphorical, as when technical processes were illustrated in summary form by angelic cherubs producing effects by means of divine intervention [12].

Transparency in measurement is a two-edged sword. Wide access to comparable measures is achieved only to the extent that technical complexities can be ignored. This point was emphasized by Whitehead [13], who observed that "Civilization advances by extending the number of important operations which we can perform without thinking about them" (p. 61). But what happens when those making these advances do not record—or do not themselves fully understand—*how* they extended the number of important operations that can be performed by persons unversed in their technicalities?

In his study of the geometric assumptions Galileo employed in his physics, Husserl [14]was sensitive to the ways in which a hidden agenda set priorities. Like Galileo, we find ourselves in a situation, in accord with the philosophical problems attending measurement, in general, where

> Metrology has not often been granted much historical significance. ... Intellectualist condescension distracts our attention from these everyday practices, from their technical staff, and from the work which makes results count outside laboratory walls [6].

Researchers in the natural sciences make use of commercially available precision tools calibrated to universally uniform reference standards, standards capitalizing on the value of invariant laws. Transparent measures communicated in a network sharing common values situates metrology's often unrecognized historical significance in a complex overall context offering important lessons for psychology and the social sciences [1], [2], [3], [4],[5], [6], [7], [8], [9], [10], [11] and [12]. The culture of science rewards a mix of convergent, divergent, and reflective thinking in ways that have proven their productivity and inform a vital culture of ongoing innovation [8], [9], [11], [12] and [15].

Consequences for Psychology and the Social Sciences

But in the social sciences, the lack of metrological institutions, methods, and traditions, and the associated absence of the intercalated disunity of distinct theoretical, experimental, and instrumental communities observed by Galison in the natural sciences[8], has been catastrophic. As social scientists have long recognized for themselves [16],[17] and [18], mainstream research methods and statistical models employ scale-dependent ordinal data in a search for a kind of significance that is often irrelevant to and even antithetical to the production of new knowledge. Even when regularities akin to natural laws are sought and found in psychological and social phenomena [19], [20], [21],[22] and [23], results are typically assessed in the language and methods of statistics rather than of measurement and metrology, meaning the focus is on data analysis and not on theory development or the calibration of instruments traceable to a standard unit. The human, social, economic, and scientific consequences of this failure to coordinate and balance convergent, divergent, and reflective field-defining activities are profound. Ideas on how such activities might be organized in education have recently been proposed [24].

The lack of institutions and traditions concerning metrological traceability and standards in psychology and the social sciences may have more to do with broad and deep cultural presuppositions than with an actual lack of a basis for them in evidence. After all, what systematic program of experimental evaluation has ever irrefutably established that uniform metrics based in lawful regularities are impossible in psychology and the social sciences? Evidence indicates that provisional possibilities exist in some circumstances[19], [20], [21], [22], [23] and [24].

METROLOGICAL TRACEABILITY FOR READING MEASUREMENT

The longstanding need to provide students with reading challenges appropriate to their reading abilities is usually approached in terms of general curricular structures, and teacher training and experience. Theory has not been of significant interest [25] and [26]. Rasch's development of a new class of measurement models in the 1950s was an important step forward in improving the quantification of reading ability [25]. This research led to improvements in the matching of readers to text.

When Rasch's concept of specific objectivity (the modelled independence of the ability and difficulty parameters, as shown in Eq. (1)) as it was obtained in local measures was combined with a general predictive theory of English text complexity in the 1980s, following the work of Stenner and colleagues [27], [28] and [29], the stage was set for the efficient creation of a network of reading measurement instruments calibrated in a common unit. By the late 1990s, all of the major high stakes English reading tests in the U.S. had been brought into the system. These are today complemented by the hundreds of thousands of books, tens of millions of short articles and hundreds of millions of readers that have been brought into the system in the intervening years.

In this system, reader abilities and text complexities are measured in the same unit. The scale ranges from below 200 for beginning readers to over 1600 for very high level readers and texts. Knowing the text measure of a book and the reader's measure predicts the degree to which the book will be comprehensible to the student.

More than 30 million measures annually are reported in the U.S. from state and commercial assessments, and from classroom reading program assessments, in a common unit of measurement [29]. The 21 U.S. state departments of education that have formally adopted this unit for use are shown in the map in Fig. 1.

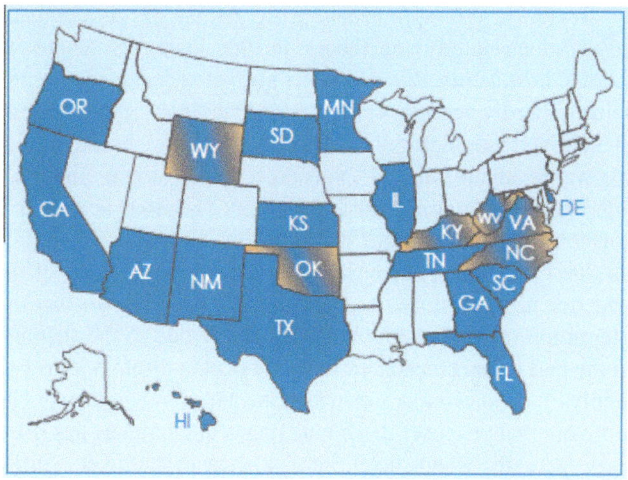

Figure 1. Map of U.S. states employing a common unit of reading measurement (two tones indicate common units in use for both reading and mathematics).

Traceability to the common unit is determined via both empirical, data-based equating studies and theory-based text analyses [29]. Additional features of the system include electronic tools integrating instruction and assessment for mass customized diagnostics[30], and others charting growth in reading ability relative to college and career readiness[31] and [32]. Establishing this network of comparable assessments required formal relationships with book and test publishers, teachers, schools and school districts, state departments of education, and psychometric researchers. Furthermore, a new array of material practices was needed to give all the parties involved ongoing and verifiable confidence in the theory. Though great efficiencies stood to be gained, credibility demanded a cautious approach to their implementation. Formal documentation of the birth of this traceability system would be a valuable contribution to the sociotechnical qualities of education.

IMPLICATIONS FOR PSYCHOLOGY AND THE SOCIAL SCIENCES

In 1965, the National Academy of Sciences published a report articulating common assumptions as to the sequence of events supposed to take place in the development of new instrumentation [33]. Four stages were identified:

- discovery of suitable means of observing some phenomenon,
- exploration of this phenomenon with special, homemade instruments or commercial prototypes,
- widespread use of commercial instruments,
- routine applications of the instrument to control industrial production as well as research.

Textbook assumptions and presentations of this sequence have indoctrinated researchers in the human sciences to believe, mistakenly, that this is the normal sequence of events. Because hardly anyone is involved in every part of the process, unexamined assumptions cohere into a "just-so" narrative that says more about cultural expectations than about historical complexities. Scientists and non-scientists alike accept this story, against the grain of actual events. Rabkin [33] points out that this scheme seems to be at variance with much of the evidence in the history of science. It has been shown that the integration of instruments has been rarely due to the demand on the part of the researcher. Rather it occurs through vigorous supply of advanced instruments on the part of the industry. The company that proposes these four stages in the report has itself had experience when stages 3 and 4 occur in the reverse order and, moreover, stage 4 is by far the most decisive factor in the development of new instrumentation.

The "vigorous supply of advanced instruments", and not demand, also characterizes the introduction of popular electronic appliances. Just as Rabkin points out has been the case in research, there was little or no clamour among the public for telephones, televisions, faxes, the Internet, microwaves, blenders, or cell phones before they were developed and introduced.

Scientists and the public both tend to think of instrumentation only as tools employed in the service of the individuals who use them. This perspective is at odds with the historical evidence as well as with philosophers' observations, such as, for instance, Thoreau's realization that humanity has become the tool of its tools [34] and Nietzsche's insight that the victory of science is better cast as a victory of method over science [35].

This alternative perspective is important because, in the history of science, theory follows from extensive experience with instruments more often than instruments are designed and built from theoretical projections. Standardized and commercially available instrumentation make possible the predictable and routine reproduction of scientific effects essential to the conduct of controlled experiments—and so also to the development of precise and accurate theoretical predictions. As stated by Price,

Historically, we have almost no examples of an increase in understanding being applied to make new advances in technical competence, but we have many cases of advances in technology being puzzled out by theoreticians and resulting in the advancement of knowledge. It is not just a clever historical aphorism, but a general truth, that 'thermodynamics owes much more to the steam engine than ever the steam engine owed to thermodynamics.' ...historically the arrow of causality is largely from the technology to the science [36].

In the context of reading measurement, the repeated reproduction of consistent results following the work of Rasch and others led to the Anchor Test Study in the 1970s [37]. This study equated seven major reading tests in the U.S. and involved over 350,000 students in all 50 states. But the purely empirical basis of the calibration and the lack of predictive theory meant that the value of the common unit of measurement was lost as soon as new items were added to the tests, which was immediately.

A plain feature of the equated test results, however, was the similarity of the items from different tests that calibrated in the same locations. The stability of this phenomenon may not surprise anyone able to read, but its practical application in a predictive theory relating text complexity, comprehension rates, and reading ability was difficult to achieve [38].

THEORY FOR READING MEASUREMENT

The ability to read is fundamental to education, and it is accordingly tested and measured more often than any other subject area. The index to the eighteenth edition of the Buros Mental Measurements Yearbook [39] includes over 140 tests with the word "reading" in their titles. This count does not include tests focused on vocabulary or word meaning, which are also numerous.

Though the issues are complex, literacy remains essential to productivity in the global economy [40]. The need for effective and efficient reading education will only intensify as communication, teamwork, and information management are increasingly demanded as basic skills [41].

And despite the longstanding fundamental importance of reading as the tool most essential to learning, reading research remained atheoretical until 1953, and interest in a unified theory of reading is a relatively new phenomenon [25] and [26]. Further, in the years since 1953, available reading theories have not generally been used to inform the design or interpretation of assessments of reading ability [25].

Though it may seem counterintuitive, this failure to apply theory in the course of empirical measurement research is not unusual, nor is it restricted to reading research. On the contrary, measurement technologies in the natural sciences have historically been developed through socially-contextualized trial-and-error solutions to practical engineering problems, such as consistent, stable results, and not directly from theoretical principles [3], [4], [5], [6], [7], [8], [33], [36] and [42]. Theory generally comes later, after researchers have had the opportunity to employ standardized technologies in the routine and repeated reproduction of a controlled phenomenon. Only then do applicable general principles emerge as useful insights that can be fed back into technical refinements.

Syntactic and Semantic Elements

In the same way putting things in words reduces an infinite variety of ways an experience might be expressed into a particular set of words expressed in a particular language, science reduces the infinite variations that phenomena exhibit to simpler models. The truth of the models is less an issue than their usefulness [43] and [44]. Simplification is usually achieved only in contexts that respect constraints and accept limited goals. The efficiency and power obtained when useful tools can be created, however, confers great value on a simplified process.

In the 1950s, Rasch's parameter separability theorem, concept of specific objectivity, and models useful in practical measurement applications combined in an important step forward in educational measurement [45]. These developments were followed by Wright's introduction of improved estimation algorithms, model fit tests, and software in the 1960s, along with his vigorous championing of Rasch's ideas [46]. By the

1970s, enough data from reading tests had been successfully fit to Rasch models in the U.S. to support the viability of the Anchor Test Study [37]. Success in this large project and additional research predicting item difficulties on the Peabody Vocabulary Test and the Knox Cube Test (a measure of short term memory and attention span) [27] and [28], led to a new effort focused on developing explanatory theory for reading.

Reading theories build on the fact that all symbol systems share two features: a semantic component and a syntactic component. In language, the semantic units are words. Words are organized according to rules of syntax into sentences [47]. Semantic units vary in familiarity and the syntactic structures vary in complexity. The readability of a text passage is dominated by the familiarity of the semantic units and by the complexity of the syntactic structures used in constructing the message. Many readability equations therefore use a two-variable equation to forecast text difficulty. The word-frequency and sentence-length measures combine to produce a regression equation, known as a construct specification equation [27] and [28]. This equation provides a theoretical model evaluated in terms of the proportion of the variance of reading comprehension task difficulties (or, more recently, the means of specification-equivalent ensembles of item difficulties, following Gibbs [48]) that can be explained as plausibly structured by causal relationships [49].

The Specification Equation

One approach to such a specification equation first employs the mean of the logarithm of the frequencies with which words in a text appear in a 550-million word corpus of K-16 texts. More specifically, the log frequency of the word family, which is more highly correlated with word difficulty, comprises one term in the equation. Word families include the stimulus word, all plurals, adverbial forms, comparatives, superlatives, verb forms, past participles, and adjectival forms. The frequencies of all words in the family are summed and the log of that sum is used in the specification equation.

The second term of the specification equation is the logarithm of the text's mean sentence length. This parameter is operationalized simply by counting and averaging the number of words in each sentence.

The theoretical logit is then a function of sentence length and word frequencies in the language stated in the specification equation:

$$\text{Reading difficulty(or readability)} = A * \log(\text{MSL}) - B * \overline{\log(\text{WF})} + C \qquad (1)$$

where MSL is the mean sentence length and WF is the word frequencies. Log(MSL) and the mean log(WF) are used as proxies for syntactic complexity and semantic demand, and the coefficients are drawn from the empirical regression study [50]. Research is continuing into the decimal place significance of the coefficients and measurement uncertainty for the values of A (9.82247), B (2.14634), and C (a constant). The resulting logits are then scaled as follows:

$$(\text{logit}+3.3)*180+200 \qquad (2)$$

The relationship of word frequency and sentence length to text readability was investigated in research that extended a previous study on semantic units [28]. The original study found success on items at about −3.3 logits as indicating the earliest reading ability, and set that level at 200. A practical top to the scale for the end of high school was at 2.3 logits, and this was set 1000 units higher, to 1200. There is no upper limit to the scale, but text measures above 1600L are rare.

In this unit, when student and text measures match, a 75% comprehension level is expected. A student with a measure of 500L is expected to answer correctly 75% of the questions on an assessment made from any text that also measures 500L, within the range of uncertainty. The 75% comprehension rate differs from the default rate of 50% comprehension usually associated with matching measures and calibrations. Though the lowest uncertainty is associated with the 50% rate, teachers find that instruction has a firmer basis in student confidence when success is more likely. For this reason, the relation of ability to difficulty was shifted from 50% to 75% comprehension.

The uncertainty (standard error) of the individual measures [51] is

$$SE=X*[L/(r(L-r))]^{(1/2)} \qquad (3)$$

which is the square root of the test length L divided by the count correct r times the L − r count incorrect, times an expansion factor X that depends on test width. This logit is then converted to the standard unit. A standard unit uncertainty for a well-targeted 36-item test measuring with an uncertainty of about .40 logits is the original logit range of 2.3 − (−3.3) = 5.6 divided into the 1000L range, times .40, which comes to about 71L.

The analysis reported in the original study [28] involved calculation of the mean word frequency and the log of the mean sentence length for each of the 66 reading comprehension passages on the Peabody Individual Achievement Test. The observed difficulty of each passage was the mean difficulty of the items associated with the passage (provided by the publisher) converted to the logit scale.

A regression analysis based on the word-frequency and sentence-length measures produced a regression equation that explained much of the variance found in the set of reading comprehension tasks. The resulting correlation between the observed logit difficulties and the theoretical calibrations was 0.97 after correction for range restriction and measurement error [50].

The regression equation was further refined based on its use in predicting the observed difficulty of the reading comprehension passages on eight other standardized tests (see Table 1). Repeated and ongoing comparisons of theoretically expected calibrations with data-based estimates produced from test data analysis provide continually updated validity evidence.

Table 1. Correlations of theory-based calibrations produced by the specification equation and data-based item difficulties

Test	# of Questions	# of Passages	$r_{(OT)}{}^{a}$	$R_{(OT)}{}^{b}$	$R^{*}_{(OT)}{}^{c}$
SRA	235	46	.95	.97	1.00
CAT-E	418	74	.91	.95	.97
Lexile	262	262	.93	.95	.97
PIAT	66	66	.93	.94	.97
CAT-C	253	43	.83	.93	.96
CTBS-U	246	50	.74	.92	.95
NAEP	189	70	.65	.92	.94
Battery	26	26	.88	.84	.87
Mastery	85	85	.74	.75	.77
Totals	1780	722			
Means			.84	.91	.93

[a] $r_{(OT)}$ = raw correlation between observed difficulties (O) and theory-based calibrations (T).

[b] $R_{(OT)}$ = correlation between observed difficulties (O) and theory-based calibrations (T) corrected for range restriction.

[c] $R^{*}_{(OT)}$ = correlation between observed difficulties (O) and theory-based calibrations (T) corrected for range restriction and measurement error.

The regression equation links the syntactic and semantic features of text to the empirically determined difficulty of text. That link, in turn, is reproduced across thousands of test items and millions of examinees.

In applications the consistent display of the link over time provides a basis for using the equation to perform theory-based calibrations of test items and texts, thus rendering empirical calibrations necessary only as checks on the system.

This specification equation joins together previously separated but analogous developments in measures of information. Hartley's [52] log(N) measure of information content (the number of signs in a message), for instance, is akin to the sentence length parameter in equation (1). Similarly, the word frequency parameter is akin to Shannon's[53] classic expression $p * \log(p)$, where more information is implied by a word's greater rarity in the language. Including Shannon's extra p (multiplying the log of the probability of observing a sign by that probability) indicates the entropy of the area under the curve in the logistic ogive [54].

BENEFITS OF METROLOGICAL COMPARABILITY

A wide range of applications for text measures have emerged in recent years [31], [32],[55] and [56]. Measures of information content are taking a wide range of forms, many involving entropy. These statistical approaches tend to be dependent on particular data sets and algorithms. Little, if any, attention is put into identifying and implementing an invariant unit of measurement, or into designing and maintaining a metrological network of instruments traceable to such a unit.

The benefits of metrological comparability for measuring reading ability extend from the advancement of education science's basis in theory to practical quality improvement methods in schools and classrooms [57]. The natural sciences and the monetary economy both enjoy a degree of efficiency in their markets for the exchange of information and prices. This efficiency stems in large part from the existence of rules, roles, and responsibilities [1], [2], [3], [4], [5], [6], [7], [8], [9], [10] and [58] associated with the institutionalization of common units of measurement, such as meters, grams, degrees Celsius, or dollars.

Suppliers, manufacturers, marketers, accountants, advocates, and customers are able to better coordinate and align their investments in physical capital when information systems employ common languages. Similar kinds of coordinations can be expected to emerge as teachers, researchers, and psychometricians establish firmer expectations for educational outcomes and the exceptions that prove (in the sense of test) the rules. For instance, quality circles will facilitate the exchange of instructional outcome information across classrooms, grades, and schools in ways not possible with test scores reported in traditional percentages correct. Curriculum publishers are already developing individualized reading instruction modules that integrate assessment information in ways that make student learning trajectories portable across proprietary tests, schools, and countries.

DISCUSSION

Projected comprehension rates should not be the only factor influencing text selection. To make the quantified measure the sole determinant of a curricular decision would be analogous to reducing a table to its physical dimensions when its color, style, or sentimental or historical value might also be relevant.

Initial efforts at deploying the unit of measurement quickly encountered a chicken and egg question from book publishers: why should they adopt the unit as a means of indicating the text complexity of their books and articles if there were no schools or students prepared to take advantage of that information? Conversely, state departments of education and school districts asked, why should they be interested in a universally uniform measure of reading ability if there were no books or articles to match with students' ability measures?

The solution arose when one publisher incorporated the unit in their own system, involving both a reading curriculum and a reading assessment system. This coordinated reader-text matching made the link to the unit more attractive to testing agencies, who could now point to an additional use for their results; to book publishers, who now were assured of a population of students with measures to match with their books; and to state departments of education and school districts, who could now effectively put the matching system to work.

The English-based system is in use in the U.S., Canada and Australia (with applications emerging in New Zealand, South Africa, and England), and in ESL applications in Korea, Japan, Malaysia, Hong Kong, and elsewhere in Asia. A Spanish system for matching readers and texts in the same unit is in use in Mexico and the Philippines. Researchers in various parts of the world are exploring possibilities for expanding the reader-text matching system to Mandarin, French, Arabic, and other languages.

Educational textbook and curriculum publishers have developed online software applications for tracking individual student growth in reading ability. A report from one such system is shown in Fig. 2. The value of repeated measures of a student over time and across texts is evident in the growth trajectory and the expected convergence of the student's ability with the reading demands of adult life.

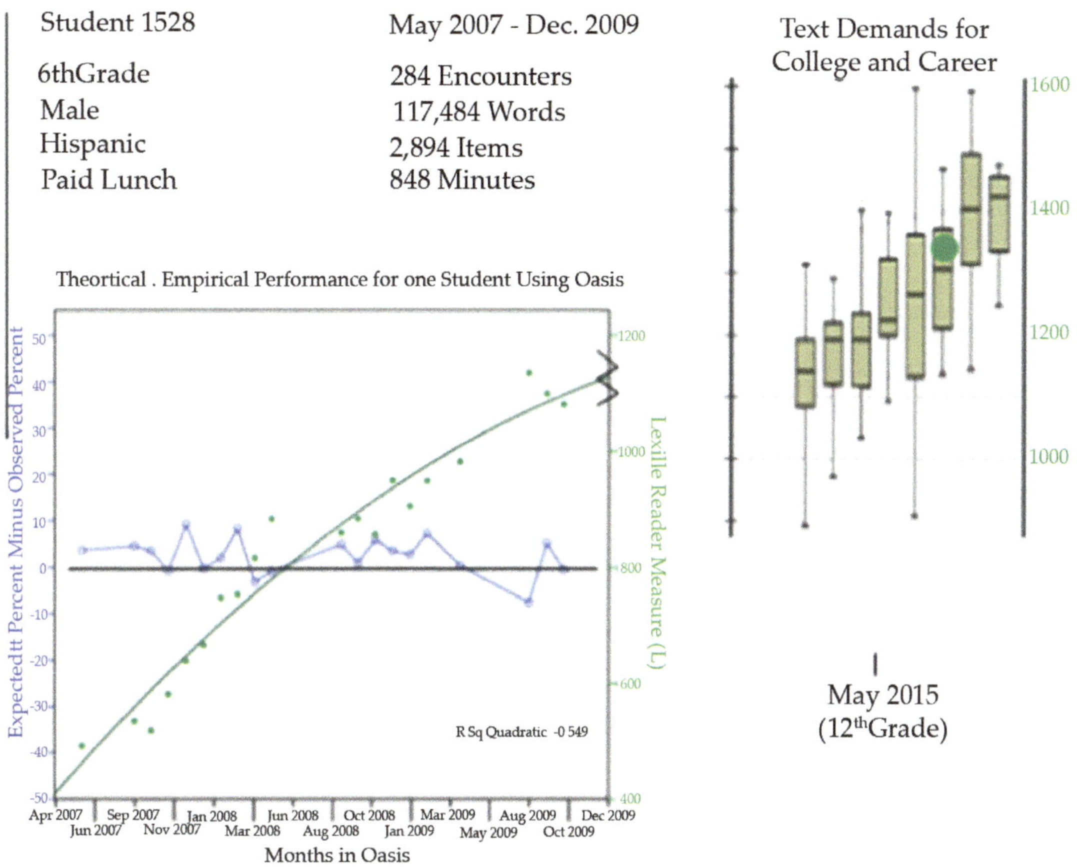

Figure 2. Individual student online reading measurement tracking system report, text domains from left to right are: High School (11–12), SAT I ACT AP, Military, Citizenship, Workplace, Community College, University, Graduate Record Exam.

Fig. 3 shows the relationship between expected and observed text complexity measures in the online system. This plot illustrates the power of theory. Traceability to the standard unit is achieved not only by estimating student reading ability measures from data, but by gauging text complexity from its syntactic and semantic makeup. Given theory-based estimates of item difficulty, items can be adaptively selected for custom-tailored individualized administration, and those students' measures may then be estimated from their comprehension rates relative to the scale values of those items.

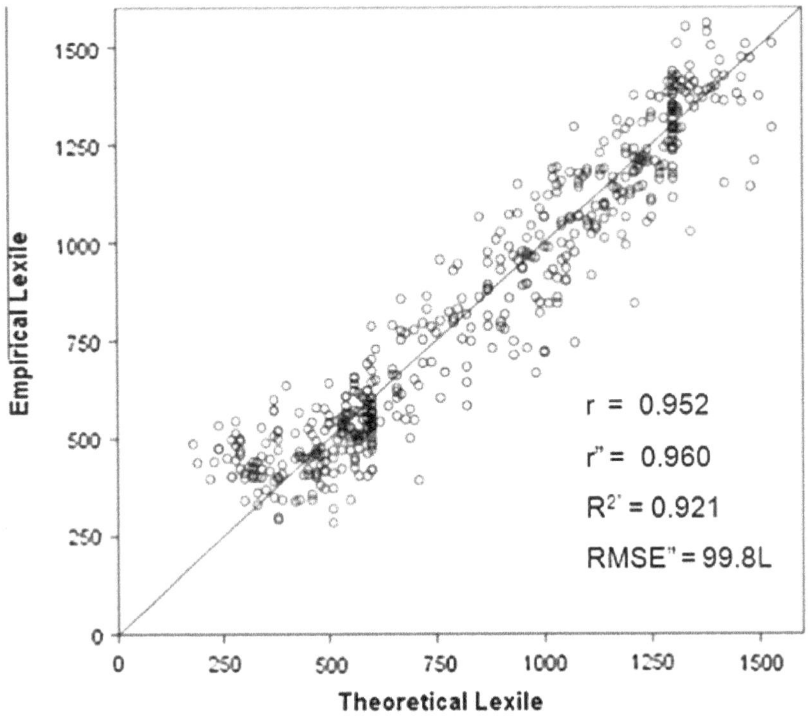

Figure 3. Empirical vs. theoretical Lexile text complexity estimates.

The specification equation operationalizes Rasch's notion of a frame of reference in a way that extends the frame beyond the specific objectivity obtained in the context of a particular test or set of equated tests to an indefinitely large collection of actual or virtual instruments, students, and texts. Theory-based instrument calibration eliminates the need to use data to both calibrate instruments and measure persons. The pay-off from using theory instead of data to calibrate instruments is large and immediate. When data fit a Rasch model, *differences* among person measures are, within the limits of uncertainty and response consistency, free of dependencies on other facets of the measurement context (i.e., the differences are specifically objective). When data fit a causal or theory-enhanced Rasch model, *absolute* person measures are free (again, within the limits of uncertainty and response consistency) of the conditions of measurement (items, occasions, etc.) making them objective beyond the limits of a specific frame of reference tied to local samples of examinees and test items [49] and [50]. In the theory-referenced context, person measures are individually-centered statistics; no reference to another person(s) figures in their estimation.

One of the most important uses of reading test scores is to predict how a reader will perform on non-test tasks. For example, imagine that first year college textbooks are virtual reading tests with item calibrations provided by the specification equation. Arbitrarily, but usefully, fixing a success rate on the virtual items for each textbook enables solving for the reader measure needed to correctly answer that percentage of those items. The individual reader's measure is then interpreted relative to the text complexity measure for each text in the freshman book bag. If the likely success rate in correctly answering the virtual items is high, so is the expectation of having the reading skills needed to complete the first year of college.

High school graduates' reading measures can thus be compared to college text demands and a reasoned prediction can be made as to the likelihood of having the reading level needed for first year completion. The efficiencies this system realizes from its use of validated predictive theory shows special promise as a tool for tracking reading readiness for post-secondary experiences in college, the work place, and the responsibilities of citizenship [31] and [32].

CONCLUSION

Historians of science have repeatedly documented the roles in theory development played by researchers with hands-on experience with instrumentation, as when Kuhn [59]notes that seven of the nine pioneers in quantifying energy conversion processes were either trained as engineers or were working with engines when they made their contributions. Indeed, this attitude that an instrument can make a science was taken from physics into economics by both Stanley Jevons and Irving Fisher in their uses of the balance scale as a model of market equilibrium [60] and [61].

But history shows that instruments alone are insufficient to the task of making a science. Furthermore, interestingly, equilibrium models have failed as guides to economic phenomena in large part because of problems in stochastic aggregation and variation in individual consumer behaviors [62]. In specific circumstances [1], [2], [3], [4], [5], [6], [7],[8], [9], [10], [11] and [12], however, instruments providing consistent information expressed in a common language throughout interconnected nodes of a network, as with the reading measurement system described here, may serve as a medium for coordinating spontaneous individual behaviors and decisions over time and space.

The historical success of science increasingly appears to stem from its embodiment of evolving ecologies of this kind of data-theory-instrument assemblage. Current conceptualizations and institutional systems prioritizing centralized design, data analysis, and policy formation stand in paradigmatic opposition to this ecologizing perspective [63],[64], [65], [66] and [67]. How will cultures of decentralized innovation, complex self-organization, and authentic engagement with substantive, meaningful processes emerge in education and the social sciences? The organic integration of theory, data, and instruments in institutional contexts sensitive to ground-up self-organizing processes requires systematic conceptualizations of measurement as a distributed process, where scientific fields, markets, and societies operate as massively parallel stochastic computers [66] and [67]. Recent comparisons of engineering and psychometric perspectives on the possibility of such systems in education suggest a viable basis for such conceptualizations [68], [69], [70], [71], [72] and [73]. Metrological traceability systems of this kind [24] will integrate qualitative progressions in learning defined by predictive theories of causal relations [49], construct maps [74], and associated item hierarchies in educational assessments generally. Systematically introduced infrastructural supports could effectively exploit the proven value of formative assessment[75] in a hopeful development for broadly enhancing educational outcomes via research and local quality improvement efforts.

ACKNOWLEDGEMENTS

This research was supported by funding from the Gates Foundation and MetaMetrics, Inc. Thanks is extended to an anonymous reviewer for multiple suggestions that improved the quality of the article.

REFERENCES

1. E. Hutchins, Cognition in the Wild, MIT Press, Cambridge, Massachusetts, 1995.
2. E. Hutchins, Mind, Culture, Activity 19 (2012) 314–323.
3. B. Latour, Science in Action: How to Follow Scientists and Engineers Through Society, Cambridge University Press, New York, 1987.
4. B. Latour, Reassembling the Social: An introduction to Actor-Network-Theory, Oxford University Press, Oxford, England, 2005.
5. N.J. Nersessian, Mind, Culture, Activity 19 (2012) 222–239.
6. S. Schaffer, in: R. Bud, S.E. Cozzens (Eds.), Invisible Connections: Instruments, Institutions, and Science, SPIE Optical Engineering Press, Bellingham, Washington, 1992, pp. 23–56.
7. M.N. Wise, The Values of Precision, Princeton University Press, Princeton, New Jersey, 1995.
8. P. Galison, Image and Logic: A Material Culture of Microphysics, University of Chicago Press, Chicago, 1997.
9. P. Dear, Osiris 27 (2012) 37–55.
10. J. O'Connell, Soc. Stud. Sci. 23 (1993) 129–173.
11. M. Sundberg, Soc. Stud. Sci. 41 (2011) 107–125.
12. S. Shapin, Am. Sci. 77 (1989) 554–563.
13. A.N. Whitehead, An Introduction to Mathematics, Henry Holt and Co., New York, 1911.
14. E. Husserl, The Crisis of European Sciences and Transcendental Phenomenology, Northwestern University Press, Evanston, Illinois, 1970.
15. A.W. Woolley, E. Fuchs, Org. Sci. 22 (2011) 1359–1367.
16. J. Cohen, Am. Psychol. 49 (1994) 997–1003.
17. T.P. Wilson, Soc. Forces 49 (1971) 432–444.
18. D.S. Salsburg, Am. Stat. 39 (1985) 220–223.
19. G. Rasch, Probabilistic Models for Some Intelligence and Attainment Tests, 1960, Danmarks Paedogogiske Institut; Copenhagen, Denmark (Reprint, with Foreword and Afterword by B.D. Wright, University of Chicago Press; Chicago).
20. R.D. Luce, Philos. Sci. 45 (1978) 1–16.
21. R.D. Luce, J.W. Tukey, J. Math. Psychol. 1 (1964) 1–27.
22. D.S. Burdick, M.H. Stone, A.J. Stenner, Rasch Meas. Trans. 20 (2006) 1059–1060.
23. W.P. Fisher, Measurement 42 (2009) 1278–1287.
24. W.P. Fisher Jr., M. Wilson, Pensamiento Educativo: Revista de Investigacion Educacional Latinoamericana 52 (2015) 55–78.
25. M. Sadoski, A. Paivio, Sci. Stud. Read. 11 (2007) 337–356.
26. G. Engelhard, J. Appl. Meas. 2 (2001) 1–26.
27. A.J. Stenner, M. Smith, Percept. Mot. Skills 55 (1982) 415–426.
28. A.J. Stenner, M. Smith, D.S. Burdick, J. Educ. Meas. 20 (1983) 305–316.
29. A.J. Stenner, H. Burdick, E.E. Sanford, D.S. Burdick, J. Appl. Meas. 7 (2006) 307– 322.
30. V. Rivero, 2011, June 10 Edtech Digest. Retrieved 23 May 2012, from http:// edtechdigest.wordpress.com/2011/06/10/interview-jack-stenner-takes-educationbeyond-the-metrics/.
31. G.L. Williamson, J. Adv. Acad. 19 (2008) 602–632.
32. G.L. Williamson, J. Fitzgerald, A.J. Stenner, Educ. Res. 42 (2013) 59–69.
33. Y.M. Rabkin, in: R. Bud, S.E. Cozzens (Eds.), Invisible Connections: Instruments, Institutions, and Science, SPIE Optical Engineering Press, Bellingham, Washington, 1992, pp. 57–82.
34. H.D. Thoreau, Walden, or Life in the Woods, Ticknor and Fields, Boston, 1854.
35. F. Nietzsche, The Will to Power, Vintage, New York, 1967.
36. D.J. Price, In Little Science, Big Science—and Beyond, Columbia University Press, New York, 1986, pp. 237–253.
37. R.R. Rentz, W.L. Bashaw, J. Educ. Meas. 14 (1977) 161–179.
38. A.J. Stenner, M. Stone, J. App. Meas. 11 (2010) 244–252.
39. R.A. Spies, J.F. Carlson, K.F. Geisinger (Eds.), The Eighteenth Mental Measurements Yearbook, University of Nebraska Press, Lincoln, Nebraska, 2010.
40. M. Hamilton, K. Pitt, Int. J. Educ. Dev. 31 (2011) 596–605.

41. S.B. Neuman, K. Roskos, Reading Teacher 66 (2012) 207–210.
42. W.E. Bijker, T.P. Hughes, T. Pinch (Eds.), The Social Construction of Technological Systems: New Directions in the Sociology and History of Technology, MIT Press, Cambridge, Massachusetts, 2012.
43. G. Rasch, Rasch Measurement Transactions, 24 (1973/2011) 1309.
44. G.E.P. Box, in: R.L. Launer, G.N. Wilkinson (Eds.), Robustness in Statistics, Academic Press, New York, 1979, pp. 201–235.
45. J. Loevinger, Psychol. Rev. 72 (1965) 143–155.
46. G. Rasch, Rasch Measurement Transactions, 2 (1972/1988) 19.
47. R.P. Carver, J. Reading Behav. 6 (1974) 249–274.
48. J.W. Gibbs, Elementary Principles in Statistical Mechanics, Yale University Press, New Haven, Connecticut, 1981.
49. A.J. Stenner, W.P. Fisher Jr., M.H. Stone, D.S. Burdick, Front. Psychol.: Quant. Psychol. Meas. 4 (2013) 1–14.
50. A.J. Stenner, D.S. Burdick, The Objective Measurement of Reading Comprehension, MetaMetrics, Inc., Durham, North Carolina, 1997. .
51. B.D. Wright, M.H. Stone, Best Test Design, MESA Press, Chicago, Illinois, 1979.
52. R.V.L. Hartley, Bell Syst. Tech. J. 7 (1928) 535–563.
53. C. Shannon, Bell Syst. Tech. J. 27 (1948) 379–423.
54. J.M. Linacre, Rasch Meas. Trans. 20 (2006) 1062–1063.
55. L. Coviello, J.H. Fowler, M. Franceschetti, Proc. IEEE 102 (2014) 1911–1921.
56. W. Zhang, T. Yoshida, X.J. Tang, Knowl.-Based Syst. 21 (2008) 879–886.
57. W.P. Fisher Jr., Assess. Learn. 2 (2013) 6–22.
58. P. Miller, T. O'Leary, Acc. Organ. Soc. 32 (2007) 701–734.
59. T.S. Kuhn, The Essential Tension: Selected Studies in Scientific Tradition and Change, University of Chicago Press, Chicago, 1977.
60. M. Boumans, in: M.S. Morgan, J. Klein (Eds.), The Age of Economic Measurement, Duke University Press, Durham, North Carolina, 2001, pp. 313–344.
61. H. Maas, in: M.S. Morgan, J. Klein (Eds.), The Age of Economic Measurement, Duke University Press, Durham, North Carolina, 2001, pp. 277–302.
62. F. Ackerman, J. Econ. Methodol. 9 (2002) 119–139.
63. F.A. Hayek, The Fatal Conceit, University of Chicago Press, Chicago, 1988.
64. B. Latour, Ecologie Politique 13 (1995) 5–27.
65. T. Weitzel, Economics of Standards in Information Networks, Physica-Verlag, New York, 2004.
66. W.B. Arthur, Complexity and the Economy, Oxford University Press, New York, 2014.
67. C. Hidalgo, Why Information Grows: The Evolution of Order, from Atoms to Economies, Basic Books, New York, 2015.
68. L. Mari, M. Wilson, J. Phys. Conf. Ser. 459 (2013) 012002.
69. L. Pendrill, W.P. Fisher Jr., J. Phys. Conf. Ser. 459 (2013) 012057.
70. L. Pendrill, W.P. Fisher Jr., Measurement 71 (2015) 46–55.
71. L. Pendrill, NCSLi Meas.: J. Meas. Sci. 9 (2014) 22–33.
72. V. Turetsky, E. Bashkansky, Measurement 78 (2016) 348–357.
73. M. Wilson, L. Mari, A. Maul, D. Torres Irribara, J. Phys. Conf. Ser. 588 (2015) 012034.
74. M.R. Wilson, J. Res. Sci. Teach. 46 (2009) 716–730.
75. J. Hattie, Visible Learning: a Synthesis of Over 800 Meta-Analyses Relating to Achievement, Routledge, New York, 2008.

CHAPTER 3 RELATIONAL MEASUREMENTS AND UNCERTAINTY

Ken Krechmer

757 Greer Road, Palo Alto, CA 94303, USA

ABSTRACT

In representational measurement theory, the current theory of all measurements, calibration and sampling processes are assumed to be a linear transformation of the coordinate system, of no effect. In this paper calibration and sampling are shown to be independent non-linear processes which do change measurement results. Relational measurement theory is developed to include calibration and sampling. The measurement changes caused by calibration and sampling are proven to be equal to the quantum measurement disturbance described by the universal uncertainty relation which has been verified by experiments. Therefore relational measurement theory explains the measurement disturbance in quantum mechanics.

INTRODUCTION

In representational measurement theory [1] a measurement result is a magnitude of equal intervals. In this paper relational measurement theory [2] defines a measurement result as a sum of intervals, where each interval's magnitude is modified by both calibration and sampling processes. The application of relational measurement theory to quantum systems explains the disturbance of one observable when measuring a second observable [3]. In this paper: sampling is the division of a continuous observable into discrete intervals [4], a measurement is the sum of the projection of all these discrete intervals onto the measuring apparatus intervals [5], and calibration is the correlation of sets of the measuring apparatus intervals to externally defined intervals [6]; together these three processes produce a measurement result.

In von Neumann's development of a measurement in Hilbert space [7], the possible discrete elements are equivalent (not necessarily equal) and of varying probability. Each such quantum element (in one dimension) may be seen as one of a measuring apparatus' minimum intervals in the same dimension. In representational measurement theory, these elements/intervals are assumed to be equal. Then the standard deviation of a distribution of these elements/intervals is zero.

The Heisenberg Uncertainty Relation (HUR) [8] identifies an inconsistency when the elements/intervals of a measurement system are assumed to be equal. In this paper the standard deviations in the HUR, shown in Eq. (7), are proved

Citation: Ken Krechmer, Relational measurements and uncertainty, Measurement, Volume 93, November 2016, Pages 36-40, ISSN 0263-2241, http://dx.doi.org/10.1016/j.measurement.2016.06.058.

to be of the magnitudes of the observable's intervals, not of the magnitude of the observable, and are never zero. In the relational measurement view of the HUR, each observable's standard deviation is a measure of a distribution of calibration corrections. These calibration corrections appear as a measurement disturbance of the second observable when both observables are correlated by calibration corrections to the same reference.

Sections 2, 3, 4 and 5 develop relational measurement theory by applying concepts from classical metrology. Sections 6 and 7 apply relational measurement theory to the HUR, proving that the standard deviation of a distribution of discrete intervals in the HUR equals the effect of calibration on experimental measurement results.

In 2003, Ozawa [3] developed the universal uncertainty relation, Eq. (12), which is shown to support relational measurement theory. His universal uncertainty relation, with a minor modification, formalizes both calibration and sampling indeterminacy [9]. In Appendix A, experiments by others [10] are presented which verify the universal uncertainty relation and therefore relational measurement theory.

RELATIONAL MEASUREMENTS

A relational measurement system defines the mean measurement result as the product of measurement magnitude and the standard deviation of the measurement intervals due to calibration and sampling processes (4).

Fig. 1 identifies m intervals of i_u, a relational measurement result, relative to m intervals of i, a representational measurement. Sampling, often treated as part of a measurement process, divides the observable into increments (Δs) indicated in Fig. 1 by short lines at right angles to the observable. A measurement is indicated by downward arrows which project the intervals (usually two or more sample increments) onto the measuring apparatus to be summed. Calibration of the measuring apparatus intervals to u (a reference unit) is indicated by the inward arrows. Calibration is represented in an orthogonal plane as it is independent of sampling and measurement. Both planes have a common basis, the measuring apparatus.

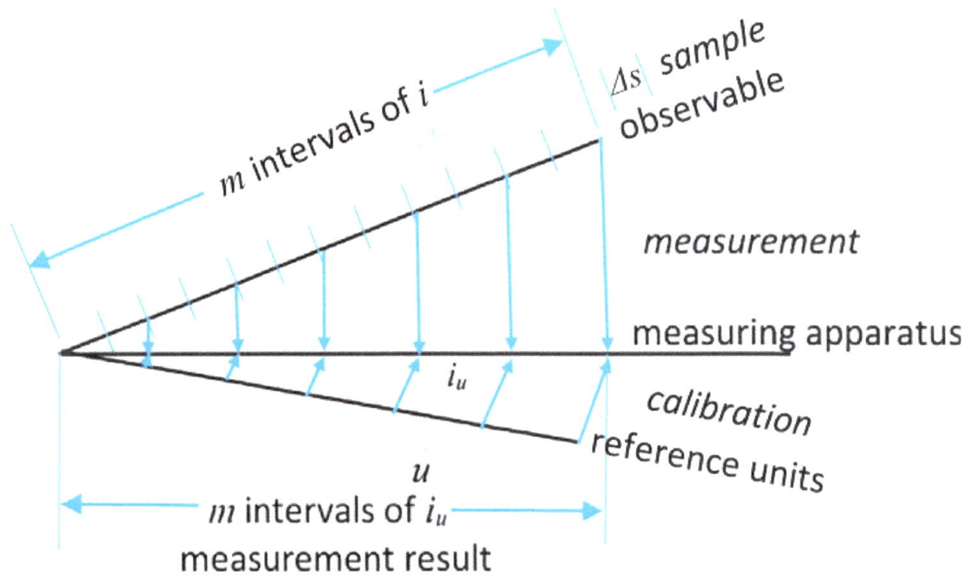

Figure 1. Relational measurement system.

Example: X is a rod's length (the observable). X is experimentally defined as a magnitude (m) of intervals each correlated to a minimum reference unit x (in this example a centimeter). The magnitude of each correlated interval is i_x. The representational measurement (normalized) magnitude of the intervals of X is i = 1/m. The experimental measurement result precision [11] (local variation) is $\pm\Delta s_{xi}$ per interval i [12]. The measuring apparatus is a meter scale which defines the magnitude of the intervals, where each interval is calibrated to x, the reference interval. A calibration process makes feasible comparisons between independent measurement results. The measured accuracy [11] (relative variation) of each i_x is determined by calibration to x. The differences between mi, mx and mi_x are not treated in representational measurement theory.

Applying Fig. 1, Δx_i = calibration variation of each interval and Δs_{xi} = sample increment of each i interval, the calibration (1) and sampling (2) operators are:

$$x\pm\Delta x_i=i_x \quad (1)$$

$$i\pm\Delta sxi=i_x \quad (2)$$

The indeterminacy (the sum of calibration and sampling variation) of i_x is:

$$\Delta i_x=(\mp\Delta x_i)+(\pm\Delta sxi) \quad (3)$$

In representational measurement theory, this indeterminacy is assumed to reduce to zero as accuracy and precision move toward perfect [13]. This is not experimentally possible. In all discrete measurements of continuous observables there is a non-cancelable minimum interval indeterminacy min Δi_x < the product of the speed of light and the reciprocal of the highest sampling frequency observed. The higher the sampling frequency, the smaller $|\min \Delta i_x|$ (the vertical bars represent an absolute value). When all other indeterminacy is nulled, the magnitude of each interval i_x randomly deviates by just less than $\pm\Delta s_{xi}$, the sampling increment. Therefore Δx_i cannot completely cancel Δs_{xi}.

Eq. (4) presents four different functions which represent the observable ⟨X⟩ (where brackets indicate the mean of X) based upon three different assumptions.

$$\int \psi^* \widehat{X} \psi dx \rightarrow \sum_{i=1}^{i=m} i \rightarrow \sum_{i_x=1}^{i_x=m} i_x \cong m[\sigma(i_x) + \sigma(\Delta s_{xi})] = \langle X \rangle \quad (4)$$

The first function is from quantum mechanics [14]: $\int \psi^* \widehat{X} \psi dx$. ψ^* is the complex conjugate of the state vector ψ (observable) of x and $\widehat{X}$ is the operator of X, therefore $\int \psi^* \widehat{X} \psi dx$ (assumes infinitesimal sampling increments) represents the mean of a continuous observable. Normalized sampling (assumes i = x) transforms $\int \psi^* \widehat{X} \psi dx$ to a discrete measurement [15], shown as the second function: $\sum_{i=1}^{i=m} i$ which is the rod's length, m intervals of i.

Calibration (1) transforms $\sum_{i=1}^{i=m} i$ to the third function: $\sum_{i_x=1}^{i_x=m} i_x$ which is the rod's length in i_x intervals and is the current classical metrology model [16]. When the sampled increment $\Delta s_{xi} \ll i_x$ (assumed in classical metrology), $\sum_{i_x=1}^{i_x=m} i_x$ is close to the fourth function: $m[\sigma(i_x)+\sigma(\Delta sxi)]$ as $\sigma(\Delta sxi)<1$ sampled increment (derived below in Section 3).

The fourth function, $m[\sigma(i_x)+\sigma(\Delta sxi)]$, does not require these three assumptions. This function sums over the common basis both the calibration (1) and sampling (2)operators and is the only function of the four which represents the mean of the experimental measurement results of a continuous observable at all experimentally possible sampling frequencies.

SAMPLING AND CALIBRATION EXAMPLES

Consider a digital voltmeter (measuring apparatus) where 00.01 is the voltmeter's display of the minimum interval. Measuring a fixed voltage (observable) multiple times produces a stochastic distribution of measurement results. The maximum indeterminacy of this distribution is specified by the manufacturer, for all voltmeters of this model, to be ±1% indeterminacy. To maintain the ±1% of a 0.01 V measurement requires ±0.0001 V precision, i.e., each interval is between 0.0099 and 0.0101 V. The ±1% indeterminacy allows a laboratory with multiple voltmeters to make comparable measurements or compare measurements with other laboratories. To achieve this precision, a sampling increment of 0.0001 (Δs_{xi}) V or less is required. The precision of a measurement result is ultimately limited by the Planck constant, the minimum possible sampled increment, and is never zero.

A stochastic distribution of voltmeter measurement results occurs when applying the 1.00000 V observable (continuous relative to the sampling increment of 0.0001 V) to many of the same model voltmeters. Indeterminacy of the continuous observable less than the sampling increment is not identifiable. The sampling process causes each of the 100 (m) 0.01 intervals to have an indeterminacy of $\Delta s_{xi} = -0.0001$, 0.0000 ($<\Delta s_{xi}$) or, +0.0001. Less than Δs_{xi} is not a zero state, but the transition between $+\Delta s_{xi}$ and $-\Delta s_{xi}$. Therefore $<\Delta s_{xi}$ statistically occurs less often than either $\pm\Delta s_{xi}$ and the standard deviation of this sampling distribution is always $< |\Delta s_{xi}|$.

In this sampling example, the three possible interval magnitudes (from (2)) are: $i \pm \Delta sxi = i_x = 0.0099i$, 0.0100i or 0.0101i. This identifies 3^{100} combinations of the 100 intervals which establish the probability of the 200 valid measurement results with values between 0.9900 to 1.0100 and within the defined ±1% precision. The distribution of the 3^{100} possible combinations of the 100 intervals will, as the number of voltmeter measurements increase, converge to a normal distribution (bell shaped curve) as described by the central limit theorem. Such a distribution (i.e., sampling noise) occurs in all measurement results and is sometimes identified in the literature as 1/f noise, where f represents the sampling frequency in [17].

As a further example, an observable's angular rotation is determined by counting the teeth of an attached gear with m teeth per 2π rotation. When counted by an observer, the gear teeth are assumed to be intervals of equal width and no indeterminacy appears. But a perfect measurement of angular rotation by using the gear teeth is not possible. As m becomes larger precision increases, but Δs always remains just less than the wavelength of the highest frequency light reflected from a gear tooth and visible to an observer's eye.

Calibration correlates each interval of the experimental measuring apparatus to external or non-local references [18] (from a representational measurement system) for the purpose of comparison. In classical metrology, calibration is assumed to be linear (very close to correct when $i_x \gg \Delta x_i$). Then the minimum interval variation cancels in normal distributions. As example: the 0.00 display of the voltmeter is adjusted, a known 3.00 V measurement standard is applied and then a different voltmeter adjustment (calibration) changes the display to: 300 × 0.01 = 3.00 V.

The effect of both the calibration and sampling processes on a theoretical measurement result is excluded in representational measurement theory. This exclusion results in the dichotomy, presented in (4), between existing measurement theories based upon representational measurement theory (second function) and classical metrology measurement results (third function). Section 4 defines the value of the fourth function.

SAMPLING ENTROPY

The division between discrete and continuous observables is determined by the sampling frequency, f. That is, when f ³ 2 times the highest frequency of the observable, the observable's intervals are often assumed to be discrete relative to the measuring apparatus. Such a measurement assumption may be described as counting. When f < 2 times the highest frequency of the observable, the observable appears continuous relative to the measuring apparatus and measurement results often have a specified precision and accuracy.

Shannon [19] identifies this entropy difference between the discrete and continuous processes that produce a measurement result: "There is one important difference between the continuous and discrete entropies. In the discrete case the entropy measures in an absolute way the randomness of the chance variable. In the continuous case the measurement is relative to the coordinate system". The italics are Shannon's.

Shannon's formal development [20] provides (5), the sampling entropy due to Δs.

$$H(\Delta s) = -(\log_e \sqrt{2\pi e} + \log_e \sigma(\Delta s)) \quad (5)$$

As was demonstrated in Section 3, σ(Δs) is near 1 in classical measurements with normal sampling distributions, so the $\log_e$σ(Δs) is near zero. This makes it practical in many classical measurements to ignore the sampling effects. In quantum measurements where σ(Δs) may not be near 1, this entropy change must be treated.

RELATIONAL STANDARD DEVIATION

Metrology establishes a hierarchy of measurement references (standards) to establish accuracy, beginning with the primary standards, then intermediate standards, and finally to the calibration of a measuring apparatus [21]. When a measuring apparatus is calibrated to x (a specific standard), its accuracy is measured by a relational standard deviation, where the mean of the intervals is replaced by a standard interval magnitude, x. Based upon (1), the relational standard deviation is defined as:

$$\sigma(i_x) = \sigma(x + \Delta x_i) = \sqrt{\frac{1}{m}\sum_{i=1}^{i=m}(x + \Delta x_i)^2} \qquad (6)$$

Considering that x > Δx_i, each Δx_i may vary, and that square roots (except of the rare perfect squares) are irrational numbers, the magnitude of (6) is almost always an irrational number, which is indeterminate. This is a formal demonstration that the calibration process, like the sampling process, is almost always indeterminate. Therefore, the magnitude of the relational standard deviation of the intervals will vary with the accuracy of calibration.

Uncertainty relations

A measurement disturbance, where the first measurement of an observable instantaneously changes the measurement of a second correlated observable is well known in theory and practice [22]. The measurement disturbance appears in the uncertainty relation:

$\sigma(i_a)\sigma(i_b)$=σ(A)σ(B) 1/2|∫ψ*(AB-BA)ψdx| (7)

The terms on either side of ³ in (7) are the Robertson form [23] of the HUR (where A and B need not be canonical conjugates). The left most term of (7) is from (6) and calculates the standard deviations of the intervals, i_a or i_b; the middle term calculates the standard deviation of the sum of each observable's intervals, which is σ(A) or σ(B). When m(magnitude of A) and m′ (magnitude of B) are constant (i.e., m and m′ are magnitudes of fixed observables), then $\sigma(i_a)$ = $\sigma(mi_a)$ = σ(A) and $\sigma(i_b) = \sigma(m'i_b)$ = σ(B). When the observables are fixed, (7) is not correlated to m or m′ as has been assumed in quantum formalisms [24].

In (7) (AB – BA), the first term AB indicates the observable A is measured first, then Band the reverse in the second term. The magnitude of the first term is almost always different than the second, identifying the first measurement changes a second differently each time. While this does not appear logical when measuring a fixed observable, it follows directly when the relational standard deviation (6) of the set of the intervals of each observable is considered.

As the scale of an experimental measurement approaches zero, Δi_a and Δi_b(indeterminacy of i_a and i_b) and i_a and i_b (intervals) both approach the Planck constant as a limit. Then the indeterminacy as a percentage of an interval increases in quantum scale measurements but is very small in measurement results above the quantum scale.

Based upon recent experiments and formal developments (see Appendix A), (7)expands into (8). The indeterminacy relation, (8) is the product of accuracy, $\sigma(x + \Delta i_a)$ and precision $\sigma(\Delta s_{ai})$, for one observable and the equivalent $\sigma(x + \Delta i_b)$ and $\sigma(\Delta s_{bi})$ for the other observable. x is the reference unit common to A and B.

$$[\sigma(x + \Delta a_i) + \sigma(\Delta s_{ai})][\sigma(x + \Delta b_i) + \sigma(\Delta s_{bi})] \geqslant 1/2\left|\int \psi * (AB - BA)\psi dx\right| \qquad (8)$$

The product of the left side of (8):

$$\sigma(x + \Delta a_i)\sigma(x + \Delta b_i) + \sigma(x + \Delta a_i)\sigma(\Delta s_{bi}) + \sigma(\Delta s_{ai})\sigma(x + \Delta b_i) + \sigma(\Delta s_{ai})\sigma(\Delta s_{bi}) \qquad (9)$$

The sum of the four variances in (9) is shown in Appendix A to be the same as the universal uncertainty relation (12). When Δs is assumed to be 0, the variance of (9) is$\sigma(x+\Delta a_i)\sigma(x+\Delta b_i)$, equal to the HUR. The sum of the four variances in (9) will rarely be equal when resampled or recalibrated. In agreement with the experimental results and theory referenced in Appendix A, the calibrations and samplings to references establish the correlation between A and B, and measurement intervals will remain within $\pm\Delta i_x$ if the relational measurement system is not changed by time evolution, resampling or recalibration.

The terms of (9) are expanded to better identify how precision (10) and accuracy (11)appear:

$$\sigma(\Delta s_{ai}) = \sqrt{\frac{1}{m}\sum_{i=1}^{i=m}\left(\Delta s_{ai} - \frac{1}{m}\sum_{i=1}^{i=m}\Delta s_{ai}\right)^2}\ \sigma(\Delta s_{bi}) = \sqrt{\frac{1}{m}\sum_{i=1}^{i=m}\left(\Delta s_{bi} - \frac{1}{m}\sum_{i=1}^{i=m}\Delta s_{bi}\right)^2} \qquad (10)$$

Applying (10) and assuming no other interval indeterminacy than sampling, the mean of all Δs_{ai} = 0. Then $\sigma(\Delta s_{ai}$ or $\Delta s_{bi})$ approaches 1, because Δs_{ai} or Δs_{bi} = 0 or ±1 increment, and Δs_{ai} or Δs_{bi} = 0 is statistically rare. This analysis is corroborated by the experiments described in Appendix A. The standard deviations (10) represent the sampling effects related to the local intervals and do not include the effects related to the non-local reference x, shown in (11).

$$\sigma(x+\Delta a_i)=\sigma(i_a)=\sqrt{\frac{1}{m}\sum_{i=1}^{i=m}(x+\Delta a_i)^2}\sigma(x+\Delta b_i)=\sigma(i_b)$$
$$=\sqrt{\frac{1}{m}\sum_{i=1}^{i=m}(x+\Delta b_i)^2} \quad (11)$$

The product of the two equations in (11) is $\sigma(i_a)\sigma(i_b)=\sigma(x+\Delta a_i)\sigma(x+\Delta b_i)$, the same as the left most variance (calibration) in (9) and the same as the left most term in(7), the HUR. This is the proof that a measurement disturbance is established by calibration. This is corroborated by the experiments in Appendix A.

SUMMARY

The result of an experimental measurement of a continuous observable is the sum of a finite set of imperfect intervals. These intervals are established by sampling, which is imperfect division, and may be partially corrected by calibration. The external (from the measurement system) calibration reference and sampling frequency reference are applied in rigorous experimental measurements. Non-linear changes correlated to these external references change the entropy of the measurement system. In classical metrology, these entropy changes appear as the indeterminacy of a measurement result, termed accuracy and precision. In current quantum mechanics, these entropy changes appear as a measurement disturbance.

Since all experimental measurements are calibrated and have finite sampling increments, representational measurement theory does not represent the experimental measurement of continuous observables. Relational measurement theory, by including the external references, supports consistent measurement calculations and experimental results at all scales.

ACKNOWLEDGEMENTS

The author would like to acknowledge: Frank Barnes, Chris Fields, Toshi Kurokawa, Elaine Baskin, Luca Mari, Yuji Hasegawa and Savely Karshenboim for their active support and assistance in the development of this work.

This research did not receive any specific grant from funding agencies in the public, commercial, or not-for-profit sectors.

APPENDIX A

Experimental verification of the universal uncertainty relation (12) is provided by neutron spin axis position experiments on a stream of neutrons (observables) [10]. This Appendix relates these experiments and the formal development [3] the experiments are based upon, to the relational measurements paper. A complete description of the experiments or the formal developments is provided in the source references. The notation in this Appendix follows the notation in each source referenced, except where noted.

Experimental Verification

In the first measurement of the experiments, the measuring apparatus **A1** is detuned (de-calibrated) in a way that it projectively measures the output operator O_A instead of A thus establishing an error ε in the A measurement. The subsequent measurement of the orthogonal observable B in the eigenstate of O_A, performed by measuring apparatus **A2**, virtually modifies B to be O_B from whose expectation values the disturbance η on B can be determined (from Fig. 5 in [10]).

The neutron spin measurement process begins with adjustments which fix at zero the azimuthal angle, ϕ_{OA} which is the detuning between A and O_A. ϕ_{OA} is the measure of calibration in relational measurements. O_A is a metric of time/distance relative to the fixed precession period, 2π radians, which is the reference unit of the measuring apparatus and appears as a distance along the y axis. Referring to Fig. 6 in [10], this preparation process establishes the physical placement of DC-2 to DC-3 relative to each other and to Analyzer 1. This adjustment process correlates the physical distance on the y axis between DC-2 and DC-3 to n × 2π radians. For the measurement of O_B relative to B, DC-4, Analyzer 2 and the final detector are similarly adjusted to Analyzer 1.

After the above adjustments are complete the detuning of O_A may be shown. The detuning process changes a distance on the y axis relative to the 2π Larmor precession which is the reference. The 2π radians represents a fixed distance on the y axis that is further divided into each detuning angle (ϕ_{OA}) shown in Fig. 8 in [10]. When DC-2/DC-3 is detuned relative to Analyzer 1 and Analyzer 2, O_A is de-calibrated from A, O_B is de-calibrated from B, and both ∈(A) and η(B) change as shown in Fig. 8. The two orthogonal dimensions are not related by a Fourier transform. Thus the measurement results of Aand B are only correlated by their common reference. Without the additional correlation of a Fourier transform, the effect of de-calibration on the measurement results (Fig. 10 in[10]) follows Ozawa's theory [3] closely.

When ϕ_{OA} is zero, O_A = A. This is the discrete case where the measurement interval and the reference interval are equal and there is no calibration effect. As this experiment has reasonable precision (~1.5°, stated in page 8 in [10]) the sampling effects, σ(A) and σ(B) are close to 1 increment which verifies the analysis in Section 3.

Formal Development

The experiments above verify the universal uncertainty relation, equation (266) in [3], as:

$$\varepsilon(A,\rho)\eta(B,\rho)+\varepsilon(A,\rho)\sigma(B,\rho)+\sigma(A,\rho)\eta(B,\rho)\geqslant\frac{1}{2}|Tr([A,B]\rho)| \quad (12)$$

Eq. (12) is equivalent to the indeterminacy relation shown in (9), the expanded version of(8). In (12)∈(A) is a disturbance of A caused by a measurement and η(B) is the noise of B that occurs due to this disturbance. ρ (applying the relational measurement vocabulary) represents a normalized possible measurement interval. In (12) the standard deviations shown as σ(A, ρ) and σ(B, ρ) represent the sampling effects the same as σ(A) and σ(B) in[10]. When comparing the left side of (9), (10), (11) and (12), (12) must be slightly modified. This is developed below.

In (12)∊(A, ρ) and η(B, ρ) are two root mean squared values each calibrated to x (termed a true value [a] in [3] page 17 bottom of the right column). Thus ∊(A, ρ) and η(B, ρ) are equal to the relational standard deviations$\sigma(x+\Delta a_i)$ and $\sigma(x+\Delta b_i)$ in (11). The next two product terms identify the variance due to the sampling effects on these two relational standard deviations.

Note that a fourth term, σ(A, ρ) σ(B, ρ), appears mathematically consistent in (12). Modifying (12) to include this fourth term does not change the inequality of (12). This modification is supported by: A and B are correlated by sampling to a common reference (time) in the neutron spin example (not uncorrelated as expressed in [3] (235)). Therefore the product of the "pre-measurement uncertainties" ([3] page 25 section C) should be included in (12). And, the test results in [10] show that both sampling standard deviations are slightly less than one, therefore their product is close to one, not zero.

The point of modifying (12) to include this fourth term is to factor the modified (12) into (8). Eq. (8) identifies that the first standard deviation term of each observable is a calibration process and the second is a sampling process. The experimental results can ignore this fourth term because the sampling entropy (5), a constant plus the log of close to one (which is close to zero), so the fourth term has little effect on the experimental results. But it does effect the understanding of indeterminacy.

REFERENCES

1. D.H. Krantz, R.D. Luce, P. Suppes, A. Tversky, Foundations of Measurement, Academic Press, New York, 1971 (This three volume work is the foundational text on representational measurement theory).
2. L. Mari, S. Sartori, A relational theory of measurement: traceability as a solution to the non-transitivity of measurement results, Measurement 40 (2007) 233–242 (Mari and Sartori include calibration in a relational measurement, but not sampling).
3. M. Ozawa, Uncertainty relations for noise and disturbance in generalized quantum measurements, Ann. Phys. 311 (2004) 350–416. .
4. C.E. Shannon, Communications in the Presence of Noise, in: N.J.A. Sloane, A.D. Wyner (Eds.), Collected Papers, IEEE Press, New York, NY, 1993, p. 161 (The Sampling Theorem).
5. J. von Neumann, Mathematical Foundations of Quantum Mechanics, Princeton University Press, Princeton, NJ, USA, 1955, p. 351 (Process 1).
6. A.E. Fridman, The Quality of Measurements: A Metrological Reference, Springer Science+Business Media, 2012, p. 132 (Calibration is the name given to a set of operations that establish a relationship between the value of the quantity obtained using the given measuring instrument, and the corresponding value of the quantity determined using the standard).
7. J. von Neumann, Mathematical Foundations, p. 34, Abstract Hilbert Space.
8. W. Heisenberg, The Physical Content of Quantum Kinematics and Mechanics, in: J.A. Wheeler, W.H. Zurek (Eds.), Quantum Theory and Measurement, Princeton University Press, Princeton, NJ, 1983, pp. 62–84.
9. In this paper quantum mechanical uncertainty is the same as accuracy in metrology and indeterminacy is accuracy plus precision in metrology. D.C. Cassidy, Am. J. Phys. 66 (1998) 278–279. Answer to Question #62: When did the indeterminacy principle become the uncertainty principle?
10. G. Sulyok, S. Sponar, J. Erhart, G. Badurek, M. Ozawa, Y. Hasegawa, Violation of Heisenberg's error-disturbance uncertainty relation in neutron-spin measurements, Phys. Rev. A 88 (2013) 022110. .
11. A.E. Fridman, The Quality of Measurements, Chapter 2, Measurement errors; page 180, correctness (accuracy) and precision.
12. A. Gersho, R.M. Grey, Vector Quantization and Signal Compression, Kluwer Academic Publishers, Boston, MA, 1992 (Chapter 3, Sampling. Nyquist sampling identifies that P2 times the highest frequency of the observable is necessary in the time dimension).
13. D.H. Krantz, et al., Foundations of Measurement. Dimensional Analysis and Numerical Laws, p. 454 (Chapter 10, All the physical measures with which we shall deal are...completely determined except for an arbitrarily chosen unit).
14. E. Schrödinger, About Heisenberg Uncertainty Relation, in: Proceedings of the Prussian Academy of Sciences, Physics-Mathematical Section, 1030, XIX, 23 May 2008, pp. 296–303. .
15. J. von Neumann, Mathematical Foundations, p. 334 (The measurement process has a discrete spectrum).
16. International Vocabulary of Metrology (VIM), third ed., BIPM JCGM 200, 2012, para. 2.1 .
17. L.M. Ward, P.E. Greenwood, 1/f noise, Scholarpedia 2 (12) (2007) 1537.
18. J.S. Bell, Speakable and unspeakable in quantum mechanics, Cambridge University Press, Cambridge, England, 1987 (On the Einstein-Podolsky-Rosen paradox, p. 20. In relational measurements calibration to a reference is the non-local variable).
19. C.E. Shannon, The Mathematical Theory of Communications, University of Illinois Press, Chicago, IL, 1963, p. 90. para 8.
20. C.E. Shannon, The Mathematical Theory of Communications, p. 89, provides the derivation of (5), the entropy of a one dimensional normal distribution.
21. A.E. Fridman, The Quality of Measurements, p. 123 on, uniformity and traceability of measurement.
22. J. von Neumann, Mathematical Foundations, p. 417 (Chapter VI, The measuring process).
23. H.P. Robertson, The Uncertainty Principle, Phys. Rev. 34 (1929) 163–164.
24. J. von Neumann, Mathematical Foundations, p. 211 (3. Simultaneous measurability).

CHAPTER 4

EFFECTS OF THE CALIBRATION PROCEDURE ON THE METROLOGICAL PERFORMANCES OF STEREOPHOTOGRAMMETRIC SYSTEMS FOR HUMAN MOVEMENT ANALYSIS

Roberto Di Marco [a,b], Stefano Rossi [c], Enrico Castelli [d], Fabrizio Patanè [e], Claudia Mazzà [b,f], Paolo Cappa [a,d]

[a]Department of Mechanical and Aerospace Engineering, 'Sapienza' University of Rome, via Eudossiana, 18, Roma 00184, Italy

[b]Department of Mechanical Engineering, The University of Sheffield, Sir Frederick Mappin Building, Mappin Street, Sheffield S1 3JD, UK

[c]Department of Economics and Management, Industrial Engineering (DEIM), University of Tuscia, Via del Paradiso, 47, 01100 Viterbo, Italy

[d] Laboratory of Movement Analysis and Robotics (MARlab), Children Hospital 'Bambino Gesù', via della torre di Palidoro, 00050 Passoscuro, Rome, Italy

[e] School of Mechanical Engineering, 'Niccolò Cusano' University, Via Don Carlo Gnocchi, 3, 00166 Rome, Italy

[f]INSIGNEO Institute for In Silico Medicine, The University of Sheffield, The Pam Liversidge Building, Sir Frederick Mappin Building, Mappin Street, Sheffield S13JD, UK

ABSTRACT

This paper proposes a methodology for evaluating the effect of different stereophotogrammetric system calibration procedures on the calculation of marker-based kinematics information. The methodology, based on calibrating the system using data recorded from capture volumes of different sizes and in trials of different durations, was applied to two different systems. The calibration data were used to reconstruct the static and dynamic position and orientation in space of a rigid wand carrying markers in known positions. The inaccuracies in the reconstruction of distances and angles from the wand markers were independent on the calibration data, with average errors lower than 1.7 mm and 0.7°, respectively. Similar results were obtained from human gait data, with the highest variations observed in the transverse plane kinematics and in the foot segment, suggesting that successful calibration procedures of different durations and performed in different volumes did not affect the metrological performance of the investigated systems.

INTRODUCTION

In the past decades, movement analysis techniques have been increasingly used to study human/animal motion [1], [2], [3] and [4]. Besides emerging techniques based on MIMU systems [5], [6], [7] and [8] and markerless approaches [9], [10], [11] and [12], the majority of the human movement analysis techniques are based on the measurement of three-dimensional position of active or passive markers attached to the body skin, as obtained using a stereophotogrammetric approach. These markers are used to track the three-dimensional pose of the subject's bones, to which they are uniquely associated through a procedure called anatomical calibration [1]. Once the pose

Citation: R. Di Marco et al., Effects of the calibration procedure on the metrological performances of stereophotogrammetric systems for human movement analysis, Measurement (2016), http://dx.doi.org/10.1016/j.measurement.2016.01.008.

of a bone is known, the joint kinematics, i.e. the relative orientation between adjacent bones, are estimated and used to quantify movement alterations and limitations and to plan and evaluate a patient's treatment.

Although stereophotogrammetric systems (SS) are routinely used in research and clinical practice, relevant data suffer from a number of inaccuracy sources that could hinder the sought information. The main sources of inaccuracy are: (i) the soft-tissue artefacts (STA) due to the relative movement between the markers attached on the skin and the underlying bones [13]; (ii) errors in the anatomical calibration due to markers' misplacement [14]; and (iii) instrumental errors [15]. Whereas the first two errors are intrinsic in the use of skin markers, the third one is due to the use of a camera-based approach and it has been found to be dependent on: the number and position of the cameras [16] and [17], their lens distortion [18], the dimension of the capture volume[19] and [20] and, last but not least, the algorithms used for the reconstruction of a marker's 3D position [21], i.e. the marker tracking.

The effect of instrumental error on marker tracking have been originally quantified by placing a goniometer equipped with retroreflective markers in different zones of the capture volume [19] and [22], imposing known static angles and random trajectories to the goniometer and then comparing its outputs with the angles measured with a SS. More recently, a T-pendulum has been used for similar purposes, and it has been shown that increased angular velocities of the body under observation can decrease the accuracy of angle measurements [23]. Shifting the problem closer to the human movement analysis, a 'walking test' was proposed in [24] and [25]: a subject was asked to walk at a self-selected speed within the capture volume holding an aluminium bar equipped with two markers and eight SSs were tested. They showed that the systems with low noise generally seem to exhibit better performances. Subsequently, the Movement Analysis Laboratory (MAL) test has been proposed [26], which is based on recording the position of a rod carrying a 2-marker cluster, manually rotated around its tip either following a pseudo circle or two orthogonal arches. The MAL test allows to quantify both precision and accuracy associated with SS-based measurements (respectively related to random and systematic errors).

The need to move cameras, the changes in light conditions, or the presence of reflecting objects call for frequent recalibration of a stereophotogrammetric system within the human movement analysis context. As a matter of fact, the SS manufacturers recommend performing a calibration before each session of data collection. This calibration procedure is performed manually by the operator, who usually has to freely move an object within the camera capture volume, and is hence dependent on the modality of its execution. The evaluation of possible errors associated with the calibration procedure of the SS has been the object of a few investigations. In [27] and [28] two different methodologies to quantify the intrinsic error of the calibration algorithms that reconstruct the marker time histories have been introduced. Despite providing interesting results concerning the quantification of the calibration algorithm errors, they are by definition not useful in quantifying the variations following the need of a system recalibration. More recently, a custom-made robot, which could be used to move a L-frame equipped with retroreflective markers to perform the calibration, has been devised[16]. The authors showed that the use of the robot can significantly improve the accuracy of the calibration. However, the robot was moved within a capture volume ($180 \times 180 \times 150$ mm^3) that is much smaller than those needed in human movement analysis. Last but not least, the effects that the calibration procedure has on the metrological performances of a SS – to the authors' knowledge – have not been fully exploited.

The aim of this paper is to propose a methodology that can be used to evaluate the effect of different calibration procedures, as executed on data capture from different acquisition volumes and of longer or shorter duration, and to use it to quantify the relevance of the effects that those calibration procedures can have on the estimate of the joint kinematics. The proposed methodology will be applied to two different SSs.

MATERIAL AND METHODS

Calibration Procedure

Two stereophotogrammetric systems were set up in two centers: an 8-camera Vicon system MX-series (SS#1 installed at the Movement Analysis and Robotics Laboratory 'MARLab' of the Children Hospital 'Bambino Gesù', Palidoro – Rome, Italy) and a 10-camera Vicon system T-series (SS#2 installed at The University of Sheffield, Sheffield – United Kingdom). The data collection was performed with a sampling frequency of 200 Hz and the marker position reconstruction was performed using the software Vicon Nexus 1.8.5 (Vicon Motion Systems, Oxford – UK). It is worth noticing that having a different number of cameras in the two systems does not affect the accuracy of the calibration, since at least six cameras have been used in both cases [17]. As shown in Fig. 1, a $2.4 \times 3.6 \times 1.6$ m^3 capture volume (Global volume, GV) was identified. Two calibration wands of the same size were used in the two centres, both equipped with five markers, placed at a known distance between each other (Fig. 2), passive for SS#1 and active for SS#2, respectively.

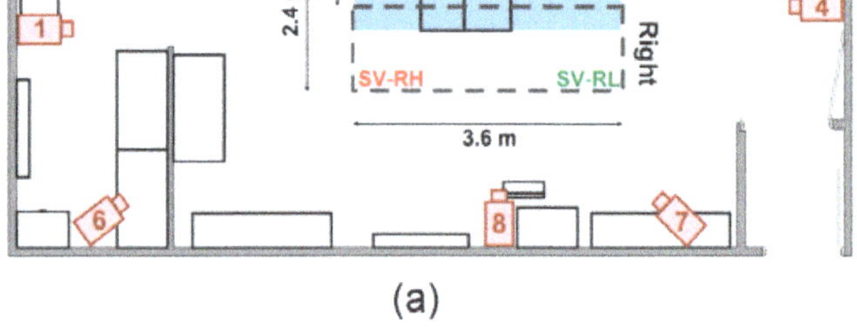

(a)

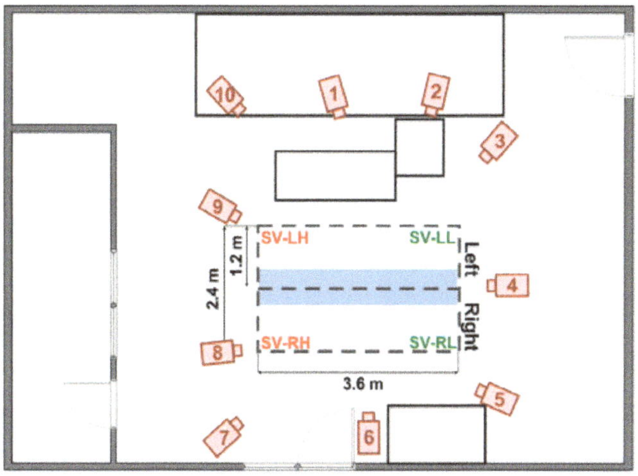

(b)

Figure 1. Maps of the considered laboratories with the highlighted volumes: (a) Movement Analysis and Robotics Laboratory 'MARLab' of the Children Hospital 'Bambino Gesù', Palidoro – Rome, Italy; (b) The University of Sheffield, Sheffield – United Kingdom. The blue areas are the areas where the subject was asked to walk on. The sub-volumes are bounded by the grey dashed-lines: the green tags indicate the SVs on the lower part of the GV, while the red ones indicate the higher. All the measures are given in meters. (For interpretation of the references to color in this figure legend, the reader is referred to the web version of this article.)

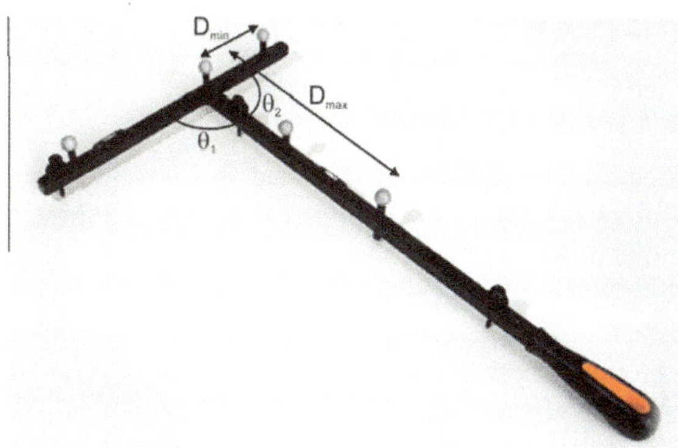

(a)

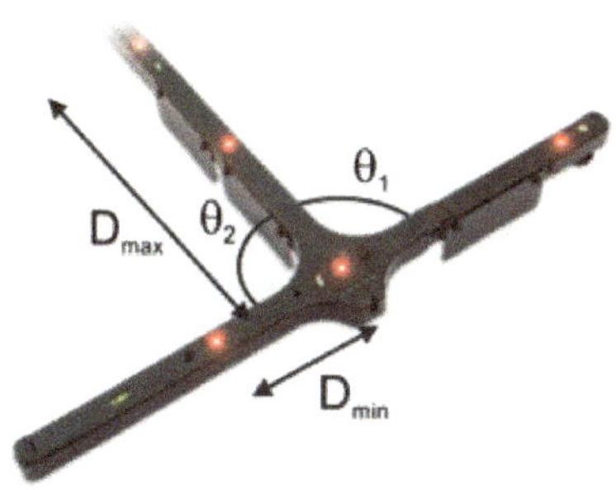

(b)

Figure 2. Rigid calibration wand (a) equipped with retroreflective passive markers at the Movement Analysis and Robotics Laboratory 'MARLab' of the Children Hospital 'Bambino Gesù', Palidoro – Rome, Italy; (b) and equipped with active markers at The University of Sheffield, Sheffield – United Kingdom.

Several calibration volumes, and consequently several calibration files, can be obtained moving the wand only in specific sections of the laboratory. The GV was hence partitioned into four 1.2 × 3.6 × 0.8 m^3 sub-volumes (SVs) defined by the intersections of the half-right, half-left, half-upper and half-lower parts of the GV: (i) left and lower sub-volume, SV-LL; (ii) left and upper sub-volume, SV-LH; (iii) right and lower sub-volume, SV-RL; and (iv) right and upper sub-volume, SV-RH.

The calibration procedure was performed in two phases, following the manufacturer recommendations: a dynamic phase, in which the rigid calibration wand was waved throughout the empty capture volume, ensuring that the markers on the wand are visible to the cameras, and a static phase, in which the wand was placed flat on the floor to identify the origin coordinates and axes of the global reference system. The number of frames (Refinement Frames, RF) used by the calibration and reconstruction algorithm to compute the calibration parameters has to be set before the calibration procedure. The manufacturer of the systems used in this study recommends setting the RF to a value higher than 1000 frames and possibly ranging between 3000 and 5000. Being the frame rate constant, the higher is the RF, the higher the time length of dynamic phase. Two sets of calibration procedures were performed: (i) the volume GV was calibrated varying RF from 1000 to 5000 in steps of 1000; and (ii) each of the SVs was calibrated setting RF = 3000 frames. In order to account for the variability related to the operator, each of the above calibration procedure was repeated three times, for a total of 27 datasets (three × five repetitions for the GV, plus three for each of the four SVs). The files containing the calibration parameters calculated by the calibration algorithm, were stored for the post-processing.

To validate the performances of the stereophotogrammetric systems, two tests were performed in each of the two centres. The first test was called low-level validation test and aimed at quantifying the error associated with measuring fixed distances and angles on a rigid body in static and dynamic conditions. The second test, named high-level validation test, aimed to assess the effect of different calibration procedures on the estimate of the joint kinematics during gait analysis. The post-processing was conducted by using Matlab (MathWorks, Natick – USA).

Test 1 – Low-Level Validation

The calibration wand was put flat on the floor and a trial of 5 s was collected. The standard deviation of the distances between each couple of markers on the wand was calculated from the static trial data. In order to evaluate their expanded uncertainty, for each of the two systems SS#1 and SS#2, the highest among the standard deviation values was multiplied by a coverage factor k = 3. The same procedure was carried out on the angles between the arms of the wand, and the relevant expanded uncertainty was calculated.

The wand was then moved at a velocity comparable with the one used in the dynamic phase of the calibration procedure, and one trial of 20 s was collected. In the perspective of resembling clinical gait analysis, we focused on two couples of markers (Fig. 2) the two closest (D_{min}) and the two most distant (D_{max}) markers on the wand; in such a manner we plan to exploit the SS performances by imposing two known values as input. The calculated distances were compared with the values declared by the manufacturer. As regards the angles, differently to the chosen rationale for the measured distances and considering the options given by the wands, we decided to verify the SS capability in recognizing as equal value two angles (θ_1 and θ_2) between the arms defined by markers differently positioned. The two measured angles were compared with the known value of 90°. For each trial and each calibration file, the RMSE of the distances and angles were computed as an accuracy index. Finally, the average of the RMSEs over the three calibration repetitions were computed. When referring to RMSE values, the following notations will be used: the analysed calibration volume and RF values will be noted as superscript and the investigated variable as subscript (e.g., the RMSE computed for the distance D_{max} considering the calibration volume GV and a number of frames with RF = 2000 is $RMSE_{Dmax}^{GV2000}$). The RF is not indicated when the RMSE was evaluated for the SVs, since it was always equal to 3000.

Test 2 – High-Level Validation

One healthy adult (age 27, height 183 cm, mass 78 kg) was involved in this part of the study after having signed an informed written consent, approved by the two local ethical boards. The subject was equipped with 16 passive markers of 9.5 mm diameter, according to the Vicon Plug-in-Gait protocol: four markers on the pelvis, two on each thigh, two on each shank and three markers on both the feet [29]. One gait trial was acquired asking the subject to walk barefoot at a self-selected speed in the middle of the capture volume (Fig. 1 highlights the walkway in blue). The subject was asked to walk back and forth along a straight line. A total of five right and five left strides, chosen among those recorded in the centre of the measurement volume, were retained for further analysis. As in the Test 1, 27 calibration files were applied to the acquired trials, and the joint kinematics were then estimated for each of them.

As a parameter for waveform similarity, we evaluated the between-calibration coefficient of multiple correlation (CMC) [30] over the five strides. The closest is the CMC to 1, the more similar the waveforms are, with CMC values between 0.85 and 0.95 indicating very good correlation between the waveforms and CMC higher than 0.95 indicating excellent correlations [31]. The maximum angular differences ($\Delta\theta$) among all the calculated average variable waveforms were also determined.

Test 2a – are Joint Kinematics Affected by the RF?

In order to test whether joint kinematics are affected by the RF, a comparison among the calibration performed within GV and changing RF was performed. Assuming that a higher number of RF can improve the calibration performances, in this step we used the values obtained with the calibration Global Volume GV at a RF of 5000 as a reference for the one obtained with other GVs at different RFs and calculated the relevant CMC and $\Delta\theta$. In the following of the paper, RF is used as superscript for the CMC and the angle for which the CMC was extracted as a subscript (e.g. $CMC_{R\text{-}A\text{-}Abd/Add}^{RF}$ stands for CMC computed among the RFs of the right (R) ankle (A) and for the Abd/Adduction). The maximum angular differences between two calibrations were defined to have the comparison as superscript and the considered kinematic variable as subscript: i.e. $\Delta\theta_{L\text{-}H\text{-}Int/Ext}^{5000/2000}$ is the maximum difference between the left (L) hip (H) internal/external rotation computed by using the calibration GV with RF equal to 5000 and 2000.

2.3.2. Test 2b – are joint kinematics affected by the dimension and the position of the calibration volume?

In order to test whether the joint kinematics are affected by the dimension and the position of the calibration volume, a series of comparisons between GV and each SV were performed. In this case, it was assumed that the calibration performances improve considering a GV rather than a SV. Then, we made a comparative analysis of the parameter CMC and $\Delta\theta$ towards the kinematic variables when applying those calibration files.

For this setting, V (Volume) was the superscript for the CMC and the angle was again the subscript: e.g. $CMC_{R\text{-}K\text{-}Flx/Ext}^{V}$ stated for CMC computed among the different volumes and for the right (R) knee (K) flexion/extension. As reported for the Test 2a, the maximum angular differences were defined to have the comparison as superscript and the considered kinematic variable as subscript: i.e. $\Delta\theta_{L\text{-}H\text{-}Int/Ext}^{GV/SV\text{-}RH}$ is the maximum difference between the left (L) hip (H) internal/external rotation computed by using the calibration GV and SV-RH.

RESULTS AND DISCUSSION

Test 1 – Low-Level Validation

Considering the static trial collected in both the centers on the calibration wand and applying the whole calibration set, the expanded uncertainty was evaluated as equal to 0.1 mm for the distances between target points and 0.1° for the angles when using SS#1, and 0.3 mm and 0.3° when using SS#2, respectively. These values were considered as references to estimate the effects that the calibration procedure can have on the performances of SSs in dynamic trials.

According to the literature [23], higher inaccuracies in measuring distances and angles on the rigid wand might be expected when comparing dynamic trials to static ones. In this study, the dynamic inaccuracies were found to be up to five times higher than the static ones. Table 1 shows the mean values of the RMSEs computed for D_{max}, D_{min}, θ_1 and $\theta_2$and for both systems SS#1 and SS#2. For SS#1, both $RMSE_{Dmax}$ and $RMSE_{Dmin}$ for each calibration condition were always less than 0.4 mm. The lowest error (0.2 mm) was obtained for $RMSE_{Dmax}^{GV3000}$ and $RMSE_{Dmax}^{GV4000}$. Changing among the calibration conditions, we always found $RMSE_{\theta 1}$ and $RMSE_{\theta 2}$ equal to 0.2° and 0.5°, respectively. Considering the system SS#2, the $RMSE_{Dmax}$ and $RMSE_{Dmin}$ were found to be always less than 1.7 mm and 1.0 mm, respectively. The reference value obtained for the distances was 0.5 mm. With regard to the angles, the lowest value was found for $RMSE_{\theta 1}^{GV5000}$, equal to 0.2°, and the highest for $RMSE_{\theta 1}^{SV\text{-}RH}$ and $RMSE_{\theta 1}^{SV\text{-}RL}$, both equal to 0.7°. For θ_2 the lowest value was found for $RMSE_{\theta 2}^{GV5000}$ (0.2°), while the highest was found for $RMSE_{\theta 2}^{SV\text{-}RH}$ (0.6°).

Table 1. RMSE values computed by considering the systems SS#1 and SS#2.

		GV1000	GV2000	GV3000	GV4000	GV5000	SV-LH	SV-LL	SV-RH	SV-RL
SS#1	$RMSE_{Dmax}$(mm)	0.3	0.3	0.2	0.2	0.3	0.4	0.3	0.3	0.3
	$RMSE_{Dmin}$(mm)	0.3	0.3	0.3	0.3	0.3	0.4	0.3	0.3	0.3
	$RMSE_{\theta 1}$ (°)	0.2	0.2	0.2	0.2	0.2	0.2	0.2	0.2	0.2
	$RMSE_{\theta 2}$ (°)	0.5	0.5	0.5	0.5	0.5	0.5	0.5	0.5	0.5
SS#2	$RMSE_{Dmax}$(mm)	0.7	0.9	0.9	0.9	0.5	0.9	0.9	1.7	1.1
	$RMSE_{Dmin}$(mm)	0.7	1.0	0.7	0.7	0.5	0.7	0.7	0.9	0.8
	$RMSE_{\theta 1}$ (°)	0.3	0.4	0.4	0.4	0.2	0.4	0.3	0.7	0.7
	$RMSE_{\theta 2}$ (°)	0.3	0.4	0.5	0.4	0.2	0.4	0.4	0.6	0.5

It is worth highlighting that the RMSE values did not vary when comparing the effect on the measurements of distances and angles obtained with the different calibrations. Moreover, despite of the fact that the cameras of the system SS#2 are technologically advanced with respect to those of the system SS#1, not only we did not obtain more accurate results when evaluating the data acquired from this system, but even slightly higher values of errors were found. As mentioned in the introduction, the accuracy of reconstructing marker time histories can depend on several aspects [13], [14], [15], [16],[17], [18], [19], [20] and [21]. It was noticed only after the experiments, that the cameras of the system SS#2 were set up with a high value of aperture than those of the system SS#1. As a matter of fact, the higher is the aperture, the noisier are the measurements. This may be the reason of the slightly increase of the inaccuracy of tracking markers and measuring distances and angles using SS#2. It has to be highlighted, however, that this factor was certainly not relevant when comparing series of data acquired with the same system.

Test 2 – High-Level Validation

Test 2a – are Joint Kinematics Affected by the RF?

As regard the test on the articular kinematics, CMC^{RF} was higher than 0.94 for both SS#1 and SS#2. It means that the waveforms were very similar to each other (Table 2). Considering SS#1 the worst case was found to be $CMC^{RF}_{R\text{-}A\text{-}Inv/Eve}$ (0.94), while for SS#2 it was $CMC^{RF}_{L\text{-}H\text{-}Int/Ext}$ (0.94). Instead, the higher values for the CMC (1.00) were obtained for $CMC^{RF}_{R\text{-}H\text{-}Flx/Ext}$, considering the SS#1, and for $CMC^{RF}_{R\text{-}H\text{-}Flx/Ext}$, $CMC^{RF}_{R\text{-}K\text{-}Flx/Ext}$ and $CMC^{RF}_{L\text{-}H\text{-}Flx/Ext}$, considering the SS#2.

Table 2. CMC values computed on the kinematics both considering the comparison between GV5000 and other GVs (CMC^{RF}), and between GV3000 and SVs (CMC^{V})

		CMC^{RF}		CMC^{V}	
		SS#1	SS#2	SS#1	SS#2
Right hip	$CMC_{R\text{-}H\text{-}Flx/Ext}$	1.00	1.00	1.00	1.00
	$CMC_{R\text{-}H\text{-}Abd/Add}$	0.99	0.98	0.99	0.98
	$CMC_{R\text{-}H\text{-}Int/Ext}$	0.98	0.98	0.98	0.97
Right knee	$CMC_{R\text{-}K\text{-}Flx/Ext}$	0.99	1.00	0.99	1.00
	$CMC_{R\text{-}K\text{-}Abd/Add}$	0.98	0.98	0.97	0.98
	$CMC_{R\text{-}K\text{-}Int/Ext}$	0.97	0.97	0.97	0.97
Right ankle	$CMC_{R\text{-}A\text{-}Plt/Drs}$	0.99	0.99	0.99	0.99
	$CMC_{R\text{-}A\text{-}Int/Ext}$	0.97	0.98	0.97	0.98
	$CMC_{R\text{-}A\text{-}Inv/Eve}$	0.94	0.95	0.93	0.94
Left hip	$CMC_{L\text{-}H\text{-}Flx/Ext}$	0.99	1.00	0.99	1.00
	$CMC_{L\text{-}H\text{-}Abd/Add}$	0.98	0.99	0.98	0.99
	$CMC_{L\text{-}H\text{-}Int/Ext}$	0.98	0.94	0.98	0.93
Left knee	$CMC_{L\text{-}K\text{-}Flx/Ext}$	0.99	0.99	0.99	0.99
	$CMC_{L\text{-}K\text{-}Abd/Add}$	0.97	0.97	0.97	0.97
	$CMC_{L\text{-}K\text{-}Int/Ext}$	0.97	0.97	0.97	0.97
Left ankle	$CMC_{L\text{-}A\text{-}Plt/Drs}$	0.98	0.97	0.98	0.97
	$CMC_{L\text{-}A\text{-}Int/Ext}$	0.98	0.96	0.98	0.96
	$CMC_{L\text{-}A\text{-}Inv/Eve}$	0.95	0.95	0.95	0.94

The first four columns of Table 3 and Table 4 show the maximum angular differences ($\Delta\theta$) between the kinematic variables when processing the static and dynamic trials with the GV calibrations. The $\Delta\theta$ was for SS#1 (Table 3) never higher than 0.3° and it was found, for most of the cases, less than 0.1°. Looking at the same results for SS#2 (Table 4), we found values lower than 0.3° for $\Delta\theta^{5000/1000}_{R\text{-}H\text{-}Flx/Ext}$, $\Delta\theta^{5000/4000}_{L\text{-}K\text{-}Flx/Ext}$, and for the entire set of comparison on the right and left hip Abd/Add; while the higher value 2.8° was reached only for $\Delta\theta^{5000/3000}_{R\text{-}K\text{-}Int/Ext}$. These results allow arguing that the number of Refinement Frames RF does not significantly affect either the waveforms or the angular values of the articular kinematic estimates during the gait cycle.

Table 3. Maximum angular differences between angles estimated by considering the comparison between GV5000 and other GVs ($\Delta\theta^{5000/RF}$), and between GV3000 and SVs ($\Delta\theta^{GV/SV}$) for the system SS#1

Comparison		**5000/1000**	**5000/2000**	**5000/3000**	**5000/4000**	**GV/SV-LH**	**GV/SV-LL**	**GV/SV-RH**	**GV/SV-RL**
Right hip	$\Delta\theta_{R\text{-}H\text{-}Flx/Ext}$ (°)	<0.1	0.1	0.1	0.1	0.1	0.1	0.1	<0.1
	$\Delta\theta_{R\text{-}H\text{-}Abd/Add}$ (°)	<0.1	<0.1	<0.1	<0.1	<0.1	<0.1	<0.1	<0.1
	$\Delta\theta_{R\text{-}H\text{-}Int/Ext}$ (°)	<0.1	<0.1	<0.1	<0.1	0.2	<0.1	0.2	0.1
Right knee	$\Delta\theta_{R\text{-}K\text{-}Flx/Ext}$ (°)	<0.1	<0.1	<0.1	<0.1	0.1	<0.1	0.1	<0.1
	$\Delta\theta_{R\text{-}K\text{-}Abd/Add}$ (°)	<0.1	<0.1	<0.1	<0.1	0.1	<0.1	0.1	0.1
	$\Delta\theta_{R\text{-}K\text{-}Int/Ext}$ (°)	0.1	0.1	0.1	0.1	0.2	0.1	0.1	0.1
Right ankle	$\Delta\theta_{R\text{-}A\text{-}Plt/Drs}$ (°)	0.1	0.1	0.1	0.1	0.1	0.1	0.1	0.1
	$\Delta\theta_{R\text{-}A\text{-}Int/Ext}$ (°)	0.1	0.1	0.1	0.1	0.2	0.1	0.2	0.2
	$\Delta\theta_{R\text{-}A\text{-}Inv/Eve}$ (°)	0.2	0.2	0.2	0.2	0.7	0.2	0.6	0.3
Left hip	$\Delta\theta_{L\text{-}H\text{-}Flx/Ext}$ (°)	<0.1	<0.1	<0.1	0.1	0.1	0.1	0.1	0.1
	$\Delta\theta_{L\text{-}H\text{-}Abd/Add}$ (°)	<0.1	<0.1	<0.1	<0.1	<0.1	<0.1	<0.1	<0.1
	$\Delta\theta_{L\text{-}H\text{-}Int/Ext}$ (°)	0.1	0.1	0.1	0.1	0.1	0.1	0.2	0.2
Left knee	$\Delta\theta_{L\text{-}K\text{-}Flx/Ext}$ (°)	<0.1	<0.1	<0.1	<0.1	0.1	<0.1	0.1	<0.1
	$\Delta\theta_{L\text{-}K\text{-}Abd/Add}$ (°)	<0.1	<0.1	<0.1	<0.1	0.1	<0.1	0.1	0.1
	$\Delta\theta_{L\text{-}K\text{-}Int/Ext}$ (°)	0.1	0.1	0.1	0.1	0.2	0.1	0.1	0.1
Left ankle	$\Delta\theta_{L\text{-}A\text{-}Plt/Drs}$ (°)	0.1	0.1	0.1	0.1	0.2	0.1	0.2	0.2
	$\Delta\theta_{L\text{-}A\text{-}Int/Ext}$ (°)	0.2	0.2	0.1	0.2	0.2	0.1	0.2	0.2
	$\Delta\theta_{L\text{-}A\text{-}Inv/Eve}$ (°)	0.2	0.3	0.3	0.3	0.6	0.2	1.0	0.4

Table 4. Maximum angular differences between angles estimated by considering the comparison between GV5000 and other GVs ($\Delta\theta^{5000/RF}$), and between GV3000 and SVs ($\Delta\theta^{GV/SV}$) for the system SS#2

		5000/1000	5000/2000	5000/3000	5000/4000	GV/SV-LH	GV/SV-LL	GV/SV-RH	GV/SV-RL
Right hip	$\Delta\theta_{R\text{-}H\text{-}Flx/Ext}$ (°)	<0.3	0.3	0.3	0.4	0.6	0.7	0.7	0.4
	$\Delta\theta_{R\text{-}H\text{-}Abd/Add}$ (°)	<0.3	<0.3	<0.3	<0.3	0.3	1.0	0.4	0.4
	$\Delta\theta_{R\text{-}H\text{-}Int/Ext}$ (°)	1.9	1.9	1.4	1.2	2.1	2.3	2.2	2.2
Right knee	$\Delta\theta_{R\text{-}K\text{-}Flx/Ext}$ (°)	0.6	0.8	0.6	0.3	0.9	0.9	0.9	0.8
	$\Delta\theta_{R\text{-}K\text{-}Abd/Add}$ (°)	1.0	0.9	0.7	0.7	1.2	1.3	1.1	1.2
	$\Delta\theta_{R\text{-}K\text{-}Int/Ext}$ (°)	0.6	1.2	2.8	2.1	3.3	2.7	2.8	2.8
Right ankle	$\Delta\theta_{R\text{-}A\text{-}Plt/Drs}$ (°)	0.3	0.5	0.6	0.7	0.8	0.5	0.6	0.6
	$\Delta\theta_{R\text{-}A\text{-}Int/Ext}$ (°)	1.1	0.9	1.0	1.1	2.0	1.4	2.1	1.4
	$\Delta\theta_{R\text{-}A\text{-}Inv/Eve}$ (°)	1.2	1.4	1.4	1.5	1.9	1.7	2.4	2.1
Left hip	$\Delta\theta_{L\text{-}H\text{-}Flx/Ext}$ (°)	0.3	0.3	0.3	0.3	0.5	0.8	0.6	0.4
	$\Delta\theta_{L\text{-}H\text{-}Abd/Add}$ (°)	<0.3	<0.3	<0.3	<0.3	<0.3	1.0	<0.3	<0.3
	$\Delta\theta_{L\text{-}H\text{-}Int/Ext}$ (°)	0.5	0.6	0.7	0.5	1.3	1.3	1.3	1.2
Left knee	$\Delta\theta_{L\text{-}K\text{-}Flx/Ext}$ (°)	0.4	0.3	0.4	<0.3	0.5	0.4	0.7	0.4
	$\Delta\theta_{L\text{-}K\text{-}Abd/Add}$ (°)	0.3	0.4	0.6	0.4	0.9	0.8	0.7	0.6
	$\Delta\theta^{x}_{L-K-Int/Ext}$ (°)	0.8	0.7	1.2	0.6	1.7	1.7	1.3	1.4
Left ankle	$\Delta\theta_{L\text{-}A\text{-}Plt/Drs}$ (°)	0.3	0.3	0.3	0.3	0.5	0.4	0.5	0.6
	$\Delta\theta_{L\text{-}A\text{-}Int/Ext}$ (°)	1.3	1.2	1.1	1.1	1.8	1.5	1.7	1.4
	$\Delta\theta_{L\text{-}A\text{-}Inv/Eve}$ (°)	0.9	0.8	0.7	0.7	1.2	1.2	1.6	1.4

Test 2b – are Joint Kinematics Affected by the Dimension and the Position of the Calibration Volume?

For the system SS#1, we obtained a CMC equal to 0.93 for the Inv/Eve of the right ankle and a CMC equal to 0.93 for the Int/Ext of the left hip (Table 2). The highest values for the CMC (1.00) were obtained for $CMC^{V}_{R\text{-}H\text{-}Flx/Ext}$ both at SS#1 and SS#2, and $CMC^{V}_{L\text{-}H\text{-}Flx/Ext}$ and $CMC^{V}_{R\text{-}K\text{-}Flx/Ext}$ at SS#2. The lowest CMC computed considering the SS#1 was obtained for $CMC^{V}_{R\text{-}A\text{-}Inv/Eve}$ (0.93), while considering SS#2 it was measured for $CMC^{V}_{L\text{-}H\text{-}Int/Ext}$ (0.93). As for the previous case, we mainly found an excellent correlation between the waveforms with each SS.

The second four columns of Table 3 and Table 4 show, instead, the maximum angular differences on kinematics when processing static and dynamic trials with the GV and SVs calibrations. The Δθ was for SS#1 (Table 3) never higher than 0.7° ($\Delta\theta^{GV/SV\text{-}LH}_{R\text{-}A\text{-}Inv/Eve}$) and, similarly to the previous test, it was found to be less than 0.1° for a few cases. Examining the Δθ for SS#2 (Table 4), we found the lowest value equal to 0.2° for $\Delta\theta^{GV/SV\text{-}LH}_{L\text{-}H\text{-}Abd/Add}$, $\Delta\theta^{GV/SV\text{-}RH}_{L\text{-}H\text{-}Abd/Add}$, and $\Delta\theta^{GV/SV\text{-}RL}_{L\text{-}H\text{-}Abd/Add}$, while the higher value 3.3° was reached for $\Delta\theta^{GV/SV\text{-}LH}_{R\text{-}K\text{-}Int/Ext}$. Looking at the results with regard to both CMCs and Δθs, we can affirm that the effect of the considered volume, in which the operator performs the calibration procedure, is negligible on the articular kinematics during the gait cycle analysis compared to those induced from other sources of error [32]. Indeed, the similarities obtained in this research, modifying either the calibration duration or the calibration volumes, are higher than those normally obtained for intra- and inter-session repeatability analyses [33]. In these studies, an operator normally performs the marker placement more than once and in different testing-days, and the stride variability is also accounted. As an example, Pinzone and colleagues [33] observed maximum Δθ of 9.7°.

Coherently with the literature [33] and [34], higher variability was found on the transverse plane and for the foot joint data, whereas the sagittal plane was confirmed to be the most reliable. CMC and Δθ values confirm this assertion both for Test 2a and Test 2b. In conclusion, we can assert that the effect of the calibration procedure on articular kinematic variables is negligible both considering the waveforms similarity and the angular differences between them [32].

CONCLUSIONS

This paper presented a methodology to evaluate the effects that a set of calibration procedures, diversified for both acquisition volumes and duration, can have on calculating distances and angles starting from trajectory data measured by a stereophotogrammetric system. The inaccuracy of the estimated distances, angles and joint kinematics was found to be higher in dynamic than in static conditions, but still negligible in both conditions and not dependent on the performed calibration procedure. Between the two investigated systems, the one with the highest performances was also the one that led to the highest inaccuracies. This apparent paradox was indeed explained by the different aperture of the camera lenses. These findings led to the conclusion that successful calibration procedures of different durations and performed in different volumes did not affect the metrological performance of the investigated systems.

ACKNOWLEDGMENTS

The authors would like to acknowledge the Seventh Framework Programme FP7 (FP7-ICT-2011-9; co-PI: P. CAPPA) and the MIUR Italian Ministry of Instruction, Research and University (PRIN 2012 – 20127XJX57; PI: P. CAPPA) for supporting the present research. The instrumentation used at The University of Sheffield was funded by the UK EPSRC, Great Technologies Capital Call, Robotics and Autonomous Systems (EP/J013714/1).

REFERENCES

1. A. Cappozzo, U. Della Croce, A. Leardini, L. Chiari, Human movement analysis using stereophotogrammetry. Part 1: Theoretical background, Gait Posture 21 (2005) 186–196, http://dx.doi.org/ 10.1016/j.gaitpost.2004.01.010.
2. G. Wu, P.R. Cavanagh, ISB recommendations in the reporting for standardization of kinematic data, J. Biomech. 28 (1995) 1257–1261.
3. G. Wu, S. Siegler, P. Allard, C. Kirtley, A. Leardini, D. Rosenbaum, et al., ISB recommendation on definitions of joint coordinate system of various joints for the reporting of human joint motion – part I: ankle, hip, and spine, J. Biomech. 35 (2002) 543–548. http:// www.sciencedirect.com/science/article/pii/S0021929001002226.
4. G. Wu, F.C.T. Van Der Helm, H.E.J. Veeger, M. Makhsous, P. Van Roy, C. Anglin, et al., ISB recommendation on definitions of joint coordinate systems of various joints for the reporting of human joint motion – part II: shoulder, elbow, wrist and hand, J. Biomech. 38 (2005) 981–992, http://dx.doi.org/10.1016/j. jbiomech.2004.05.042.
5. E. Bergamini, G. Ligorio, a. Summa, G. Vannozzi, a. Cappozzo, a.M. Sabatini, Estimating orientation using magnetic and inertial sensors and different sensor fusion approaches: accuracy assessment in manual and locomotion tasks, Sensors 14 (2014) 18625–18649, http://dx.doi.org/10.3390/s141018625.
6. E. Palermo, S. Rossi, F. Marini, F. Patanè, P. Cappa, Experimental evaluation of accuracy and repeatability of a novel body-to-sensor calibration procedure for inertial sensor-based gait analysis, Measurement 52 (2014) 145–155, http://dx.doi.org/10.1016/j.measurement.2014.03.004.
7. E. Palermo, S. Rossi, F. Patanè, P. Cappa, Experimental evaluation of indoor magnetic distortion effects on gait analysis performed with wearable inertial sensors, Physiol. Meas. 35 (2014) 399–415, http:// dx.doi.org/10.1088/0967-3334/35/3/399.
8. P. Picerno, A. Cereatti, A. Cappozzo, Joint kinematics estimate using wearable inertial and magnetic sensing modules, Gait Posture 28 (2008) 588–595, http://dx.doi.org/10.1016/j.gaitpost.2008.04.003.
9. A. Fernández-Baena, A. Susín, X. Lligadas, Biomechanical validation of upper-body and lower-body joint movements of kinect motion capture data for rehabilitation treatments, in: Proc. 2012 4th Int. Conf. Intell. Netw. Collab. Syst. INCoS 2012, 2012, pp. 656–661.
10. R.A. Clark, Y.H. Pua, K. Fortin, C. Ritchie, K.E. Webster, L. Denehy, et al., Validity of the Microsoft Kinect for assessment of postural control, Gait Posture 36 (2012) 372–377.
11. S. Obdrzalek, G. Kurillo, F. Ofli, R. Bajcsy, E. Seto, H. Jimison, et al., Accuracy and robustness of Kinect pose estimation in the context of coaching of elderly population, in: Proc. Annu. Int. Conf. IEEE Eng. Med. Biol. Soc. EMBS, 2012, pp. 1188–1193.
12. B. Bonnechère, B. Jansen, P. Salvia, H. Bouzahouene, L. Omelina, F. Moiseev, et al., Validity and reliability of the Kinect within functional assessment activities: comparison with standard stereophotogrammetry, Gait Posture 39 (2014) 593–598.
13. A. Leardini, L. Chiari, U. Della Croce, A. Cappozzo, Human movement analysis using stereophotogrammetry. Part 3. Soft tissue artifact assessment and compensation, Gait Posture 21 (2005) 212–225, http://dx.doi.org/10.1016/j.gaitpost.2004.05.002.

14. U. Della Croce, A. Leardini, L. Chiari, A. Cappozzo, Human movement analysis using stereophotogrammetry: Part 4: assessment of anatomical landmark misplacement and its effects on joint kinematics, Gait Posture 21 (2005) 226–237.

15. L. Chiari, U. Della Croce, A. Leardini, A. Cappozzo, Human movement analysis using stereophotogrammetry. Part 2: Instrumental errors, Gait Posture 21 (2005) 197–211, http://dx.doi.org/10.1016/ j.gaitpost.2004.04.004.

16. M. Windolf, N. Götzen, M. Morlock, Systematic accuracy and precision analysis of video motion capturing systems–exemplified on the Vicon-460 system, J. Biomech. 41 (2008) 2776–2780, http:// dx.doi.org/10.1016/j.jbiomech.2008.06.024.

17. R. Di Marco, S. Rossi, F. Patanè, P. Cappa, Technical quality assessment of an optoelectronic system for movement analysis, J. Phys: Conf. Ser. 588 (2015) 012030, http://dx.doi.org/10.1088/1742- 6596/588/1/012030.

18. J. Weng, P. Cohen, M. Herniou, Camera calibration with distortion models and accuracy evaluation, IEEE Trans. Pattern Anal. Mach. Intell. 14 (1992) 965–980.

19. D.W. Vander Linden, S.J. Carlson, R.L. Hubbard, Reproducibility and accuracy of angle measurements obtained under static conditions with the Motion Analysis(TM) video system, Phys. Ther. 72 (1992) 300–305.

20. F.P. Branca, P. Cappa, An experimental study of the accuracy and precision associated to an opto-electronic system utilized for gait analysis, in: 12th Trienn. World Congr. Int. Meas. Confed., Beijin, 1991.

21. Y.I. Abdel-Aziz, H.M. Karara, Direct linear transformation from comparator coordinates in close-range photogrammetry, in: ASP Symp. Close-Range Photogramm. Illinois, Urbana, 1971.

22. J.P. Scholz, Reliability and validity of the WATSMART threedimensional optoelectric motion analysis system, Phys. Ther. 69 (1989) 679–689.

23. D.J. Wilson, B.K. Smith, J.K. Gibson, Accuracy of reconstructed angular estimates obtained with the Ariel performance analysis system TM, Phys. Ther. 77 (1997) 1741–1746.

24. Y. Ehara, H. Fujimotob, S. Miyazaki, S. Tanaka, S. Yamamoto, Comparison of performance of 3D camera systems, Gait Posture 3 (1995) 166–169.

25. Y. Ehara, H. Fujimoto, S. Miyazaki, M. Mochimaru, S. Tanaka, S. Yamamoto, Comparison of the performance of 3D camera systems II, Gait Posture 5 (1997) 251–255, http://dx.doi.org/10.1016/S0966- 6362(96)01093-4.

26. U. Della Croce, A. Cappozzo, A spot check for estimating stereophotogrammetric errors, Med. Biol. Eng. Comput. 38 (2000) 260–266, http://dx.doi.org/10.1007/ BF02347045.

27. L. Chen, C.W. Armstrong, D.D. Raftopoulos, An investigation on the accuracy of three-dimensional space reconstruction using the direct linear transformation technique, J. Biomech. 27 (1994) 493–500.

28. P.J. Klein, J.J. Dehaven, Accuracy of three-dimensional linear and angular estimates obtained with the ariel performance analysis system, in: Am. Congr. Rehabil. Med. Am. Acad. Phys. Med. Rehabil., 1995, pp. 183–189.

29. R.B. Davis, S. Ounpuu, D. Tyburski, J.R.R. Gage, A gait analysis data collection and reduction technique, Hum. Mov. Sci. 10 (1991) 575– 587. http://www.sciencedirect.com/science/article/pii/0167945791 90046Z.

30. M.P. Kadaba, H.K. Ramakrishnan, M.E. Wootten, J. Gainey, G. Gorton, G.V. Cochran, Repeatability of kinematic, kinetic, and electromyographic data in normal adult gait, J. Orthop. Res. 7 (1989) 849–860.

31. A. Ferrari, A.G. Cutti, P. Garofalo, M. Raggi, M. Heijboer, A. Cappello, et al., First in vivo assessment of "outwalk": a novel protocol for clinical gait analysis based on inertial and magnetic sensors, Med. Biol. Eng. Comput. 48 (2010) 1–15, http://dx.doi.org/10.1007/ s11517-009-0544-y.

32. A. Cappozzo, Minimum measured-input models for the assessment of motor ability, J. Biomech. 35 (2002) 437–446, http://dx.doi.org/ 10.1016/S0021-9290(01)00186-5.

33. O. Pinzone, M.H. Schwartz, P. Thomason, R. Baker, The comparison of normative reference data from different gait analysis services, Gait Posture 40 (2014) 286–290, http://dx.doi.org/10.1016/ j.gaitpost.2014.03.185.

34. A. Ferrari, M.G. Benedetti, E. Pavan, C. Frigo, D. Bettinelli, M. Rabuffetti, et al., Quantitative comparison of five current protocols in gait analysis, Gait Posture 28 (2008) 207–216, http://dx.doi.org/ 10.1016/j.gaitpost.2007.11.009.

CHAPTER 5 VEHICLE EMISSION FACTORS OF SOLID NANOPARTICLES IN THE LABORATORY AND ON THE ROAD USING PORTABLE EMISSION MEASUREMENT SYSTEMS (PEMS)

Barouch Giechaskiel, Francesco Riccobono, Theodoros Vlachos, Pablo Mendoza-Villafuerte, Ricardo Suarez-Bertoa, Georgios Fontaras, Pierre Bonnel and Martin Weiss

Sustainable Transport Unit, Joint Research Centre, Institute for Energy and Transport, European Commission, Varese, Italy

Emission inventories are used to quantify sources and identify trends in the emissions of air pollutants. They use vehicle-specific emission factors that are typically determined in the laboratory, through remote-sensing, vehicle chasing experiments and, more recently, on-board measurements with Portable Emission Measurement Systems (PEMS). Although PEMS is widely applied to measure gaseous pollutants, their application to Solid Particle Number (SPN) emissions is new. In this paper, we discuss the current status of determining SPN emission factors both on the chassis dynamometer in the laboratory and on the road using PEMS-SPN. First, we determine through dedicated tests in the laboratory the influence of the measurement equipment, ambient temperature, driving style, and cycle characteristics, and the extra mass of the PEMS equipment on the SPN emissions of vehicles. Afterward, we present the SPN emissions under type-approval conditions as well as on the road of two heavy-duty diesel vehicles equipped with Diesel Particulate Filter (DPF; one of them Euro VI), two light-duty diesel vehicles equipped with DPF, one light-duty vehicle equipped with a Port Fuel Injection engine (PFI), and seven Gasoline Direct Injection (GDI) passenger cars (two of them Euro 6). We find that cold-start and strong accelerations tend to substantially increase SPN emissions. The two heavy-duty vehicles showed on-road emissions around 2×10^{13} p/km (Euro V truck) and 6×10^{10} p/km (Euro VI truck), respectively. One of the DPF-equipped light-duty vehicles showed emissions of 8×10^{11} p/km, while the other one had one order of magnitude lower emissions. The PFI car had SPN emissions slightly higher than 1×10^{12}p/km. The on-road emissions of GDI passenger cars spanned approximately from 8×10^{11} p/km to 8×10^{12} p/km. For the cars not equipped with a DPF, the on-road SPN emissions remained within a factor of two of the laboratory results. This factor was on average around 0.8 for the Euro 6 and 1.6 for the Euro 5 GDIs. The DPF equipped vehicles showed a difference of almost one order of magnitude between laboratory and on-road tests due to the different DPF fill state and passive regeneration during the tests. The findings of this study can (i) help improving the inventories on SPN emissions and (ii) assist policy makers in designing effective test procedures for measuring the SPN emissions of vehicles under real-world driving conditions.

Citation: Giechaskiel B, Riccobono F, Vlachos T, Mendoza-Villafuerte P, Suarez-Bertoa R, Fontaras G, Bonnel P and Weiss M (2015) Vehicle Emission Factors of Solid Nanoparticles in the Laboratory and on the Road Using Portable Emission Measurement Systems (PEMS). Front. Environ. Sci. 3:82. doi: 10.3389/fenvs.2015.00082

INTRODUCTION

Particulate Matter (PM) has been shown to have a negative impact on the human health. Epidemiological studies (e.g., Pope, 2000) and toxicological studies (e.g., Oberdörster, 2000)

have associated urban PM pollution with carcinogenic and adverse respiratory and cardio-vascular effects. Recent reviews (EPA, 2009;HEI, 2013; WHO, 2013) concluded that adverse health effects are best linked to an exposure to PM with an equivalent aerodynamic diameter below 2.5 μm. However, a number of studies also suggest that the inflammatory response to particles from non-toxic materials correlates more closely with particle surface area rather than particle mass (e.g., Sager and Castranova, 2009). Nanoparticles of a diameter below 2.5 μm are considered more harmful than larger particles because they have a larger specific surface area and higher deposition fraction in the human respiratory tract. Moreover, inhaled ultrafine solid nanoparticles (< 100 nm) may pass through the lungs› alveolar membrane and via the blood to the heart and other organs where they can directly exert a toxic effect (translocation effect; Oberdörster et al., 2004). Thus, ultrafine Particle Number (PN) emissions have recently gained the attention of toxicologists and environmental policy makers.

Around 20% of the European Union›s (EU) urban population lives in areas where the daily air quality standard for PM with a diameter smaller than 10 μm (50 μg/m^3) was exceeded in 2010 (Sundvor et al., 2012; EEA, 2014). Exceedances of the PM standard (< 10 μm) occur most frequently in urban locations with dense road traffic. Quantification of the traffic contribution to PM levels in urban areas is complex, due to the many sources of primary PM emissions and due to the formation of secondary PM from precursor gases such as nitrogen oxides and volatile organic compounds emitted by vehicles and other sources. Road traffic contributes some 9–66% to the urban PM concentrations, with an average estimate of 34% in the vicinity of urban traffic sites and 15% at urban background sites. The contribution of road traffic to PN pollution is larger (60%) due to the high number of nuclei particles formed from the unburnt fuel and lubricant or secondary formation, reaching 90% in cases of busy roads (Kumar et al., 2014 and references therein).

The contribution of vehicle emissions to air pollution is typically quantified with emission inventories (EMEP/EEA, 2013). An emission inventory lists the total emissions of a specific component, disaggregated into source categories relevant in a certain geographical area and within a specified time span. Emission inventories are used since the 1970s for scientific research but also for air quality management and environmental policy making. In the context of road transport, emission inventories are specifically used to monitor the effectiveness of the vehicle emissions legislation and the emissions reduction resulting from new vehicles entering the fleet. The vehicle emission factors are a key input for emission inventories relative to on road transportation. These are typically implemented as default values in the various calculation tools (e.g., COPERT, MOVES, HBEFA) and frequently updated using experimental results (e.g., Boulter and McCrae, 2007). However, while data on gaseous emissions are generally readily available from laboratory and on-road measurements (e.g.,Ntziachristos and Galassi, 2014), experimental data for PN emissions are limited. In addition, PN emission measurements are not always comparable due to differences in sampling conditions and instruments used (Keogh and Sonntag, 2011).

The European regulation on particle emissions was initially based solely on the smoke levels recorded by opacity meters (Giechaskiel et al., 2014a). However, over the past 35 years, the regulatory method has adopted the gravimetric quantification of PM collected on filters while the vehicle is driven on a prescribed speed profile (test cycle). With the introduction of Diesel Particulate Filters (DPFs) around the year 2000, the gravimetric method was no longer sensitive enough to measure very low PM emissions. To address the situation, in 2001, the United Nations Economic Commission for Europe (UN-ECE) launched the Particle Measurement Programme (PMP) to develop a new and sufficiently sensitive, precise, and representative measurement methodology (Martini et al., 2009). In addition to an improved PM filter method, the Solid Particle Number (SPN) measurement method proposed by PMP in 2007, based on the counting of solid particles with a diameter larger than 23 nm, was integrated into the European emissions regulation in 2011 for diesel light-duty vehicles (Euro 5b)[1] (EC, 2007; Giechaskiel et al., 2008a), in 2014 for Gasoline Direct Injection (GDI) light-duty vehicles (Euro 6) and in 2013 for heavy-duty engines (Euro VI; EC, 2009; Giechaskiel et al., 2012a). A minimum diameter of 23 nm size was selected in order to include the primary soot particles and exclude volatile nucleation mode particles (see Giechaskiel et al., 2014a). Volatile particles were excluded in order to improve the repeatability and reproducibility of the method to levels acceptable for legislative purposes. The limit is 6×10^{11} p/km (or p/kWh for heavy-duty engines) for the transient type approval test cycle. The limit for GDIs was set to 6×10^{12} p/km in 2014 but it will decrease to 6×10^{11} p/km in 2017.

The vehicle regulations in both the United States of America (US) and the European Union (EU), also address emissions from in-use vehicles. In-service conformity (or in-use compliance in the US) requires laboratory testing of vehicles recruited from the vehicle fleet that is actually operated on the road. Recently this has been augmented by regulations requiring on-board in-service conformity emissions testing of heavy-duty vehicles (EC, 2011) with Portable Emission Measurement Systems (PEMS; Bonnel et al., 2006; EPA, 2008; Perujo and Mendoza-Villafuerte, 2015). Moreover, the European Commission (EC) is currently introducing a complementary and mandatory PEMS on-road test procedure, the so-called Real-Driving Emissions (RDE) procedure for light-duty vehicles (EC, 2012; Weiss et al., 2013; Vlachos et al., 2014). These initiatives aim at reducing the large differences that are found between laboratory and on-road measurements, particularly with respect to the NO_x emissions of light-duty diesel vehicles (Weiss et al., 2011a, 2012; Fontaras et al., 2014; Franco et al., 2014; Vlachos et al., 2014). A detailed discussion of in-use emissions testing with PEMS can be found in Vlachos et al. (2014). SPN measurements with PEMS is at the final stages of investigation for light duty vehicles and recently a program started for the applicability of SPN measurements for the in-service conformity testing of heavy-duty vehicles. However, for SPN, information about the difference between laboratory and on-road emissions is limited (Andersson et al., 2014).

The main objective of this article is to give an overview of the status of SPN emissions from trucks and passenger cars under real-driving conditions based on experimental data of 2 heavy-duty vehicles and 10 light-duty vehicles. The measured SPN emission levels are compared with the literature and the regulated emission limits. A second objective of this study is to analyze and discuss differences in the SPN emissions between chassis dynamometer studies and on-road measurements using the same vehicles. We focus our analysis on the effect of some parameters on particle number measurements (i.e., uncertainty of SPN instruments for on-board application, DPF load level, ambient temperature, vehicle mass, driving style, and driving cycle). The findings of this study can (i) help improving the emission inventories on SPN emissions and (ii) assist policy makers in designing effective test procedures for measuring SPN on-board of vehicles.

BACKGROUND INFORMATION

The PN emissions of vehicles can be determined in the same ways that are used to determine gaseous emissions, namely through remote-sensing measurements at the roadside, tunnel studies, vehicle chasing experiments, chassis dynamometer measurements, and on-board measurements with PEMS during the actual vehicle use (for details see Franco et al. (2013) and references therein). Here only studies focusing on PN will be discussed.

Remote-sensing measurements, also referred to as "roadside measurement," determine the instantaneous ratios of pollutant concentrations as vehicles pass by a measurement station on the roadway (Pant and Harrison, 2013). Remote-sensing enables monitoring the emissions of large numbers of vehicles (up to thousands per day) and offers the necessary resolution to identify emission levels of single vehicles, whose emission technology may be determined by video recording of license plates and cross-referencing with vehicle registration databases. On the other hand, remote-sensing only gives an instantaneous (snapshot) estimate of emissions at a specific location under specific operating conditions, and cannot be used across multiple lanes of heavy traffic. Thus, it is difficult to derive emission factors that are representative of the wide range of vehicle operations from remote-sensing measurements only. Remote-sensing particle measurements include emissions from combustion, tires, brakes and the secondary particles formed in the atmosphere (Pant and Harrison, 2013), although sometimes tracers can be used to apportion the contribution of each source. Moreover, remote-sensing needs to carefully deal with unstable weather and background pollution levels, that hinder the evaluation of the actual vehicle emissions.

Tunnel studies involve measuring the total flux of pollutants from vehicles passing through a tunnel and correlating the pollutant flux to the traffic flow. An advantage of tunnel studies is that, contrary to remote-sensing measurements, the wind conditions are well defined. The average absolute levels of emissions can be obtained. Tunnel studies are thus able to capture a cross-section of the on-road vehicle fleet under real-world operation conditions at the location of the tunnel. These, in turn, may not be typical of the wide range of real-world driving conditions (specifically those in urban environments), since tunnels are usually traversed at steady speed. As a notable disadvantage, tunnel studies usually do not allow allocating emissions to specific vehicle classes unless different tunnel bores are dedicated to them. PN measurements from both roadside and tunnel studies typically include solid and volatile particles and non-exhaust emissions particles from tire wear and road dust (see reviews of Morawska et al., 2008; Keogh et al., 2010).

During vehicle chasing experiments, individual vehicles are followed by a mobile laboratory on a van or trailer (Vogt et al., 2003; Giechaskiel et al., 2005; Ježek et al., 2015) that capture the exhaust plume of the vehicle being followed, thus providing real-world emissions data under a wide range of operating conditions (Kittelson et al., 2000, 2006a,b). Vehicle chasing experiments make it possible to study a statistically representative sample of vehicles for fleet characterization. However, such experiments are best conducted on a test track to avoid obstructing traffic safety. Comparisons between laboratory and chasing experiments at constant speed showed good agreement for particles in the accumulation mode regime (peak around 50–100 nm; Casati et al., 2007; Giechaskiel et al., 2007a) and qualitative good agreement for particles in the (volatile) nucleation mode regime (peak < 30 nm; Rönkkö et al., 2006; Casati et al., 2007; Giechaskiel et al., 2007a). However, the diluted exhaust of low emitting vehicles (i.e., those with particulate filters) will be at the levels of the ambient air particle background (Bergmann et al., 2009); thus the results will have high uncertainty and variability.

During chassis dynamometer testing, the vehicle remains stationary as a driver operates it in a test cell according to a predetermined time-speed profile and gear shift pattern shown on a monitor. In vehicle regulations around the world, chassis dynamometer testing is specified as the primary measurement method for verifying vehicle emissions [e.g., Regulation 83, UN/ECE (2015)] due to its high accuracy and repeatability compared to other methods. Climatically controlled test cells can simulate driving under a wide range of temperatures, including sub-zero tests (currently, the European emissions regulation covers the temperature range of down to −7°C); however, to date only a few laboratories can simulate driving at high altitude, i.e., at low ambient pressure. Chassis dynamometer testing does not capture the wide dynamicity range of real-world driving and may thus underestimate the on-road emissions of vehicles (Pelkmans and Debal, 2006). Moreover, the driving resistance that simulates road load is obtained from vehicle coast-down tests under artificially favorable conditions, thus frequently yielding lower fuel consumption and emission results as compared to real-world on-road driving (Mellios et al., 2011). In addition, especially for particles, the sampling conditions in the laboratory (e.g., dilution ratio, dilution temperature, etc.) can affect the measurement result (e.g., Khalek et al., 1999) to the extent that the measured particle mass and number emissions might not be representative of the actual emissions on the road (Vogt et al., 2003). Secondary particle formation from volatile species in the atmosphere (see e.g., Keogh and Sonntag, 2011) can lead to significant differences between particle emissions measured by roadside studies and on the chassis dynamometer, although in many cases there is a good agreement between both methods (Beddows and Harrison, 2008 and references therein). Chassis dynamometer tests are still the most important source of SPN emission factors (see e.g., review of Giechaskiel et al., 2012a and references therein).

On-board measurements are conducted with PEMS, mounted inside the vehicle cabin or outside thereof on a tow bar. PEMS on-board measurements can provide long series of emission data of a particular vehicle driven under a wide range of operating conditions, including those that would otherwise be difficult to replicate in the laboratory (e.g., large road gradients, strong accelerations). However, with PEMS only a relatively limited number of vehicles compared to remote-sensing can be tested with a given effort. In the past decade, extensive programs have confirmed the ability of PEMS to measure gaseous pollutants at a quality similar to that achieved by certification laboratories (Rubino et al., 2007; EPA, 2008; Johnson et al., 2009a; Bonnel et al., 2011). PEMS for PM measurement have been developed in the past 5 years with satisfactory results (Johnson et al., 2009b, 2011; Giechaskiel et al., 2011; Mamakos et al., 2013a; Mendoza-Villafuerte et al., 2015).

PEMS for PN have been used in a few studies (Bergmann et al., 2009; Li et al., 2013). However, PEMS that measure solid PN (PEMS-SPN) similarly to the regulated SPN systems became available only recently and were assessed in a comprehensive program by Giechaskiel et al. (2014b, 2015). The technical specifications of PEMS-SPN for regulatory purposes have not been finalized yet. To date, there are two measurement principles available to measure SPN emissions with PEMS: (i) diffusion charging (Fierz et al., 2008) and (ii) condensation particle counting that represents the measurement principle applied in the regulated PMP systems (Giechaskiel et al., 2009). Both principles measure PN emissions after thermal pre-treatment of the exhaust. One particular issue of mobile SPN measurements refers to emissions sampling directly from the tailpipe; the resulting SPN counts might not be identical to those measured by following the regulated procedure at the full dilution tunnel that alters particle properties due to e.g., coagulation, thermophoresis, diffusion etc. Comparisons of PEMS-SPN with reference PMP systems, especially during on-board operation, as well as of emissions of vehicles both in the laboratory and during on-road driving are still scarce (e.g., Merkisz et al., 2009; Andersson et al., 2014). Quite often high differences between laboratory and on-road testing are found (Andersson et al., 2014). Other reasons of differences between PEMS-SPN and chassis dynamometer SPN results include the typical effect of on-road operating conditions (extra vehicle mass, auxiliaries, aerodynamics), the driver, the route, the traffic, and the road conditions (road morphology, altitude, road gradients, etc.; see also Zacharof et al., 2015 and references therein). Although the effect of the above parameters has been discussed for fuel consumption and to some extent gaseous pollutant emissions (Weiss et al., 2011a, 2012), their effect on SPN has only been touched recently (e.g., Clark et al., 2002; Wang et al., 2006; Barrios et al., 2012; Huang et al., 2013; Li et al., 2013). Additional research is needed.

EXPERIMENTAL METHODS AND MATERIALS

The heavy-duty and light-duty vehicles were tested both in the laboratory and on the road using the same SPN equipment.

Test Vehicles and Fuels

We tested two heavy-duty vehicles and 10 light-duty vehicles (Table 1). All test vehicles complied during laboratory testing under type-approval conditions with their respective emissions limit. The test fleet included diesel vehicles equipped with a DPF and gasoline vehicles with Port Fuel injection (PFI) or direct injection (GDI) engines equipped with Three-Way Catalyst (TWC) (Table 1). The light-duty vehicles were mainly Euro 5 (two Euro 6). The tested light-duty vehicles mainly comprised passenger cars of Category M1. Exceptions constitute vehicles DPF #1 and #2 which were vans of Category N1. The heavy-duty vehicles were of Category N3. One of the heavy-duty vehicles was certified as Euro V and the other as Euro VI; both vehicles were equipped with a DPF. The heavy-duty trucks weighted ~18 tons; an extra load of 8 tons was added during the on-road tests.

Table 1. Characteristics of test vehicles and extra evaluation tests in the laboratory

Vehicle ID	Engine and after-treatment technology	Emissions limit	Engine capacity [l]	Rated Power [kW]	Test mass [kg] NEDC/WLTC	Odometer [km]	Extra tests
HD #1	Diesel; DPF	Euro V	11.0	326	18,000	53,000	-
HD #2	Diesel; DPF	Euro VI	13.0	335	-	6200	-
DPF #1	Diesel; DPF	Euro 5	3.0	140	2960/3100	32,700	regeneration
DPF #2	Diesel; DPF	Euro 5	3.0	125	2270/2440	4300	-
PFI	PFI; TWC	Euro 5	1.6	86	1250/1310	4200	-
GDI #1	GDI; TWC	Euro 6	1.0	59	1020/1135	500	temperature, mass, EFM
GDI #2	GDI; TWC	Euro 6	2.0	155	1700/1760	11,200	temperature, cycles
GDI #3	GDI; TWC	Euro 5	1.2	63	1250/1360	2500	mass, driving style, EFM
GDI #4	GDI; TWC	Euro 5	1.6	100	1360/1450	15,300	
GDI #5	GDI; TWC	Euro 5	1.6	140	1470/1530	8000	-
GDI #6	GDI; TWC	Euro 5	1.2	59	–/1190	5000	cycles
GDI #7	GDI; TWC (flex fuel)	Euro 5	1.6	132	1470/1560	28,500	temperature, mass

All chassis dynamometer and on-road tests were conducted with market fuels compliant with the standards EN 228 and EN 590 for gasoline and diesel fuels, respectively. The ethanol content of the gasoline fuel was < 7%; the sulfur content of the diesel fuel was < 10 ppm. Due to the low mileage of all tested vehicles, we assume that the lubricants were the manufacturer's factory-fill products.

Laboratory

The laboratory tests were conducted in the Vehicle Emissions Laboratory (VELA 2 for light-duty and VELA 7 for the heavy-duty vehicle) of the Joint Research Centre (JRC) in Italy. All light-duty vehicles were tested on a 2 axis roller dynamometer. The exhaust gas was connected to the full dilution tunnel with a 6 m tube heated at 100°C. Typical flow rates of 9–12 and 100 m^3/min were used at the full dilution tunnel with constant volume sampling (CVS; Giechaskiel et al., 2007b) for the light-duty and heavy-duty vehicles, respectively. SPN instruments (see description below) were used both at the dilution tunnel and at the tailpipe of the vehicles.

Test Cycles in the Laboratory

The Euro V heavy-duty vehicle was tested under cold and hot start conditions with the Worldwide Harmonized Vehicle Cycle (WHVC), i.e., a chassis dynamometer test developed based on the same set of data used for the development of the World Harmonized Transient Cycle (WHTC) for heavy-duty engines [which replaced the European Transient Cycle (ETC) since Euro VI]. Unlike the WHTC, however, testing with the WHVC does not follow a standardized procedure; the WHVC is not used for regulatory testing. The dyno settings were not available from the vehicle manufacturers. We thus used the values from similar trucks.

The light-duty vehicles were tested over three different driving cycles, each consisting of several phases (Table2):

- New European Driving Cycle (NEDC) consisting of an Urban Driving Cycle (UDC) and an Extra Urban Driving Cycle (EUDC); the NEDC cycle is currently used for type-approval within the EU;
- Worldwide Harmonized Light vehicles Test Cycle (WLTC) consisting of a low-speed, medium-speed, high-speed, and extra high-speed phase; this cycle is considered more representative of real-world driving than the NEDC and will likely be used for type-approval within the EU from 2017 onward;
- Random Cycles (RC) consisting of three phases of ~900 s each that resemble soft (RC_A), medium (RC_B), and severe (RC_C) driving on European roads. Each phase consists of a low-speed, medium-speed and high-speed portion. The RC was developed by the RDE working group of the European Commission during the development of the RDE on-road test procedure (Weiss et al., 2013).

Table 2. Characteristics of driving cycles

Cycle	Duration [s]	Distance [km]	Speed max [km/h]	Speed mean [km/h]	Idling [%]	Temp. [°C]
WHVC	1800	20.1	87.8	40.2	15	25
UDC	780	4.0	50.0	18.4	23	25
EUDC	400	7.0	120.0	62.6	9	25
NEDC	1180	11.0	120.0	33.6	21	25
WLTC_A (low)	589	3.1	56.5	18.9	26	23
WLTC_B (medium)	433	4.8	76.6	39.5	11	23
WLTC_C (high)	455	7.2	97.4	56.6	7	23
WLTC_D (extra high)	323	8.3	131.3	92.0	2	23
WLTC	1800	23.3	131.3	46.5	13	23
RC_A	900	10.6	112.5	42.2	3	23
RC_B	900	13.2	114.0	53.2	4	23
RC_C	900	13.0	134.0	52.8	7	23

Unless specified otherwise, the NEDC and WLTC tests were conducted with a cold engine as required by the regulation. At least two repetitions of the NEDC and WLTC were conducted with each vehicle. Tests with more cycles were conducted with Vehicles GDI #1, #2, #3, #6, and #7 and Vehicle DPF #1. All tests were conducted with the same driver. One vehicle (GDI #3) was driven by four drivers to analyze the effect of the driving style on the SPN emissions. Vehicle DPF #1 was tested with various cycles (i.e., under cold and hot start conditions, as well as at various ambient temperatures) until the DPF was regenerated. Then cycles were run again until a second regeneration. A third sequence of cycles was initiated but the tests were stopped before the regeneration.

All test masses and dyno coefficients for NEDC testing were (i) used as given by the vehicle manufacturers (Vehicles DPF#1, DPF #2, PFI, GDI #5, GDI #6) or (ii) taken from Regulation 83 (UN/ECE, 2015) after weighing the vehicle and adding 100 kg for driver and fuel when vehicle manufacturer's data were not available. The test masses for WLTC and the RC testing were chosen according to the Worldwide Harmonized Light vehicles Test Procedure (WLTP); the implemented values exceeded those used for NEDC testing by 6% on average (see Table 1).

The WLTC tests with Vehicles GDI #1 (three repetitions) and GDI #3 (two repetitions) were conducted with a higher dyno inertia to match the elevated vehicle mass during on-road tests, caused by the PEMS equipment and a co-driver. The dyno coefficients were adjusted to the new inertia.

The laboratory tests were in general conducted at room temperature of 23–25°C. Some tests with Vehicles GDI #1, #2, and #7 were conducted at temperatures of 8° and −7°C (three repetitions at each temperature). For the tests at −7°C, the chassis dyno coefficients were increased by 10% as required by Regulation 83 (UN/ECE, 2015). It should be noted that Regulation 83 requires to conduct the low-temperature test at −7°C only over the UDC phase of the NEDC; for this test, there is no limit for SPN emissions defined.

All vehicles were pre-conditioned with the required cycle the day before as stated in Regulation 83 (UN/ECE, 2015).

On-Board Testing with PEMS

The procedures described in Regulation 582/2011 for in-service conformity testing of heavy-duty vehicles (EC, 2011) and the proposal on the RDE test procedure for the type-approval for light-duty vehicles (Proposal for Annex IIIA, Regulation 692/2008) were followed as close as possible. However, instead of employing an analyzer for gaseous components, our PEMS consisted of a Global Positioning System (GPS), an exhaust flowmeter (EFM), and the respective SPN instrument, i.e., a PEMS-SPN (NanoMet 3, Testo) and/or a PMP-TP (NanoMet 1, Testo; see description below). The total weight of the PEMS, including batteries and the flowmeter, was estimated to be 100 kg (160 kg when both PEMS-SPN and PMP-TP systems were on-board). Vehicles GDI #1, #3, and #5 were also tested with the analyzer for gaseous components in order to measure CO_2 emissions and apply the future RDE data evaluation method.

The vehicles were tested at routes which capture urban, rural and highway driving, with an average trip length of ~100 km and an average speed of 50 km/h (see Table 3). The DPF and the PFI vehicles were tested only on urban and rural routes with an average distance of 50 km. Vehicle GDI #7 was not tested on the road. Route 1 and 4 consist mainly of rural and motorway driving, Route 2 captures urban and rural driving, while Route 3 is a rural route with steep uphill and downhill parts (up to 1100 m.a.s.l.). The other routes are located at an elevation between 200 and 400 m.a.s.l. Table 3 summarizes the number of routes and their characteristics for each vehicle. Details can be found in Weiss et al. (2011b). The driver for the on-road tests was different from the drivers in the laboratory.

Table 3. Characteristics of on-road test routes

Vehicle ID	Route*	Number of tests	Duration [h]	Distance [km]	Speed max [km/h]	Speed average [km/h]	Idle duration %	Temp. [°C]
HD #1	4	2	2.5–3.0	115	95	45	10	10–15
HD #2	4	6	2.5–3.0	130	95	50	2	10–15
DPF #1	2	2	1.0	50	130	50	17	20
DPF #2	2, 4	3	2.0	100	130	45	22	27
PFI	–	1	0.5	20	100	45	2	20
GDI #1	1, 2	2	2.0	130	130	60	9	21
GDI #2	1, 2	6	1.0–2.0	50–100	130	55	8	5–10
GDI #3	2, 4	3	1.5	75–180	130	50	15	22
GDI #4	1, 2	3	1.0–2.0	130	130	50	14	5
GDI #5	1	1	2.0	110	130	50	20	22
GDI #6	1	2	2.0–2.5	130	130	55	8	7
GDI #6	3	3	2.0	60	65	30	4	6

*Identification numbers refer to the test routes described in Weiss et al. (2011b) or Figure S4 in the Supplementary Material A.

SPN Equipment

A SPN measurement system (AVL APC 489) compliant with the light-duty vehicles Regulation 83 (UN/ECE, 2015) and regulation 49 for heavy-duty engines (UN/ECE, 2013) was used at the full dilution tunnel of the laboratory (Giechaskiel et al., 2010a). The system consists of a hot dilution at 150°C, an evaporation tube at 350°C and a secondary dilution at ambient temperature. The system was calibrated by the manufacturer. The chosen dilution including the particle losses (as average of sizes 30, 50, and 100 nm) is referred to as Particle number Concentration Reduction Factor (PCRF). In our experiments, we typically, used a primary PCRF of 100 and a secondary of 10. Downstream of the thermal pre-treatment system, a Condensation Particle Counter (CPC; TSI 3790) with 50% counting efficiency at 23 nm ($d_{50\%}$ = 23 nm) was connected. The application of hot dilution and the evaporation tube with a CPC of $d_{50\%}$ = 23 nm follows the protocol defined in the Particle Measurement Program (PMP) and is used accordingly for regulatory SPN measurements (Giechaskiel et al., 2008b). We will refer to this system as PMP-CVS.

For a few laboratory tests, another PMP compliant system (NanoMet 1, Testo) was connected to the tailpipe of the vehicles in order to investigate the effect of different sampling locations (tailpipe versus full dilution tunnel) on the measured SPN emissions. This system also consists of a hot dilution at 150°C, an evaporation tube at 350°C, a secondary dilution at ambient temperature and a CPC (TSI 3790) with $d_{50\%}$ = 23 nm. The PCRF chosen was 540 (110 × 4). The Nanomet 1 system was also used for some PEMS on-board tests (Vehicles HD #1, HD #2, DPF #2, PFI, GDI#1, GDI #3, GDI #5). In the following, we will refer to this system as PMP-TP.

Moreover, we conducted laboratory and on-board SPN emission measurements with a portable PEMS-SPN (NanoMet 3, Testo). The system samples the raw exhaust gas through a 2 m long line kept at 110°C. The raw gas is then diluted by a factor of 10 or 100 in a rotating disk diluter (Giechaskiel et al., 2014b). The diluted sample then passes through an evaporation tube set at 300°C. The concentration of SPN is estimated by measuring the current through a mini Diffusion Size Classifier (DiSC mini, Testo; Fierz et al., 2008). The DiSC charges the particles with a unipolar diffusion charger. Subsequently, the charged particles pass through a diffusion stage where the smallest particles are deposited by diffusion and detected as an electric current. The remaining particles end up in a second stage, the so called "filter stage," where a second current is measured. The ratio of the two currents allows estimating the average particle diameter. Taken together the two currents, the average particle diameter, the exhaust flow rate and the calibration parameters allow estimating the SPN concentration in the exhaust. This estimation includes particles < 23 nm (the 50% detection limit is around 15 nm). Future devices will have hardware and software "filters" to mimic the PMP efficiency curve. These features were not available in our device. Thus we used the total current from the two stages (proportional to the Lung Deposited Surface Area—LDSA) and we calibrated it with a PMP system (thus >23 nm); the calibrated signal was used for our calculations (Giechaskiel et al., 2014b, 2015). The results presented below refer to the calibrated LDSA to give SPN >23 nm, unless otherwise specified.

Exhaust Flow Determination

We estimated the vehicle's exhaust flow in all laboratory tests from the difference between the total flow of the dilution tunnel and the dilution air flow. In some tests of Vehicle GDI#1 (two cold start WLTCs and two hot start WLTCs) and Vehicle GDI #3 (two NEDCs, two WLTCs, one test with constant speeds) the exhaust passed through an exhaust gas flow meter (EFM; Sensors Inc., 2 inches diameter, accuracy 1.5 kg/h) connected at the exit of the vehicle's tailpipe before entering the dilution tunnel. The same flow meter was also used for the on-road PEMS measurements. The correlation between the EFM flow rate and the estimated flow rate from the dilution tunnel for steady state points was $R^2 > 0.99$ with a slope of 1.13. At transient operation, the slope was about 1.03 ($R^2 > 0.90$). Thus, the on-road exhaust mass flow as determined with the EFM could be subject to uncertainty in the range of 3–13%. We did not apply correction to the flowmeter signal because the exhaust flow determined by the difference of two big flows (dilution air and total flow in the dilution tunnel) has an uncertainty too. For the typical flow rates of the dilution tunnel that have an uncertainty of 0.5%, an uncertainty of 7% was estimated for the calculated mean vehicle exhaust gas flow. For the heavy-duty on-board tests a 5 inches exhaust flowmeter was used (reported accuracy 10 kg/h) and similar overestimation of the flow rate was found when evaluated in the laboratory.

Calculations

The CVS flow rate and the distance traveled by the vehicle were used to calculate the emissions in particles per kilometer [p/km] from the SPN systems at the dilution tunnel (CVS). For the systems at the tailpipe (both in the laboratory and on the road), the instantaneous particle concentration was time-aligned with the exhaust flow rate and then both parameters were multiplied with each other. The sum of all instantaneous particle number emissions was then divided by the test distance to obtain the distance-specific particle number emissions over a give sub-cycle, cycle, or route [p/km] (integrated emissions).

For the on-board tests, all emission events from the whole trip, including cold start, idling, and regeneration events, were included in the calculations.

Note that the approach for the on-road tests is not fully compliant with the upcoming RDE regulation that excludes cold start and idling from the emission calculations and requires that the trip remains within specific boundary conditions, e.g., with respect to acceleration and elevation gain. We regard our approach justified as the target of this paper is not to check the conformity of vehicles with respect to legislative emission limits, but to compare laboratory-based results with on-board measurements. Nevertheless, the evaluation method and its application will be presented in the "Results and Discussion" Section.

RESULTS AND DISCUSSION

Before presenting the on-road SPN emission results for the tested vehicles, the importance of some parameters that affect the level of measured emissions will be discussed.

Sampling Location and Instruments Evaluation

Measuring particle number emissions after the dilution tunnel or directly from the tailpipe can lead to substantial differences due to processes like thermophoresis, agglomeration, evaporation/condensation, and diffusion (Giechaskiel et al., 2012b). Comparing two PMP systems, one measuring at the tailpipe (PMP-TP) and one at the dilution tunnel (PMP-CVS), we found mean deviations ranging from −15 to +15%, with a standard deviation of the mean differences of ~ 15% (Figure 1, blue bars and error bars). These deviations are slightly higher compared to those we found for the two PMP systems measuring in parallel (5–10%; Figure not shown).

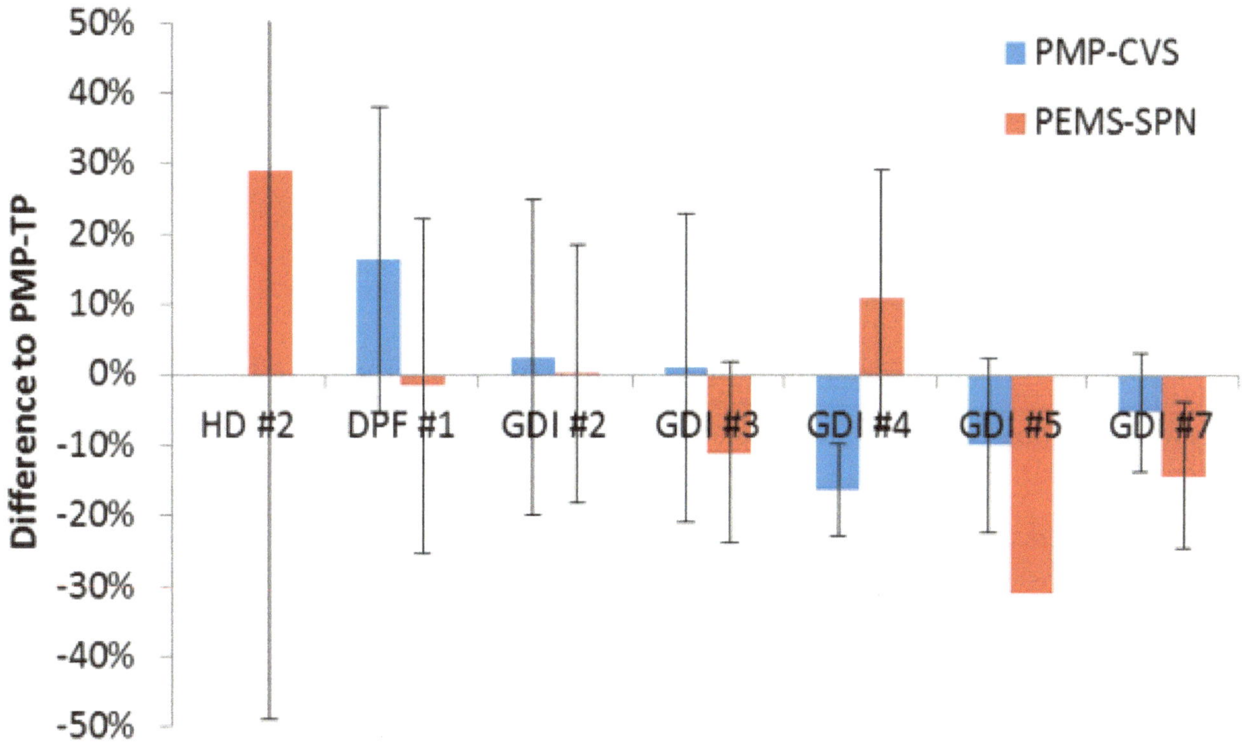

Figure 1. Deviations between SPN emissions measured with various systems at various sampling locations. Error bars show one standard deviation of SPN emissions measured over >10 phases of various driving cycles. All tests were conducted in the laboratory, except the comparison of PMP-TP with the PEMS-SPN for HD #2 and GDI #5 that were conducted on the road (on-board).

The results are in agreement with Giechaskiel et al. (2015) who compared two reference PMP systems that were compliant to Regulation 83 (UN/ECE, 2015) and installed at the tailpipe and at the dilution tunnel, respectively. Based on a sample of >130 tests with >15 vehicles, the average differences were within ±20% with a standard deviation of 15%. Giechaskiel et al. (2015) attributed the differences to agglomeration (0–15%), thermophoresis (0–5%), exhaust flow estimation (±10%), and time alignment issues (0–10%).

These observations suggest that particle dynamics play a minor role in the determination of SPN emissions; the differences in the measurements taken at the tailpipe and after the dilution tunnel are relatively small. This conclusion could be different for (i) older, high emitting vehicles where coagulation is important (Isella et al., 2008) or (ii) heavy-duty engine testing where the exhaust gas temperatures and thus thermophoretic losses are higher (Giechaskiel et al., 2010b).

The mean differences between the measurements taken at the tailpipe in the laboratory by the PEMS-SPN and the PMP-TP systems are within ±15% (Figure 1, red bars, except HD #2 and GDI #5) with a standard deviation of the differences of 25%. These differences are similar with those obtained for the two PMP systems measuring at the tailpipe and at the dilution tunnel, respectively, but it should be taken into account that the calibration of the PEMS-SPN was conducted with these vehicles. When the on-road comparisons are included the differences are within ±30% (Figure 1, red bars, HD #2 and GDI #5). The larger differences can be explained by the different principle of operation, i.e., diffusion charging of the PEMS-SPN that is sensitive to the size of particles measured versus condensation particle counting of the PMP system (see Giechaskiel et al., 2014b). For example, for Vehicle GDI #5, the concentration of the PEMS-SPN as estimated by the system using the mean size information would be within 4% of the PMP system. The results scatter over a wider range for the DPF-equipped Vehicle HD #2, where the particle number emissions were low and close to the detection limit of the PEMS-SPN. A relatively high dilution was used for these tests because the target was to catch any regeneration event that would happen. Nevertheless, although the relative difference in the measured SPN emissions seems high, in absolute terms, it was very low.

The relatively good agreement between the PEMS-SPN and the PMP-TP, both connected to the tailpipe, holds true even when emission levels span almost three orders of magnitude (from 5×10^{10} to 2×10^{13} p/km, Figure2). Almost all points lie within the range of ±35% depicted by the dotted lines in Figure 2. Note that the tests of Vehicle GDI #5 (on-road) had similar differences with Vehicle DPF #1 (laboratory) and GDI #7 (laboratory).

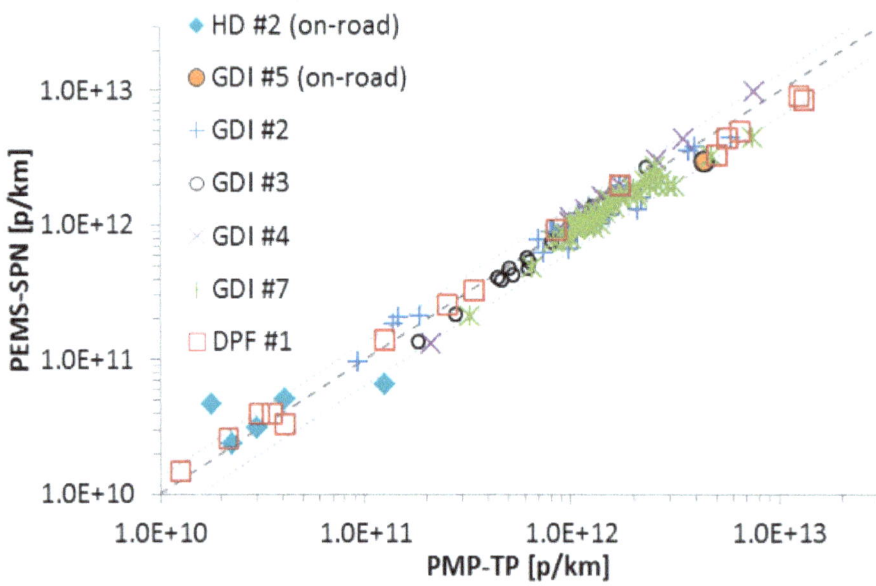

Figure 2. Deviations between the SPN emissions measured at the tailpipe with the PEMS-SPN and the PMP-TP system; each point represents a phase of a test cycle. Tests are conducted in the laboratory, unless otherwise specified (on-road). The dotted lines represent deviations of −35% (0.65:1) and +35% (1.35:1) from the 1:1 line.

The real-time response of the PEMS-SPN is very close to that of the PMP-TP (Figure 3), even though it is based on diffusion charging and not optical counting. A similar behavior was found for the other test vehicles as well. In most cases there was no significant effect of the mean size of the particles on the particle number measurement. Partly, the reason was due to the similar Geometric Mean Diameter (GMD) of the particles emitted by the tested vehicles, which ranged approximately from 37 to 59 nm. However, during, e.g., cold-starts and accelerations when the instantaneous mean particle size is different than the mean size over the whole test, the instantaneous emissions can be over- or underestimated (see details in Giechaskiel et al., 2015). For example, the spikes usually have differences on the order of ±60% between the PEMS-SPN and the PMP-TP. Partly this is due to the change of the mean particle size and partly due to the difference in the response time of the two instruments.

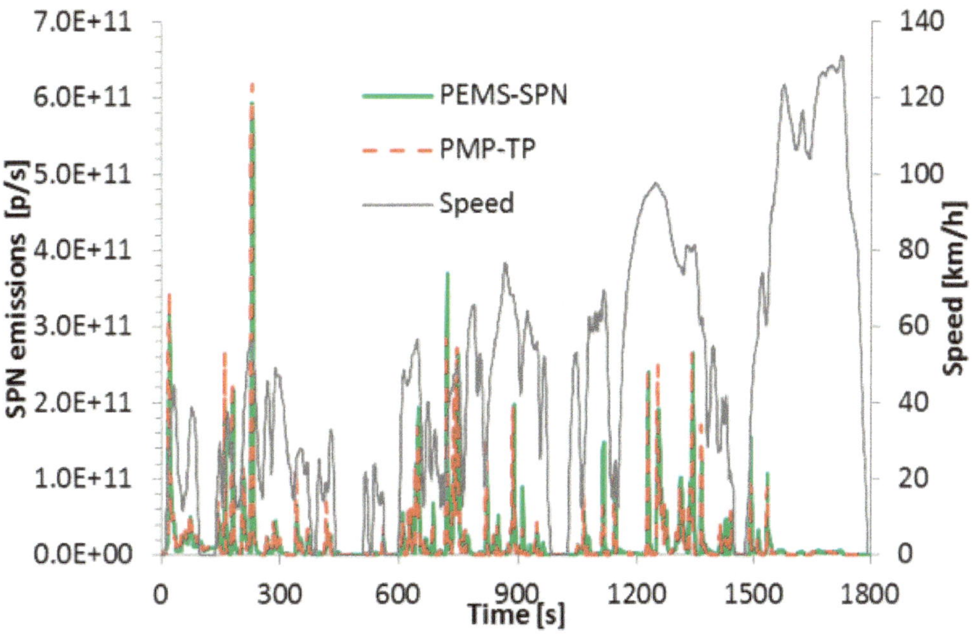

Figure 3. Real-time comparison of PEMS-SPN and PMP-TP measurements of Vehicle GDI #2 over one WLTC in the laboratory. Both systems were connected to the tailpipe.

As a final note, although the deviations between the PEMS-SPN and the PMP-TP measurements are higher than those between the two PMP systems, it should be kept in mind that the PEMS-SPN is designed for on-board use, thus it is light and relatively energy efficient; and thus simplified approaches (e.g., the sensor) are acceptable. In addition, the uncertainty is relatively constant for a wide range of emission levels. Nevertheless, we argue that the measurement uncertainty has to be taken into account in the Conformity Factors of the future PEMS-SPN regulation.

DPF Load and Regeneration

DPFs reduce tailpipe SPN emissions by more than 95% but lead to highly variable tailpipe emissions due to variations in the particle removal efficiency during cold-start and in dependence of the DPF fill state. The efficiency of a DPF can thus vary from 90 to >99%. In addition, the particle number emissions during regeneration can be significantly elevated for aggressive cycles including prolonged operation at motorway driving conditions (Mamakos et al., 2013b). The SPN emissions (>23 nm) during phases with regeneration can exceed by several factors the Euro 5b and Euro 6 SPN limit, due to the decreased efficiency of the DPF after the oxidation and removal of the accumulated soot.

Figure 4 shows three sequences of measurements that were conducted with Vehicle DPF#1 until regeneration events occurred. Each sequence depicts the SPN emissions measured over individual phases of the NEDC and WLTC at various conditions (including cold or hot start, different ambient temperatures, mixed sequence of cycles). Immediately after a regeneration event (0 km) the particle number emissions are high because the DPF is empty and its efficiency is relatively low (typically 90%). The emissions decrease over time to very low levels with a few exceptions (e.g., cold-start cycles at low ambient temperatures). A similar behavior was observed at different sequences of tests between regeneration events.

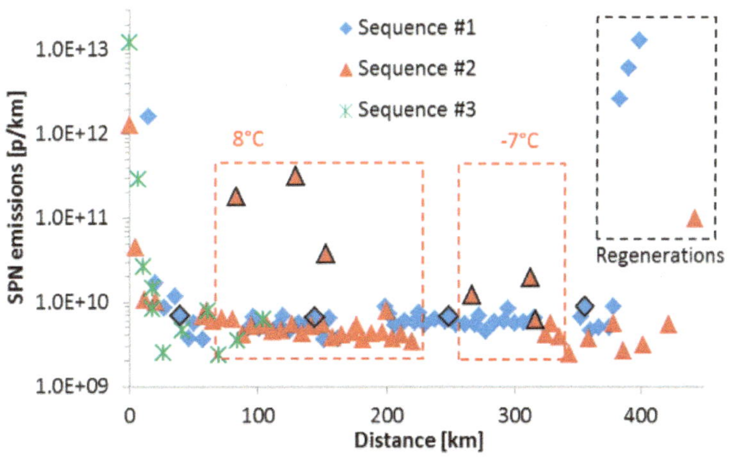

Figure 4. SPN emissions of Vehicle DPF#1 over the distance between regenerations. Each point represents a phase of a test cycle (mix of NEDCs and WLTCs). Cold-start cycles are indicated by black border line. All tests were conducted in the laboratory with PMP-CVS equipment. Sequence #3 was stopped before regeneration.

The particles during cold-start could be (i) semi-volatile material escaping oxidation as the catalytic converters have not yet reached the light-off temperature and cannot be evaporated in the SPN system or (ii) blow-out of loose non-volatile particle deposits, as the filter is exposed to highly transient operation with respect to thermal and flow conditions (Giechaskiel et al., 2007b). Another explanation is that small filter defects, which affect porosity, reduce DPF filtration efficiency at low temperatures but tend to close as the temperature rises, thus causing the DPF efficiency to increase with temperature (Braisher et al., 2010; Mamakos et al., 2013b). In conclusion, DPF equipped vehicles have low, but variable particle number emissions depending on the operating conditions of the vehicle.

Ambient Temperature

The SPN emissions tend to increase as the ambient temperature decreases (Figure 5). The emissions of Vehicle GDI #2 increased by 160% when tested over the WLTC at −7°C instead of the prescribed 23°C. The temperature effect was smaller for Vehicle GDI #1 (70%). The findings are in agreement with the literature (Aakko and Nylund, 2003; Giechaskiel et al., 2007b; Mamakos et al., 2012, 2013b; Chan et al., 2013; EPA, 2014) and can be explained by two effects. First, the enrichment of the air-fuel mixture during cold-start engine operation, which compensates for the reduced fuel vaporization and elevated friction of engine components, leads to incomplete fuel combustion. Second, at low ambient temperature, catalytic after-treatment systems need more time to reach their light-off temperature. By comparison, the increase in SPN emissions of a hot engine and after-treatment system at low ambient temperature is small (< 50%, Figure 5). This observation is expected because solid particles are being formed in the engine and are therefore relatively unaffected by ambient conditions after engine warm-up (Aakko and Nylund, 2003; Ristimäki et al., 2005). Mamakos et al. (2013b) have mentioned that the opposite effect of lower SPN emissions with decreasing ambient temperatures is also possible in cases where the Exhaust Gas Recirculation (EGR) in diesel engines is reduced at low temperatures, thus decreasing the soot-out emissions of the engine.

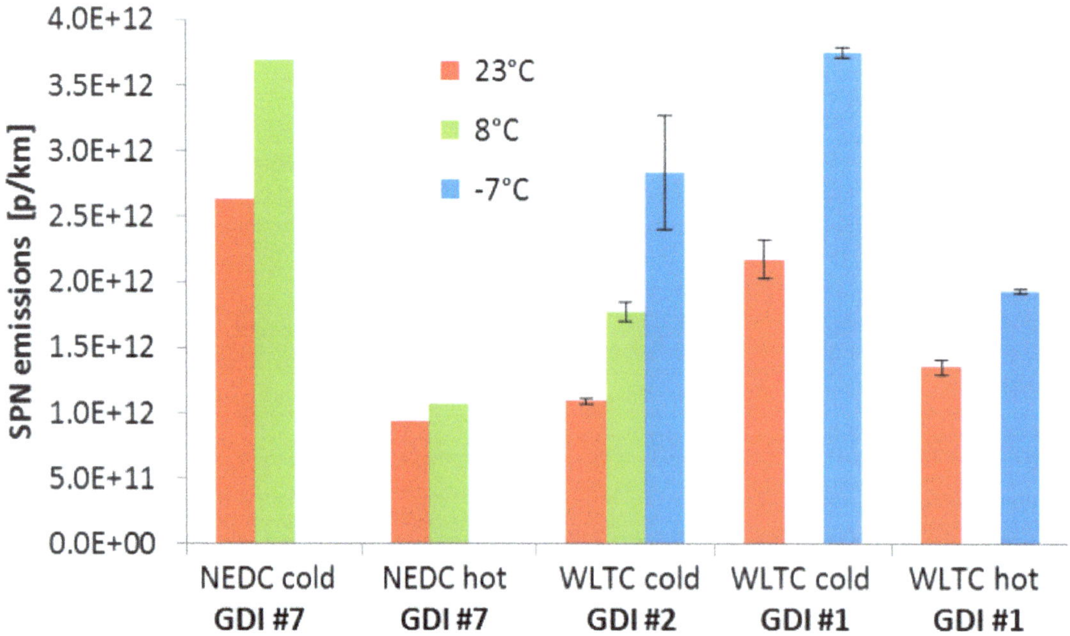

Figure 5. The effect of ambient temperature on SPN emissions. Error bars indicate the minimum and maximum emissions of three test repetitions. Tests were conducted in the laboratory with PMP-CVS equipment.

The instantaneous real-time SPN emissions and the exhaust gas temperatures of Vehicle GDI #1 are shown. During cold start, SPN emissions are higher at lower ambient temperature. The cold start effect is significant in the first 300 s but remains visible until almost 750 s. For the hot cycles at both ambient temperatures (−7° and 23°C) the SPN emissions at the beginning of the cycle are similar as the exhaust gas temperature is also similar. However, the emissions diverge after 300 s and correlate with the exhaust gas temperature. The cold and hot cycles at the same ambient temperature have similar emissions at the last part of the cycle as the exhaust gas temperature stabilizes at the same levels.

Test Mass

The mass added by the PEMS equipment to the vehicle can sometimes exceed 150 kg (especially for the first generation of PEMS). Accounting for the mass of one co-driver of 80 kg, PEMS-SPN testing may result in a vehicle load of 230 kg. This additional load may be negligible for buses or trucks; however for passenger cars, the added mass can affect engine load and thus particle number emissions. We found that increasing the chassis dyno inertia to account for an additional vehicle load of 230 kg increases the SPN emissions of Vehicle GDI #3 by 5–15% (Figure 6). The same tests suggested an increase in the CO_2 emissions by 10–20% as well, which is in line with the expected CO_2 increase of 6–7% for every 100 kg of vehicle mass (Zacharof et al., 2015). A 10% increase of the inertia of Vehicle GDI #1 resulted in an 18–27% increase of SPN emissions. A similar trend was found by May et al. (2014) when testing vehicles with different inertia in the laboratory.

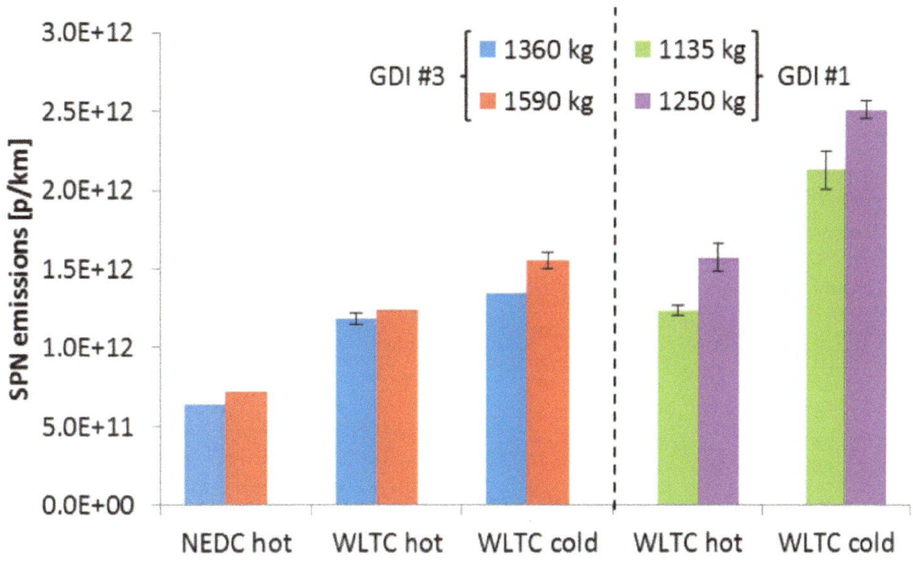

Figure 6. Effect of vehicle load (simulated here by chassis dynamometer inertia) on the SPN emissions of Vehicles GDI #1 and #3. Error bars indicate the minimum and maximum emissions of three (Vehicle GDI #1) and two (Vehicle GDI #3) test repetitions. Tests were conducted in the laboratory with PMP-CVS equipment.

Driving Style

Driving the WLTC speed profile without any gear shift indication on the display by four different but experienced drivers causes deviations in the SPN emissions of less than 10% (Figure 7). However, deliberately driving at higher engine speed, thus at lower gear, and changing the gears often can double the SPN emissions. Dedicated constant speed tests showed that the gear change has the biggest effect on SPN emissions, while a smaller effect can be attributed to the engine speed; i.e., the transition phase has higher contribution probably due to incomplete combustion. Thus, even with a well-defined speed profile, the effect of driving style on the SPN emissions can be significant. This observation is in agreement with the dynamics of gaseous pollutants and CO_2 observed during on-road driving (Jackson et al., 2006). Aggressive driving can increase trip-average fuel consumption by up to 24%, while eco–driving can provide estimated benefits in the order of 6% compared to normal operation, thus directly impacting particle formation (Zacharof et al., 2015).

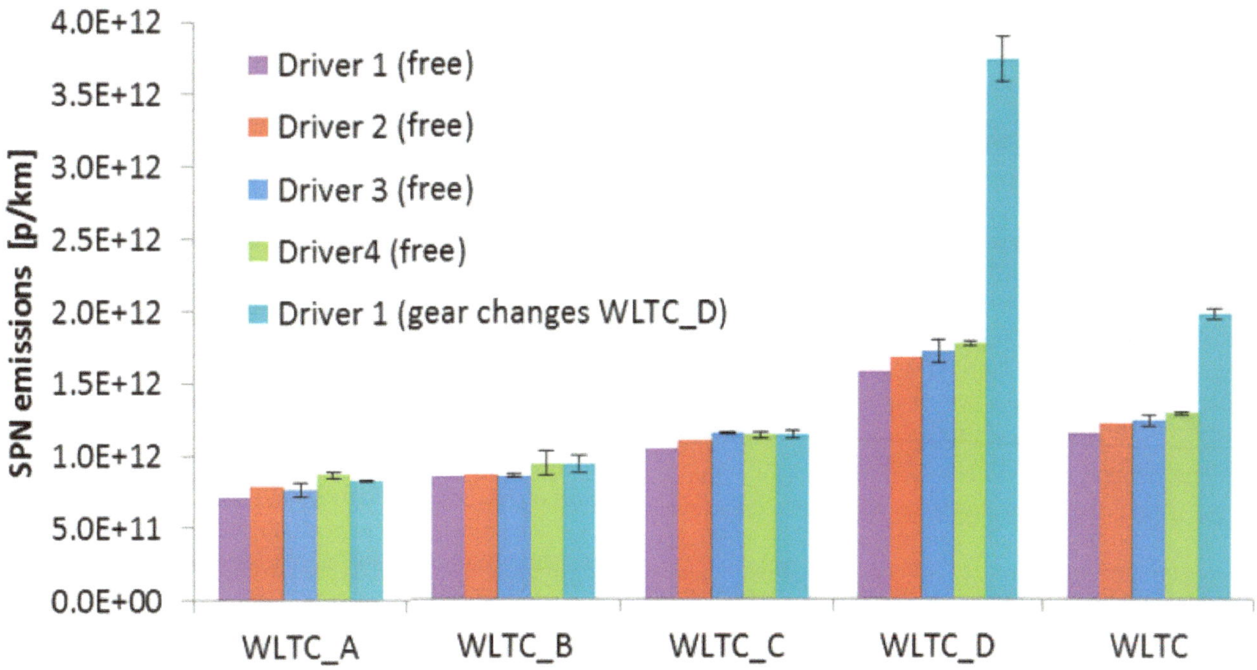

Figure 7. Effect of driving style on the SPN emissions of Vehicle GDI #3. Error bars indicate the minimum and maximum SPN emissions of two test repetitions conducted with 3 months difference between them. All drivers were free to choose their gear shifting, except Driver 1 who followed in one test sequence of WLTC_D a gear shift strategy at higher rpm. Tests in the laboratory were conducted with PMP-CVS equipment.

Driving Cycle

In line with the sensitivity to driving style, SPN emissions vary substantially depending on the driven cycles and their phases (Figure 8). For Vehicle GDI #6, the SPN emissions were on average ~ 4 × 10^{12} p/km with the engine working in hot conditions, but lower emissions were measured over the high-speed phase of the WLTC (WLTC_C; 2 × 10^{11} p/km) and higher emissions over the low-speed phase of the WLTC (WLTC_A at cold-start conditions; 8.5 × 10^{12} p/km). Aggressive accelerations during the severe portion of the random cycle (RC_C) led to elevated SPN emissions. The on-board results are in qualitative agreement with the laboratory results. The higher SPN emissions over PEMS Route 1 as compared to Route 3 can be explained by the higher average speeds (55 vs. 30 km/h). Qualitatively similar results were observed for the other GDI vehicles. In most cases, cold start contributed the largest portion to the differences in SPN emissions observed between different sub-cycles.

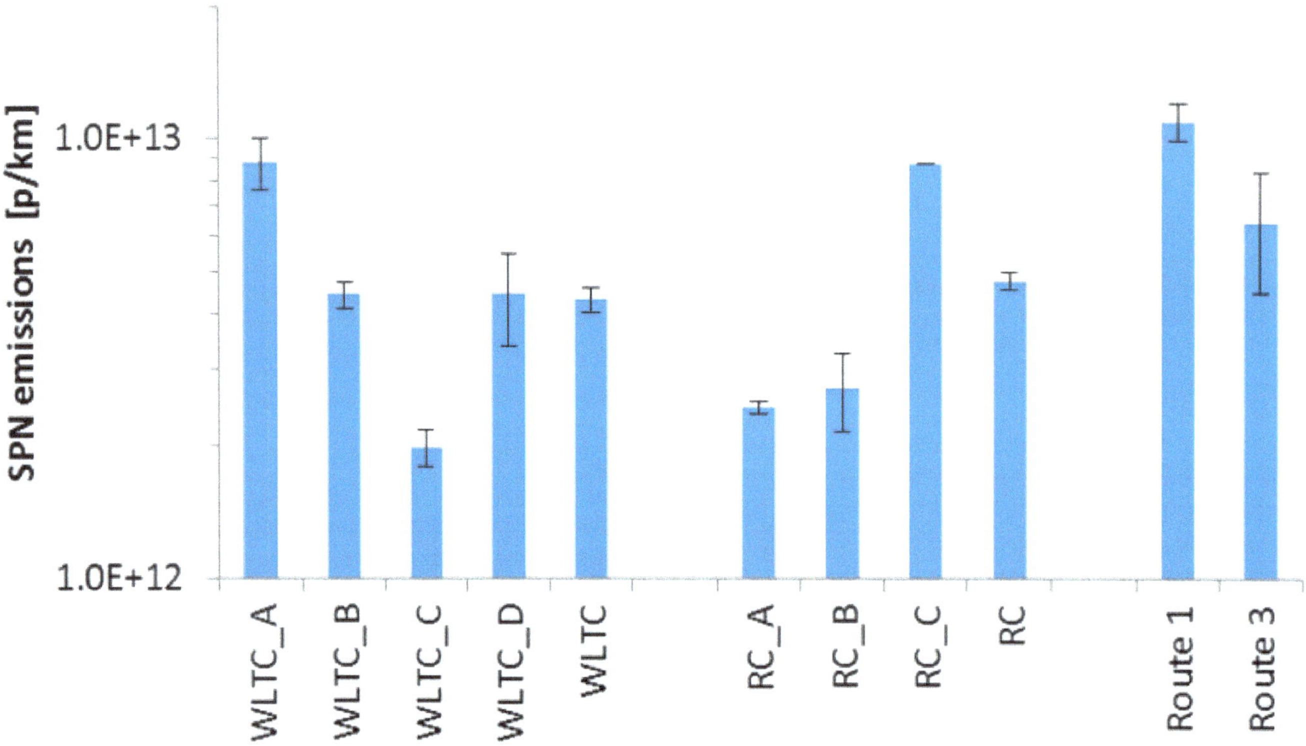

Figure 8. SPN emissions of Vehicle GDI #6 over various phases of the WLTC and the random driving cycle as well as two on-road trips.

As mentioned in Section Ambient Temperature, SPN emissions at cold-start are high due to engine enrichment resulting in incomplete combustion. The SPN emissions decrease as the engine gets warmer due to better combustion. In addition, the increasing surface temperatures of cylinder walls aid better fuel vaporization and avoid "pool fires" (Witze and Green, 2005; Peckham et al., 2011). The higher SPN emissions during accelerations correlate with the air/fuel ratio; emissions increase when the mixture is rich (Maricq et al., 1999;Wang et al., 2006).

A similar tendency was observed for the DPF equipped vehicles, when the emissions were in measurable levels. However, due to the high efficiency of the DPF, in most cases absolute emission levels were so low that it is difficult to draw robust conclusions.

Evaluation of On-Road Data

The on-road SPN emission results are presented as integrated emissions over the whole trip. (Figures 8, 10). However, the trip share (urban, rural, motorway) and the conditions (e.g., vehicle mass, ambient conditions, road grade, wind, accelerations etc.) have a big impact on the final result.

For heavy-duty in-service conformity testing, the trip requirements and the Moving Average Window (MAW) evaluation method are described in Regulation 582/2011 (EC, 2011). The MAW is based on data evaluation of windows of on-road operation whose length is defined by the work produced over the reference laboratory WHTC. However, the method contains several boundaries; for example, operation points with coolant temperature < 70°C (i.e., cold-start) and low power operation, where the engine power is < 20% of the maximum rated power are excluded. For N3 category trucks (i.e., those tested in this study) the trip should consist of ~20% urban (< 50 km/h), 25% rural, and 55% motorway (>75 km/h) operation (Annex II of Regulation 582/2011; EC, 2011). These boundary conditions are currently under evaluation (Perujo and Mendoza-Villafuerte, 2015).

For the type approval of light-duty vehicles, the share of urban (< 60 km/h), rural (60–90 km/h), and motorway (>90 km/h) driving should be of equal duration. The data evaluation is based on the MAW method. The lengths of windows match half of the mass of CO_2 emitted by the respective vehicle over the WLTC. Additional boundary conditions limit the instantaneous acceleration at a given speed and the cumulative elevation gain during a trip. However, there are still some open points regarding the boundary conditions for the analysis of the results (e.g., exclusion of high acceleration rates or certain ambient conditions).

To illustrate the variability of SPN emissions over a test, Figure 9 shows the MAW SPN results over the mean speed of each MAW as well as the average results for urban, rural, and motorway driving during an on-road trip. The WLTC laboratory results are also depicted for comparison. Figure 9 confirms the findings of Figure 8, suggesting that cold-start and high accelerations increase SPN emissions. Control of parameters such as fuel injection, spark timing, and air-to-fuel ratio during cold-start and acceleration may be important in limiting the extent of particles formation.

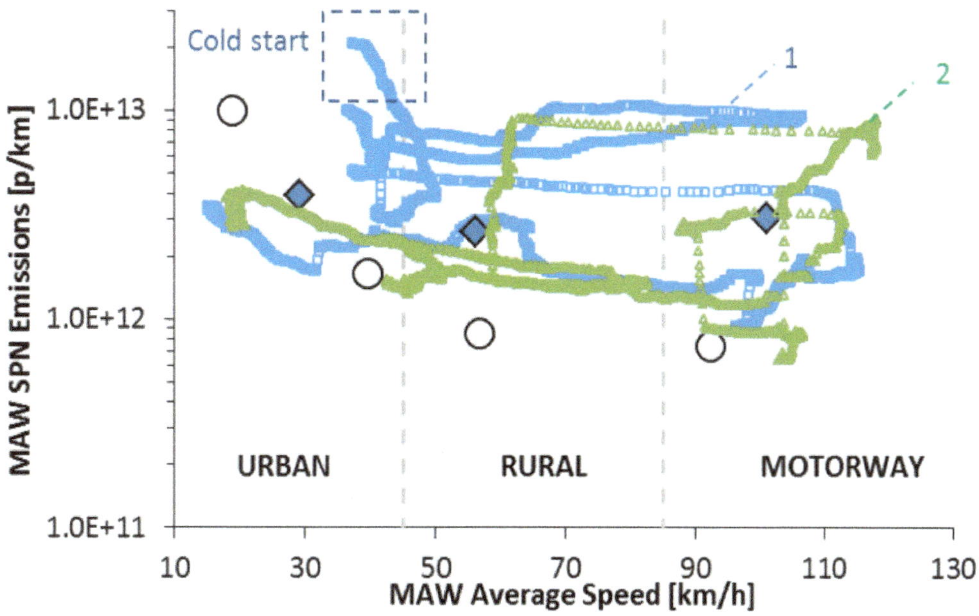

Figure 9. MAW SPN emissions of a road trip with Vehicle GDI #5 in function of the MAW speed. The blue small squares (1) show the first half way of the trip and the small green triangles (2) the return trip. Solid diamonds depict the mean on-road SPN emissions for urban, rural, and motorway driving. White circles depict the laboratory SPN emissions over the four phases of the WLTC.

The on-road PEMS-SPN emissions are calculated without any data exclusions (4.1×10^{12} p/km) and alternatively by following the RDE regulation with the exclusion of cold start and idling and the weighing of emissions if these belong to windows whose CO_2 emissions deviate by more than 25% from the emissions measured over the WLTC (3.2×10^{12} p/km). The on-road SPN emissions are higher than those measured in the laboratory over the cold-start WLTC (2.2×10^{12} p/km).However, it is important to keep in mind when reporting emissions using the evaluation tools (like MAW or power binning method) because they are partly normalized emissions to the type approval cycle.

SPN Emission Factors

Experimental Results

The results of the on-road PEMS tests obtained with the PMP-TP or PEMS-SPN and those of the laboratory tests obtained with the PMP-CVS are shown in Figure 10. Beginning with the PEMS results, the Euro V heavy-duty diesel Vehicle HD #1 had high emissions (2×10^{13} p/km), while the Euro VI heavy-duty diesel Vehicle HD#2 showed very low SPN emissions (6×10^{10} p/km). Assuming 1.2 kWh/km for the WHVC (Ntziachristos et al., 2006), then the 6×10^{11} p/kWh Euro VI SPN limit would translate to 7.2×10^{11} p/km, and the on-road emissions of the Euro VI truck are one order of magnitude lower than the limit. As far as the compression ignition light-duty vehicles is concerned, Vehicle DPF #1 had emissions of 8×10^{11} p/km, while the emissions of Vehicle DPF#2 were one order of magnitude lower, thus close or even below the Euro 6 limit (6×10^{11}p/km). The Vehicle PFI had SPN emissions slightly higher than 1×10^{12} p/km (note that the trip comprised urban driving only). The SPN emissions of the GDI vehicles spanned ~ from 8×10^{11} to almost 8×10^{12} p/km. Most GDIs complied with the current Euro 6 limit (6×10^{12} p/km) but not with the future Euro 6c limit (6×10^{11} p/km, to be applied from 2017 onward).

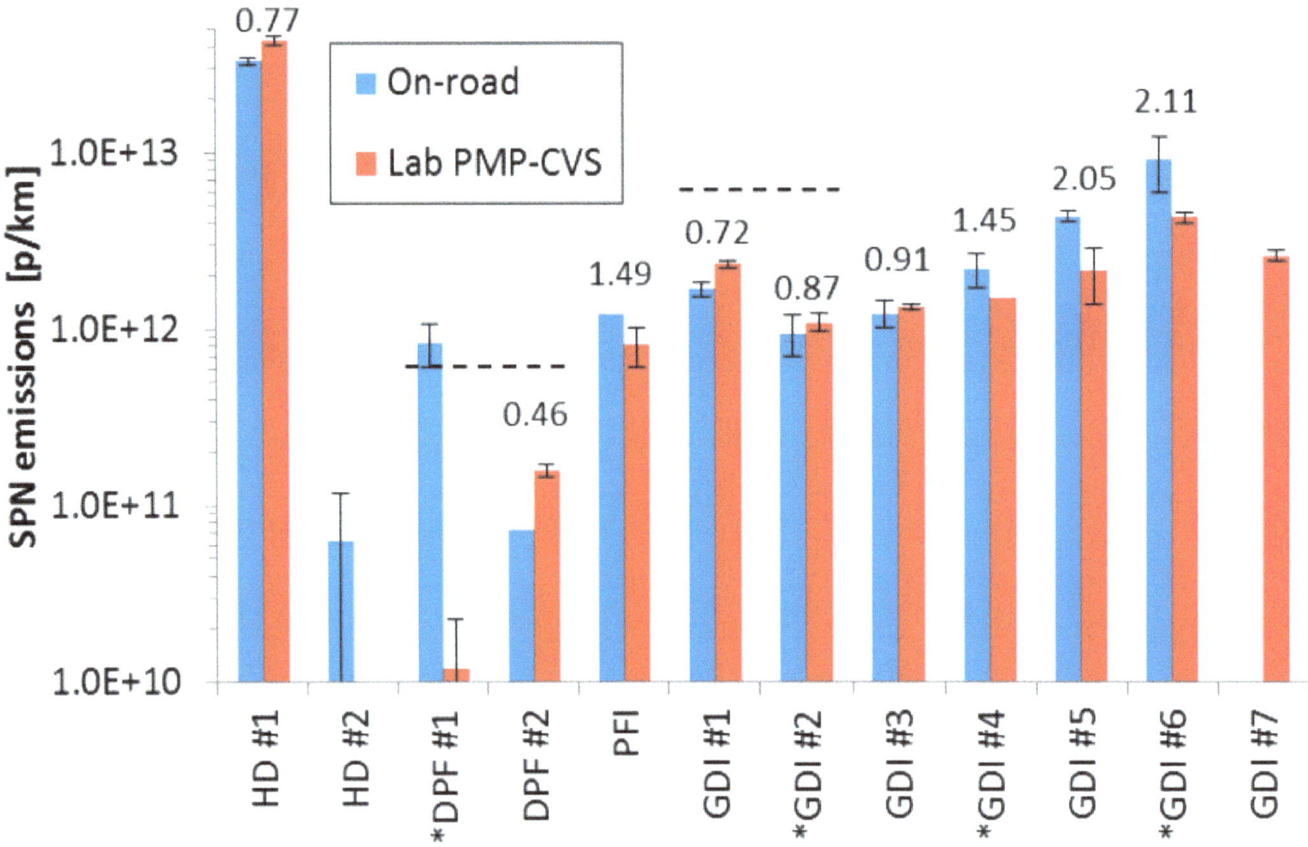

Figure 10. SPN emissions of heavy-duty and light-duty vehicles on the road (PMP-TP) and in the laboratory (PMP-CVS). Asterisk indicates that PEMS-SPN was used instead of PMP-TP system for on-road measurements. DPF regeneration has been excluded from the emission measurements of Vehicles DPF #1 and DPF#2 when tested in the laboratory. On-road results are averages of all tested routes (see Table3). Error bars indicate the range between minimum and maximum SPN emissions. Dashed lines depict the applicable emission limits. Numbers indicate the ratio of on-road to lab emissions.

The laboratory results of the (future type approval) cold-start 23°C WLTCs (or WHVC for the Euro V truck, which is not a type approval cycle) were in relatively good agreement with the on-road PEMS results, with the exception of the DPF equipped light-duty vehicles. As discussed in the previous sections, the DPF fill state and the passive regeneration during the on-road tests most likely contributed to the high, though unsystematic differences (see also discussion of Figure 4). The on-road tests of Vehicle DPF #1 were conducted almost immediately after a regeneration event in the laboratory and for this reason they are elevated. For the remaining vehicles, deviations may be explained by: (i) uncertainties of the measurement equipment (e.g., Vehicles GDI #2, #4, #6), (ii) the lower ambient temperatures on the road (Vehicles GDI #2, #4, #6), (iii) the differences in driving conditions on the road and in the laboratory, (iv) the uncertainty of the chassis dynamometer parameters to simulate on-road driving, and (v) the mass of the PEMS.

A deviation of about a factor of two was observed when comparing the on-road and laboratory tests of Vehicles GDI #5 and GDI #6. A closer examination of the real time signals in the laboratory tests revealed that these vehicles had very high emissions during accelerations. In addition, vehicles were running under rich conditions during most part of the on-road tests. The Euro 6 Vehicles GDI #1 and #2 showed on the road lower SPN emissions than in the laboratory. This observation may probably be explained by the relatively high contribution of cold-start to the average SPN emissions in the laboratory. The cold start (around 5 min) has a higher impact on a 30 min test cycle (WLTC) than on a 120 min on-road test. To illustrate the effect, we compared the SPN emissions measured on the road with those measured over a hot WLTC test. The ratio of SPN emissions changed from 0.72 to 1.2 for Vehicle GDI #1 and from 0.87 to 1.27 for Vehicle GDI #2.

It should be mentioned that the light-duty on-board emissions were compared with the WLTC emissions which are considered more representative of the real-life operation of the vehicles than the NEDC. Comparisons with SPN emissions measured over the NEDC would probably result in higher differences between on-road and laboratory measurements. Similarly the Euro V truck on-road emissions were compared with the WHVC and not the ETC. As a side note, the fuel consumption of the three GDI vehicles (#1, #3, #5) was 15–35% higher on the road than in the lab; this observation is in line with the ranges reported in the literature (Mock et al., 2013). For example, Weiss et al. (2012) found 24 ± 15% differences at seven diesel Euro 4–6 vehicles. Another study of 600,000 vehicles found 40% differences (Tietge et al., 2015).

Literature Overview on SPN Emission Levels

The SPN emissions measured in this study are in general within the ranges reported in the literature on *chassis dynamometer studies* (see Giechaskiel et al., 2012a; Ntziachristos and Galassi, 2014). The substantial decrease in the SPN emissions between Euro V and Euro VI trucks shows the potential of engine and after-treatment technologies following the introduction of a Euro VI SPN limit. Until Euro V a mass limit of 20 mg/kWh was only required and any DPF was optimized for such emission levels (which are equivalent to 4×10^{13} p/km assuming a correlation of 2×10^{12} p/mg, see Giechaskiel et al., 2012a). The SPN emissions of the PFI vehicle (1.2×10^{12} p/km) are at the high end of reported values, probably because this vehicle was tested in "urban" conditions only. The emissions of the seven GDI vehicles (2×10^{12} p/km) were at the lower end of reported values. This observation may be attributed to the improvements of the new generation of GDIs.

On-road measurements of Euro III and IV diesel buses showed levels of 9×10^{13} and 3×10^{13} p/km, respectively (Liu et al., 2011). Huang et al. (2013) found much higher emissions for a Euro IV diesel bus using PEMS equipment (7×10^{14} p/km). The Euro V truck of our study showed lower SPN emissions than the previously mentioned Euro III and Euro IV buses. Johnson et al. (2009b) and Zheng et al. (2014) measured trucks with a Continuously Regenerating Trap (CRT) and found SPN emissions below 6×10^{11} p/km. However,Zheng et al. (2014) found that SPN emissions during uphill driving could increase by one order of magnitude compared to the test without uphill dribing. These findings are in agreement with the SPN emissions measured in our study with the Euro VI heavy-duty vehicle on the road (Vehicle HD #2).

The on-road PN emissions of diesel cars that are not equipped with a DPF range approximately from 1×10^{14}p/km (Vogt et al., 2003) to 6×10^{14} p/km (Huang et al., 2013), depending on the specific test route and operating conditions. The SPN emissions of diesel cars equipped with a DPF are two orders of magnitude lower (5×10^{12} p/km; Armas et al., 2013) and can be even close to the ambient background levels under favorable operating conditions, e.g., optimum DPF fill state (Bergmann et al., 2009). Our results are within the reported range (see Vehicles DPF #1 and #2 in Figure 10).

The PN levels measured on-road for light-duty gasoline vehicles vary over several orders of magnitude. Wehner et al. (2009) found total PN emissions (including volatile and semi-volatile particles) of 1.8×10^{12} p/km for a Euro 4 gasoline car. Huang et al. (2013) found PN emissions two orders of magnitude higher (1.5×10^{14}) for another Euro 4 gasoline car. Due to the low temperature at the thermal pre-treatment (80°C) of the sampling system, it is possible that these emissions included semi-volatile particles as nucleation mode. Our results are in agreement with Wehner et al. (2009).

The on-road SPN emission levels of the two heavy-duty vehicles are lower than those found by the *roadside measurements*. Total PN emissions of 1.4×10^{14} p/km have been measured for DPF equipped buses and 6.5×10^{14} p/km for non-DPF buses (see reviews of Morawska et al., 2008; Hallquist et al., 2013). The Euro V DPF equipped truck had emissions of 2×10^{13} p/km which is lower but close to the reported value. However, the Euro VI truck had emissions $< 1 \times 10^{11}$ p/km indicating the capabilities of high efficiency DPFs that have to fulfill the Euro VI SPN limits. The reported *roadside emission levels* of light-duty vehicles are on the order of 3×10^{13} p/km, which is much higher than the emissions reported here ($1–5 \times 10^{12}$ p/km; see Figure 10). The discrepancy may be explained by (i) the share of diesel cars without DPF in the vehicle fleet, (ii) the observation that roadside measurements often include nucleation mode particles and secondary particles, and (iii) the overall higher uncertainty, thus variability, of roadside measurements.

Literature Overview on Comparisons between Laboratory and On-Road Tests

Direct comparisons of SPN emission measurements conducted in the laboratory and on the road are still scarce. Vogt et al. (2003) and Giechaskiel et al. (2007a) found very good agreement (within 15%) between laboratory and chase steady state tests for light-duty diesel vehicles. Bergmann et al. (2009) measured lower emissions in the laboratory compared to on-road tests for a DPF equipped light-duty vehicle. Similarly,Andersson et al. (2014) found almost one order of magnitude lower SPN emissions in the laboratory than on the road with PEMS for DPF-equipped diesel cars. These results are in agreement with our findings.

To our knowledge, there are no comparisons of SPN emissions from GDI vehicles in the literature. Here, we show that SPN emissions on the road are generally higher by a factor < 2 than those found in the laboratory. The Euro 6 vehicles had on-road emissions close to the laboratory values (values $<$ 1). This observation indicates that for current technologies, the on-road SPN emissions are roughly in line with the applicable emissions limit. We have not observed exceedances comparable to those found for NO_x emissions, where factors of seven have been reported (Franco et al., 2014). In addition, the on-road SPN emissions of the GDI vehicles were lower than the current limit of 6×10^{12} p/km. However, it remains to be investigated whether the relatively good agreement between laboratory and on-road SPN emissions may still hold for future GDI vehicles that have to emit less than 6×10^{11} p/km and may be equipped with new technologies (i.e., Gasoline Particle Filters).

In general, the results of the present study are at the low end of PN emissions reported in the literature. We attribute this observation to technological improvements, especially related to GDI vehicles, and the fact that we examined only solid particles >23 nm (as currently regulated in the EU). Previous studies also investigated the emissions of sub-23 nm particles in the laboratory. In agreement with the literature, Giechaskiel et al. (2014c) found that sub-23 nm particles over a test cycle (WLTC) accounted for 30–40% of the total PN emissions. The sub-23 nm fraction during on-board tests was not measured. However, an estimation can be made based on our PEMS-SPN results: The difference of the SPN particles (>15 nm) as estimated by the instrument versus the >23 nm SPN emissions as calibrated by us reached some 30–50%, which is in line with the expected percentages. However, this estimation is subject to high uncertainty as the actual on-road mean particle size could be different compared to the one observed in lab tests used for the >23 nm calibration of the PEMS-SPN instrument. The total emissions (solids and volatiles) were not evaluated.

CONCLUSIONS

Limits for SPN emissions have been introduced in the European Union in 2011 for light-duty vehicles and 2013 for heavy-duty engines, respectively. However, quite often, differences in SPN emissions are found between laboratory and on-road tests. European policy makers address this situation with the recently approved RDE on-road test procedure for light-duty vehicles.

On-road emissions testing with PEMS is well established for determining the in-service conformity of heavy-duty vehicles with respect to gaseous pollutants. The application of PEMS for measuring particulate mass emissions is already accepted in the USA and might be included in the European regulation too. However, the application of PEMS for SPN emissions for the type-approval testing of light-duty and heavy-duty vehicles is new and presents technical challenges.

Particle sampling at the tailpipe requires robust systems that can handle high temperatures. Agglomeration (0–15%), thermophoresis (0–5%), uncertainty in exhaust flow measurements (±10%), and time alignment issues (0–10%) can result in differences between SPN emissions measured at the tailpipe and after dilution tunnel sampling (as prescribed in the current regulations). The application of two different principles (i.e., diffusion charging and electrical detection vs. optical counting) for determining particle number concentration with PEMS-SPN introduces an extra measurement uncertainty which was found to be on the order of 30% (with a standard deviation 25%). Nevertheless, this uncertainty was similar for emission levels that span over almost three orders of magnitude (from 5×10^{10} to 2×10^{13} p/km).

This article has also analyzed parameters that can affect SPN emission levels. Among these are the status and pre-conditioning of the after-treatment systems, namely the DPF. Here it was demonstrated that the DPF status can significantly affect the repeatability of SPN measurements. Differences of more than one order of magnitude were found between regenerating and non-regenerating cycles. Low ambient temperature testing (−7°C) increased the emissions by 70–160% compared to testing at 23°C. However, tests conducted with already warm engines at different ambient temperatures typically varied by less than 50% with respect to the resulting SPN emissions. A 10–15% increase of the vehicle weight resulted in a 10–25% increase of SPN emissions. Aggressive driving that includes strong acceleration and frequent changes in engine rpm resulted in elevated SPN emissions. Subsequently, SPN emissions can increase by a factor 2–5 if cold-start and strong accelerations are covered by a driving cycle.

We also conducted on-road emission measurements of two heavy-duty vehicles equipped with diesel particle filter (DPF), two light-duty diesel vehicles equipped with DPF, one fuel port injection (PFI), and six gasoline direct injection (GDI) passenger cars. The Euro V heavy-duty diesel truck showed SPN emissions of around 2×10^{13} p/km, while the Euro VI diesel truck emitted only 6×10^{10} p/km. One of the DPF equipped light-duty vehicle had emissions of 8×10^{11} p/km, while the other one had one order of magnitude lower emissions. The PFI showed SPN emissions slightly higher than 1×10^{12} p/km on an urban trip. The SPN emissions of the GDI vehicles spanned ~ from 8×10^{11} to almost 8×10^{12} p/km. For the non-DPF vehicles the on-road PEMS results were in relatively good agreement with those obtained in the laboratory, with differences of a factor < 2.1. However, the SPN emissions of the DPF equipped vehicles differed by almost one order of magnitude when driven in the laboratory and on the road.

Concluding, SPN measurements on-board are possible with a relative small increase of the measurement uncertainty. This uncertainty is smaller than the differences observed between laboratory and on-road measurements. Reasons that can partly explain the differences are DPF fill state, ambient air temperature, chassis dynamometer settings, extra mass of PEMS and driving conditions/style. Even higher differences can be observed compared to roadside measurements, but there total PN measurements are typically conducted that include formation of secondary particles.

AUTHOR CONTRIBUTIONS

BG had the concept of the work, did the analysis of the data and wrote the paper. FR organized the light-duty vehicles work. TV did the analysis of the MAW method. PM and RS organized the heavy-duty vehicles testing. GF and MW contributed in the structure of the paper. PB and MW were responsible for the project. All authors supported in the experiments, and critically commented, and revised the paper.

ACKNOWLEDGMENTS

The views expressed here are those of the authors and may not be considered as the official position of the European Commission. The authors would like to thank R. Colombo, M. Cadario, M. Sculati, P. Lelijour, D. Lesueur, F. Forni, and M. Carriero for executing the laboratory and on-road tests.

REFERENCES

1. Aakko, P., and Nylund, N. (2003). Particle Emissions at Moderate and Cold Temperatures using Different Fuels. *SAE Technical Paper* doi: 10.4271/2014-01-2826
2. Andersson, J., May, J., Favre, C., Bosteels, D., de Vries, S., Heaney, M., et al. (2014). On-road and chassis dynamometer evaluations of emissions from two Euro 6 diesel vehicles. *SAE Int. J. Fuel Lubr.* 7, 919–934.
3. Armas, O., Garcia-Contreras, R., and Ramos, A. (2013). Emissions of light duty vehicle tested under urban and extra-urban real-world driving conditions with diesel, animal fat biodiesel and GTL fuels. *SAE Technical Paper* 2013-24-0176.
4. Barrios, C., Domínguez-Sáez, A., Rubio, J., and Pujadas, M. (2012). Factors influencing the number distribution and size of the particles emitted from a modern diesel vehicle in real urban traffic. *Atmos. Environ.* 56, 16–25. doi: 10.1016/j.atmosenv.2012.03.078
5. Beddows, D., and Harrison, R. (2008). Comparison of average particle number emission factors for heavy and light duty vehicles derived from rolling chassis dynamometer and field studies. *Atmos. Environ.* 42, 7954–7966. doi: 10.1016/j.atmosenv.2008.06.021
6. Bergmann, M., Kirchner, U., Vogt, R., and Benter, T. (2009). On-road and laboratory investigation of low-level PM emissions of a modern diesel particulate filter equipped diesel passenger car. *Atmos. Environ.* 43, 1908–1916. doi: 10.1016/j.atmosenv.2008.12.039
7. Bonnel, P., Brunella, A., Carriero, M., Fumagalli, I., Kubelt, J., Manfredi, U., et al. (2006). *European Project on Portable Emissions Measurement Systems: EU PEMS.* Ispra: EU Report EUR 22143 EN.
8. Bonnel, P., Kubelt, J., and Provenza, A. (2011). *Heavy-Duty Engines Conformity Testing Based on PEMS -*

Lessons Learned from the European Pilot Program. Ispra: EU Report 24921.

9. Boulter, P., and McCrae, I. (2007). *The Links Between Micro-Scale Traffic, Emission and Air Pollution Models.* Transport Research Laboratory Project Report PPR 269.

10. Braisher, M., Stone, R., and Price, P. (2010). Particle number emissions from a range of european vehicles. *SAE Technical Paper* doi: 10.4271/2010-01-0786

11. Casati, R., Scheer, V., Vogt, R., and Benter, T. (2007). Measurement of nucleation and soot mode particle emission from a diesel passenger car in real world and laboratory in situ dilution. *Atmos. Environ.* 41, 2125–2135. doi: 10.1016/j.atmosenv.2006.10.078

12. Chan, T., Meloche, E., Kubsh, J., Brezny, R., Rosenblatt, D., and Rideout, G. (2013). Impact of ambient temperature on gaseous and particle emissions from a direct injection gasoline vehicle and its implications on particle filtration. *SAE Int. J. Fuel Lubr.* 6, 350–371. doi: 10.4271/2013-01-0527

13. Clark, N., Kern, J., Atkinson, C., and Nine, R. (2002). Factors affecting heavy-duty diesel vehicle emissions. *J. Air Waste Manag. Assoc.* 52, 84–94. doi: 10.1080/10473289.2002.10470755

14. EC (2007). Regulation (EC) No 715/2007 of the European parliament and of the council of 20 June 2007 "on type-approval of motor vehicles with respect to emissions from light passenger and commercial vehicles (Euro5 and Euro6) and on access to vehicle repair and maintenance information." *Off. J. Eur. Union L* 171, 1–16.

15. EC (2009). Regulation (EC) no 595/2009 of the European parliament and of the Council of 18 June 2009 "on type-approval of motor vehicles and engines with respect to emissions from heavy-duty vehicles (EuroVI) and on access to vehicle repair and maintenance information and amending Regulation (EC) no. 715/2007 and directive 2007/46/EC and repealing directives 80/1269/EEC, 2005/55/EC and 2005/78/EC". *Off. J. Eur. Union L* 188, 1–13.

16. EC (2011). Commission Regulation (EU) No 582/2011 of 25 May 2011 "implementing and amending Regulation (EC) No 595/2009 of the European Parliament and of the Council with respect to emissions from heavy-duty vehicles (EuroVI) and amending Annexes I and III to directive 2007/46/ECof the European parliament and of the council". *Off. J. Eur. Union L* 167, 1–168.

17. EC (2012). *Communication from the Commission to the European Parliament, the Council, the European Economic and Social Committee and the Committee of the Regions. Cars 2020: Action Plan for a competitive and sustainable automotive industry in Europe. (2012), COM.* Available online at: http://eur-lex.europa.eu/LexUriServ/LexUriServ.do?uri=COM:2012:0636:FIN:EN:PDF

18. EEA (2014). *Air Quality in Europe - 2014.* Copenhagen: EEA Report No, 5/2014.

19. EMEP/EEA (2013). *Air Pollutant Emission Inventory Guidebook.* Copenhagen: European Monitoring and Evaluation Programme/European Environment Agency, 2013.

20. EPA (2008). *Determination of PEMS Measurement Allowances for Gaseous Emissions Regulated Under the Heavy-Duty Diesel Engine in-use Testing Program.* United States Environmental Protection Agency, Revised Final report. Available online at: http://www.epa.gov/otaq/regs/hd-hwy/inuse/420r08005.pdf

21. EPA (2009). *Integrated Science Assessment for Particulate Matter (Final Report).* Washington, DC: U.S. Environmental Protection Agency. EPA/600/R-08/139F, 2009.

22. EPA (2014). *Emission Adjustments for Temperature, Humidity, Air Conditioning, and Inspection and Maintenance for on-road Vehicles in MOVES2014.* Washington, DC. Report EPA-420-R-14-012.

23. Fierz, M., Burtscher, H., Steigmeier, P., and Kasper, M. (2008). Field measurement of particle size and number concentration with the diffusion size classifier (DISC). *SAE Technical Paper* 2008-01-1179.

24. Fontaras, G., Franco, V., Dilara, P., Martini, G., and Manfredi, U. (2014). Development and review of Euro 5 passenger car emission factors based on experimental results over various driving cycles. *Sci. Total Environ.* 468–469, 1034–1042. doi: 10.1016/j.scitotenv.2013.09.043

25. Franco, V., Kousoulidou, M., Muntean, M., Ntziachristos, L., Hausberger, S., and Dilara, P. (2013). Road vehicle emission factors development: a review. *Atmos. Environ.* 70, 84–97. doi: 10.1016/j.atmosenv.2013.01.006

26. Franco, V., Sanchez, F., German, J., and Mock, P. (2014). *Real-World Exhaust Emissions from Modern Diesel Cars. A meta-Analysis of PEMS Emissions Data from EU (Euro 6) and US (Tier 2 bin 5/ULEV ii) Diesel Passenger Cars.* Berlin: ICCT–The International Council on Clean Transportation.

27. Giechaskiel, B., Arndt, M., Schindler, W., Bergmann, A., Silvis, W., and Drossinos, Y. (2012b). Sampling of non-volatile vehicle exhaust particles: a simplified guide. *SAE Int. J. Engines* 5, 379–399. doi: 10.4271/2012-01-0443

28. Giechaskiel, B., Carriero, M., Martini, G., Bergmann, A., Pongratz, H., and Jörgl, H. (2010b). Comparison of particle number measurements from the full dilution tunnel, the tailpipe and partial flow systems. *SAE Int. J. Fuel Lubr.* 2, 512–530.

29. Giechaskiel, B., Cresnoverh, M., Jörgl, H., and Bergmann, A. (2010a). Calibration and accuracy of a particle number measurement system. *Meas. Sci. Technol.* 21, 045102. doi: 10.1088/0957-0233/21/4/045102

30. Giechaskiel, B., Dilara, P., and Andersson, J. (2008a). Particle Measurement Programme (PMP) light-duty inter-laboratory exercise: repeatability and reproducibility of the particle number method. *Aerosol Sci. Technol.* 42, 528–543. doi: 10.1080/02786820802220241

31. Giechaskiel, B., Dilara, P., Sandbach, E., and Andersson, J. (2008b). Particle Measurement Programme (PMP) light-duty inter-laboratory exercise: comparison of different particle number measurement systems. *Meas. Sci. Technol.* 19:095401. doi: 10.1088/0957-0233/19/9/095401

32. Giechaskiel, B., Mamakos, A., Andersson, J., Dilara, P., Martini, G., Schindler, W., et al. (2012a). Measurement of automotive non-volatile particle number emissions within the European legislative framework: a review. *Aerosol Sci. Technol.* 46, 719–749. doi: 10.1080/02786826.2012.661103

33. Giechaskiel, B., Manfredi, U., and Martini, G. (2014c). Engine exhaust solid sub-23 nm particles: I. literature survey. *SAE Int. J. Fuels Lubr.* 7, 950–964. doi: 10.4271/2014-01-2834

34. Giechaskiel, B., Maricq, M., Ntziachristos, L., Dardiotis, C., Wang, X., Axmann, H., et al. (2014a). Review of motor vehicle particulate matter emissions sampling and measurement: from smoke and filter mass to particle number. *J. Aerosol Sci.* 67, 48–86. doi: 10.1016/j.jaerosci.2013.09.003

35. Giechaskiel, B., Massimo, C., Bonnel, P., Schindler, W., Scheder, D., Bassoli, C., et al. (2011). PEMS feasibility of particulate mass and number measurement with portable emission measurement systems (PEMS) for in-use testing. *SAE Technical Paper* 2011-24-0199.

36. Giechaskiel, B., Munoz-Bueno, R., Rubino, L., Manfredi, U., Dilara, P., De Santi, G., et al. (2007b). Particle measurement programme (PMP): particle size and number emissions before, during and after regeneration events of a euro 4 DPF equipped light-duty diesel vehicle. *SAE Technical Paper* 2007-01-1944.

37. Giechaskiel, B., Ntziachristos, L., Samaras, Z., Casati, R., Scheer, V., and Vogt, R. (2005). Formation potential of vehicle exhaust nucleation mode particles onroad and in the laboratory. *Atmos. Environ.* 39, 3191–3198. doi: 10.1016/j.atmosenv.2005.02.019

38. Giechaskiel, B., Ntziachristos, L., Samaras, Z., Casati, R., Volker, S., and Rainer, V. (2007a). Effect of speed and speed transition on the formation of nucleation mode particles from a light-duty diesel vehicle. *SAE Technical Paper* 2007-01-1110.

39. Giechaskiel, B., Riccobono, F., and Bonnel, P. (2014b). *Feasibility Study on the Extension of the Real Driving Emissions (RDE) Procedure to Particle Number (PN): Experimental Evaluation of Portable Emission Measurement Systems (PEMS) with Diffusion Chargers (DCs) to Measure Particle Number (PN) Concentration.* JRC Report 26997, Ispra.

40. Giechaskiel, B., Riccobono, F., and Bonnel, P. (2015). *Feasibility Study on the Extension of the Real Driving Emissions (RDE) Procedure to Particle Number (PN): Chassis Dynamometer Evaluation of Portable Emission Measurement Systems (PEMS) to Measure Particle Number (PN) Concentration:* Phase II. Ispra: EU Report 27451.

41. Giechaskiel, B., Wang, X., Horn, H., Spielvogel, J., Gerhart, C., Southgate, J., et al. (2009). Calibration of condensation particle counters for legislated vehicle number emission measurements. *Aerosol Sci. Technol.* 43, 1164–1173. doi: 10.1080/02786820903242029

42. Hallquist, Å., Jerksjö, M., Fallgren, H., Westerlund, J., and Sjödin., Å (2013). Particle and gaseous emissions from individual diesel and CNG buses. *Atmos. Chem. Phys.* 13, 5337–5350. doi: 10.5194/acp-13-5337-2013

43. HEI (2013). *Understanding the Health Effects of Ambient Ultrafine Particles. HEI Perspectives 3.* Health Effects Institute Review Panel on Ultrafine Particles.

44. Huang, C., Lou, D., Hu, Z., Feng, H., Chen, Y., Chen, C., et al. (2013). A PEMS study of the emissions of gaseous pollutants and ultrafine particles from gasoline- and diesel-fueled vehicles. *Atmos. Environ.* 77, 703–710. doi: 10.1016/j.atmosenv.2013.05.059

45. Isella, L., Giechaskiel, B., and Drossinos, Y. (2008). Diesel-exhaust aerosol dynamics from the tailpipe to the dilution tunnel. *J. Aerosol Sci.* 39, 737–758. doi: 10.1016/j.jaerosci.2008.04.006

46. Jackson, E., Qu, Y., Holmén, B., and Aultman-Hall, L. (2006). Driver and road type effects on light-duty gas and particulate emissions. *Transport. Res. Rec. J. Transport. Res. Board* 1987, 118–127. doi: 10.3141/1987-13

47. Ježek, I., Drinovec, L., Ferrero, L., Carriero, M., and Moènik, G. (2015). Determination of car on-road black carbon and particle number emission factors and comparison between mobile and stationary measurements. *Atmos. Meas. Tech.* 8, 43–55. doi: 10.5194/amt-8-43-2015

48. Johnson, K. C., Durbin, T. D., Jung, H., Chaudhary, A., Cocker, D. R., Herner, J. D., et al. (2009b). Evaluation of the European PMP methodologies during on-road and chassis dynamometer testing for DPF equipped heavy-duty diesel vehicles. *Aerosol Sci. Technol.* 43, 962–969. doi: 10.1080/02786820903074810

49. Johnson, K., Durbin, T., Cocker, D., Miller, W., Bishnu, D., Maldonado, H., et al. (2009a). On-road comparison of a portable emission measurement system with a mobile reference laboratory for a heavy-duty diesel vehicle. *Atmos. Environ.* 43, 2877–2883. doi: 10.1016/j.atmosenv.2009.03.019

50. Johnson, K., Durbin, T., Jung, H., Cocker, D. R. III, Bishnu, D., and Giannelli, R. (2011). Quantifying in-use PM measurements for heavy duty diesel vehicles. *Environ. Sci. Technol.* 45, 6073–6079. doi: 10.1021/es104151v

51. Keogh, D., and Sonntag, D. (2011). Challenges and approaches for developing ultrafine particle emission inventories for motor vehicle and bus fleets. *Atmosphere* 2, 36–56. doi: 10.3390/atmos2020036

52. Keogh, D., Kelly, J., Mengersen, K., Jayaratne, R., Ferreira, L., and Morawska, L. (2010). Derivation of motor vehicle tailpipe particle emission factors suitable for modelling urban fleet emissions and air quality assessments. *Environ. Sci. Pollut. Res.* 17, 724–739. doi: 10.1007/s11356-009-0210-9

53. Khalek, I., Kittelson, D., and Brear, F. (1999). The influence of dilution conditions on diesel exhaust particle size distribution measurements. *SAE Technical Paper* 1999-01-1142.

54. Kittelson, D., Johnson, J., Watts, W., Wei, Q., Drayton, M., and Paulsen, D. (2000). Diesel aerosol sampling in the atmosphere. *SAE Technical Paper* 2000-01-2212.

55. Kittelson, D., Watts, W., and Johnson, J. (2006a). On-road and laboratory evaluation of combustion aerosols. Part 1: summary of diesel engine results. *J. Aerosol Sci.* 37, 913–930. doi: 10.1016/j.jaerosci.2005.08.005

56. Kittelson, D., Watts, W., Johnson, J., Schauer, J., and Lawson, D. (2006b). On-road and laboratory evaluation of combustion aerosols Part 2: summary of spark ignition engine results. *J. Aerosol Sci.* 37, 931–949. doi: 10.1016/j.jaerosci.2005.08.008

57. Kumar, P., Morawska, L., Birmili, W., Paasonen, P., Hu, M., Kulmala, M., et al. (2014). Ultrafine particles in cities. *Environ. Int.* 66, 1–10. doi: 10.1016/j.envint.2014.01.013

58. Li, T., Chen, X., and Yan, Z. (2013). Comparison of fine particles emissions of light-duty gasoline vehicles from chassis dynamometer tests and on-road measurements. *Atmos. Environ.* 68, 82–91. doi: 10.1016/j.atmosenv.2012.11.031

59. Liu, Z., Ge, Y., Johnson, K., Shah, A., Tan, J., Wang, C., et al. (2011). Real-world operation conditions and on-road emissions of Beijing diesel buses measured by using portable emission measurement system and electric low-pressure impactor. *Sci. Total Environ.* 409, 1476–1480. doi: 10.1016/j.scitotenv.2010.12.042

60. Mamakos, A., Bonnel, P., Perujo, A., and Carriero, M. (2013a). Assessment of portable emission measurement systems (PEMS) for heavy-duty diesel engines with respect to particulate matter. *J. Aerosol Sci.* 57, 54–70. doi: 10.1016/j.jaerosci.2012.10.004

61. Mamakos, A., Dardiotis, C., and Martini, G. (2012). *Assessment of Particle Number Limits for Petrol Vehicles.* Ispra: EU Report 25592 EN.

62. Mamakos, A., Martini, G., and Manfredi, U. (2013b). Assessment of the legislated particle number measurement procedure for a Euro5 and a Euro6 compliant diesel passenger cars under regulated and unregulated conditions. *J. Aerosol Sci.* 55, 31–47. doi: 10.1016/j.jaerosci.2012.07.012

63. Maricq, M., Podsiadlik, D., and Chase, R. (1999). Examination of the size-resolved and transient nature of motor vehicle particle emissions.*Environ. Sci. Technol.* 33, 1618–1626. doi: 10.1021/es9808806

64. Martini, G., Giechaskiel, B., and Dilara, P. (2009). Future European emission standards for vehicles: the importance of the UN-ECE particle measurement program. *Biomarkers* 14, 29–33. doi: 10.1080/13547500902965393

65. May, J., Bosteels, D., and Favre, C. (2014). An assessment of emissions from light-duty vehicles using PEMS and chassis dynamometer testing.*SAE Int. J. Engines* 7, 1326–1335. doi: 10.4271/2014-01-1581

66. Mellios, G., Hausberger, S., Keller, M., Samaras, Z., and Ntziachristos, L. (2011). *Parameterisation of Fuel Consumption and CO_2 Emissions of Passenger Cars and Light Commercial Vehicles for Modelling Purposes.* JRC Report 24927, Ispra.

67. Mendoza-Villafuerte, P., Perujo, A., Bonnel, P., Carriero, M., Forni, F., and Riccobono, F. (2015). *EU-PEMS PM Pilot Program: Testing, Data Analysis and Results.* JRC Report 27326, Ispra.

68. Merkisz, J., Pielecha, J., and Gis, W. (2009). Gaseous and particle emissions results from light duty vehicle with diesel particle filter. *SAE Technical Paper* 2009-01-2630.

69. Mock, P., German, J., Bandivadekar, A., Riemersma, I., Ligterink, N., and Lambrecht, U. (2013). *From Laboratory to Road: A Comparison of Official and 'Real-World' Fuel Consumption and CO2 Values for Cars in Europe and the United States.* ICCT White Paper. Berlin: The International Council on Clean Transportation.

70. Morawska, L., Ristovski, Z., Jayaratne, E., Keogh, D., and Ling, X. (2008). Ambient nano and ultrafine particles from motor vehicle emissions: characteristics, ambient processing and implications on human exposure. *Atmosp. Environ.* 42, 8113–8138. doi: 10.1016/j.atmosenv.2008.07.050

71. Ntziachristos, L., and Galassi, M. (2014). *Emission Factors for New and Upcoming Technologies in Road Transport.* JRC Report 26952, Ispra.

72. Ntziachristos, L., Mamakos, A., Samaras, Z., Rexeis, M., and Hausberger, S. (2006). Diesel particle exhaust emissions from light duty vehicles and heavy duty engines. *SAE Technical Paper* 2006-01-0866.

73. Oberdörster, G. (2000). Toxicology of ultrafine particles: *in vivo* studies. *Philos. Trans. R. Soc. A* 258, 2719–2740. doi: 10.1098/rsta.2000.0680

74. Oberdörster, G., Sharp, Z., Atudorei, V., Elder, A., Gelein, R., Kreyling, W., et al. (2004). Translocation of inhaled ultrafine particles to the brain.*Inhal. Toxicol.* 16, 437–445. doi: 10.1080/08958370490439597

75. Pant, P., and Harrison, R. (2013). Estimation of the contribution of road traffic emissions to particulate matter concentrations from field measurements: a review. *Atmos. Environ.* 77, 78–97. doi: 10.1016/j.atmosenv.2013.04.028

76. Peckham, M., Finch, A., Cambell, B., Price, P., and Davies, M. (2011). Study of Particle Number Emissions from a Turbocharged Gasoline Direct Injection (GDI) Engine Including Data from a Fast Response Particle Size Spectrometer. *SAE Technical Paper* 2011-01-1224.

77. Pelkmans, L., and Debal, P. (2006). Comparison of on-road emissions with emissions measured on chassis dynamometer test cycles. *Transport. Res. D* 11, 233–241. doi: 10.1016/j.trd.2006.04.001

78. Perujo, A., and Mendoza-Villafuerte, P. (2015). *PEMS Emissions Testing of Heavy Duty Vehicles/Engines: Assessment of PEMS Procedures in Fulfilment of Article 14(3) to Regulation (EU) 582/2011.* JRC Report 27251, Ispra.

79. Pope, C. A. III. (2000). Epidemiology of fine particulate air pollution and human health: biologic mechanisms and who's at risk? *Environ. Health Perspect.* 108, 713–723. doi: 10.2307/3454408

80. Ristimäki, J., Keskinen, J., Virtanen, A., Maricq, M., and Aakko, P. (2005). Cold temperature PM emissions measurement: method evaluation and application to light

duty vehicles. *Environ. Sci. Technol.* 39, 9424–9430. doi: 10.1021/es050578e

81. Rönkkö, T., Virtanen, A., Vaaraslahti, K., Keskinen, J., Pirjola, L., and Lappi, M. (2006). Effect of dilution conditions and driving parameters on nucleation mode particles in diesel exhaust: laboratory and on-road study. *Atmos. Environ.* 40, 2893–2901. doi: 10.1016/j.atmosenv.2006.01.002
82. Rubino, L., Bonnel, P., Hummel, R., Krasenbrink, A., Manfredi, U., De Santi, G., et al. (2007). PEMS light duty vehicles application: experiences in downtown milan. *SAE Technical Paper* 2007-24-0113.
83. Sager, T., and Castranova, V. (2009). Surface area of particle administered versus mass in determining the pulmonary toxicity of ultrafine and fine carbon black: comparison to ultrafine titanium dioxide. *Part. Fibre Toxicol.* 6:15. doi: 10.1186/1743-8977-6-15
84. Sundvor, I., Castell Balaguer, N., Viana, M., Querol, X., Reche, C., Amato, F., et al. (2012). *Road Traffic's Contribution to Air Quality in European Cities.* Bilthoven: The European Topic Centre on Air Pollution and Climate Change Mitigation, ETC/ACM Technical Paper, 2012/14.
85. Tietge, U., Zacharof, N., Mock, P., Franco, V., German, J., Bandivadekar, A., et al. (2015). *From laboratory to road: a 2015 update of official and "real-world" fuel consumption and CO2 values for passenger cars in Europe.* ICCT white paper.
86. UN/ECE, Regulation 49. (2013). *Addendum 48: Regulation No. 49, Revision 6. UNECE - United Nations – Economic Commissions for Europe.* Available online at: http://www.unece.org/fileadmin/DAM/trans/main/wp29/wp29regs/2013/R049r6e.pdf
87. UN/ECE (2015). *Addendum 82: Regulation No. 83, Revision 5. UNECE – United Nations - Economic Commission for Europe.* Available online at: http://www.unece.org/fileadmin/DAM/trans/main/wp29/wp29regs/R083r5e.pdf.
88. Vlachos, T., Bonnel, P., Perujo, A., Weiss, M., Mendoza-Villafuerte, P., and Riccobono, F. (2014). In-use emissions testing with portable emissions measurement systems (PEMS) in the current and future European vehicle emissions legislation: overview, underlying principles and expected benefits. *SAE Int. J. Commer. Veh.* 7, 199–215. doi: 10.4271/2014-01-1549
89. Vogt, R., Scheer, V., Casati, R., and Bender, T. (2003). On-road measurement of particle emission in the exhaust plume of a diesel passenger car.*Environ. Sci. Technol.* 37, 4070–4076. doi: 10.1021/es0300315
90. Wang, J., Storey, J., Domingo, N., Huff, S., Thomas, J., and West, B. (2006). Studies of diesel engine particle emissions during transient operations using an engine exhaust particle sizer. *Aerosol Sci. Technol.* 40, 1002–1015. doi: 10.1080/02786820600919408
91. Wehner, B., Uhrner, U., von Löwis, S., Zallinger, M., and Wiedensohler, A. (2009). Aerosol number size distributions within the exhaust plume of a diesel and a gasoline passenger car under on-road conditions and determination of emission factors. *Atmos. Environ.* 43, 1235–1245. doi: 10.1016/j.atmosenv.2008.11.023
92. Weiss, M., Bonnel, P., Hummel, R., Manfredi, U., Colombo, R., Lanappe, G., et al. (2011b). *Analyzing On-Road Emissions of Light-Duty Vehicles with Portable Emission Measurement Systems (PEMS).* JRC report 24697, Ispra.
93. Weiss, M., Bonnel, P., Hummel, R., Provenza, A., and Manfredi, U. (2011a). On-road emissions of light-duty vehicles in Europe. *Environ. Sci. Technol.* 45, 8575–8581. doi: 10.1021/es2008424
94. Weiss, M., Bonnel, P., Hummel, R., and Steininger, N. (2013). *A Complementary Emissions Test for Light-Duty Vehicles: Assessing the Technical Feasibility of Candidate Procedures. Conclusions of the Real Driving Emissions – Light Duty Vehicles (RDE-LDV) Working Group.* JRC Report 25572, Ispra.
95. Weiss, M., Bonnel, P., Provenza, A., Lambrecht, U., Alessandrini, S., Carriero, M., et al. (2012): Will Euro 6 reduce the NO_x emissions of new diesel cars? Preliminary conclusions from on-road emissions tests with Portable Emission Measurement Systems (PEMS). *Atmos. Environ.* 62, 657–665. doi: 10.1016/j.atmosenv.2012.08.056.
96. WHO (2013). *Review of Evidence on Health Aspects of Air Pollution - REVIHAAP Project.* WHO Technical Report, World Health Organization, Regional Office for Europe, Copenhagen, Denmark.
97. Witze, P., and Green, R. (2005). Comparison of single and dual spray fuel injectors during cold start of a PFI spark ignition engine using visualization of liquid fuel films and pool fires. *SAE Technical Paper* 2005-01-3863.
98. Zacharof, N., Fontaras, G., Tsiakmakis, S., Ciuffo, B., Codrea, C., and Anagnostopoulos, K. (2015). *Review of in use Factors Affecting the Fuel Consumption and CO2 Emissions of Passenger Cars.* Ispra: EU Report.
99. Zheng, Z., Durbin, T., Xue, J., Johnson, K., Li, Y., Hu, S., et al. (2014). Comparison of particle mass and solid particle number (SPN) emissions from a heavy-duty diesel vehicle under on-road driving conditions and a standard testing cycle. *Environ. Sci. Technol.* 48, 1779-1786. doi: 10.1021/es403578b

CHAPTER 6

COMPARATIVE MEASUREMENT OF THE BREAKUP LENGTH OF LIQUID JETS IN AIRBLAST ATOMISERS USING OPTICAL CONNECTIVITY, ELECTRICAL CONNECTIVITY AND SHADOWGRAPHY

Georgios Charalampous, Constantinos Hadjiyiannis, Yannis Hardalupas

Department of Mechanical Engineering, Imperial College London, London SW7 2AZ, UK

ABSTRACT

A comparative examination of the optical connectivity, electrical connectivity and shadowgraphy techniques for the measurement of the break-up length of atomizing liquid jets from a co-axial airblast atomizer is presented. The atomizer was operated over air-to-liquid Momentum Ratios between 27 and 335 and Momentum Flux Ratios between 0.67 and 8.27. Shadowgraphy records instantaneous images of the shadow of the atomizing liquid jet when it is back-illuminated by a light source. The electrical connectivity uses the continuity of an electrically conducting atomizing liquid jet to measure the potential during the presence of an electrical connection between the spray nozzle and a probe further downstream. The optical connectivity visualizes the atomizing liquid jet, doped with a fluorescing dye, as it is illuminated from within the nozzle using a laser beam. Comparison of the measured breakup lengths with time resolved shadowgraphy, optical connectivity and electrical connectivity, following the proposed novel processing of the time-dependent potential, showed that the mean values are all within ±15% of each other. The advantages and limitations of each technique are discussed.

INTRODUCTION

The atomization of liquid jets by injection in a gaseous environment is a process, which is utilized in many practical applications, such as automotive and aviation engines, agricultural sprays, dust control, firefighting and nuclear reactor core cooling. The atomization process is complex and several parameters, such as the physical properties and the flow characteristics of both the injected liquid and of the surrounding air and the geometry of the atomizer, contribute to the formation of the final spray. However, the importance of all the influencing parameters is not known. Yet, there is high demand for precise control of the spray characteristics. For example, the injected liquid fuel in a burner must atomize quickly near the nozzle to form small droplets, which disperse well into the surrounding air, so that they evaporate quickly and mix with the air to improve fuel consumption and reduce combustion emissions. A lubricating liquid jet on the other hand needs to maintain coherence for some distance beyond the nozzle, so that the lubricant is delivered to the precise

Citation: Georgios Charalampous, Constantinos Hadjiyiannis, Yannis Hardalupas, Comparative measurement of the breakup length of liquid jets in airblast atomizers using optical connectivity, electrical connectivity and shadowgraphy, Measurement, Volume 89, July 2016, Pages 288-299, ISSN 0263-2241, http://dx.doi.org/10.1016/j.measurement.2016.03.062.

location it is required in order to reduce its consumption and avoid contamination of surrounding areas.

The control of the spray characteristics requires a deeper understanding of the atomization process, which is broken down in smaller sub-processes that can be independently examined and understood. The most fundamental description of the atomization process follows a two stage approach, namely the primary and the secondary atomization [1] and [2]. Primary atomization begins immediately at the nozzle exit, where destabilizing forces operate on the gas–liquid interface to break up the continuous liquid stream into discrete liquid fragments. The liquid jet usually maintains continuity for several liquid jet diameters downstream of the nozzle exit. After the continuity of the liquid jet is interrupted, secondary atomization takes place, which involves further breakup of the liquid fragments into progressively smaller sizes until stable droplets are obtained. This complex process involves many spatial and temporal scales. One of the key spatial scales is the length of the continuous liquid jet past the nozzle exit, which is known as the "breakup length". The breakup length is critical because it defines the spatial extent of the primary atomization region. The measurement of the breakup length is therefore crucial in spray studies, because it is essential for the physics of atomization and the performance of atomizers [3], [4], [5] and [6] and for the evaluation of computational models for atomization [7].

A number of techniques have been developed for the measurement of the break-up length. The most prominent is photography. It is the most widely used method [6], [8], [9],[10] and [11], because it is straightforward to apply, it places only moderate demands on equipment and the photographic images provide detailed information of the continuous liquid geometry. In this method, the atomizing liquid jet is imaged directly by a camera. An intense and fast light source is required to illuminate the liquid jet, so that the exposure time is short compared to the time scales of atomization in order for the jet image to remain sharp. The liquid jet may be illuminated from different directions, but usually back illumination is preferred (Shadowgraphy). In this way, the shadow of the jet is recorded against a bright background, delivering high contrast images where the liquid jet contour is well defined (Figure 1). The shadowgraphy technique is extensively presented in [12] where a theoretical analysis on the formation of the shadow of the visualized object is provided and the capabilities of shadowgraphic visualization are highlighted. Shadowgraphy facilitates the visualization of the whole liquid jet from which many morphological characteristics can be extracted, including the breakup length. While this method is easy to implement and gives good results under certain flow conditions, it has the disadvantage that, when atomization becomes intense and the density of droplets around the liquid jet increases, the optical access to the liquid jet core is reduced. In this case, there is poor jet visualization and, therefore, ambiguity in the estimate of the break-up length and other liquid jet characteristics.

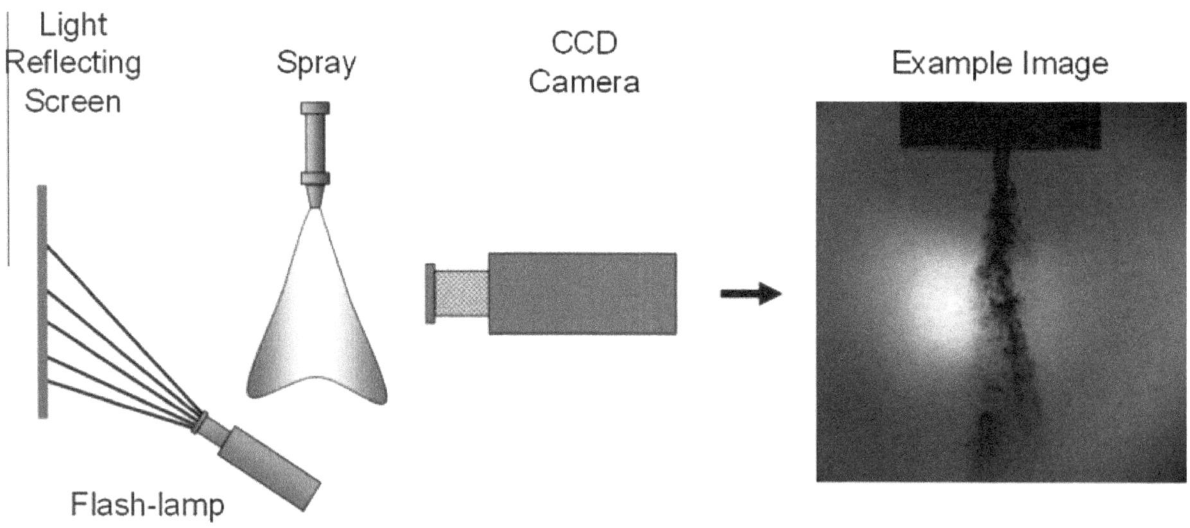

Figure 1. Principle of photographic spray imaging using shadowgraphy.

A different approach for the measurement of the breakup length is the electrical connectivity technique, which does not rely on the visualization of the continuous liquid jet and, therefore, does not share the same difficulties. This technique is based on the conduction of electricity along the length of the continuous liquid jet downstream of the nozzle. A potential is applied between the spray nozzle and a probe downstream. If there is continuity of the liquid phase between the nozzle and the probe, a closed electrical circuit will ensue. The probe can be moved across different positions to determine the continuity of the liquid jet as a function of the downstream distance. If the measured potential between the nozzle and the probe is low, then there is electrical connectivity between the two, indicating continuity of the liquid jet core. Discontinuity of the liquid jet can be inferred when the measured potential between the nozzle and the probe is high. The theoretical basis of the electrical connectivity method was developed in the seminal work of Hiroyasu et al. [13]. A theoretical analysis of the expected measured potential between the needle probe and the nozzle was performed by Chehroudi et al. [14] and Yule and Salters [15]. These authors consider the continuous liquid jet as a resistor. Therefore, the jet resistance increases with the jet length and decreases with the jet cross sectional area and consequently the measured potential across the nozzle and the probe. Various morphologies of the continuous liquid jet core were examined from which its resistance is estimated.

However, no consensus exists on the expected output signal behavior among authors, since different assumptions were made on the shape of the jet. Additionally, there are differences in the implementation of the technique between different researchers. Hiroyasu et al. studied the breakup length of a high-speed liquid jet by measuring an electrical resistance between the nozzle and a fine wire screen detector located in a spray jet [13]. Chehroudi et al. tried to determine the shape and length of the intact liquid core by applying a voltage between the nozzle unit and fine needles, rods and screens [14]. Their results indicate that electrical current can be transmitted, not only by continuous liquid cores, but, also, by atomized unconnected liquid fragments, which would represent a serious limitation for the technique. Yule and Salters investigated the breakup zone of a transient diesel spray as a function of time and position employing a wire probe [15]. For this reason a visual inspection of the process is necessary in order to assess how the measured potential should be interpreted.

A novel technique for the measurement of the break-up length was recently proposed by Charalampous et al. [16] and [17], which is based on the optical connectivity of a continuous liquid jet and aims at addressing some of the shortcomings of shadowgraphy and electrical connectivity. In this approach, the liquid jet is illuminated from within the nozzle by a laser beam. The beam propagates along the liquid jet by reflecting at the gas–liquid interface. Due to the higher index of refraction of the liquid jet to that of the surrounding gas, when the incident angle of the laser beam rays on the gas–liquid interface, defined from the normal to the interface, is greater than the angle for total internal reflection (about 48° for water), the laser beam is reflected completely back into the liquid stream and propagates downstream for a long distance, in the same way that light propagates along optical fibers. This phenomenon was demonstrated as early as the 19th century [18] and [19]. The addition of a fluorescent dye, such as Rhodamine WT [20]in the liquid jet, causes some of the intensity of the laser beam to be absorbed and re-emitted at a longer wavelength as fluorescence. This causes the volume of the continuous jet to become luminous, which allows the evaluation of the break-up length. Beyond the point of liquid discontinuity, the laser beam is diffused and its intensity is significantly reduced. A schematic of the process is shown in Figure 3. An extension of the theoretical principle is related to the propagation of the laser beam within a continuous liquid jet and the subsequent fluorescent emission from the continuous liquid jet. This has been performed numerically for ideal jet geometries [21]. It is found that in straight jets, such as those observed in coaxial airblast atomizers considered here, the laser beam can propagate sufficiently to illuminate liquid jets over 10 liquid nozzle diameters. This numerical investigation was expanded for instantaneous liquid jet geometries derived from Large Eddy Simulations of liquid jets in an air coflow [22], which were close to observed jet geometries under intense atomization regime (We = 1040 and MR = 336). It was found that the laser beam can illuminate the full length of the liquid jet with some fluorescent intensity decrease close to the tip, where the jets become narrower. This technique offers the unique advantage that the continuous liquid jet is visualized without significant interference from the surrounding droplets and, therefore, does not suffer from the same issues as shadowgraphy. As such, the breakup length can be resolved in cases where atomization is intense and shadowgraphy fails. These are the cases of significant interest as in most practical applications intense atomization for the generation of fine droplets is required. There is a limitation of the jet length that can be visualized using optical connectivity, as light intensity losses by refraction occur at the jet interface. Nevertheless, this issue is mitigated by the fact that optical connectivity is useful for cases where atomization is intense. In such cases, the continuous liquid jet length is short and the visualization of the full jet length with optical connectivity is successful.

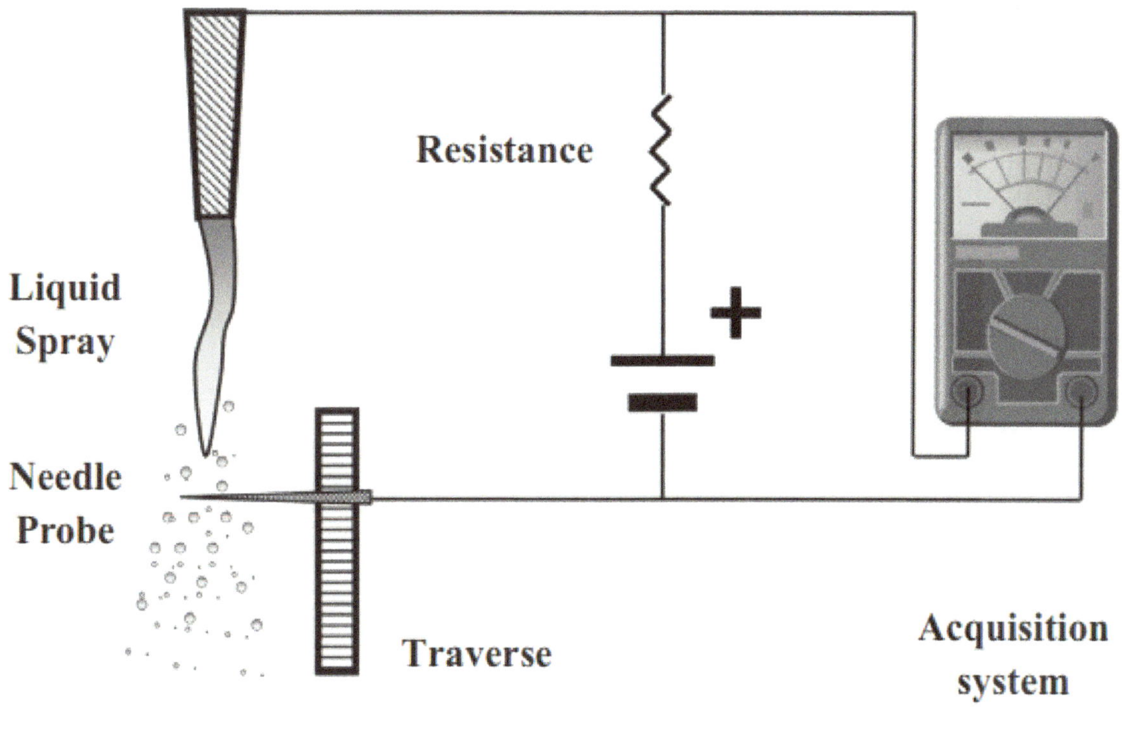

Figure 2. Principle of the electrical connectivity technique.

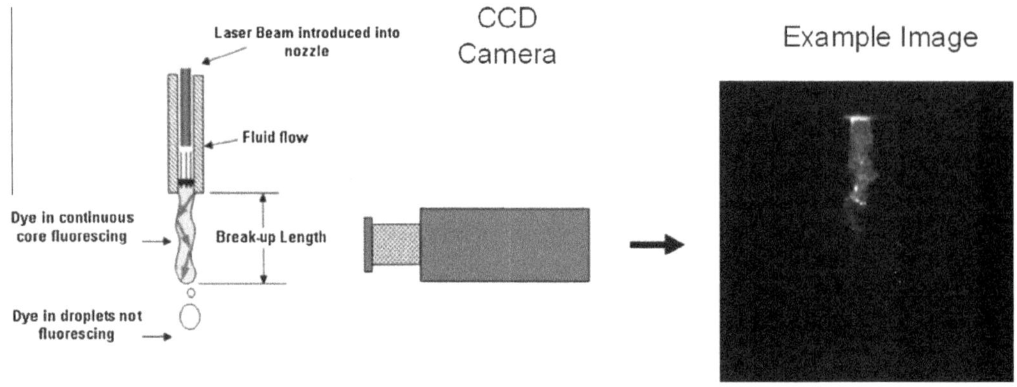

Figure 3. Principle of the optical connectivity technique.

Other techniques exist for the measurement of the breakup length, such as X-ray absorption [23], [24] and [25] and ballistic imaging [26], [27], [28] and [29]. While these techniques are powerful, they require advanced optical instrumentation, which is not commonly available in an engineering laboratory. Additionally, complicated experimental arrangements make these techniques unavailable to most researchers at present, while additional evaluation of their measurement accuracy is required. It is the purpose of this paper to provide a comparison between the measurement of the breakup length of a liquid jet by shadowgraphic imaging, electrical connectivity and optical connectivity, which can be implemented in most research laboratories. Different processing methods of measured data using the electrical connectivity technique are evaluated and the best approach is proposed.

The remaining of the paper is structured in the following sections. In the next section, the details of the experimental configuration of the tested co-axial airblast atomizer are presented along with the description of the implementation of each of the three considered techniques. The evaluation of data processing approaches of the electrical connectivity signals and the description of the results and their interpretation follows in Section 3. Section 4 presents a discussion of the results where the merits and weaknesses of each of the tested techniques are exposed. The paper ends with a summary of the conclusions.

EXPERIMENTAL SET-UP

The investigation was conducted on the liquid jet injected by the air-blast atomizer of Figure 4. The internal and external diameters of the liquid nozzle were $D_L = 2.3$ mm and 2.95 mm respectively and the internal diameter of the gas nozzle was $D_G = 14.95$ mm. The length of the liquid nozzle was over $20D_L$ which ensures that the mean velocity profile of the liquid jet is sufficiently developed, while the gas nozzle had a contraction with an area ratio of about 30. The two working fluids were water and air. The atomizer was set up vertically and exhausted downwards in order to avoid complexities due to gravitational effects that would cause the liquid jet to deflect from a straight path. The flow rate of the water and the air was controlled by and measured using rotameters. The resolution of the water rotameter was 0.1 l/min while the resolution of the air rotameter was 50 l/min. Corrections were applied to the measured air flow rate to account for compressibility effects due to the increased pressure of the air supply. All measurements were obtained at room temperature and pressure.

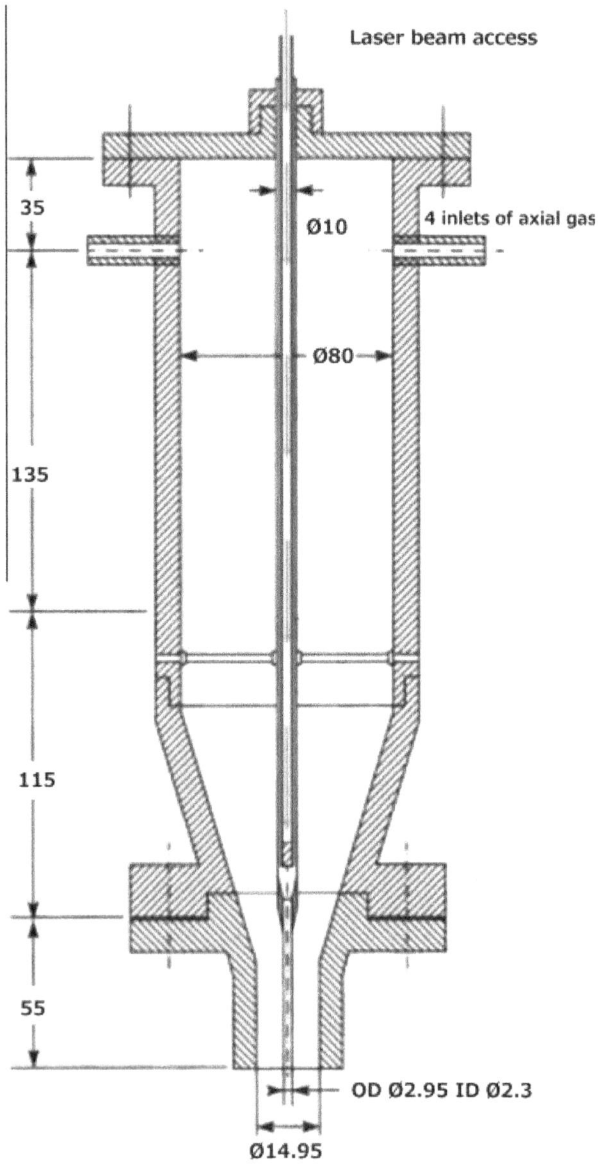

Figure 4. Geometry of coaxial air-blast atomizer.

The shadowgraphic imaging measurements were performed with a 10-bit Photron Fastcam-APX RS high speed CMOS camera. The camera was fitted with a 105 mm lens and the imaged region was resolved within approximately 30 μm per pixel, resulting to a spatial resolution of about 75px over the diameter of the liquid jet at the nozzle exit. Therefore, the liquid jet breakup length can be resolved in increments smaller than 5% ofD_L. The camera frame acquisition rate was set to 20,000 frames/s for all flow conditions. This frequency is greater than the breakup frequency of the liquid jet core for all tested flow conditions. As such, temporally resolved image sequences were recorded for each flow condition. The purpose of acquiring temporally resolved image sequences was to examine the possibility of improving the accuracy of the determination of the location of the liquid break-up in cases where the full length of the atomizing jet is not clearly visible. This corresponds to flow conditions for which the products of atomization obstruct the direct optical access to the liquid jet. If the liquid jet images are uncorrelated, the ambiguity cannot be overcome. However, the breaking point of the liquid jet could be traced upstream from a temporal sequence of correlated images. The source of illumination was a 100W lamp that was placed behind the liquid jet, so that the shadow of the liquid jet was imaged in the forward scattering direction.

The measurements with the electrical connectivity technique were performed with a stainless needle probe with 1.1 mm diameter, which crossed the axis of the atomizer. The probe was sufficiently robust to withstand the impact of the liquid without deflection as verified by shadowgraphy. The needle probe was supported on a robust beam to minimize vibrations. The distance between the needle probe and the nozzle was adjusted by a vertical traverse in steps of 1 mm from the nozzle exit. This is sufficient to resolve the breakup length in increments of up to 40% of D_L. The spatial resolution of the electrical connectivity was not as fine as that of shadowgraphy. The current setup was appropriate to deliver spatial resolution of the breakup length measurements better than half the liquid jet diameter, which is more than adequate for the purposes of this investigation and a good compromise between spatial resolution and measurement time. An electrical circuit was designed (Figure 2) to measure the electrical potential between the nozzle and the probe. A constant voltage of 10 V from a stabilized battery supply was applied between the liquid nozzle of the atomizer and the probe. The 10 V electrical potential was selected because it offers a natural scale to compare the measurements and, as shown previously, the results of the technique do not depend on the applied electrical potential in the range between 6 V and 60 V [15]. The positive pole of the power supply was connected to the needle probe in series with a 1 MΩ resistance to regulate the current in the circuit. The negative pole of the power supply was wired straight to the nozzle of the atomizer. A low potential between the nozzle and the probe indicates good electrical connectivity and, therefore, continuity of the liquid jet. Conversely, a high potential between the two indicates poor electrical connectivity and, therefore, marginal continuity or complete discontinuity of the liquid jet. An analogue oscilloscope was used to monitor the potential between the nozzle and the probe, while acquisition of the potential was performed by a 12-bit analogue to digital converter (National Instruments PCI-6023E) capable of sampling at 200 kHz. However, the sampling rate of the electrical connectivity was set at 20 kHz, so that the electrical connectivity measurements could be synchronized to the camera of the high speed shadowgraphic measurements for simultaneous measurements.

The measurements of the breakup length with the optical connectivity technique were reported elsewhere [16]. However, they are used here for comparison purposes. The details of the technique have been described elsewhere, however, we provide here a brief summary of the implementation of the technique. A stainless steel light guiding tube with a quartz window was inserted into the tube that delivers the liquid jet to the nozzle. The beam from a Nd:YAG laser, operating at 532 nm, was steered into the light guiding tube by a system of mirrors and lenses, so that it entered in a direction parallel to the tube axis to minimize reflections on the internal surfaces of the tube. The laser beam exited just before the contraction of the liquid jet nozzle (see Figure 4), in a direction parallel to the nozzle axis, so that it follows the liquid stream. The atomizing liquid was doped with Rhodamine WT dye, which is excited by laser light at 532 nm and fluoresces strongly with a peak at 587 nm. As a consequence, the liquid jet became luminous along the length of propagation of the beam. An Andor Classic iCCD camera was used to record images of the fluorescent liquid jet. A long pass optical filter that absorbs the laser wavelength at 532 nm and transmits the longer wavelengths of the fluorescent light was placed in front of the camera lens to suppress scattered light. The optical connectivity measurements were not temporally resolved.

Simultaneous measurements of electrical connectivity and high speed shadowgraphic imaging were obtained for conditions with liquid jet area-averaged velocity of U_L = 2.0 m/s and annular flow area-averaged air velocities of U_G = 47 m/s, 71 m/s, 95 m/s, 119 m/s, 142 m/s and 166 m/s. For the optical connectivity technique, the same liquid jet of U_L = 2.0 m/s was examined under U_G = 47 m/s, 71 m/s, 119 m/s and 166 m/s. The operating conditions of the atomizer are considered in terms of the Reynolds number, Re, the Weber number, We, the momentum ratio, MR, and the momentum flux ratio, which are defined as:

$$Re = \frac{\rho_L U_L D_L}{\mu} \tag{1}$$

$$We = \frac{\rho_G (U_L - U_G)^2 D_L}{\sigma} \tag{2}$$

$$MR = \frac{\rho_G}{\rho_L} \cdot \frac{U_G^2}{U_L^2} \cdot \frac{(D_G^2 - D_L^2)}{D_L^2} \tag{3}$$

$$MFR = \frac{\rho_G}{\rho_L} \cdot \frac{U_G^2}{U_L^2} \tag{4}$$

where ρ_L is the liquid density, ρ_G is the gas density, D_L is the internal diameter of the liquid nozzle, D_G is the internal diameter of the gas nozzle exit, U_L is the area-averaged velocity of the liquid at the nozzle exit, U_G is the area-average velocity of the annular gas flow at the nozzle exit, μ is the liquid dynamic viscosity and σ is the surface tension of the liquid. It should be noted that since the dimensions of the nozzle have not been changed here, the effects of MR and MFR cannot be distinguished. However, there are cases where the momentum flux may not suffice to describe the flow. If the continuous liquid jet extends beyond the potential core of the annular gas jet, the final stages of breakup occur under a decreased gas velocity. This has been pointed out by Engelbert et al. [9] and, more recently, by [30]. In such cases, the diameter of the gas nozzle will have an effect on the resulting sprays, as confirmed by Matas and Cartellier [9]. For completeness, we have presented both the MR and MFR values.

The resulting liquid Reynolds number was 5442, while the Weber number ranged between 80 and 1041, the momentum flow ratio was between 27 and 335 and the momentum flux ratios between 0.67 and 8.27. For this range of flow conditions the breakup frequency of the liquid core of this atomizer is below 2000 KHz [9] which is ten times lower than the sampling rate used in the present investigation and the sampling rate used for the high speed measurements is more than adequate to temporally resolve the breakup of the liquid jet. A summary of the tested conditions is presented in Table 1.

Table 1. Considered flow conditions of the co-axial airblast atomizer.

U_L (m/s)	U_G (m/s)	Re	We	MR	MFR
2.0	47	5442	80	27	0.67
2.0	71	5442	185	61	1.52
2.0	95	5442	334	109	2.70
2.0	119	5442	526	171	4.22
2.0	142	5442	762	246	6.07
2.0	166	5442	1041	335	8.27

EVALUATION OF DATA PROCESSING APPROACHES AND RESULTS

The results from each of the considered techniques are presented in this section. The methodology for the evaluation of the breakup length for the shadowgraphic imaging measurements and the electrical connectivity is also presented here. The methodology for the optical connectivity has been presented at a previous publication [16].

Shadowgraphic Imaging

Example images from each of the flow conditions considered here are shown in Figure 5. Atomization can be observed to progress from a destabilization regime of the main jet at low MR to the stripping of droplets from the jet surface at high MR. It is possible to evaluate many aspects of the morphology of the liquid jet, such as the waves on the surface of the jet, the shape of the detached liquid, the flapping of the liquid jet and the breakup length.

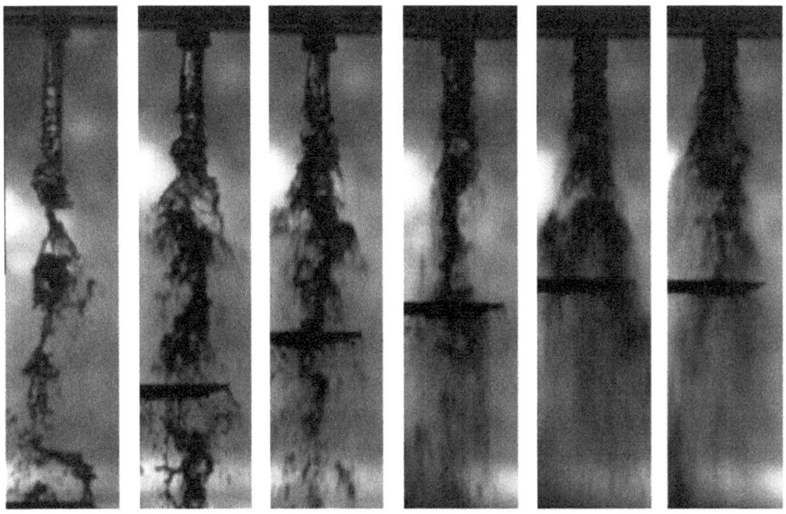

Figure 5. Examples of the shadowgraphic jet morphology for all the considered flow conditions. (a) MR = 27, (b)MR = 61, (c) MR = 109, (d) MR = 171, (e) MR = 246 and (f) MR = 335. The horizontal dark line is the electrical connectivity probe.

The evaluation of the breakup point from shadowgraphic images is a matter of interpretation. This is because it is not always possible to identify clearly the location of the breaking point. One issue is the presence of ligaments and lamella that can connect detached masses of liquid with the main liquid jet that is injected from the nozzle. This can be observed in all cases in Figure 5. For example in Figure 5a, the liquid jet breaks up in a succession of liquid fragments. While the continuity of the liquid jet is interrupted at about the middle of the image, there is a fine webbing of liquid ligaments that connect the main jet with the liquid fragment directly downstream. The same is true for the case of the most intense atomization of Figure 5f, where, at the location of the breakup point half way between the nozzle exit and the needle probe, fine ligaments connect the main jet with the detached segment downstream. This fine interconnection between successive liquid masses not only causes problems in the interpretation of the breakup length, but also makes the automated treatment of the liquid jet images unreliable, since thresholding of the image intensity will not correctly identify the main jet and lead to longer measured breakup lengths.

Another difficulty with shadowgraphy is that the detached liquid droplets and ligaments can obscure the observation of the breakup point. This type of ambiguity is demonstrated in Figure 6. If the jet is illuminated from behind the page and imaged from the front, the observer should see the gap between the continuous liquid jet and the segment of detached liquid immediately downstream, to identify the breakup location correctly. However, if the jet is illuminated from the left and imaged from the right, the optical path between the illumination source and the camera is blocked by a mass of detached liquid immediately to the right of the breakup location and the gap will not be seen. Instead, a continuous liquid jet will appear, which, in fact, comprises of three distinct liquid masses and the measured breakup length will be considerably longer. For the current example, the measurement error is about 50%.

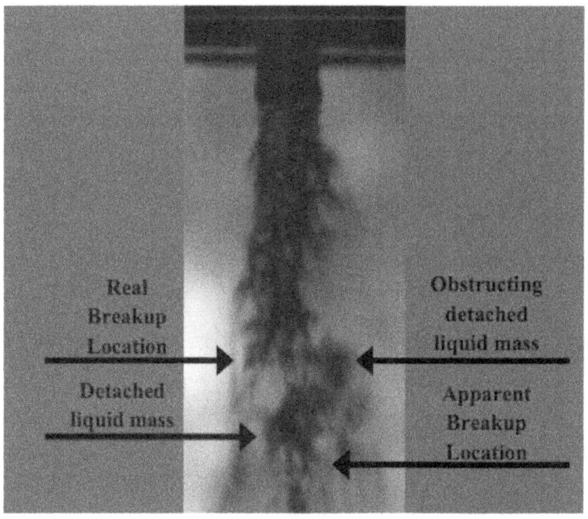

Figure 6. Demonstration of ambiguity of shadowgraphic measurements of breakup length. If the liquid jet is imaged from the left side and recorded at the right side, the detached liquid mass will obstruct the observation of the breakup point and the breakup length will appear about 50% longer than the real value.

The difficulty in the measurement of the breakup length from single images can be alleviated from temporally resolved shadowgraphic images. The uncertainty of the breakup location can be reduced by examining the sequence of the time-dependent images frame by frame. For example, in the first frame of Figure 7, the liquid jet appears to be continuous. However, after 50 μs, it is revealed that there is a discontinuity upstream of the location of the needle, which becomes greater in the subsequent frames, revealing that the liquid is divided in two distinct segments leading to a measured breakup length considerably shorter than identified from the first image.

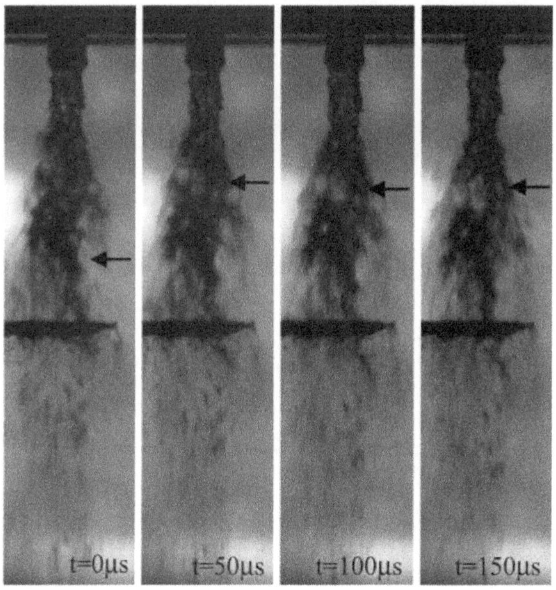

Figure 7. Time resolved imaging reveals the location of liquid jet breaking point, which could be missed by single frame shadowgraphy. The horizontal dark line in the middle of the images is the electrical connectivity probe. The black arrows indicate the location of the liquid jet breakup on each image. (The example shown is for Re = 5442, We = 526 and MR = 171).

In summary, shadowgraphy can give accurate measurements of the breakup length provided a temporal record exists. However, the evaluation of the breakup length requires considerable input from the investigator as a frame by frame analysis is required, which is not fully automated and contains a degree of subjectivity.

Electrical Connectivity

During the application of the electrical connectivity method, the electrical potential is measured between the nozzle and the probe at distinct locations downstream of the nozzle. The location of the liquid jet breakup is determined by analyzing the measured potential against the distance from the nozzle. The separation of the continuous liquid jet and the downstream liquid fragments is not always complete. This can be observed in Figure 8 for two flow conditions with different morphological development of the liquid core. In the first case (Figure 8a), close to the breaking point there is a liquid membrane that connects the main jet to the liquid fragment downstream. In the second case (Figure 8b), the vicinity of the breaking point is dominated by ligaments. This has also been observed by [15] for diesel sprays and it was proposed that diesel sprays do not breakup in a single column, but in a number of interconnecting liquid ligaments. While there are differences between the diesel sprays of that study and the air-blast atomizer considered here, the connection between the liquid jet and downstream liquid fragments conducting electricity applies to both cases. For the jets in Figure 8, while the physical connectivity between the liquid jet fragments and the main liquid jet is small, the measured electrical potential is significant (4.1 V and 3.9 V respectively, with 10 V indicating complete discontinuity). Therefore, an appropriate criterion of the measured potential needs to be defined that indicates the location of the breakup even if there is electrical connectivity between the nozzle and the probe.

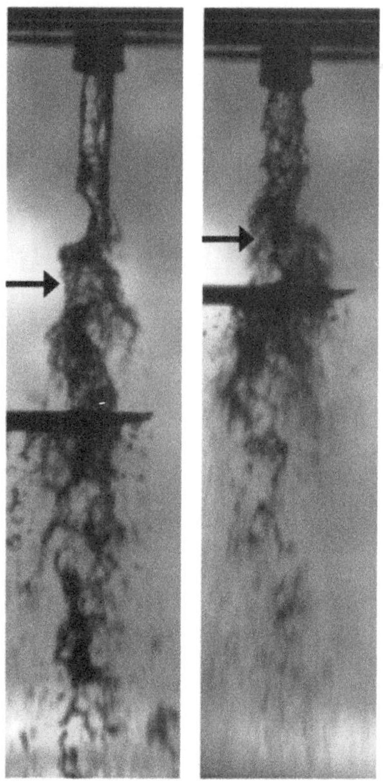

(a) (b)

Figure 8. Instantaneous images of the needle probe in contact with the main liquid jet core through minor physical connectivity. (a) We

= 185 and (b) We = 762. The black arrow shows the location of the breakup, which occurs upstream of the location of the needle probe. The respective measured voltages of 4.1 and 3.9 V indicate the presence of remaining electrical connectivity.

The breakup criterion is also affected by the change of the measured potential with the apparent contact of the liquid jet on the electrical probe. This is illustrated in Figure 9, which shows a sequence of images of the atomizing liquid jet along with the measured voltage drop across the jet nozzle and the probe. The progressively decreasing area of contact between the liquid jet and the needle probe in frames 312–320 of Figure 9 shows an increasing trend in the measured voltages between the nozzle and the probe, while in frames 321–322 the increasing contact between the jet and the probe decreases the value of the measured voltage. From a low voltage of about 4.5 V in frame 312, the measured potential rises to a peak value of about 6 V in image 320, where the apparent contact between the jet and the probe is very small.

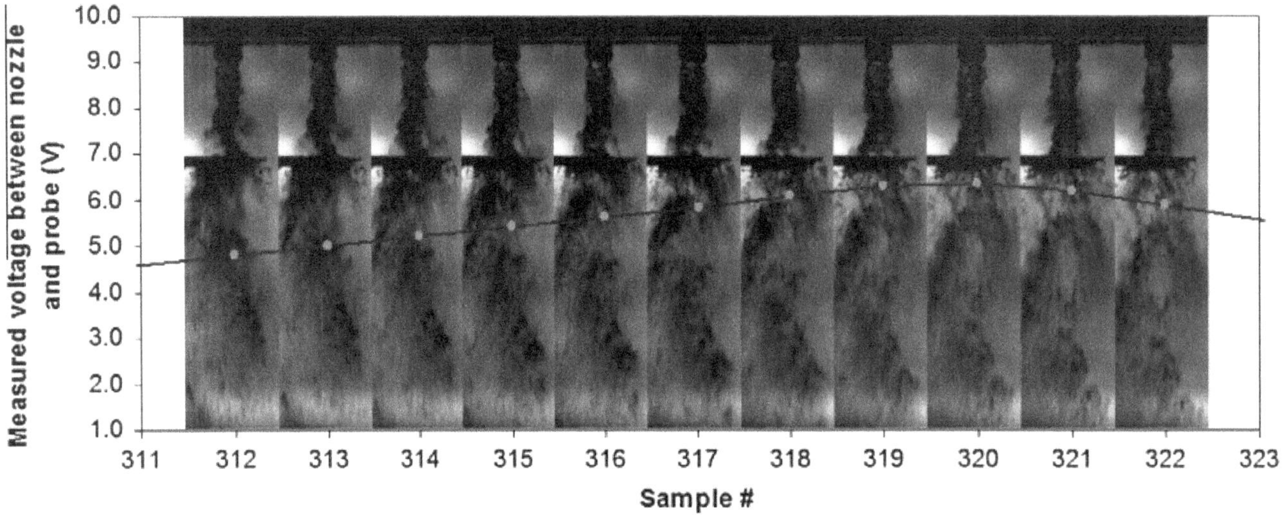

Figure 9. Sequence of temporally resolved shadowgraphic images (We = 526 and MR = 171) with the simultaneously measured voltage between the nozzle and the probe (black horizontal line on images) superimposed. The higher voltages indicate lower electrical connectivity.

Since, neither the contact area between the liquid jet and the probe nor the presence of interconnecting ligaments between the main liquid jet and the downstream fragments can be accounted for in the practical application of the technique, the probability function of the measured voltages needs to be considered as a function of the distance from the nozzle. The shape of the probability function is presented in Figure 10, and, as the probe is placed farther from the nozzle, the mean value of the measured voltage increases, the probability peaks at an intermediate location over the range of measured values and the skewness of the histogram shifts from right to left. Therefore, various statistical parameters can be considered to identify the appropriate breaking point criterion for the electrical connectivity technique.

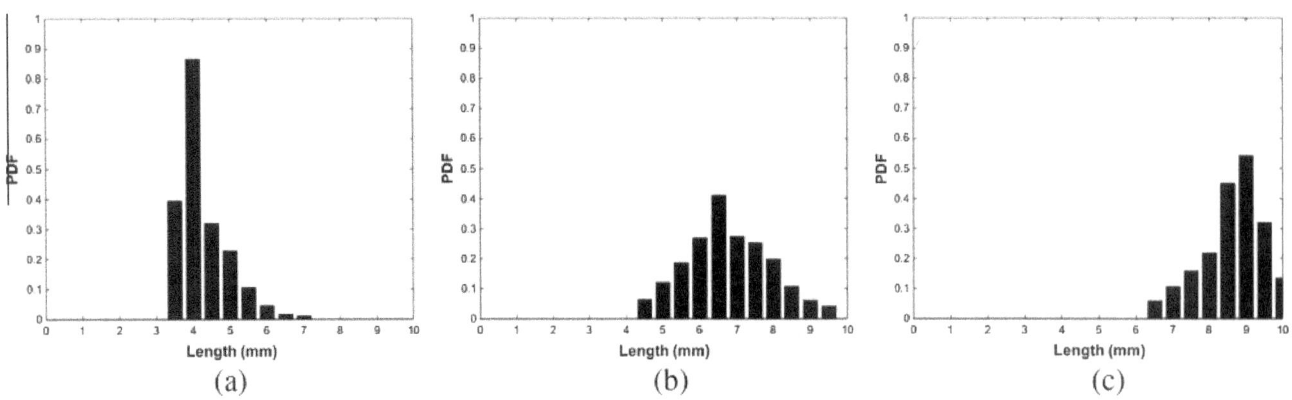

Figure 10. Example of the evolution of the Probability Density Function of the electrical connectivity signal for the flow condition of We = 526 and MR = 171 when the voltage is measured at (a) 11 mm, (b) 14 mm and (c) 16 mm downstream of the nozzle exit.

The first considered parameter is the mean value of the measured potential between the nozzle and the probe. The mean value of measured voltage is presented in Figure 11 as a function of the distance from the nozzle exit for all the considered flow conditions. In the near nozzle region, the mean measured potential is the same for all cases and its value increases linearly with the distance from the nozzle. The linear increase of the electrical resistance with length is consistent with that of a conductor of constant cross section and indicates that the liquid jet cross section is largely intact for the full length of that region. A linear baseline of the measured potential of the unbroken liquid jet with distance from the nozzle is thus established and is indicated in Figure 11 by a dashed line and is compatible with the behavior of the potential across a straight conductor with the resistivity of water. The baseline is then used as a guide for the detection of the length of the continuous liquid jet. The results for most flow conditions indicate an initial linear increase of the measured potential with distance from nozzle exit, followed by a sharp increase of the measured potential at a certain distance. This sharp increase can be explained by a sharp decrease of the mean cross section of the liquid jet and is indicative of liquid jet discontinuity. This axial location is marked in Figure 11 with a dot for each flow condition and occurs progressively closer to the nozzle exit as the atomization improves. All the dots that indicate the jet breakup fall on the baseline established along the length on the intact jet, which indicates that this method of identifying the breakup location with electrical connectivity is independent of the considered flow regimes. Beyond the breakup region, the slope of the mean potential is reduced. This is due to the increasing number of instances, where the maximum value of 10V potential is measured. The advantage of this method for identifying the breakup length is that the average potential provides a clearly defined criterion that is applicable across many flow regimes. The drawback is that the threshold is not fixed and the slope of the voltage with distance needs to be analyzed. This may pose difficulties when the slope of the potential does not change dramatically with distance from the nozzle.

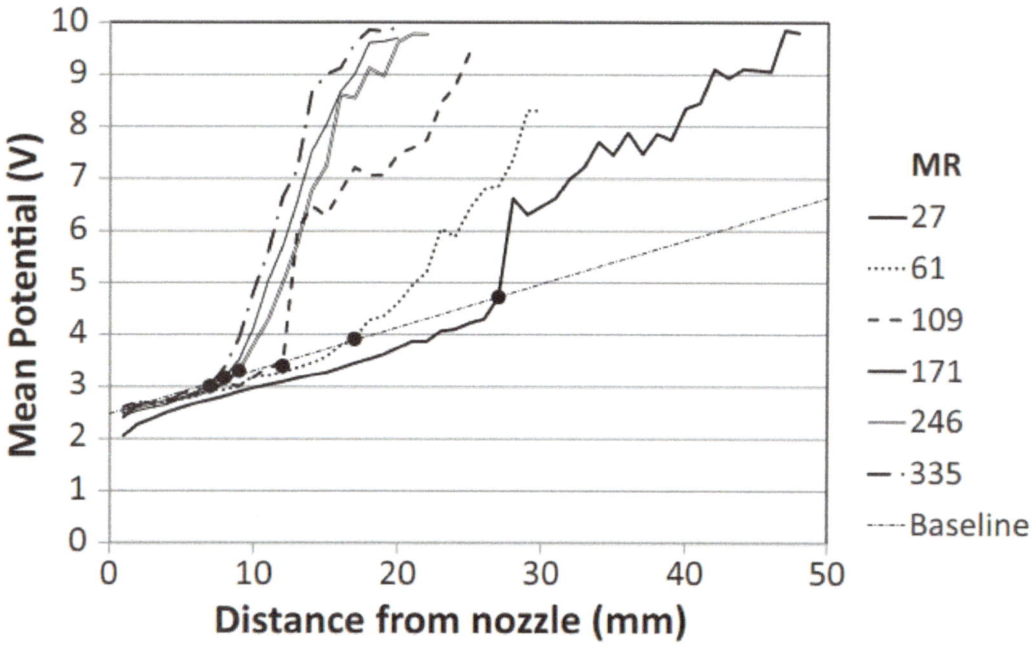

Figure 11. Average values of the measured potential against the distance of the probe from the nozzle exit. The added 'dot' on each line indicates the considered location of the break-up length of the liquid jet for that operating condition.

The second considered parameter is the standard deviation of the fluctuations of the measured potential, which is indicative of the spread of the measured potential values. The evolution of the measured standard deviation with the distance from the nozzle exit is presented in Figure 12 for all flow conditions. The observed general trend is that the standard deviation begins to increase at some location downstream of the nozzle exit and reaches a peak and then decreases again. This behavior can be explained as follows. Close to the location of the liquid jet breakup, as evaluated from the plot of average measured potential of Figure 11, and marked with a black dot on the standard deviation plots in Figure 12, there is an increase of the fluctuations of the measured voltages due to the formation of a discontinuity of the liquid jet. The black dots fall in the range of 0.2–0.4 V, which appears to be the threshold value for breakup when the standard deviation of the fluctuations of the potential is considered as a criterion. As the probe moves further downstream, greater fluctuations of the measured voltages are measured, since the interconnecting ligaments between the main liquid jet and the downstream liquid fragments become finer. The peak value of the standard deviation identifies the distance where the probe starts to detect complete discontinuity of the liquid jet, meaning that there is no connectivity between the main jet and detached liquid masses even through thin ligaments. As the liquid jet becomes completely discontinuous for most of the time further downstream from the nozzle exit, the fluctuations of the measured potential decrease and tend to zero. Since the threshold value of the standard deviation is not obvious from the trends of Figure 12, the standard deviation of the fluctuations of the measured potential is not a good criterion for detection of the liquid breakup.

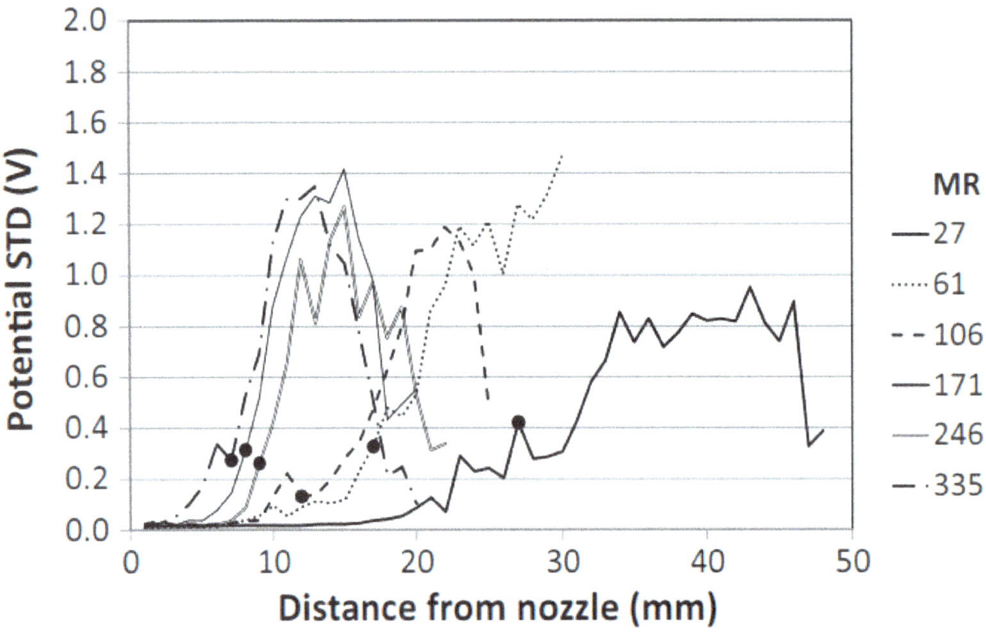

Figure 12. Standard deviation of the fluctuations of the measured potential plotted as a function of the distance of the probe from the nozzle exit. The added 'dot' on each line indicates the considered location of the break-up length of the liquid jet for that operating condition.

The third considered statistical parameter for the interpretation of the measured electrical potential is the skewness of the histogram of the measured potential. While the probe is traversed downstream from the nozzle exit, the skewness of the measured samples changes from positive to negative. This can be observed in Figure 10 for flow We = 526 and MR = 171 and is similar for all other tested flow conditions. The development of the skewness of the samples of the measured potential is shown in Figure 13 as a function of the distance from the nozzle exit for all flow conditions. Closer to the nozzle, the values of the measured potential are skewed to the right (Figure 10a). This means that, when the breakup of the jet is initiated, it develops fast and there are many samples, where the reduced electrical connectivity of the breaking liquid jet results to high values of measured potential. As the probe is traversed downstream, the probabilities of larger and lower electrical connectivity than that of the liquid jet become equal and the skewness becomes zero (Figure 10b). As the probe is traversed further downstream, the tail of the measured samples will be at low potential values as the electrical connectivity of the liquid jet is diminished (Figure 10c). From the above, it can be asserted that when the liquid jet is discontinuous on average, the skewness of the histogram of the measured values of the potential will be zero. This break-up criterion is easy to apply, because the evaluation of the breakup location is unambiguous and does not require any interpretation of the signal. However, as it will be discussed later the estimated breakup length with this method is considerably longer than what is evaluated when using the other two methods.

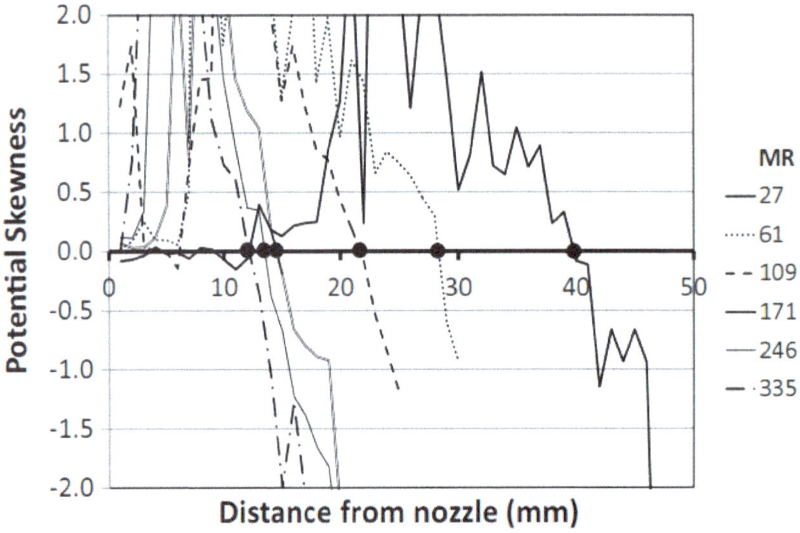

Figure 13. Skewness of the Probability Density Function of the measured potential plotted as a function of the distance of the probe from the nozzle exit for different flow conditions indicated by the MR value. The added 'dot' on each line indicates the considered location of the break-up length of the liquid jet for that operating condition.

Another approach for the evaluation of the average breakup length of the liquid jet may be attempted by considering the fraction of the samples, where the measured potential falls below a certain threshold. In Figure 14, the Cumulative probability Distribution Function (CDF) of the measured potential being below 3.0 V, 4.0 V, 5.0 V, 6.0 V, 7.0 V 8.0 V, 9.0 V and 9.9 V respectively is presented. Very close to the nozzle exit, the measured potential between the nozzle and the probe is always below the threshold values and therefore the CDF is 0. As the probe moves further away from the nozzle, the CDF increases, signifying the proportion of time that the measured potential is below the selected threshold. Naturally, higher threshold values result to the CDF shifting to greater distances from the nozzle exit. It is difficult to estimate an appropriate threshold just from examining Figure 14. For this reason the value of the potential that corresponds to 50% of the cumulative density function is selected. This represents the median value of the samples and is presented in Figure 15 as a function of the distance from the nozzle exit. Figure 15 shows considerably similar behavior with Figure 11, which shows that there is a sudden change of the slope of the mean value of the potential with distance from the nozzle exit. The location of the change in the slope is almost identical to that of the mean value of Figure 11. This occurs because the distribution of the measured potential is mono-modal and the mean and median values are similar. Therefore, there is no additional information that can be gained from the CDF approach.

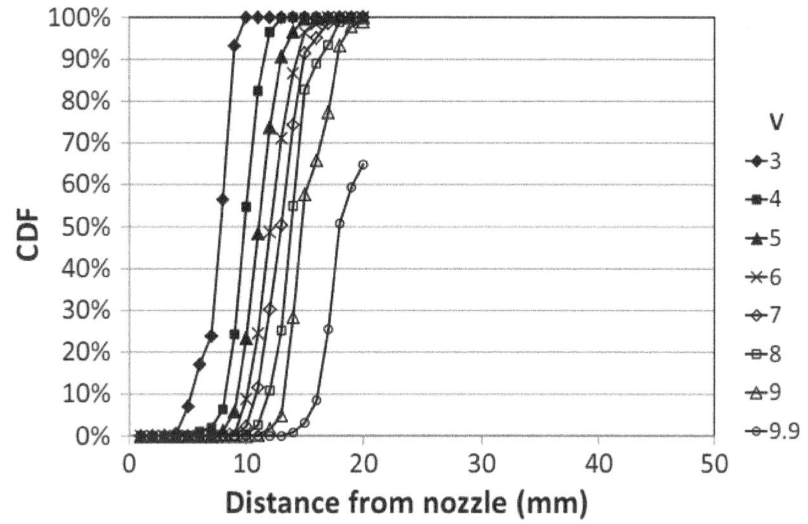

Figure 14. Cumulative Distribution Function (CDF) of the time the measured potential is below a threshold value.

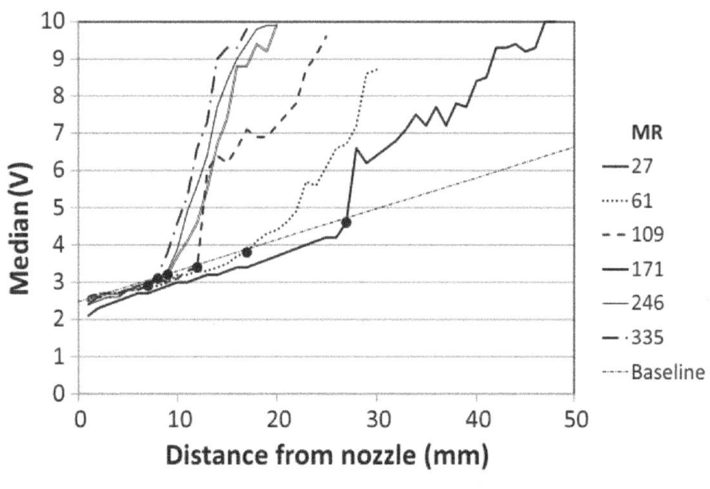

Figure 15. Evaluation of the liquid jet breakup length based on the median value of the measured potential. The added 'dot' on each line indicates the considered location of the break-up length of the liquid jet for that operating condition.

Optical Connectivity

In the optical connectivity technique, unlike shadowgraphy, the main liquid jet can be directly imaged without too much interference from the detached fragments of liquid [16]. This is demonstrated in Figure 16, where an atomizing liquid jet was imaged using shadowgraphy and optical connectivity simultaneously. In the shadowgraphic image, all of the liquid phase along the line of sight between the light source and the imaging camera is visualized without distinction between the continuous liquid jet and the surrounding detached liquid. In the image of the optical connectivity technique, only the continuous liquid jet is visualized through which the laser beam propagates.

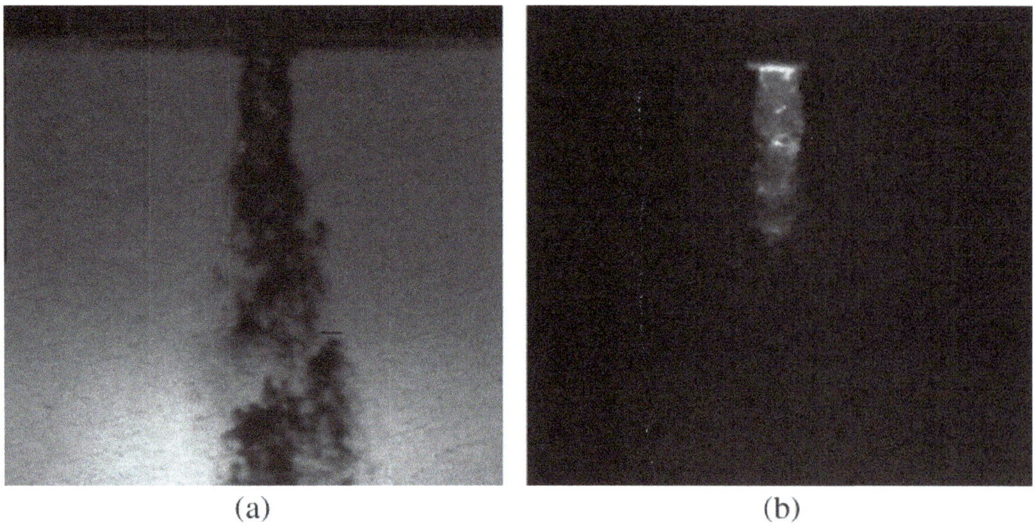

Figure 16. Example of liquid jet visualization with (a) shadowgraphy and (b) optical connectivity for operating condition We = 526 and MR = 171. The optical connectivity technique is less influenced by the products of atomization around the continuous jet, making the determination of the breakup length easier.

The interpretation of the shadowgraphic image for the determination of the breakup length is challenging, since both the background and parts of the continuous liquid jet have the same intensity. The interpretation must be based on the change of the image intensity along the boundary of the liquid jet in order to decide which bright image regions are within and which are outside the continuous liquid jet. Such a procedure is complex and automatic image processing is difficult. In the optical connectivity, the liquid phase is clearly seen as bright intensity and the gas phase is clearly seen as dark. Therefore, there is negligible uncertainty at the detection of the location of the continuous liquid phase. Consequently, the breakup length of the liquid jet can be clearly identified. In addition to the breakup length, other useful characteristics of the liquid jet morphology can be identified, such as the wavelength of the interfacial waves and the midline of the liquid jet[31] and [32].

The procedure for the processing of the optical connectivity images, a discussion of the advantages and the limitations of the technique for the determination of the breakup length as well as the structure of the liquid jet have been presented elsewhere[16] and [33] and only a summary is provided below. The intensity of the fluorescent light is tracked downstream of the liquid nozzle exit. Close to the nozzle exit, the fluorescent intensity remains fairly constant. In the vicinity of the breaking point, there is a sharp reduction of the fluorescent intensity of the liquid jet. The breaking point is identified by considering a threshold value that is a fraction of the intensity at the nozzle exit. This value is in the region of 10–20% of the fluorescent intensity at the nozzle exit. Since the measured breakup length does not change by more than 10% when changing the value of the above threshold, the determination of the breaking point of the liquid jet is relatively easy and can be performed through automatic image processing which is not susceptible to large uncertainty.

DISCUSSION

The comparison of the measurements of the average liquid jet breakup length, obtained with the electrical connectivity technique, high and low speed imaging with shadowgraphy and optical connectivity are presented in Figure 17 as a function of the air-to-liquid momentum ratio (MR).

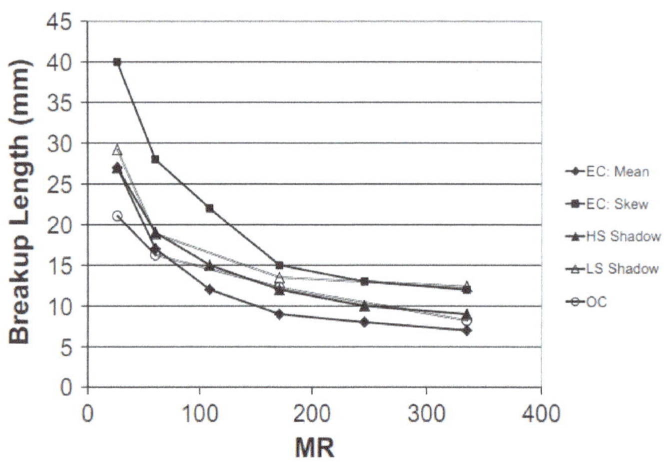

Figure 17. Comparison of the breakup length measurements obtained by the three experimental techniques. Each technique is indicated as EC Mean: Electrical Connectivity using the mean criterion (♦), EC Skew: Electrical Connectivity using the Skewness Criterion (■), HS Shadow: High Speed Shadowgraphy (▲), LS Shadow: Low Speed Shadowgraphy (Δ) and OC: Optical Connectivity (○).

For all measurements approaches and different processing approaches of each method, with the exemption of the skewness criterion for electrical connectivity, the measurement of the breakup length is within 8–10 mm, which corresponds to about $4D_L$. When considering that the average breakup length varies in the range of 25 mm to 10 mm, depending on flow conditions, the relative uncertainty [34] and [35] among the different measuring techniques ranges from ±20% to ±50%, with the uncertainty being more significant for the high MR range, where the average breakup length is shorter. The interpretation of the electrical connectivity measurements with the skewness criterion results to breakup length measurements, which are longer by about 50% than the other methods in most cases. Therefore, this criterion does not provide an accurate evaluation of the breakup length. The interpretation of the results of the remaining methods should be considered with respect to the MR of the measured flow conditions, where the merits and shortcoming of each technique can be highlighted.

For the lower values of the MR range (Figure 17), the atomization is less intense. In this flow regime, the continuous liquid jet is longer, the shedding of liquid from the surface of the jet is not intense and close to the breakup location there is formation of liquid lamella (Figure 5a–c). For this range of MR, there is good agreement of the measured breakup length among the high and low speed shadowgraphic imaging and the electrical connectivity. The high and low speed shadowgraphic measurements agree well, because the morphology of the jets is such that the liquid jet can be visualized adequately. The good agreement of the electrical connectivity with shadowgraphy for this range of MR values can be hypothesized to be due to the morphology of the constriction of the liquid jet that develops at the breaking point, which is different than the constriction at higher MR values. However, this is a hypothesis and more evidence is required, which can be obtained with high magnification imaging measurements that resolve the morphology of the breaking location better than the resolution used in this study. Considering the optical connectivity technique, the measured breakup length is shorter than for the other techniques. This is explained by Charalampous et al. [16] and is a result of the optical connectivity technique being based on the propagation of a laser beam within the length of a liquid jet in the same way that light propagates in optical fibers. Close to the breaking point the laser beam cannot propagate considerably within the ligaments that connect the continuous core and the downstream liquid jet fragments. As a result, the breakup length measured by the optical connectivity technique will be weighted towards the beginning of the breakup point, while the shadowgraphic techniques will measure a somewhat longer breakup length, since they are more likely to detect the breakup point at a later stage.

For higher values of MR (Figure 17), the atomization is more intense (Figure 5d–f), and the results show that there is good agreement between optical connectivity and high speed shadowgraphic imaging. Low speed shadowgraphy gives longer breakup lengths and electrical connectivity gives shorter breakup lengths. For low speed shadowgraphy under intense atomization, due to visual interference of the breakup location from droplets and liquid fragments, it is more difficult to identify the breakup location correctly. Therefore, the measured breakup length can appear longer. However, when using high speed shadowgraphy, the temporal record of the acquired images can help to overcome the interference from the detached liquid segments by tracking the discontinuity upstream in the liquid jet. In the optical connectivity method, since the laser beam is introduced through the nozzle and only the main liquid jet is illuminated, optical interference with the surrounding droplets is avoided. Additionally, as the breakup length is naturally short under intense atomization, the losses of the laser beam intensity along the interface of the continuous liquid jet due to refraction are small and the illumination of the continuous jet is complete. The shorter breakup length detected by electrical connectivity might be attributed to the nature of the technique. As the probe is physically inserted into the flow, electrical connectivity will be able to detect the narrowing of the continuous jet close to the nozzle. However, at high MR, the flow is fast and the narrowing does not have sufficient time to develop to a discontinuity within the probe location indicating a short breakup length. Overall, confidence in the measurement of the breakup length for the region of higher MR is higher for the high speed shadowgraphy and the optical connectivity, while electrical connectivity has greater uncertainties.

Overall, when considering the 3 most reliable measurement methods of the breakup length, namely optical connectivity, high speed shadowgraphy and electrical connectivity interpreted according to the mean value of the measured potential, the maximum measurement error among them is within ±15% of the mean measured value. This range is much narrower than the ±50% range for all the measurement techniques and processing approaches considered.

From an application point of view, the selection of the appropriate technique depends on the implementation of the experiment and the results, which are sought. The most common way to measure the breakup length is low speed shadowgraphy. It has the advantages that it is non-intrusive, very easy to set up and requires equipment that is available at all laboratories. There is no limit on the length of the breakup length that can be measured and, in addition, valuable information can be extracted regarding the overall atomization process and the behavior of the detached liquid segments. On the downside, shadowgraphy requires optical access to the atomizing spray from two opposite sides, one for the illumination and one for the imaging system, which could be a problem when the spray is contained. Also, the analysis of the images is susceptible to the subjective interpretation of the researchers, since the location of the breakup point must be determined visually. This is a significant issue when atomization is intense and there is no clear view of the liquid jet and therefore there is ambiguity in the measurements. This ambiguity can be overcome by using high speed shadowgraphy. The setup remains essentially the same, but the cost of the required imaging system is much higher and may not be available at all laboratories. In this case, however, the analysis of the shadowgraphic record requires substantial user effort, as the breakup location needs to be traced upstream which needs to be done manually. In general, shadowgraphy is a good tool for the overall investigation of the liquid jet breakup with some limitations depending on the atomization intensity, which can be overcome with temporally resolved measurements.

The optical connectivity approach for the measurement of the liquid breakup length is most appropriate when there is optical obstruction of the liquid jet from surrounding detached liquid fragments, since it visualizes only the main jet. This is particularly important for cases where many droplets surround the liquid jet core. The optical connectivity images of the liquid jet present clearly the continuous liquid jet. This makes the determination of the breakup length straightforward by tracing the fluorescent intensity along the jet length, which is an advantage over shadowgraphy and can be performed with automatic image processing. Additional information can also be extracted for the overall morphology of the liquid jet, since the full image of the continuous jet is obtained [31] and [32]. The technique does pose increased demands on equipment compared to shadowgraphy because a laser source is required for illumination. However, lasers are available at most laboratories studying atomization. Another difficulty is that the atomizer must be modified, so that the laser beam can be guided through the nozzle. In case of complex atomizer designs, this might not be possible. Finally, it has been shown that there will be a decrease of the laser beam intensity as it propagates along the liquid jet length due to refraction [16] at the liquid interface. Therefore, this technique is recommended for cases where the breakup length is expected to be short, which are related to intense atomization and, therefore, for most practical applications. This is also the case when low speed shadowgraphy becomes problematic. Therefore, the two techniques can be complementary.

Finally, the electrical connectivity technique is an alternative method to optical approaches and is dedicated to the measurement of the breakup length. This technique is attractive, since it is easy to set up, has little demand on equipment and produces small datasets. Like shadowgraphy, there is no limitation of the maximum breakup length that can be measured. The drawbacks of the technique are that the estimated breakup length can be somewhat shorter than what is measured by optical techniques. This is due to the interpretation of the measured potential between the nozzle and the probe. The requirement for physical access to the near nozzle region may be a problem for some applications, especially in limited spaces and in physically hazardous environments, where the direct contact of the spray with the probe can cause damage or contamination of the surface that would alter the electrical resistance of the probe and interfere with the measurement. The time required to characterize a single flow condition is also longer compared to other techniques, since the probe has to be traversed along the jet. Finally, a requirement of the technique is that the liquid must be electrically conductive. Overall this technique is promising when only the breakup length is required and demands on equipment needs are minimal.

The advantages and disadvantages of the considered techniques are summarized in Table 2.

Table 2. Summary of advantages and disadvantages of examined methods for measurements of liquid jet breakup length

Method	Advantages	Disadvantages
Shadowgraphy	• Non-intrusive • Easy to set up • Can measure jet morphology and spray patterns	• Measurement bias when not time-resolved • Measurement is susceptible to interpretation • Two way optical access (one for camera, one for illumination)
Electrical connectivity	• Easy to set up • Economic • Small datasets	• Time consuming scanning required • Physically intrusive probe • Only measures breakup length
Optical connectivity	• Clear separation of continuous liquid jet and gas phase • Suppresses interference of liquid droplets and fragments around continuous liquid jet • Can measure continuous liquid jet morphology • Feasible with equipment available in most laboratories	• Modification of atomizer is required • Limitation of maximum breakup length • Two way optical access (one for camera, one for illumination)

CONCLUSIONS

A comparative investigation of the measurement of liquid jet breakup length of a co-axial airblast atomizer was conducted using electrical connectivity, optical connectivity and shadowgraphy for a range of flow conditions with air-to-liquid Momentum Ratios (MR) between 27 and 335 and momentum flux ratios between 0.67 and 8.27. The findings are as follows:

During shadowgraphic measurements, the optical access to a central liquid jet is obstructed by the presence of detached liquid segments along the line of sight between the jet and the camera resulting to bias towards long breakup length measurements. High speed shadowgraphy provides the temporal evolution of the liquid jet, which can alleviate some of the uncertainty in the interpretation of the shadowgraphs and produce more reliable breakup length measurements.

The electrical connectivity technique was affected by the presence of thin liquid ligaments that interconnect the detached liquid fragments and are sufficient to allow the conduction of electricity. Different criteria for processing the measured values of potential in order to identify the breakup length were considered, which were assisted from the availability of understanding gained from simultaneous electrical connectivity and shadowgraphic measurements. The criteria included the mean value of the measured potential, its standard deviation, its skewness and its median value. The methods, based on the mean and the median value of the potential, were appropriately reliable to measure the mean value of the breakup length, but they have to be evaluated at different injector geometries. Therefore, if an appropriate criterion to process the electrical connectivity potential measurements is used, the measured breakup length is within the same range of the other techniques.

The optical connectivity technique can overcome the optical obstructions caused by droplets surrounding the liquid jet, which become more significant with the increase of Momentum Ratio of the flow conditions.

Overall, it is found that the three methods have merits and should be used complementary depending on the requirements of the investigation and the morphology of the investigated liquid jets. If applied correctly, the maximum measurement error between the techniques is within ±15% of the mean value of the breakup length.

ACKNOWLEDGEMENTS

The research was supported by the UK Engineering and Physical Sciences Research Council (EPSRC) Grant EP/G01597X/1 and the European Office of Aerospace Research and Development (EOARD) Contract FA8655-09-1-3036. Data supporting this publication can be obtained by request from MultiphaseFlows@imperial.ac.uk.

REFERENCES

1. A.H. Lefebvre, Atomization and Sprays, Hemisphere Publishing Corporation, 1989.
2. L. Bayvel, Z. Orzechowski, Liquid Atomization, Taylor & Francis, 1993.
3. P.H. Marmottant, E. Villermaux, On spray formation, J. Fluid Mech. 498 (2004) 73–111.
4. G.N. Laryea, S.Y. No, Spray angle and breakup length of charge-injected electrostatic pressure-swirl nozzle, J. Electrostat. 60 (2004) 37–47.
5. K. Sallam, C. Aalburg, G. Faeth, Breakup of round nonturbulent liquid jets in gaseous crossflow, AIAA J. 42 (2004) 2529–2540.
6. H. Eroglu, N. Chigier, Z. Farago, Coaxial atomizer liquid intact lengths, Phys. Fluids A – Fluid Dyn. 3 (1991) 303–308.
7. F. Xiao, M. Dianat, J.J. McGuirk, LES of turbulent liquid jet primary breakup in turbulent coaxial air flow, Int. J. Multiphas Flow 60 (2014) 103–118.
8. J.C. Lasheras, E. Villermaux, E.J. Hopfinger, Break-up and atomization of a round water jet by a high-speed annular air jet, J. Fluid Mech. 357 (1998) 351–379.
9. C. Engelbert, Y. Hardalupas, J.H. Whitelaw, Breakup phenomena in coaxial airblast atomizers, Proceed. Roy. Soc. Lond. Ser. A, Math. Phys. Sci. 451 (1995) 189–229.
10. A.J. Yule, I. Filipovic, On the break-up times and lengths of diesel sprays, Int. J. Heat Fluid Flow 13 (1992) 197–206.
11. M. Ochowiak, L. Broniarz-Press, S. Rozanska, Analysis of liquid jet breakup in one- and two-phase flows, Chem. Eng. Technol. 35 (2012) 1685–1691.
12. G.S. Settles, Schlieren and Shadowgraph Techniques: Visualizing Phenomena in Transparent Media, Springer, Berlin, Heidelberg, 2001.
13. H. Hiroyasu, M. Shimizu, M. Arai, The Breakup of High Speed Jet in a High Pressure Gaseous Atmosphere, ICLASS-82, 1982.
14. B. Chehroudi, S.H. Chen, F.V. Bracco, On the Intact Core of Full-cone Sprays, SAE Technical Papers 850126, 1985.
15. A.J. Yule, D.G. Salters, A conductivity probe technique for investigating the breakup of diesel sprays, Atom. Spray 4 (1994) 41–63.
16. G. Charalampous, Y. Hardalupas, A.M.K.P. Taylor, Novel technique for measurements of continuous liquid jet core in an atomizer, AIAA J 47 (2009) 2605–2615.
17. G. Charalampous, Y. Hardalupas, A. Taylor, Structure of the continuous liquid jet core during coaxial air-blast atomization, Int. J. Spray Combust. 1 (2009) 389–415.
18. D. Colladon, On the reflections of a ray of light inside a parabolic liquid stream, Comptes Rendus (1842) 800–802.
19. D. Colladon, La fontain colladon, La Nature (1884) 325–326.
20. L.A. Melton, C.W. Lipp, Criteria for quantitative PLIF experiments using highpower lasers, Exp. Fluids 35 (2003) 310–316.
21. G. Charalampous, C. Hadjiyiannis, Y. Hardalupas, Numerical and experimental investigation of the optical connectivity technique in cross flow atomization, in: 12th International Conference on Liquid Atomization and Spray Systems, Heidelberg, Germany, 2012.
22. G. Charalampous, N. Soulopoulos, Y. Hardalupas, Ray tracing analysis of realistic atomizing jet geometries for optical connectivity applications, in: 53rd AIAA Aerospace Sciences Meeting, American Institute of Aeronautics and Astronautics, AIAA 2015–0164, 2015.
23. W.Y. Cai, C.F. Powell, Y. Yue, S. Narayanan, J. Wang, M.W. Tate, M.J. Renzi, A. Ercan, E. Fontes, S.M. Gruner, Quantitative analysis of highly transient fuel sprays by time-resolved X-radiography, Appl. Phys. Lett. 83 (2003) 1671– 1673.
24. M.J. Renzi, M.W. Tate, A. Ercan, S.M. Gruner, E. Fontes, C.F. Powell, A.G. MacPhee, S. Narayanan, J. Wang, Y. Yue, R. Cuenca, Pixel array detectors for time resolved radiography (invited), Rev. Sci. Instrum. 73 (2002) 1621–1624.
25. Y. Yue, C.F. Powell, R. Poola, J. Wang, J.K. Schaller, Quantitative measurements of diesel fuel spray characteristics in the near-nozzle region using X-ray absorption, Atomization Spray 11 (2001) 471–490.

26. M. Linne, M. Paciaroni, T. Hall, T. Parker, Ballistic imaging of the near field in a diesel spray, Exp. Fluids 40 (2006) 836–846.
27. M.A. Linne, M. Paciaroni, J.R. Gord, T.R. Meyer, Ballistic imaging of the liquid core for a steady jet in crossflow, Appl. Optics 44 (2005) 6627–6634.
28. M. Paciaroni, M. Linne, Single-shot, two-dimensional ballistic imaging through scattering media, Appl. Optics 43 (2004) 5100–5109.
29. M. Paciaroni, M. Linne, T. Hall, J.P. Delplanque, T. Parker, Single-shot twodimensional ballistic imaging of the liquid core in an atomizing spray, Atomization Spray 16 (2006) 51–69.
30. J.-P. Matas, A. Cartellier, Flapping instability of a liquid jet, Comptes Rendus Mécanique 341 (2013) 35–43.
31. C. Hadjiyiannis, S. Sahu, G. Charalampous, Y. Hardalupas, A.M.K.P. Taylor, correlation between the primary break-up of liquid jets and downstream spray characteristics in air-blast atomizers, in: 6th European Combustion Meeting ECM 2013, Lund, Sweden, 2013.
32. G. Charalampous, Y. Hardalupas, C. Brown, U. Mondragon, V. McDonell, Investigation of injection characteristics of alternative aviation fuels by LaserInduced Fluorescence imaging, in: 51st AIAA Aerospace Sciences Meeting including the New Horizons Forum and Aerospace Exposition, AIAA 2013– 0162, American Institute of Aeronautics and Astronautics, 2013.
33. G. Charalampous, C. Hadjiyiannis, Y. Hardalupas, Proper Orthogonal Decomposition analysis of photographic and optical connectivity time resolved images of an atomizing liquid jet, in: 24th Annual Conference on Liquid Atomization and Spray Systems, Estoril, Portugal, 2011.
34. S. Tavoularis, Measurement in Fluid Mechanics, Cambridge University Press, 2005.
35. L. Kirkup, Experimental Methods: An Introduction to the Analysis and Presentation of Data, Wiley, 1996.

CHAPTER 7 LASER BASED MEASUREMENT FOR THE MONITORING OF SHAFT MISALIGNMENT

Anthony Simm [a], Qing Wang [a], Songling Huang [b], Wei Zhao [b]

[a] School of Engineering and Computing Sciences, Durham University, United Kingdom

[b] State Key Laboratory of Power Systems, Department of Electrical Engineering, Tsinghua University, Beijing, China

ABSTRACT

This paper presents a method for real-time online monitoring of shaft misalignment, which is a common problem in rotating machinery, such as the drive train of wind turbines. A non-contact laser based measurement method is used to monitor positional changes of a rotating shaft in real time while in operation. The results are then used to detect the presence of shaft misalignment. An experimental test rig is designed to measure shaft misalignment and the results from the work show that the technique can be used for the monitoring of both offset and angular shaft misalignment, which will have applications in the condition monitoring and maintenance of various types of rotating machinery.

INTRODUCTION

When performing maintenance on machinery, the following techniques are common: (1) run to failure where a piece of equipment must fail before any maintenance is performed; (2) preventative (or periodic) maintenance where maintenance is based on the length of the operating period, using criteria such as the mean time to failure (MTTF) measurement for the machine; (3) predictive maintenance where the operating condition of the machine is monitored to identify the need for repairs through data analysis and diagnosis [1].

Citation: Anthony Simm, Qing Wang, Songling Huang, Wei Zhao, Laser based measurement for the monitoring of shaft misalignment, Measurement, Volume 87, June 2016, Pages 104-116, ISSN 0263-2241, http://dx.doi.org/10.1016/j.measurement.2016.02.034.

Predictive maintenance strategies have led to the need for machinery condition monitoring. Condition monitoring can be defined as monitoring the physical parameters associated with the operation of the machine, such as vibration, temperature or pressure, to determine the operational condition of the machine. Improvements in maintenance strategies have economic benefits through improved production and less downtime, as well as indirect benefits through the need for fewer spare parts [2]. Examples of previous research on condition monitoring for rotating machines use vibration sensors for the monitoring of bearings to identify shaft rub and shaft misalignment [3], fault diagnosis using the empirical mode decomposition method [4] and acoustic noise measurements to predict the remaining useful life of a machine [5].

A typical application of condition monitoring for rotating machines is in wind turbines [6], specifically in offshore wind turbines, due to inaccessible locations [7], which may be expensive or difficult to access, with variable operating conditions. The drive train of the wind turbine consists of typically, a low speed shaft (on the rotor side), a gearbox, and a high-speed shaft (on the generator side) as well as support bearings, one or more couplings (between the shaft and the gearbox) and a mechanical brake [8]. A study [9] showed that 25% of the operational expenditure for an offshore wind farm is for operation and maintenance activities so any improvement in monitoring, leading to a reduction in maintenance costs, could eventually drive down the cost of energy.

SHAFT MISALIGNMENT

A shaft is an essential part of the rotating machine; it is used to transmit power and motion. A common problem (estimated to cause over 70% of vibration problems [10]) in rotating machinery is shaft misalignment. Shaft misalignment occurs when the center lines of rotation of two (or more) machinery shafts are not in line with each other [11]. This increases axial and radial forces on bearings, seals and couplings, increasing the amount of wear in these components, leading to an increase in vibration in the machine and bearings, it also increases bending of the shaft, increasing the risk of shaft failure and reducing the amount of power transmitted through the shaft [12]. Even if initially, or after adjustment, the shaft is aligned, during operation various factors such as thermal growth, piping pressure and foundation movements will alter the alignment [13].

Previous work on shaft misalignment monitoring has mainly been focused on looking at the vibration response. Arebi et al. [14] developed an on-line misalignment monitoring system using a wireless accelerometer mounted directly to the shaft, to measure the acceleration due to vibration. Work in rotor-dynamics shows that shaft misalignment caused by rotor imbalance leads to synchronous vibrations (frequency of vibration at twice the shaft speed) [15]. Dewell and Mitchell showed that the response to a misaligned coupling contained vibration frequencies of two times and four times the rotation speed[16], Xu and Marangoni also showed vibration frequencies at multiples of rotation speed[17] and [18]. Patel and Darpe [10] describe a drawback in using vibration monitoring to monitor for shaft alignment, as shaft damage, shaft stiffness and the type of coupling used can also affect vibration response. Shaft misalignment should cause vibration in both the connected machines, if it is only on a single machine, this could indicate other problems such as a cracked case [19]. As well as using vibration measurements, analysis of the motor current has been applied to shaft misalignment monitoring, Chaudhury and Gupta[20] and Verma et al. [21] used spectral characteristics of the stator current to identify shaft misalignment, Thomson and Fenger used motor current signature analysis to identify faults such as misalignment [22]. Similarly to the work on vibration measurements, Bossio et al. showed that angular misalignment has an effect on current at frequencies of two times the rotation frequency [23].

As well as measuring the response of the motor (current) or shaft (vibration) directly, indirect methods have been used to identify shaft misalignment. Rameshkumar et al. investigated the effect of misalignment on "coast down time" (time between the power cut off and the machine stopping rotating). They found that as shaft misalignment increased, the coast down time decreased, due to the increased power loss caused by the shaft misalignment and increased torque on the bearings [24]. Strain gauges have also been used to measure the presence of misalignment on turbine rotors [25] and to measure the increased gear loading caused by shaft misalignment [26]. Fulzele et al. used an optical sensor to measure shaft vibration by measuring the fluctuation of reflected light from the shaft [27]. Various methods have been developed to model or predict shaft misalignment; Sekhar and Prabhu used Finite Element Method (FEM) modelling to investigate the effect of coupling misalignment on the vibration response of a rotor-bearing system [28]. Cho and Jeong [29] and Fang et al. [12] used principal component regression (PCR) and partial least squares (PLS) to predict shaft parallel and angular misalignment. Yang and Tavner used empirical mode decomposition to reconstruct shaft orbit measurements to identify shaft misalignment [30].

Monitoring of shaft alignment during operation is needed as an effective tool in maintenance. A survey on rotating machinery in industry [31] showed that fewer than 10% of 160 machines examined were within acceptable shaft alignment, also 30% of a machine's down time is due to poor alignment [32]. Shaft misalignment consists of three types [33]:

- Offset where the two shafts are on two separate parallel centerlines (Figure 1a).
- Angular where the two shafts are coaxial but at an angle to each other (Figure 1b).
- In reality, shaft misalignment would be a combination of both of these effects (Figure 1c).

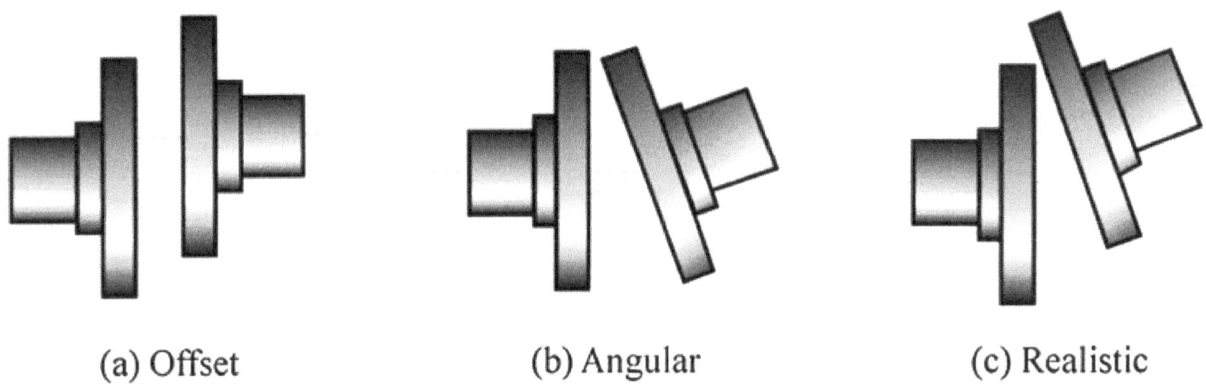

Figure 1. Types of shaft misalignment.

Offset misalignment affects power consumption more than angular misalignment and the components of misalignment are additive irrespective of whether they are horizontal or vertical [31]. The process of shaft alignment is the positioning of the shaft center lines of the driver machine and driven machine to create collinear shafts, where the rotational center lines of the coupled shafts are parallel and intersect (like a single shaft), this is accomplished through either shimming or moving the machine.

To measure the amount of shaft misalignment, the following methods are commonly used:

- Straight edge (Figure 2a).
- Dial indicator (reverse indicator method shown in Figure 2b).
- Laser indicator (Figure 2c).

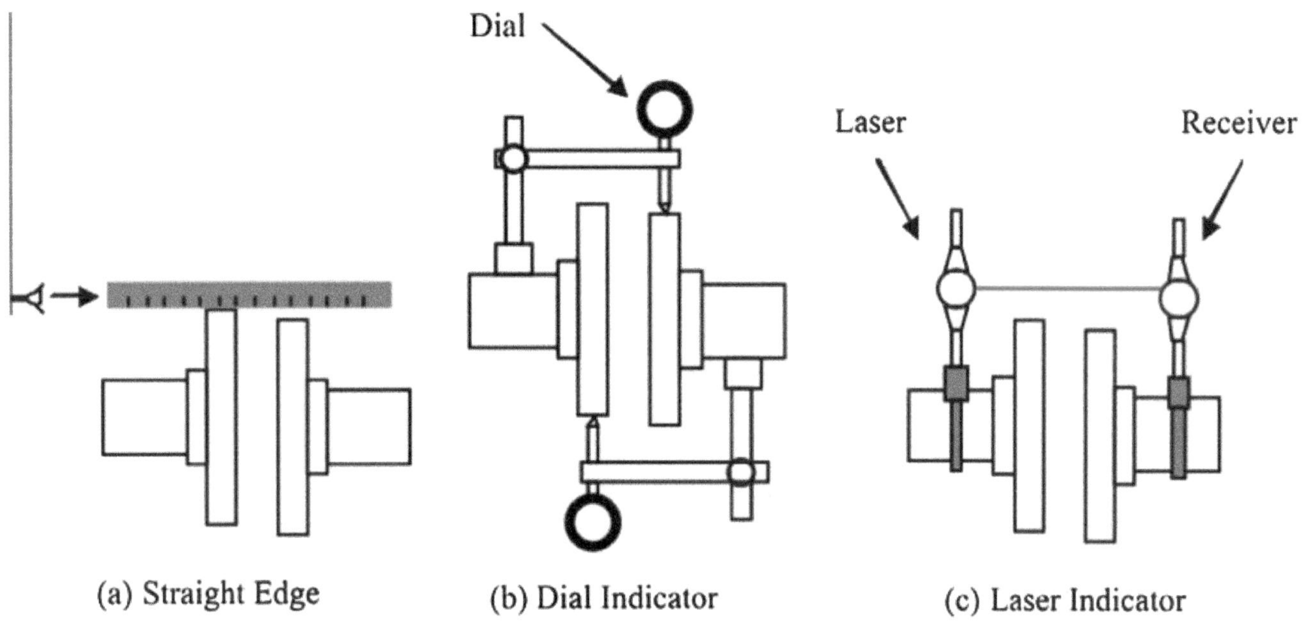

Figure 2. Misalignment measurement methods.

The straight edge is usually the first stage of inspection to get an approximate reading before moving onto the more accurate methods. The common method is to use dial indicators. Recently, these have been replaced by laser based misalignment measurement methods, which are more accurate and can be automated to calculate the amount of shaft misalignment automatically.

Monitoring and predicting the shaft alignment condition is important for making decisions on when to perform maintenance of the rotating machine, by counting misalignment events or using the measurements to identify other fault types such as bearing damage. This work proposes an alternative to the previously mentioned shaft misalignment measurement methods: a non-contact laser based measurement technique to capture the on-line positional changes of a shaft, for improved shaft misalignment monitoring.

EXPERIMENTAL STUDY

The most common method to counteract the presence of shaft misalignment is to use a coupling, where the coupling is inserted between the driving and driven shafts. Couplings are chosen based on various factors such as environment, vibration and stiffness [34]. The two main types of coupling are: (1) metallic, such as, chain, gear and disc couplings, which have high stiffness and tolerance to extreme environments and (2) elastomeric, such as, pin and bushing, jaw and sleeve couplings, which are torsionally soft and have vibration damping/shock absorbing qualities. In this work a set of huco elastomeric d-loop couplings are used, due to a large range of permissible operating misalignment (10° angular misalignment and 2.6 mm offset misalignment) [35], allowing for a range of shaft misalignment to be examined.

Test Rig Design

To simulate a rotating drive train, an experimental test rig has been designed, as shown in Figure 3. The laser based measuring method is designed to be non-contact, so it can be easily applied to various types of rotating machinery. Laser distance measurements have engineering applications in areas such as vibrometry [36], coordinate measurement systems [37] and micro displacement measurement [38]. Laser mouse sensors have also been used for non-contact shaft speed measurements [39], as an alternative to electromagnetic sensors.

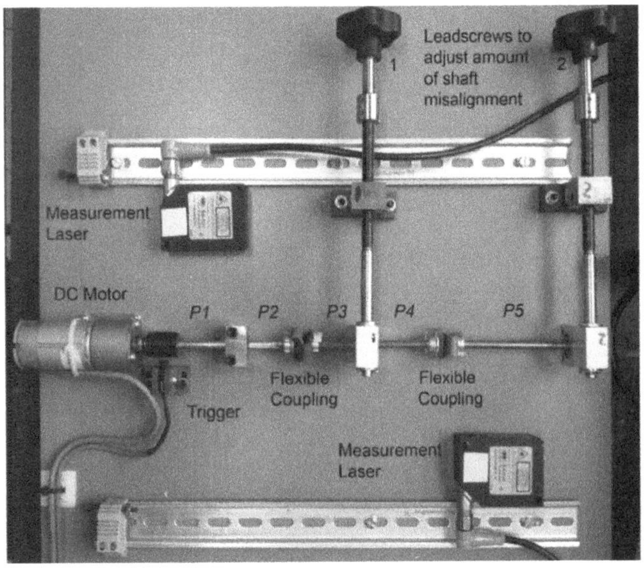

Figure 3. Experimental test rig.

A Baumer photoelectric distance measuring sensor with a measuring range of 30 mm to 130 mm, a resolution of 0.06 mm and a response time of less than 10 ms, is used to measure the amount of shaft misalignment present by measuring shaft positional changes (the distance from the laser to the shaft). The laser uses optical triangulation, where a pulsed red laser line is projected onto the surface of the shaft and part of the reflected light is measured using a photodiode detector, where the angle of incidence is used to calculate the distance.

The motor is used to rotate the shaft (~37 rpm) and an optoelectronic trigger is used to measure shaft rotation for ensemble averaging. The 6 mm diameter steel shaft is supported by support bearings that are attached to lead screws (~1.25 mm pitch) to give adjustable amounts of shaft misalignment (both offset and angular). Note that *P1–P5* are shaft misalignment measuring positions.

The first stage in a condition monitoring system is to acquire and process the data related to the condition of the machine [40]. The data acquisition is performed using a National Instruments USB-6218 card. The data acquisition software is developed using DAQmx drivers built into LabVIEW 2012. This software allows control of the motor speed and acquisition of the trigger signal and output from the two measurement lasers. Data from these measurements will allow the monitoring of any ongoing shaft misalignment from the changes in the distance measurement over time, an improvement over existing offline maintenance methods to monitor shaft misalignment [11]. The data is sampled at 300 Hz to avoid Nyquist issues with the 10 ms laser response time and 10 s of data are recorded. The data is then processed to give an estimation of real time shaft misalignment.

Shaft Misalignment Tests

- To investigate the monitoring of shaft misalignment using the experimental test rig, the following tests were performed:
- Test one: Increase misalignment using lead screw one by one revolution (offset measured at 1.46 mm).
- Test two: Increase misalignment using lead screw one by two revolutions (offset measured at 2.36 mm).
- Test three: Increase misalignment using lead screw two by one revolution (offset measured at 0.57 mm).
- Test four: Increase misalignment using lead screw two by two revolutions (offset measured at 1.18 mm).
- Test five: Increase misalignment using both lead screw one and lead screw two both by one revolution.
- Test six: Increase misalignment using both lead screw one and lead screw two both by two revolutions.
- Test seven: Continuously increase the amount of misalignment using lead screw one and measure as it changes dynamically over time.

Data Processing

The data is acquired as a voltage measurement from the lasers. It is converted to a distance using the relationship, *Distance (mm)* = *23.2* × *Voltage (V)* + *4.92*, from the line fitted to the measurement data shown in Figure 4. For example, a laser output of 1.5 V equates to a distance of 39.72 mm.

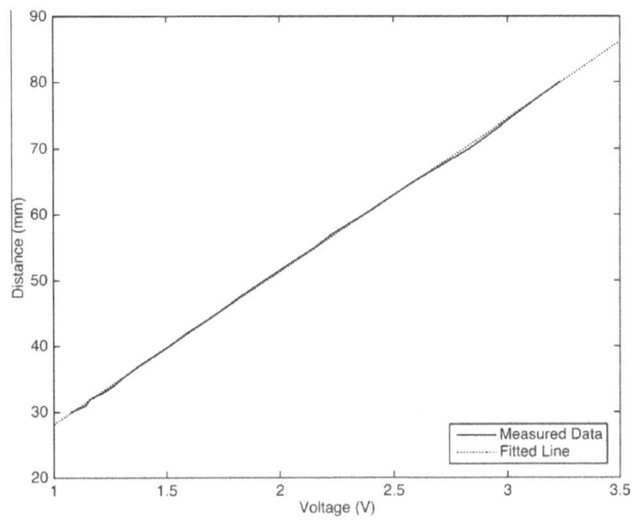

Figure 4. Laser distance calibration.

To remove noise from the data, the measurement is filtered using a weighted moving average filter from the LabVIEW Advanced Signal Processing Toolkit [41]. In this case, all of the data is read in and then the moving average filter operation is applied to smooth the data as shown in (1).

$$y(n) = a_0x(n) + a_1x(n+1) + \ldots + a_qx(n+q) = \sum_{r=0}^{q} a_r x(n+r), \quad (1)$$

where x is the vector of input data, y is the vector of filtered data and a_r is a set of user-defined weights. In this case a Henderson 23-term moving average filter is used [42], here the number of weights must be odd, the weight array must be symmetric and $\sum a_r=1$ as shown in (2).

$$\begin{aligned} y(n) = {} & -0.00428x(n) - 0.01092x(n+1) - 0.01569x(n+2) \\ & -0.01453x(n+3) - 0.00495x(n+4) \\ & +0.01343x(n+5) + 0.03893x(n+6) \\ & +0.06830x(n+7) + 0.09740x(n+8) \\ & +0.12195x(n+9) + 0.13832x(n+10) \\ & +0.14406x(n+11) + 0.13832x(n+12) \\ & +0.12195x(n+13) + 0.09740x(n+14) \\ & +0.06830x(n+15) + 0.03893x(n+16) \\ & +0.01343x(n+17) - 0.00495x(n+18) \\ & -0.01453x(n+19) - 0.01569x(n+20) \\ & -0.01092x(n+21) - 0.00428x(n+22) \qquad (2) \end{aligned}$$

The result of filtering the measurement data to smooth the data and remove outliers is shown in Figure 5. The dotted lines show the limit for outlier data (unusual values of data), in this case 1.5 times the interquartile range (middle 50% of the values in the data) [43].

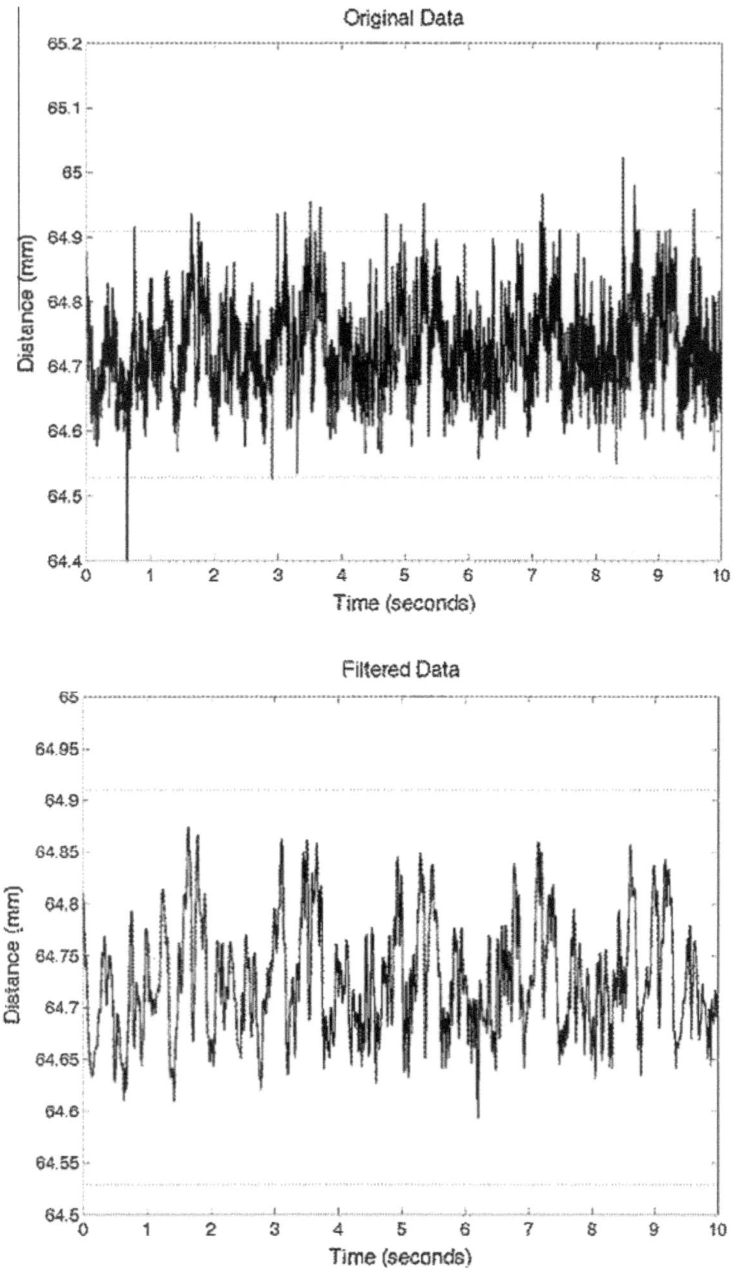

Figure 5. Filtered data.

The mean value will then be used for distance calculations, in the example in Figure 5, the mean of both the original data and filtered data is 64.72 mm, showing that in this case filtering removes extreme values with no change to the mean.

When the data has been processed the mean over the periodic results (the revolutions of the shaft) can be calculated using the ensemble average. The difference between time average and ensemble average is shown in Figure 6.

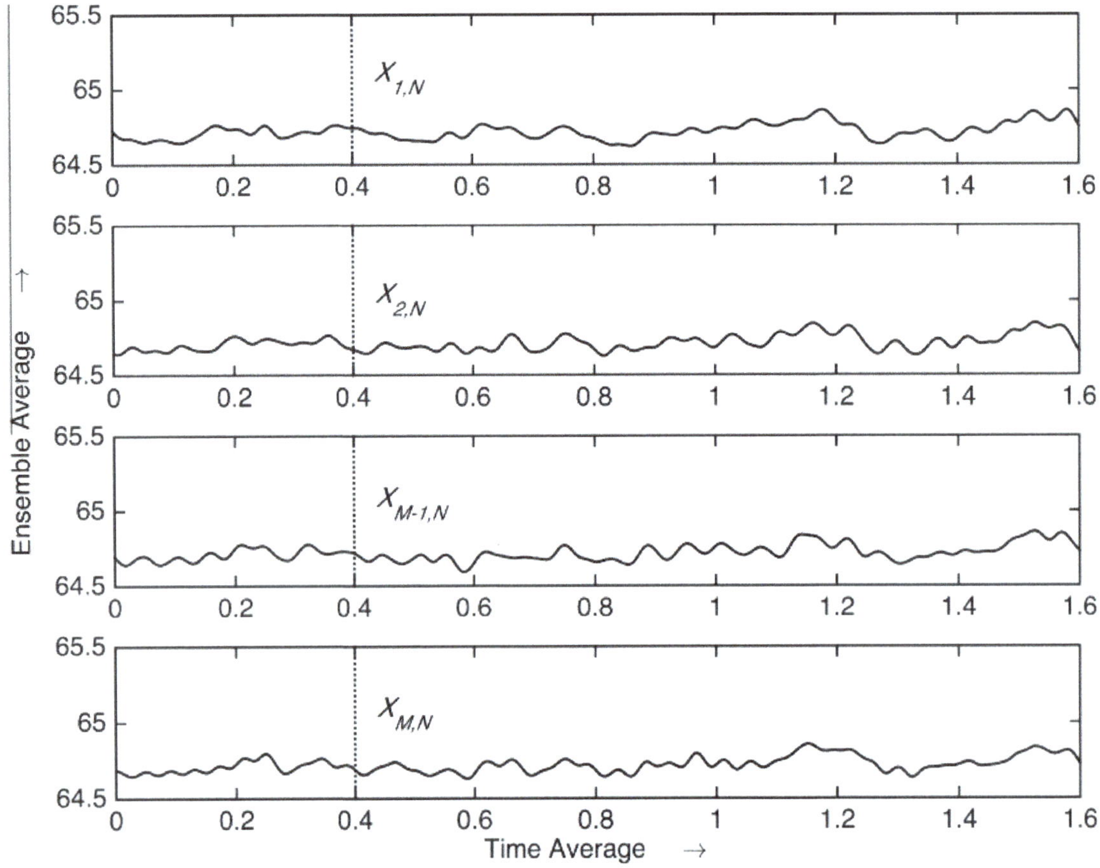

Figure 6. Types of averaging.

If $X_{i,j}$ is the *j*th time element in the *i*th revolution the ensemble can be represented using the matrix in (3) [44].

$$\text{revs} \downarrow \overset{\text{time} \atop \longrightarrow}{\begin{array}{ccccc} \mathbf{X}_{1,1} & \mathbf{X}_{1,2} & \cdots & \mathbf{X}_{1,N-1} & \mathbf{X}_{1,N} \\ \mathbf{X}_{2,1} & \mathbf{X}_{2,2} & \cdots & \mathbf{X}_{2,N-1} & \mathbf{X}_{2,N} \\ \vdots & \vdots & \ddots & \vdots & \vdots \\ \mathbf{X}_{M-1,1} & \mathbf{X}_{M-1,2} & \cdots & \mathbf{X}_{M-1,N-1} & \mathbf{X}_{M-1,N} \\ \mathbf{X}_{M,1} & \mathbf{X}_{M,2} & \cdots & \mathbf{X}_{M,N-1} & \mathbf{X}_{M,N} \end{array}} \quad (3)$$

With *N* columns of evenly spaced time samples (300 Hz = 3.33 ms) and *M* rows of time series for each shaft revolution. For the *i* th revolution it is possible to compute the time average denoted by $\overline{X}$, given by (4) [44].

$$\overline{X}_{i,j} = \frac{1}{N}\sum_{j=1}^{N} X_{i,j} \quad (4)$$

Similarly for the *j* th sample, the ensemble average, denoted by $\langle X \rangle$, can be calculated over all rotations given by (5) [44].

$$\langle X_{i,j} \rangle = \frac{1}{M}\sum_{i=1}^{M} X_{i,j} \quad (5)$$

To identify the start of each shaft revolution, a trigger signal is used as shown in Figure 7, where the start and end of each rotation is denoted with an "X".

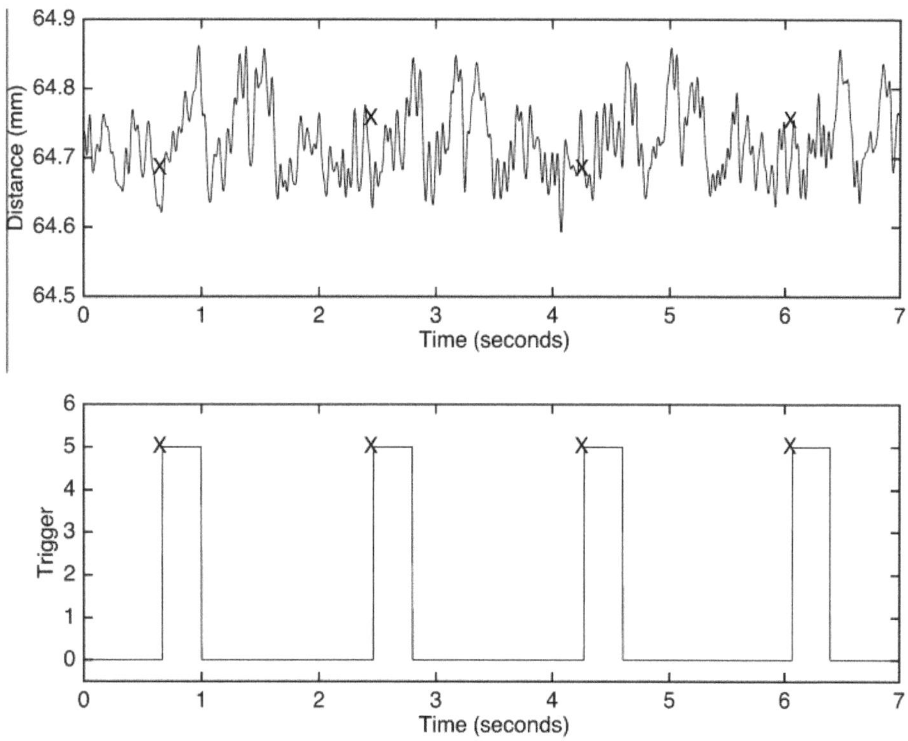

Figure 7. Trigger signals.

The triggering is based on an optoelectronic sensor measuring the change from a dark area to a light area on the shaft as it rotates. In this case, each shaft rotation takes 1.6 s (37.5 rpm). The data for each of the shaft revolutions acquired over the period can be separated as shown in Figure 8.

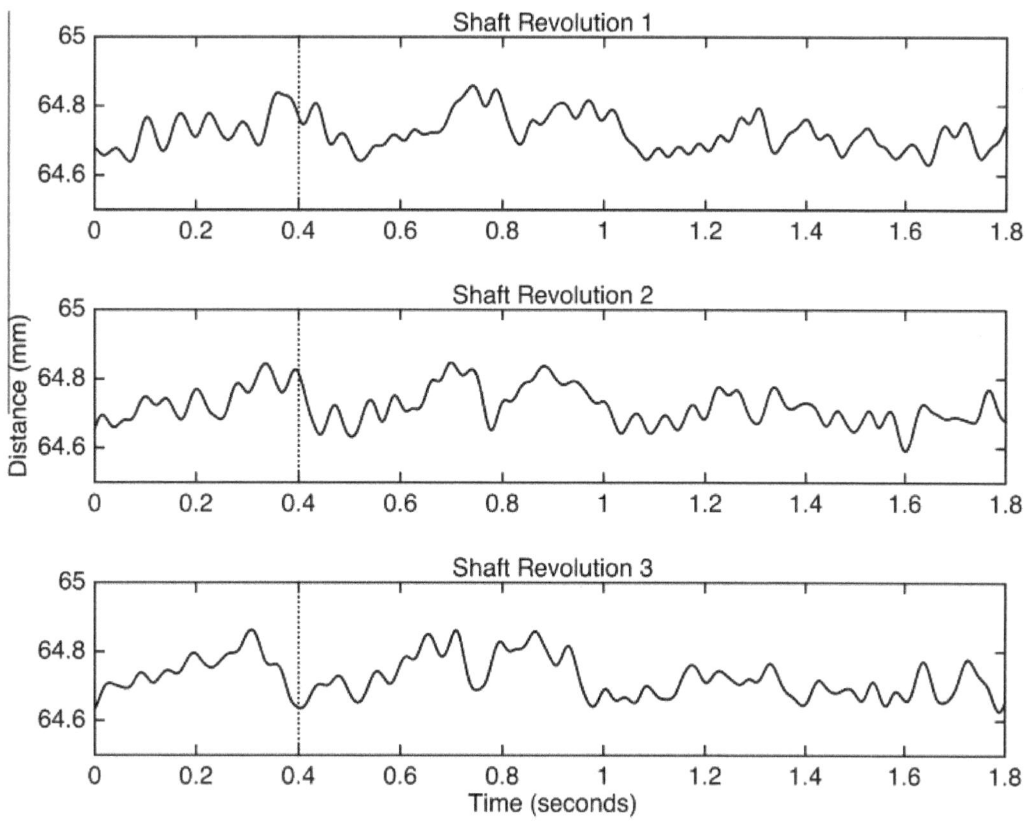

Figure 8. Shaft revolution measurements.

Finally, the ensemble average is performed to calculate the average measurement per shaft revolution, for example the average value of the three points on each revolution at time 0.4 s in Figure 8, would give the ensemble average value at 0.4 s (denoted by “X” in Figure 9), reducing the multiple periods of data for each shaft revolution to a single period.

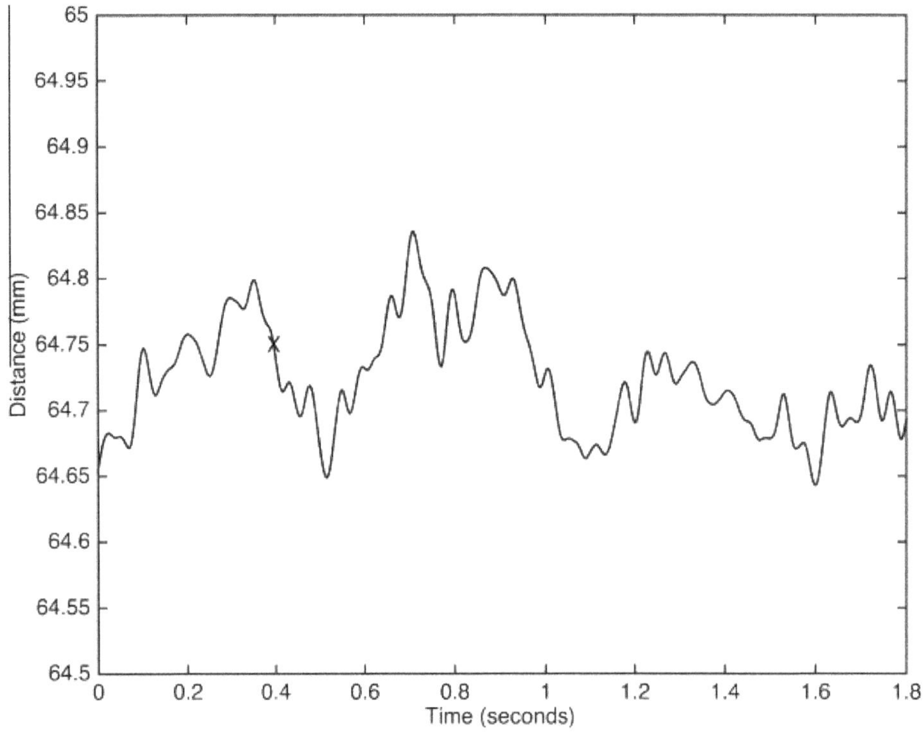

Figure 9. Shaft revolution ensemble average.

The mean of the ensemble-averaged data, the measurement over a single revolution, is then used to calculate the distance from the laser to the shaft, 64.72 mm in Figure 9. The difference in the mean value from a baseline (shaft aligned) value will then be used to estimate the amount of shaft misalignment as shown in Figure 10.

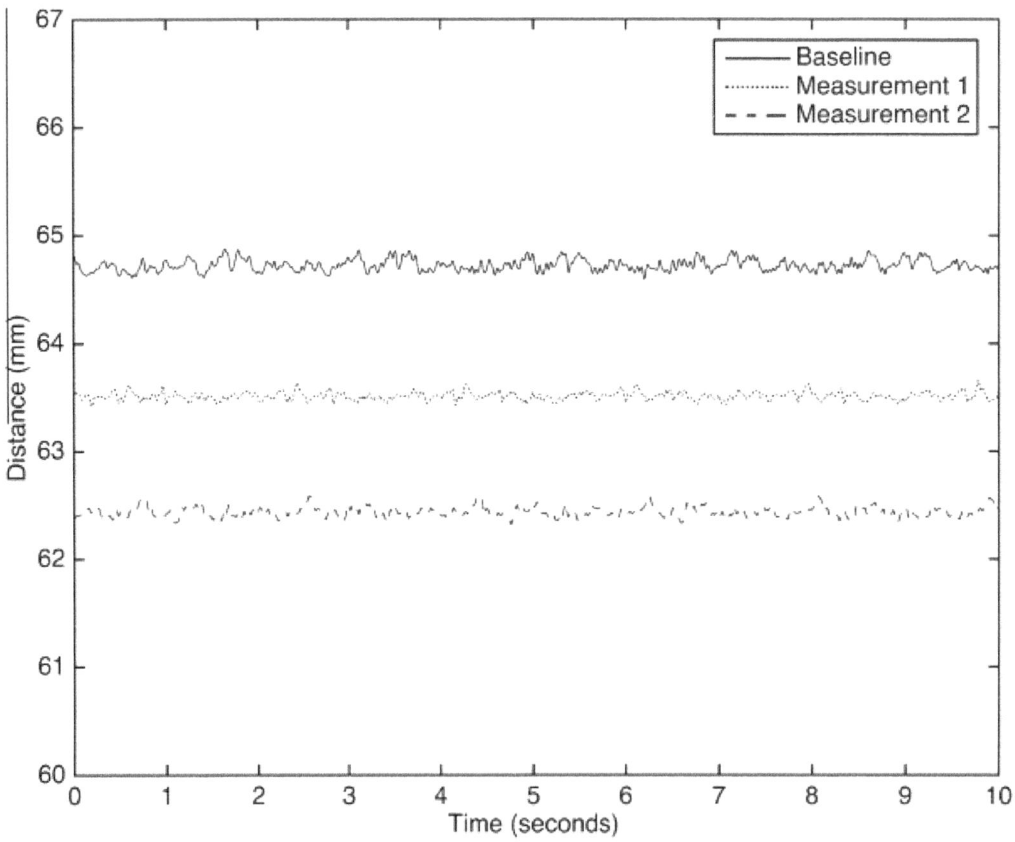

Figure 10. Measurement change from baseline value.

The measurement can be converted to cylindrical coordinates and displayed as the distance to the surface of the 6 mm diameter shaft as shown in Figure 11.

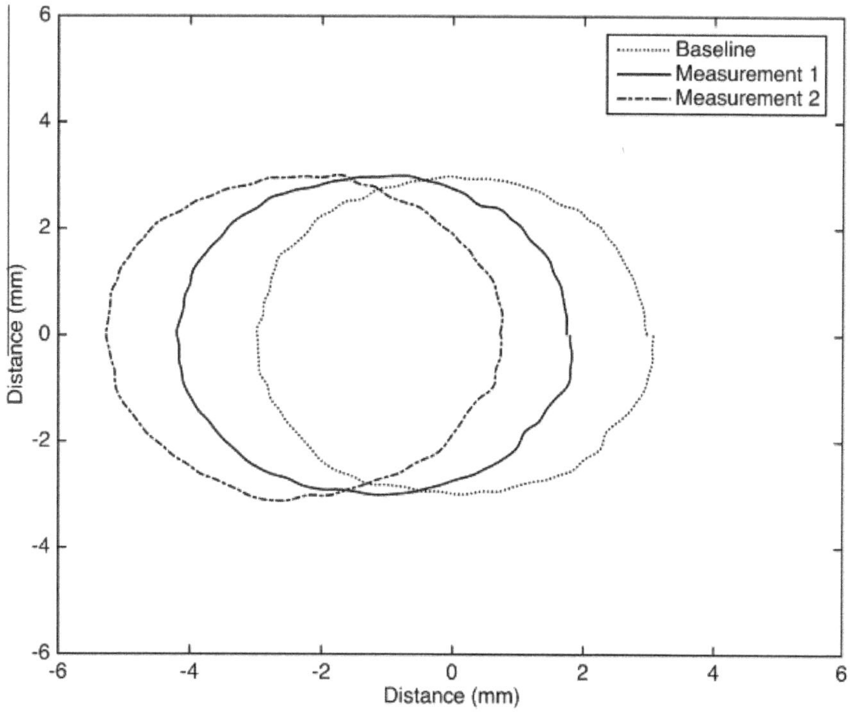

Figure 11. Shaft change from baseline value.

The difference in shaft centers show the amount of shaft misalignment and fluctuations in shaft circumference are variations in the rotation. The negative values indicate that the misalignment is towards the position of the measurement laser (the distance is decreasing).

Results

After a baseline measurement was performed (on the aligned shaft), the level of shaft misalignment was increased by adjusting the lead screws on the test rig and the distance to the shaft was measured using the distance measurement lasers.

The results for the shaft misalignment tests will now be given. The results for misalignment test 1 and misalignment test 2 (measured at position *P3* in Figure 3) are shown in Figure 12 and the results for misalignment test 3 and misalignment test 4 (measured at position *P5* in Figure 3) are shown in Figure 13.

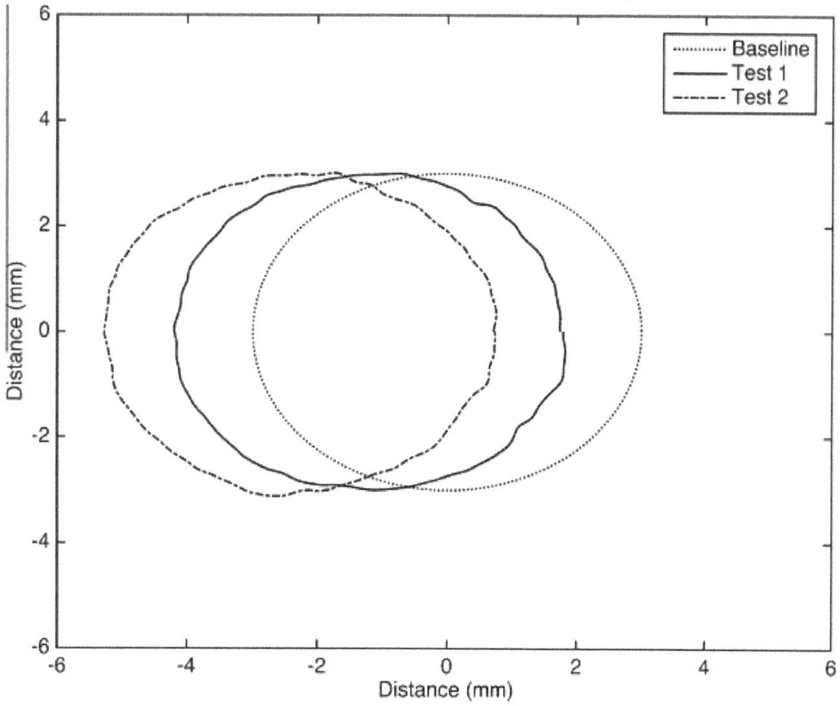

Figure 12. Shaft position change for misalignment test 1 and misalignment test 2.

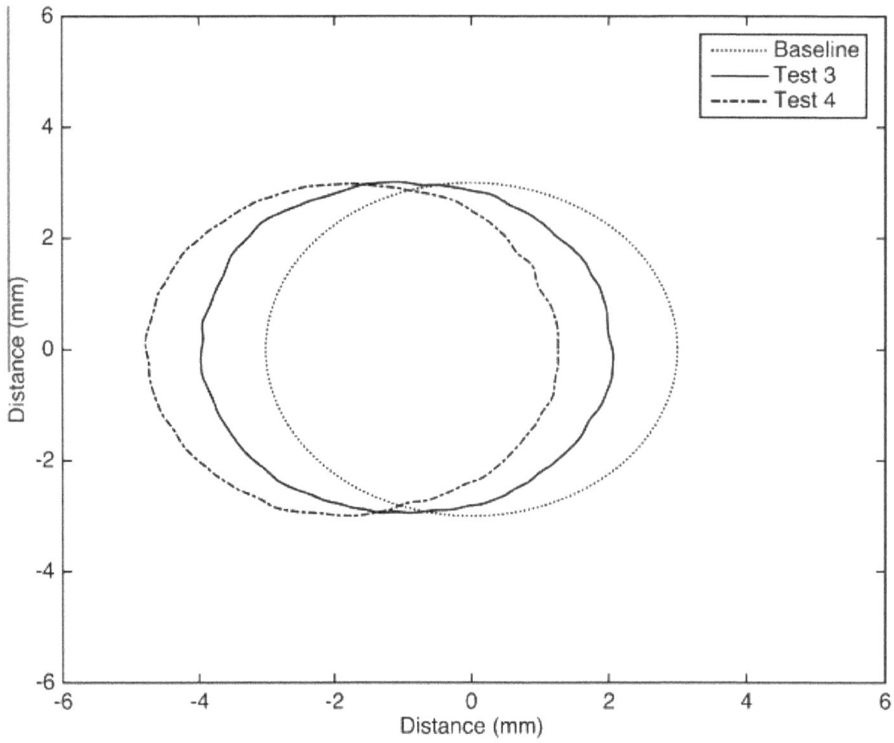

Figure 13. Shaft position change for misalignment test 3 and misalignment test 4.

The results from these tests show that the increasing amounts of shaft misalignment can be identified, as the circles are shifted from the baseline value. As well as measuring the position at a single point on the shaft, by measuring the amount of misalignment at various positions on the shaft (moving a single distance measurement laser or using multiple distance measurement lasers), the type of shaft misalignment as shown in Figure 1can be identified. The results for misalignment test 1 and misalignment test 2 (offset misalignment) are shown in Figure 14 and the results for misalignment test 3 and misalignment test 4 (angular misalignment) are shown in Figure 15.

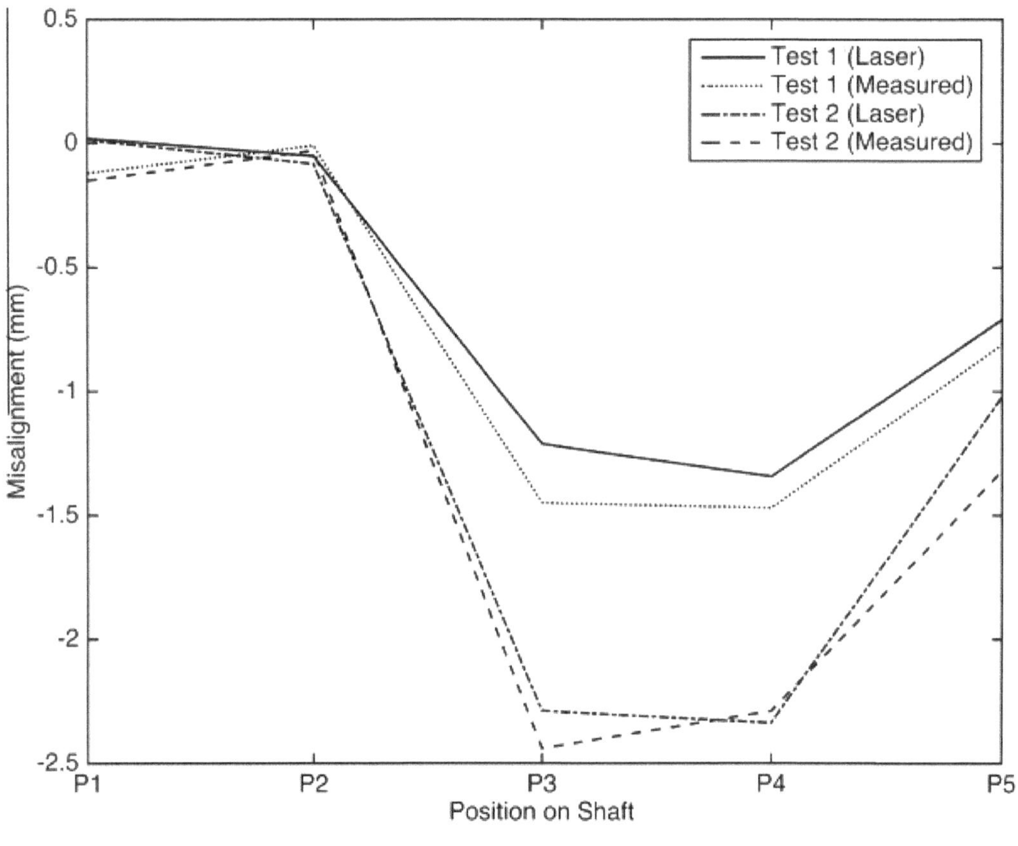

Figure 14. Shaft misalignment results for misalignment test 1 and misalignment test 2 (offset misalignment).

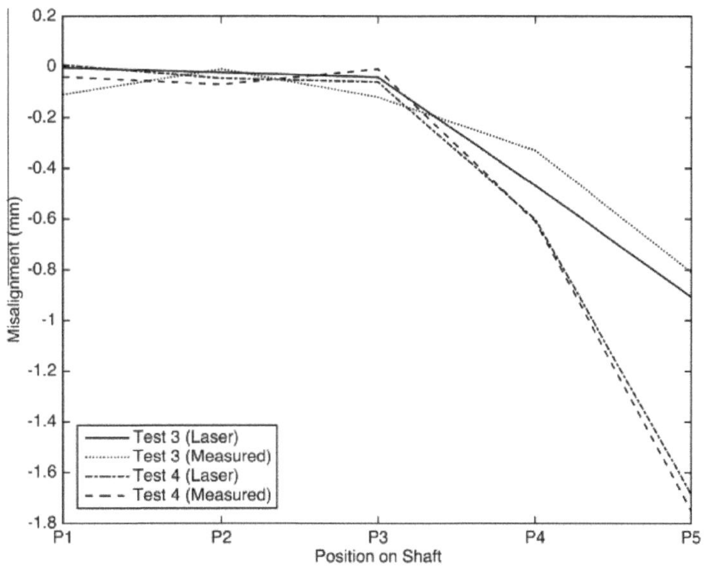

Figure 15. Shaft misalignment results for misalignment test 3 and misalignment test 4 (angular misalignment).

Figure 14 shows that the area of the shaft between positions *P3* and *P4*, attached to the first leadscrew, has been offset from the rest of the shaft as the amount of misalignment has been increased. Figure 15 shows that the angle of position *P5* has been increased as the second leadscrew is adjusted.

The distance measurements from the laser are also compared against manual distance measurements made using an electronic vernier caliper, as shown in Table 1, where both the measurement error and percentage error (in brackets) are given.

Table 1. Error between laser measurement and manual measurement of shaft misalignment

Test	*P1* (mm)	*P2* (mm)	*P3* (mm)	*P4* (mm)	*P5* (mm)
Misalignment test 1	−0.14 (116%)	0.04 (419%)	−0.24 (16%)	−0.13 (9%)	−0.10 (12%)
Misalignment test 2	−0.16 (108%)	0.05 (177%)	−0.15 (6%)	0.05 (2%)	−0.30 (23%)
Misalignment test 3	−0.11 (96%)	0.01 (135%)	−0.08 (66%)	0.14 (42%)	0.10 (12%)
Misalignment test 4	−0.05 (117%)	−0.02 (34%)	0.05 (500%)	−0.01 (1%)	−0.06 (4%)

As well as studying offset and angular types of shaft misalignment individually, the more realistic shaft misalignment can be studied by adjusting both leadscrews simultaneously, in misalignment test 5 and misalignment test 6, to add both types of shaft misalignment as shown in Figure 16. For comparison, the results of adding the individual amounts of offset and angular shaft misalignments are shown, indicating good agreement with this test result.

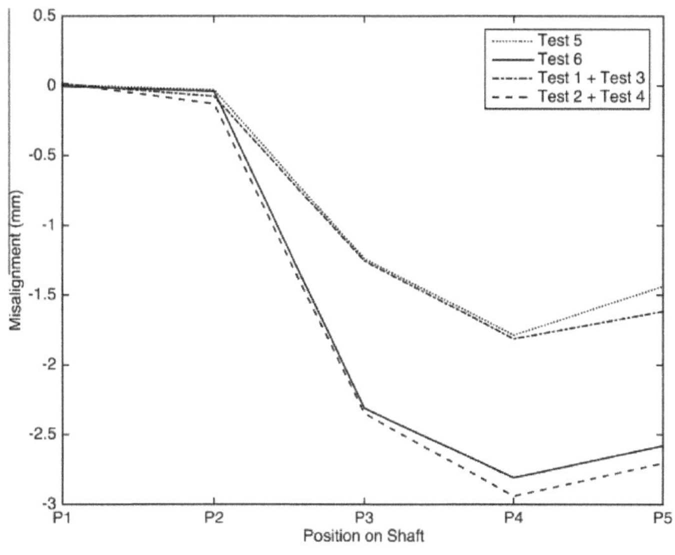

Figure 16. Shaft misalignment results for misalignment test 5 and misalignment test 6 (offset and angular misalignment combined).

These measurements give a visual representation of both the type and amount of shaft misalignment. The distance measurements can be used to perform the required machine movements to return the systems to the baseline aligned condition. In addition, in condition monitoring applications, the measurements can be used to trigger an alarm if they exceed a threshold value.

As well as measuring static values of shaft misalignment, the level of shaft misalignment was recorded as it was being adjusted over time (misalignment test 7). The result for the measured shaft misalignment (measured at position *P3* in Figure 3) is shown in Figure 17 and the corresponding shaft position changes over time (at one, two and three seconds) is shown in Figure 18.

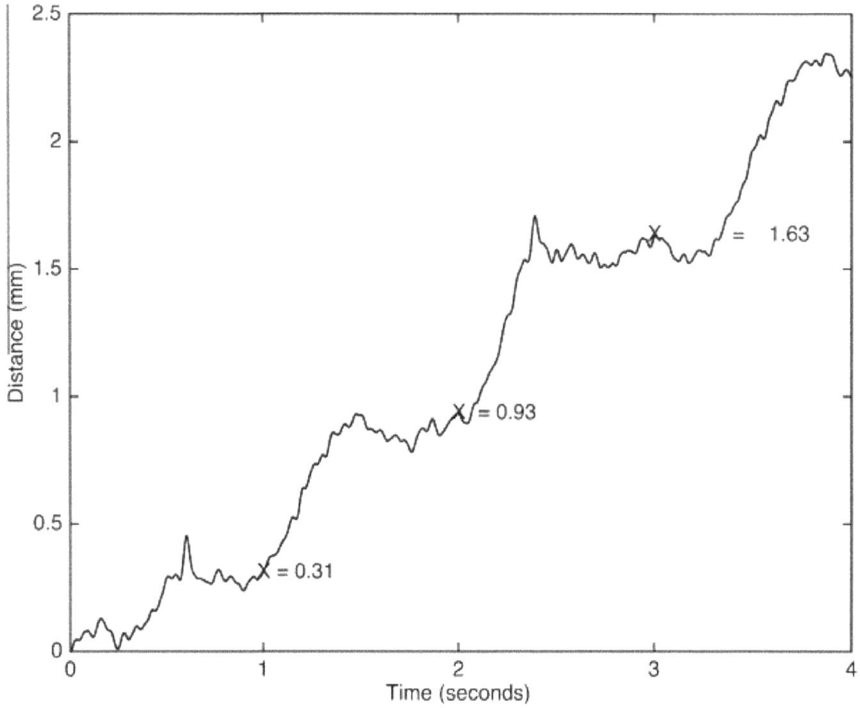

Figure 17. Shaft misalignment change over time for misalignment test 7.

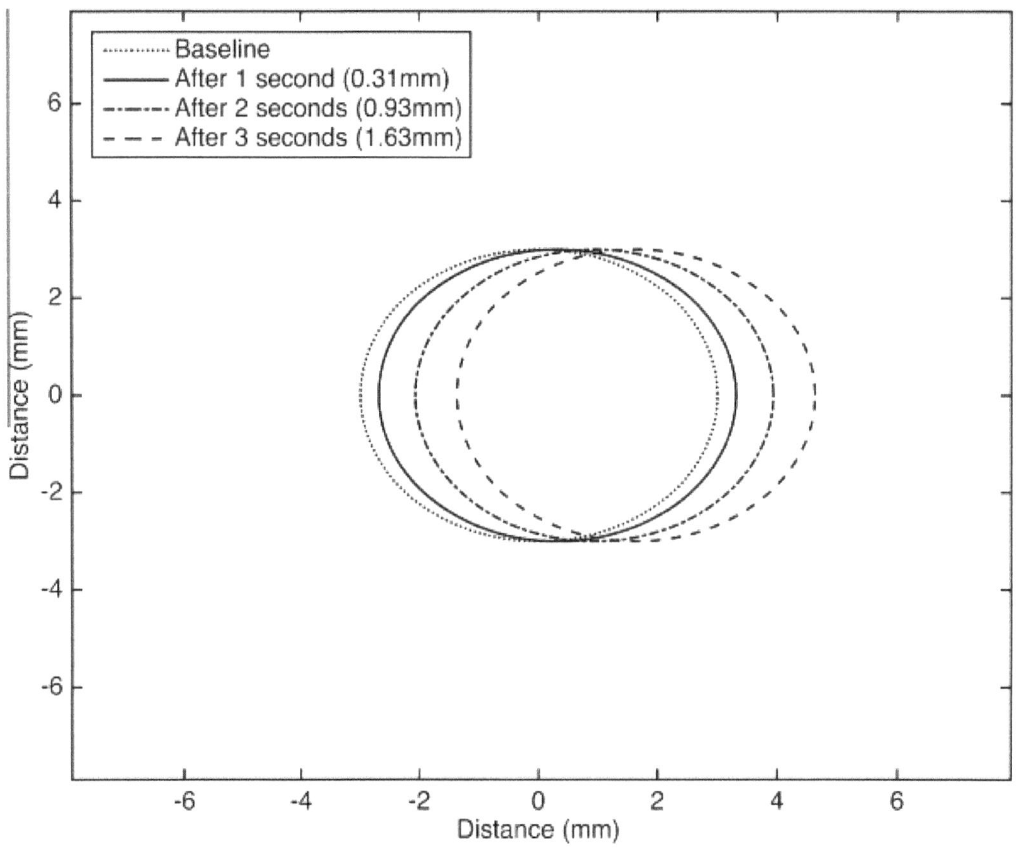

Figure 18. Shaft position change over time for misalignment test 7.

This shows the main advantage over existing laser based shaft misalignment measurement methods, this is a non-contact method and it can be used while the rotating machine is in operation (unlike existing shaft misalignment measuring tools, which are attached to the shaft so the machine should be stopped). This on-line measurement procedure is a requirement for condition monitoring, as it can detect changes in shaft misalignment over time.

CONCLUSIONS AND FUTURE WORK

A requirement exists in industry for rapid and reliable techniques to measure shaft misalignment, so that the condition of the rotating machines can be monitored and the shafts can be adjusted to achieve proper alignment. To achieve this, this work has presented a novel laser based measurement method for shaft misalignment monitoring. This uses a commercially available laser distance measuring device to measure the change in distance to a rotating shaft caused by the presence of shaft misalignment. This has advantages over existing misalignment measurement methods of being non-contact and suitable for on-line operation. Several misalignment tests have been performed to demonstrate the feasibility of the technique. Variations in misalignment have been identified as well as identification of the two common types of misalignment (offset and angular) and the measurement of changes in shaft misalignment over time.

This technique has been used to measure shaft misalignment distances of between 0.5 mm and 2.5 mm on an experimental test rig. These are larger than practical shaft misalignment tolerances, which are usually based on the rotational speed of the machine, varying between 0.02 mm and 0.3 mm [45], future work could be performed to measure alignment at this greater precision with higher specification laser measurement equipment (the resolution of the current laser measurement system is 0.06 mm) as the feasibility of the technique has been demonstrated. It would also be useful to extend this work to data acquisition over long periods and machine diagnostics based on the acquired shaft misalignment measurements, plus investigation of the system performance in harsh environments.

ACKNOWLEDGEMENTS

This work was supported by UK Engineering and Physical Sciences Research Council(EPSRC) Center for Through-life Engineering Services (EP/1033246/1, Project SC006) and EPSRC Impact Acceleration Award (EP/K50336811), Royal Academy of Engineering Newton Collaborative Research Programme (NRCP 1415 91) and Royal Academy of Engineering Distinguished Visiting Fellowship (DVF 1415 2 17).

REFERENCES

1. R.K. Mobley, Maintenance Fundamentals, Elsevier Inc., 2004.
2. S. Ebersbach, Artificial intelligence system for integrated wear debris and vibration analysis in machine condition monitoring, School of Engineering, James Cook University, 2007.
3. J.K. Sinha, K. Elbhbah, A future possibility of vibration based condition monitoring, Mech. Syst. Signal Process. 34 (2013) 231– 240.
4. Y. Lei, J. Lin, Z. He, M.J. Zuo, A review on empirical mode decomposition in fault diagnosis of rotating machinery, Mech. Syst. Signal Process. 35 (2013) 108–126.
5. P. Scanlon, D.F. Kavanagh, F.M. Boland, Residual life prediction of rotating machines using acoustic noise signals, IEEE Trans. Instrum. Meas. 62 (2013) 95–108.
6. P. Tavner, L. Ran, J. Penman, H. Sedding, Condition Monitoring of Rotating Electrical Machines, The Institution of Engineering and Technology, London, 2008.
7. Y. Amirat, M.E.H. Benbouzid, E. Al-Ahmar, B. Bensaker, S. Turri, A brief status on condition monitoring and fault diagnosis in wind energy, Renew. Sustain. Energy Rev. 13 (2009) 2629–2636.
8. J.F. Manwell, J.G. McGowan, A.L. Rogers, Wind Energy Explained Theory, Design and Application, John Wiley & Sons Ltd, Chichester, 2009.
9. G.L. Garrad Hassan, in: G.L. Garrad Hassan (Ed.), Offshore Wind – Operations & Maintenance opportunities in Scotland – An insight into opportunities for Scottish ports and the O&M supply chain, Scottish Enterprise, 2013.
10. T.H. Patel, A.K. Darpe, Experimental investigations on vibration response of misaligned rotors, Mech. Syst. Signal Process. 23 (2009) 2236–2252.
11. J. Piotrowski, Shaft Alignment Handbook, CRC Press, 2006.
12. Y. Fang, H. Cho, M. Jeong, Health monitoring of a shaft transmission system via hybrid models of PCR and PLS, in: J. Ghosh, D. Lambert, D. Skillicorn, J. Srivastava (Eds.), Sixth SIAM International Conference on Data Mining, Bethesda, Maryland, 2006.
13. A. Kr Jalan, A.R. Mohanty, Model based fault diagnosis of a rotorbearing system for misalignment and unbalance under steady-state condition, J. Sound Vib. 327 (2009) 604–622.
14. L. Arebi, F. Gu, A. Ball, A comparative study of misalignment detection using a novel Wireless Sensor with conventional Wired Sensors, in: 25th International Congress on Condition Monitoring and Diagnostic Engineering, IOP Publishing, Huddersfield, UK, 2012.
15. J.M. Vance, Rotordynamics of Turbomachinery, John Wiley & Sons, Inc., 1988.
16. D.L. Dewell, L.D. Mitchell, Detection of a misaligned disk coupling using spectrum analysis, J. Vib. Acoust. Stress Reliability Des. 106 (1984) 9–16.
17. M. Xu, R.D. Marangoni, Vibration analysis of a motor-flexible coupling-rotor system subject to misalignment and unbalance, Part I: theoretical model and analysis, J. Sound Vib. 176 (1994) 663–679.
18. M. Xu, R.D. Marangoni, Vibration analysis of a motor-flexible coupling-rotor system subject to misalignment and unbalance, Part II: experimental validation, J. Sound Vib. 176 (1994) 681–691.

19. P.N. Saavedra, D.E. Ramírez, Vibration analysis of rotors for the identification of shaft misalignment Part 2: experimental validation, Proc. Inst. Mech. Eng., Part C: J. Mech. Eng. Sci. 218 (2004) 987–999.
20. S.B. Chaudhury, S. Gupta, Online identification of AC motor misalignment using current signature analysis and modified Kmean clustering technique, in: IEEE International Conference on Industrial Technology, 2006, Mumbai, 2006, pp. 2331–2336.
21. A.K. Verma, S. Sarangi, M.H. Kolekar, Shaft misalignment detection using stator current monitoring, Int. J. Adv. Comput. Res. 3 (2013) 305–309.
22. W.T. Thomson, M. Fenger, Current signature analysis to detect induction motor faults, IEEE Ind. Appl. Mag. 7 (2001) 26–34.
23. J.M. Bossio, G.R. Bossio, C.H. De Angelo, Angular misalignment in induction motors with flexible coupling, in: 35th Annual Conference of IEEE Industrial Electronics, 2009, Porto, 2009, pp. 1033–1038.
24. G.R. Rameshkumar, B.V.A. Rao, K.P. Ramachandran, Coast down time analysis to analyze the effect of misalignment in rotating machinery, Int. J. Eng. Adv. Technol. 1 (2012) 149–156.
25. S.M. Kim, J.H. Suh, J.S. Im, S.B. Kim, S.B. Kim, A smart memory type of data acquisition system for shaft misalignment maintenance, J. Mech. Sci. Technol. 19 (2005) 15–27.
26. M.A. Hotait, D. Talbot, A. Kahraman, An investigation of the influence of shaft misalignment on bending stresses of helical gears with lead crown, Gear Technol. (2008) 54–62.
27. A.G. Fulzele, V.G. Arajpure, P.P. Holay, N.M. Patil, Condition monitoring of shaft of single-phase induction motor using optical sensor, Mech. Syst. Signal Process. 29 (2012) 428–435.
28. A.S. Sekhar, B.S. Prabhu, Effects of coupling misalignment on vibrations of rotating machinery, J. Sound Vib. 185 (1995) 655–671.
29. H.-W. Cho, M.K. Jeong, Enhanced prediction of misalignment conditions from spectral data using feature selection and filtering, Expert Syst. Appl. 35 (2008) 451–458.
30. W. Yang, P.J. Tavner, Empirical mode decomposition, an adaptive approach for interpreting shaft vibratory signals of large rotating machinery, J. Sound Vib. 321 (2009) 1144–1170.
31. J. Giannone, Laser-optical shaft alignment cuts energy costs, World Pumps 1995 (1995) 22–24.
32. V. Hariharan, P.S.S. Srinivasan, Vibration analysis of misaligned shaft-ball bearing system, Indian J. Sci. Technol. 2 (2009) 45–50.
33. R.B. McMillan, Rotating Machinery: Practical Solutions to Unbalance and Misalignment, The Fairmont Press, Inc., 2004.
34. M.M. Hodowanec, Effects of coupling installation on motor performance, IEEE Ind. Appl. Mag. (1997) 70–77.
35. Huco Engineering Industries Ltd, Flex-P Double Loop Couplings with Stainless Steel Hubs.
36. M.-A. Beeck, W. Hentschel, Laser metrology – a diagnostic tool in automotive development processes, Opt. Lasers Eng. 34 (2000) 101–120.
37. I. Besic, N. van Gestel, J.-P. Kruth, P. Bleys, J. Hodolic, Accuracy improvement of laser line scanning for feature measurements on CMM, Opt. Lasers Eng. 49 (2011) 1274–1280.
38. S.H. Wang, C.J. Jin, C.J. Tay, C. Quan, H.M. Shang, Design of an optical probe for testing surface roughness and micro-displacement, Precis. Eng. 25 (2001) 258–265.
39. P. Cheng, M.S.M. Mustafa, B. Oelmann, Contactless rotor RPM measurement using laser mouse sensors, IEEE Trans. Instrum. Meas. 61 (2012) 740–748.
40. A.K.S. Jardine, D. Lin, D. Banjevic, A review on machinery diagnostics and prognostics implementing condition-based maintenance, Mech. Syst. Signal Process. 20 (2006) 1483–1510.
41. National Instruments Corporation, TSA Moving Average VI, LabVIEW Advanced Signal Processing Toolkit, 2015.
42. R. Henderson, Note on graduation by adjusted average, Trans. Actuarial Soc. Am. 17 (1916) 43–48.
43. S. Boslaugh, P.A. Watters, Statistics in a Nutshell, O'Reilly Media, Inc., Sebastopol, CA, 2008.
44. R.M. Chervin, Estimates of first- and second-moment climate statistics in GCM simulated climate ensembles, J. Atmos. Sci. 37 (1980) 1889–1902.
45. Pruftechnik Ltd, A Practical Guide to Shaft Misalignment, Pruftechnik, 2002.

CHAPTER 8

ENABLING THE MEASUREMENT OF PARTICLE SIZES IN STIRRED COLLOIDAL SUSPENSIONS BY EMBEDDING DYNAMIC LIGHT SCATTERING INTO AN AUTOMATED PROBE HEAD

Martinus de Kanter [a], Julian Meyer-Kirschner [b], Jörn Viell [b], Alexander Mitsos [b], Michael Kather [c], Andrij Pich [c], Christoph Janzen [d]

[a] Chair for Laser Technology, RWTH Aachen University, Aachen, Germany

[b] Aachener Verfahrenstechnik – Process Systems Engineering, RWTH Aachen University, Aachen, Germany

[c] Institute of Technical and Macromolecular Chemistry, RWTH Aachen University and DWI Leibniz Institute for Interactive Materials e.V., Aachen, Germany

[d] Fraunhofer Institute for Laser Technology (ILT), Aachen, Germany

ABSTRACT

A novel probe head design is introduced, which enables in-line monitoring of particle sizes in undiluted stirred fluids using dynamic light scattering. The novel probe head separates a small sample volume of 0.65 ml from the bulk liquid by means of an impeller. In this sample volume, particle sizing is performed using a commercially available fiber-optical backscatter probe. While conventional light scattering measurements in stirred media fail due to the superposition of Brownian' motion and forced convection, undistorted measurements are possible with the proposed probe head. One measurement takes approximately 30 s used for liquid exchange by rotation of the impeller and for collection of scattered light. The probe head is applied for in-line monitoring of the particle growth during microgel synthesis by precipitation polymerization in a one liter laboratory reactor. The in-line measurements are compared to off-line measurements and show a good agreement.

INTRODUCTION

Measuring particle sizes and size distributions in colloidal suspensions is of great importance in many technical processes. These processes are necessary for the production of polymers, pharmaceutical active ingredients and additives for the food-, cosmetic- and paint-industry [1]. For a better process monitoring of these technical processes, online measurements of the particle size without time-consuming sampling are needed.

Dynamic light scattering (DLS) is one of the standard methods for measuring particle sizes in fluids and has been established for many years [2]. This method is based on the examination of random particle movement due to constant Brownian motion. The collision of particles with surrounding liquid molecules results in a diffusional process with small particles moving faster than large particles. To monitor this diffusion, the sample is illuminated with a monochromatic laser beam. Depending on the position of the particles relative to each other, light scattered by the particles undergoes constructive or destructive interference. The resulting intensity fluctuations are detected time-resolved by a photomultiplier

Citation: Martinus de Kanter, Julian Meyer-Kirschner, Jörn Viell, Alexander Mitsos, Michael Kather, Andrij Pich, Christoph Janzen, Enabling the measurement of particle sizes in stirred colloidal suspensions by embedding dynamic light scattering into an automated probe head, Measurement, Volume 80, February 2016, Pages 92-98, ISSN 0263-2241, http://dx.doi.org/10.1016/j.measurement.2015.11.024.

positioned at a certain angle to the incident light. The decay of the autocorrelation of the measured intensity is correlated to the translational diffusion coefficient, which in turn is used to calculate the hydrodynamic radius of the particles using the Stokes–Einstein equation. For accurate determination of the diffusion coefficient, multiple light scattering has to be avoided. In concentrated suspensions, which are typical in industrial polymer production, the incident light is scattered multiple times before it is detected, which directly influences the intensity autocorrelation. To circumvent this issue, there are different enhancements of DLS, such as two color-cross correlation [3] and [4] or 3D cross correlation [5]. Another demand for industrially applicable particle sizing is the possibility of direct in-line measurements. While able to measure in concentrated suspensions, both cross-correlation DLS methods involve complex optical setups and are not suitable for in-line measurements. In contrast, fiber-optic-quasi-elastic light scattering (FOQELS) [6] or fiber optic dynamic light scattering (FODLS) [7] methods use an immersion probe with a simple and robust optical setup and are therefore ideally suited for in-line applications. Both techniques are fiber-coupled and collect the backscattered light for particle size determination. With small penetration depths, concentrated dispersions with high solid content up to 40 wt% [6] can be measured [8]. Nevertheless, in-line DLS measurements are challenging, since laboratory as well as industrial reactors typically are stirred. Accurate particle sizing using DLS necessitates a resting fluid in the measured sample. In actively-mixed fluids the diffusion is overlaid by a turbulent convection, which prohibits diffusion measurements. Hence, it is not possible to apply DLS in fluids that exhibit forced convection either due to stirring or even due to sufficiently large pressure- or temperature gradients.

Apart from DLS, there are several other methods to determine the particle size in fluids[9], such as angular resolved static light scattering (SLS) of particles. This method is capable of sizing of particles with diameters of at least a size several tens of nanometers. To avoid multiple scattering, SLS measurements are usually performed in highly-diluted samples Therefore, SLS is not applicable for in-line monitoring of samples with high solids content. Also commonly used for size determination of industrial polymers or macromolecules is size exclusion chromatography (SEC), a method that involves sample drawing and uses small flow rates [10] and hence is not suitable for in-line monitoring.

In practice, the aforementioned methods are used either off-line by taking samples from the production line to the laboratory or by installing fully automated on-line trains in bypasses or loops [11]. Sampling and further sample preparation by dilution or separation are time-consuming and induce a delay of up to several minutes between sampling and size information. In contrast, in-line measurements can provide real-time data or only have a delay of a few seconds, depending on the applied method, facilitating possible improvements on process- and quality control. Hence, a particle sizing method capable of measuring in-line in undiluted and stirred suspensions and enabling a close to real-time monitoring of technical processes is desired.

By measuring the bulk turbidity of a sample, one can calculate the particle size for monodisperse particles with known optical constants [12]. Since these optical constants can vary during polymerization processes, turbidity measurements are not sufficient for reliable particle size measurements. Optical imaging methods have also been applied in combination with in-line probes in stirred vessels [13]. These methods are based on capturing images using a CCD camera, and as such restricted by the optical diffraction limit and cannot detect nanometer sized particles. A promising approach for in-line particle sizing of stirred turbid colloidal suspensions is Photon Density Wave (PDW) spectroscopy [14]. Using a special in-line probe, intensity-modulated light is detected at multiple distances between excitation and collection fiber. Since the method relies on strong multiple scattering, it can only be applied for dispersions exhibiting at least a certain level of turbidity [15] and hence cannot be used to monitor the initial stage of particle growth in processes that still exhibit single scattering.

This paper introduces a novel probe head to enable in-line FOQELS measurements in stirred colloidal suspensions. To achieve this, the novel probe head and the probe of a commercial DLS device were mounted together and then immersed directly into the reactor.

A possible application of such an in-line particle sizing method is monitoring of microgel particle growth in precipitation polymerization process. Microgels are soft colloidal particles formed by cross-linked polymer chains [16]. The particle growth during precipitation polymerization depends on the reaction conditions like concentration of main reagents (monomer(s), crosslinker, surfactant and initiator) as well as reaction temperature. Higher polymerization temperatures and initiator concentrations result in a faster particle growth. A typical microgel synthesis takes between 5 and 20 min from initiation to final conversion [17]. For highly diluted syntheses at low temperatures, the growth of microgel particles inside a cuvette was successfully monitored using static light scattering [18]. Kara et al. conducted microgel syntheses inside quartz cells at room temperature [19] and [20] and for temperatures up to 50°C [21]. Placing the cell inside a UV–VIS spectrometer, they observed a decrease in transmitted intensity during the gelation process. This decrease was attributed to an increase in light scattered by the particles. The scattered light intensity was correlated to the particle volume using Rayleigh's equation, which is only valid for particles smaller than 0.1 of the laser wavelength [19]. To monitor microgel particle growth at lab production scale in undiluted and stirred systems without being restricted to particles smaller than 40–70 nm [19], this method is not suitable.

The remaining article is structured as follows. In the experimental section the design of the probe head is presented together with commercially available DLS devices used for the in-line measurements as well as off-line validation measurements. Furthermore the experimental setups involving the probe head are described, which consist of stirred and unstirred measurements of a colloidal suspension using particles of a fixed size as well as the application of the probe head for in-line monitoring of the particle growth during microgel syntheses. The result section reports the particles sizes measured using the probe head for both experimental setups along with the off-line reference measurements taken during particle synthesis.

MATERIAL AND METHODS

Particle Metrix/Microtrac Nanotrac 250

The commercial device, which is equipped with the novel probe head is a Nanotrac 250 (Particle Metrix/Microtrac Europe GmbH, Germany). The measurement principle is shown in Figure 1. This device is factory-equipped with an immersion probe, which allows an easy access to the sample due to the probe's small form factor. The light emitted by a laser (780 nm) is guided to the optical probe by a fiber and is focused into the sample close to the protective sapphire window, allowing measurements in highly concentrated suspensions of up to 40 wt% solids. The 180° backscattered light is collected and guided through a fiber to a detector. The laser light reflected by the protective window is also guided to the detector. The scattered light of the particles and the reflected light from the protective window are overlaid on the detector and cause interferences. The time-resolved intensity fluctuations correlate with the motion of the particles from which the particle size distribution is calculated by suitable mathematical models [22]. This device covers a particle size range from 0.8 nm to 6.5 µm. The above described heterodyne method improves the signal to noise ratio especially for small particles (<100 nm), low particle concentrations and enables accurate particle sizing over a large concentration area [22] and [23]. The software delivered from the manufacturer (Microtrac FLEX 11.0.0.1) controls the device and calculates the particle sizes.

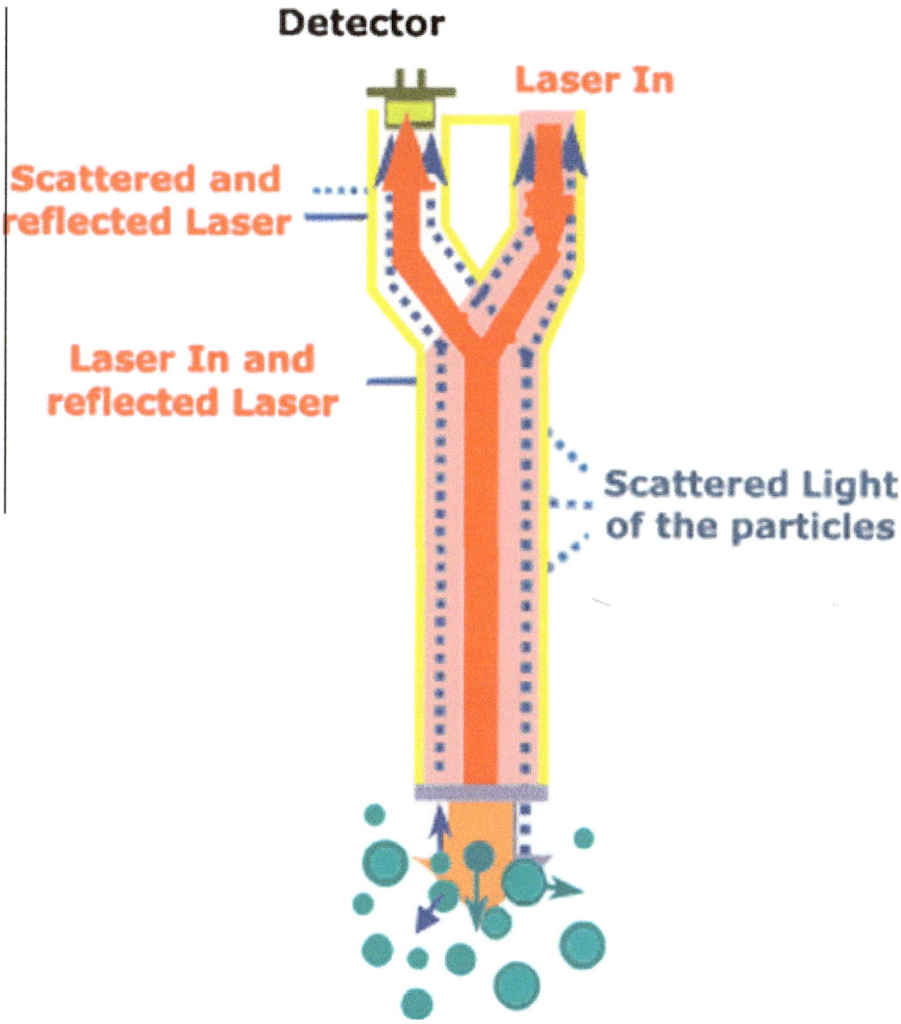

Figure 1. Overview of the heterodyne FOQELS measurement principle. (Reprinted with permission from Particle-Metrix, Germany.)

Probe Head

To perform measurements in stirred fluids with the Particle Metrix/Microtrac Nanotrac 250, a custom designed probe head pictured in Figure 2 was developed. The probe head was tailored to be attached directly to the commercial probe, which is then immersed together with the probe head into the reaction fluid. The task of this probe head is to separate a small amount of volume from the bulk fluid, which can then be measured using the commercial probe. This is further called compartmentalization. Additionally, the probe head has to actively exchange the sample volume between two measurements. The requirement for this probe head were compactness, so it can be applied in common laboratory reactors and it had to be robust enough to be used in industrial stirred reaction vessels. The separated sample volume should be as small as possible, it should be exchanged sufficiently fast between each measurement, and most importantly it has to be protected from external motion and stray light.

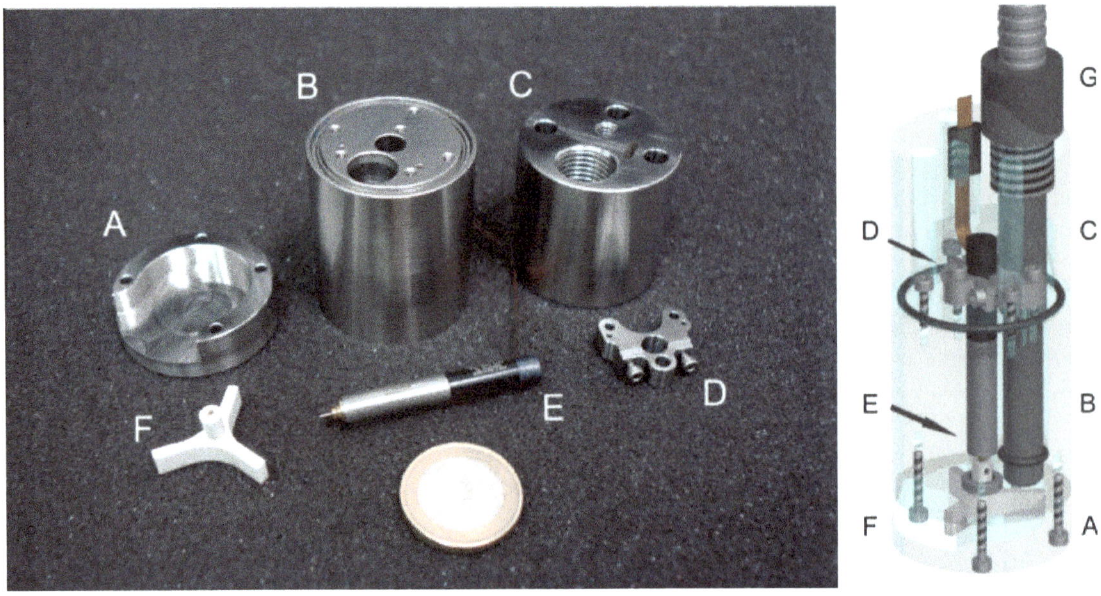

Figure 2. Left: Components of disassembled probe head for DLS measurements in stirred solutions. 1€-coin is shown for size reference. Right: CAD image of assembled probe head with DLS measurement probe.

The probe heads outer diameter is 35 mm and its length is 82 mm. The probe head consists of three parts for the enclosure (A, B, C) made from stainless steel. A miniature stepper motor (0620C006B K1855, Dr. Fritz Faulhaber GmbH & Co.KG, Germany) (E) with a diameter of 6 mm is secured by a stainless steel holder (D). The rotor made from polytetrafluorethylen (PTFE) (F) is attached to the stepper motor. There are several sealings in the probe head to keep moisture away from the stepper motor.

The custom probe head (A, B, C) surrounds the optical DLS-probe (G) of the DLS device and provides a small enclosed sample chamber (~0.65 ml) for the DLS-measurements. The exchange of the sample fluid is carried out by the miniature stepper motor. The miniature stepper motor stops the rotor in a defined position after exchanging the sample so the optical path for the DLS-probe is free and the measurement chamber is properly capsulated. Because of the small sample chamber size the fluid stalls very fast and does not disturb the measurement due to any overlaid motion. The stepper motor is controlled by a motion controller (MCBL3002, Dr. Fritz Faulhaber GmbH & Co.KG, Germany) and software to program the motion profile (Motion Manager 5.0.0, Dr. Fritz Faulhaber GmbH & Co.KG, Germany).

Malvern Zetasizer Nano ZS

A Zetasizer Nano ZS (Malvern Instruments Ltd, England) was used as an off-line reference device. This device uses classical single-scattering DLS methods to calculate particle sizes in a range from 0.3 nm to 10 μm with a 633 nm laser and detects the backscattered light at 173°. The device features a temperature-controlled holder for a cuvette. The sample in the cuvette needs to be diluted to avoid multiple scattering.

Experiments in Open Beaker

Before the probe head is applied for in-line measurements during microgel synthesis, the suitability of the compartmentalization is tested on previously synthesized microgel particles. Therefore, the probe head was immersed into a beaker filled with the colloidal microgel suspension and a magnetic stir bar was added. The beaker was placed on top of a magnetic stirrer. Measurements were conducted with and without stirring. For comparison, the same measurements were repeated without the probe head. The integration time of the scattered light for each DLS measurement was 30 s with six repeated measurements per setup.

Experiments in a One Liter Reactor

In order to show that the new probe head can be used for an in-line monitoring of particle growth during a polymerization reaction, a one liter double-walled reactor with temperature controlled oil heating mantle was equipped with the new device.

As exemplary application for in-line monitoring of particle growth a microgel synthesis was selected. Two different polymerization reactions were analyzed, namely the particle formation of poly-N-vinylcaprolactam (PVCL) and of poly-N-isopropylacrylamide (PNIPAM). The following chemicals were used for the synthesis of the microgels. As monomers N-isopropylacrylamide (NIPAM) (Sigma–Aldrich) and N-vinylcaprolactam (VCL) (Sigma Aldrich) were employed. N,N′-Methylenebisacrylamide (BIS) (Sigma Aldrich), Cetyltrimethylammoniumbromid (CTAB) (Merck) and 2,2′-Azobis(2-methylpropionamidine) dihydrochloride (AMPA) (Sigma Aldrich) were used as crosslinker, surfactant and initiator, respectively. NIPAM and VCL were purified by recrystallization in hexane with subsequent high-vacuum distillation at 80 °C. All other chemicals were used as received. Deionized water was used as solvent. The synthesized microgels have a polymer content of about 1.5 wt%. The exact amount of each chemical used in the two discussed experiments can be found in Table 1. Each microgel synthesis was performed following the procedure described in Pich et al. [16]. After the deionized water was heated to the required temperature, the components monomer, cross-linker and surfactant were added. During the entire process, the reactor was under constant nitrogen atmosphere. After 30 min, the initiator was added to start the polymerization. From this point, the light scattering measurements were started. The integration time of the Nanotrac was set to 10 s. Once a DLS measurement is finished, the sample is exchanged by movement of the impeller, which stops in the pre-defined position (see Section 2.2) before triggering a new DLS measurement. The whole measuring cycle including data processing takes about 30 s.

Table 1. Mass of reactants (g) used in the two experiments.

Component	1 l PVCL	1 l PNIPAM
Water	1050.000	1050.000
VCL	15.505	N/A
NIPAM	N/A	12.600
BIS	0.420	0.420
AMPA	0.350	0.350
CTAB	0.154	0.126

In addition to the in-line measurements, 5 ml samples were taken at certain intervals during polymerization using a syringe with a long needle. For the first ten minutes after initiation, these samples were taken each minute. Between minute 10 and 30, the samples were taken every five minutes. From then on, the samples were taken every ten minutes. The samples were directly cooled down in numbered sample vials placed in liquid nitrogen and an inhibitor (Hydrochinon, Merck) was added to prevent further polymerization and particle growth. To confirm the particle sizes measured with the novel probe head, these samples were measured off-line in a Malvern Zetasizer Nano ZS in a fused silica cuvette. Before, the samples were diluted to avoid multiple scattering and filtered through a 1.2 μm PET filter to remove dust. An additional surfactant (CTAB) was added to prevent particle aggregation due to the instability of the particles at the early stage of reaction. The surfactant concentration was below the critical micelle concentration, to prevent any influence on the DLS measurements. The cuvette temperature of the Zetasizer Nano ZS was set to 50°C and the integration time was automatically set to get a good signal-to-noise ratio. Every measurement was repeated three times and the resulting arithmetic average size was used for comparison to the in-line measurements.

RESULTS AND DISCUSSION

Test of Compartmentalization

For the test of compartmentalization the open beaker setup was used. Four different conditions were examined. First, the bare optical probe of the DLS device was used in stirred and unstirred microgel suspension. The magnetic stirrer was set to 1000 rpm for the stirred measurements. Afterwards, the optical probe of the DLS device was equipped with the new probe head and the measurements in stirred and unstirred microgel were repeated.

Figure 3 shows the results of these measurements. Zone 1 shows the particle size determined from repeated measurements with the bare probe in the unstirred suspension for a size reference. Zone 2 displays the sizes resulting from measurements taken during active stirring. Due to the convection, the particle sizes are erroneously identified to be remarkably smaller. The calculated sizes differ between 70 nm and 350 nm and a reliable value cannot be evaluated. Zone 3 contains particle sizes from measurements using the custom probe head, albeit without stirring. The deviation between the detected sizes is slightly smaller than the deviation without the custom probe head (zone 1). This smaller distribution of measured particle sizes could be caused by the compartmentalization of the sample, which effectively keeps away ambient light from the detector and restricts even small liquid movements. The particle sizes of measurements with the new probe head under stirred conditions are depicted in zone 4. The particle sizes determined under agitated condition match the values determined in unstirred condition, which ultimately proofs the operational capability of the developed probe head.

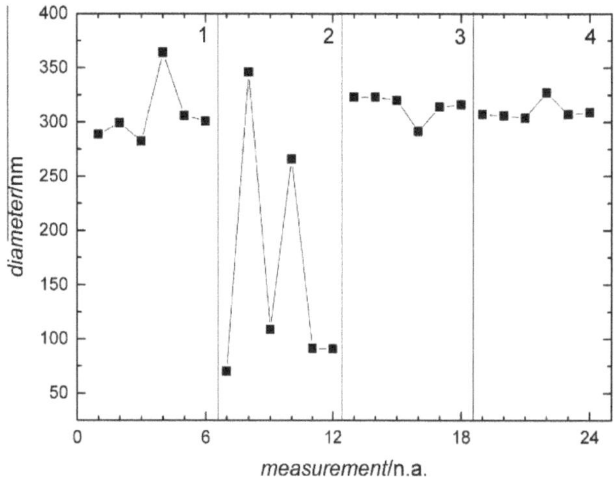

Figure 3. Performance tests of the new probe head with PVCL microgel. Comparison of the measurements with bare and custom probe in stirred and unstirred sample. 1: Unstirred liquid, bare optical DLS probe; 2: Stirred liquid, bare optical DLS probe; 3: Unstirred liquid, custom probe head; 4: Stirred liquid; custom probe head.

Table 2 shows the calculated values of the average and standard deviation for each zone. The standard deviation for the measurements with the attached custom probe head are improved by a factor of 2.4–3.2 over the measurements with the bare probe of the DLS-device. The deviations of the averages from zone 3 and zone 4 are within the deviation of zone 1.

Table 2. Average and deviation of the comparison measurements from Figure 5.

Zone	Average size (nm)	Standard deviation (nm)
1	307	29
2	N/A	115
3	330	12
4	316	9

The results of the measurements of the colloidal microgel suspension show that the design of the custom probe head is suitable to isolate and measure a small volume from the system, thereby preventing any interference of the encapsulated sample with the stirred bulk fluid. The accuracy of the custom probe under stirred conditions is at least as good as the bare probe in unstirred conditions. This renders reliable DLS measurements even in vigorously stirred environments possible.

Monitoring of Polymerization

Following the test of compartmentalization in the open beaker, the probe head was applied to monitor microgel syntheses in the temperature-controlled one liter reactor under nitrogen atmosphere. Figure 4 displays the particle size monitoring during polymerization of PVCL at a reaction temperature of 60 °C. This temperature was chosen since it is the lowest temperature that reliably starts the polymerization of PVCL using AMPA as initiator and hence permits most measurements during the particle growth phase. Each dot in the figure represents one particle size measurement, of which the red[4]triangles were visually identified as outliers and excluded for computation of the floating mean, which is determined piecewise over three successive values. The two outliers might be caused by small air bubbles that reached the measurement chamber.

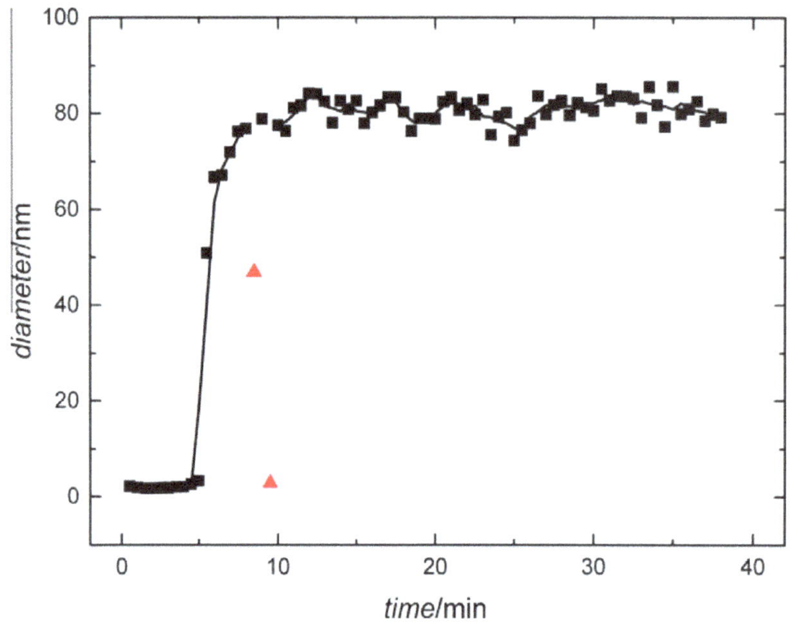

Figure 4. Monitoring of particle size during polymerization of PVCL at 60 °C in 1 L laboratory reactor. Triangles show visually identified outliers.

The first measurements after the addition of the initiator at time zero show a particle size of 2–3 nm due to background scattering. From minute five onwards, the growth of the particles can be seen very well. Approximately twelve minutes after initiation, the particle sizes reach a plateau at ~81 nm ± 3 nm SD diameter. The time between the first noticeable change in particle size and the onset of the plateau is about 8 min.

The second system monitored using the probe head was a polymerization of PNIPAM at a reaction temperature of 65 °C. The corresponding in-line recorded particle sizes are displayed in Figure 5.

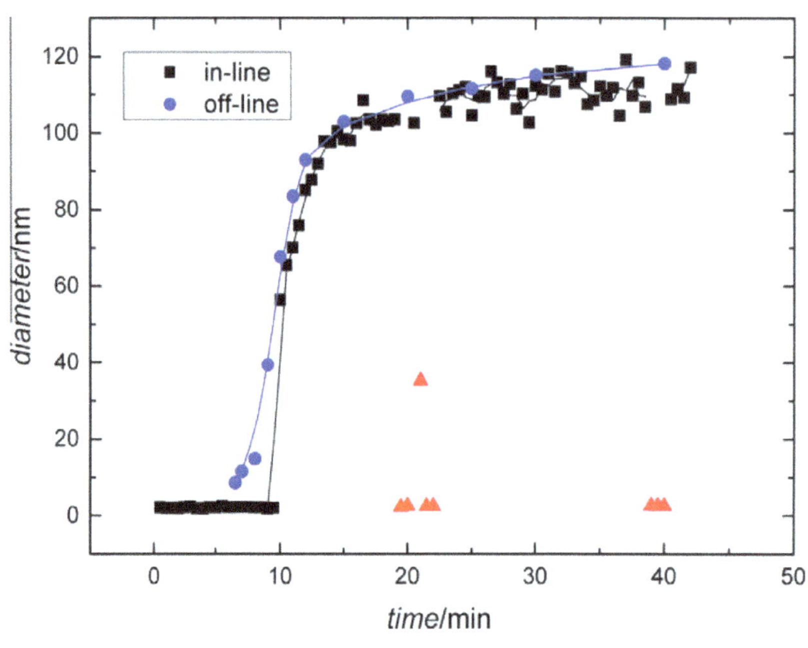

Figure 5. Monitoring of particle size taken in-line (squares) during polymerization of PNIPAM at 65 °C in 1 L laboratory reactor together with off-line DLS reference measurements (circles). Triangles show visually identified outliers.

To assess the validity of the particle sizes determined using the novel probe head, samples were taken during a microgel synthesis, directly quenched to stop further reaction and subsequently analyzed off-line. In Figure 5, the particle sizes measured using the in-line probe head are depicted using black squares and the particle sizes measured off-line using blue dots. Again, red triangles represent visually selected outliers. The slopes of both measurements are in good agreement with each other. While the in-line measured sizes remain at their initial level until minute ten, the off-line measurements are able to depict small particle growth from minute six. Starting with the significant jump to 55 nm at minute 10, the in-line measurements display a constant particle growth, which is in accordance with the off-line measurements. The differences between both measurement techniques present at early reaction times decrease as the reaction progresses. The particle sizes from both techniques reach a plateau of 111 ± 4 nm SD for the in-line and 114 nm for the off-line measurements, respectively. A reason for the decreasing differences between in-line and off-line data could be the integration time of the Particle Metrix/ Microtrac Nanotrac 250. Since the particles continue to grow during the acquisition time, the scattering pattern changes as well. The DLS evaluation method then determines a particle size that best approximates the measured signal, which might be distorted due to the shifting scattering pattern. The slower the reaction becomes, the smaller are the differences between both methods. The handling time from taking the sample and stopping the reaction with fluid nitrogen also takes a certain amount of time in which the particles keep growing.

After the syntheses, a minor contamination from chemical residue is visible on the surface of the probe head and on the protective window of the optical DLS probe. Examination of the DLS measurements shows no remarkable influence of these contaminations on the detected particle sizes. The residue rather causes a static reduction of the mean scattering intensity and thus might increase the measurement time necessary to retain the targeted measurement uncertainty. Nevertheless, the accuracy reached using the custom probe head in stirred surroundings is in a region that is at least as good as measurements performed with a bare DLS probe in unstirred fluids.

CONCLUSIONS

A novel probe design has been introduced to enable in-line monitoring of particle sizes inside a stirred fluid based on dynamic light scattering without the need for sampling and dilution. It was shown that the fluid compartmentalization in the probe head is sufficient to exclude any influences of the stirred bulk liquid. Further experiments showed that the growth of microgel particles could be suitably followed during stirred polymerization. A comparison between the direct in-line measurements and off-line determined particle sizes using a state-of-the-art device showed that both measurement techniques are in good agreement with each other. The application of the probe design in industrial environments that currently rely on tedious off-line analysis of drawn samples could vastly facilitate product quality control. In batch productions it could lead to faster reaction times, since the end of particle growth can automatically be detected. It renders possible the discovery of unexpected product conditions and triggering of counteractive measures that might be untraced for considerably longer times with conventional monitoring techniques.

ACKNOWLEDGEMENTS

This work was supported by the German Research Foundation and is part of the Collaborative Research Centre "Functional Microgels and Microgel Systems (SFB 985)" located at the RWTH Aachen University.

REFERENCES

1. D.L. Black, M.Q. McQuay, M.P. Bonin, Prog. Energy Combust. Sci. 22 (3) (1996) 267.
2. R. Pecora, J. Nanopart. Res. 2 (2000) 123.
3. P.N. Segre, W. van Megen, P.N. Pusey, W. Peters, J. Mod. Opt. 42 (9) (1995) 1929.
4. P.N. Pusey, Curr. Opin. Colloid Interface 4 (1999) 177.
5. C. Sinn, R. Niehüsser, E. Overbeck, T. Palberg, Part. Part. Syst. Charact. 16 (1999) 95.
6. F. Stieber, W. Richtering, Langmuir 2 (1995) 4724.
7. J.C. Thomas, V. Dimonie, Appl. Opt. (1990) 5332.
8. F.M. Horn, W. Richtering, J. Bergenholtz, et al., J. Colloid Interface Sci. 225 (2000) 166.
9. H.G. Barth, R.B. Flippen, Anal. Chem. 67 (1995) 257.
10. C.-S. Wu (Ed.), Handbook of Size Exclusion Chromatography and Related Techniques: Revised and Expanded, vol. 91, CRC Press, 2003.
11. A. Alb, W. Reed, Macromol. React. Eng. 4 (2010) 470.
12. G. Crawley, M. Cournil, D. DiBenedetto, Powder Technol. 91 (1997) 197.
13. N. Wei, J. You, K. Friehs, E. Flaschel, T.W. Nattkemper, Biotechnol. Bioeng. 97 (2007) 1489.
14. R. Hass, O. Reich, ChemPhysChem 12 (2011) 2572.
15. R. Hass, D. Munzke, S.V. Ruiz, J. Tippmann, O. Reich, Anal. Bioanal. Chem. 407 (2015) 2791.
16. A. Pich, A. Tessier, V. Boyko, Y. Lu, H.-J.P. Adler, Macromolecules 39 (2006) 7701.
17. J. Meyer-Kirschner, M. Kather, A. Pich, D. Engel, W. Marquardt, J. Viell, A. Mitsos, Appl. Spectrosc. 70 (2) (2016) (accepted for publication).
18. O.L.J. Virtanen, W. Richtering, Colloid Polym. Sci. 292 (2014) 1743.
19. S. Kara, Ö. Pekcan, Polymer 41 (2000) 3093.
20. S. Kara, Ö. Pekcan, J. Appl. Polym. Sci. 80 (2001) 823.
21. S. Kara, O. Okay, O. Pekcan, J. Appl. Polym. Sci. 86 (2002) 3589.
22. P.J. Freud, Nanoparticle Sizing. Dynamic Light Scattering Analysis in the Frequency Spectrum Mode, vol. 7, Microtrac Inc., 2007.
23. M.N. Trainer, P.J. Freud, E.M. Leonard, High Concentration Submicron Particle Size Distribution by Dynamic Light Scattering, vol. 34, American Laboratory, Fairfield, Connecticut, 1992.

CHAPTER 9

GLOBAL MONITORING OF FLUIDIZED-BED PROCESSES BY MEANS OF MICROWAVE CAVITY RESONANCES

Johan Nohlert [a], Livia Cerullo [a], Johan Winges [a], Thomas Rylander [a], Tomas McKelvey [a], Anders Holmgren [b], Lubomir Gradinarsky [b], Staffan Folestad [b], Mats Viberg [a], Anders Rasmuson [c]

[a]Department of Signals and Systems, Chalmers University of Technology, SE-41296 Göteborg, Sweden

[b] AstraZeneca R&D, Pharmaceutical Development, SE-431 83 Mölndal, Sweden

[c]Department of Chemical and Biological Engineering, Chalmers University of Technology, SE-41296 Göteborg, Sweden

ABSTRACT

We present an electromagnetic measurement system for monitoring of the effective permittivity in closed metal vessels, which are commonly used in the process industry. The measurement system exploits the process vessel as a microwave cavity resonator and the relative change in its complex resonance frequencies is related to the complex effective permittivity inside the vessel. Also, thermal expansion of the process vessel is taken into account and we compensate for its influence on the resonance frequencies by means of a priori information derived from a set of temperature measurements. The sensitivities, that relate the process state to the measured resonance frequencies, are computed by means of a detailed finite element model. The usefulness of the proposed measurement system is successfully demonstrated for a pharmaceutical fluidized-bed process, where the water and solid contents inside the process vessel is of interest.

INTRODUCTION

In the pharmaceutical industry, monitoring and control of manufacturing processes are important in order to achieve high-quality products to a low production-cost. A common step in the manufacturing process for many pharmaceutical products is coating and drying of particles in fluidized-bed processes. Measurement techniques that are capable of monitoring the material distribution and moisture content in this type of process are important tools for control, optimization and improved understanding of the process.

Currently available methods for monitoring of moisture content in pharmaceutical fluidized-beds include spectroscopic methods such as near infra-red (NIR) and Raman spectroscopy [1] and [2]. Although these methods yield important information on moisture content as well as physical and chemical properties of the material, they mainly provide local information for the region in the very proximity of the measurement probe, which may be insufficient or misleading for the estimation of the global process state. Recently, Buschmüller et al. [3] developed a stray field microwave resonator technique for in-line moisture measurements during fluidized-bed drying.

Citation: Johan Nohlert, Livia Cerullo, Johan Winges, Thomas Rylander, Tomas McKelvey, Anders Holmgren, Lubomir Gradinarsky, Staffan Folestad, Mats Viberg, Anders Rasmuson, Global monitoring of fluidized-bed processes by means of microwave cavity resonances, Measurement, Volume 55, September 2014, Pages 520-535, ISSN 0263-2241, http://dx.doi.org/10.1016/j.measurement.2014.05.034.

However, their approach is based on local fields in the vicinity of the resonator probe and, therefore, it does not yield global information on the process state.

At low frequencies, electric capacitance tomography (ECT) can be applied to regions with high particle density [4]. Although ECT yields important information on the particle distribution on a global scale, it suffers from limited spatial resolution and it is not suited for the estimation of dielectric losses, which leads to difficulties in distinguishing between material density and moisture content.

In this article, we present a microwave-based measurement technique that exploits electromagnetic eigenmodes that are supported by the process vessel bounded by its metal walls. The process vessel is considered as a cavity resonator and the permittivity distribution inside the vessel is estimated based on measurements of the complex resonance frequencies of the cavity. This technique resembles the well-established cavity perturbation measurement technique, which is regularly used for accurate measurements of dielectric and magnetic properties of materials at microwave frequencies [5] and [6]. In contrast to traditional cavity-perturbation measurements, our technique exploits several different eigenmodes, which yields three-dimensional spatial resolution and measurement information at multiple frequencies.

The process vessel is equipped with a set of probes that are connected to a network analyzer. Given the scattering parameters measured with the network analyzer, we estimate the complex resonance frequencies of the process vessel, where the real part corresponds to the frequency of oscillation and the imaginary part corresponds to the damping. We exploit a linearization of the electromagnetic response of the process vessel and it yields a relation between (i) a perturbation of the resonance frequencies and (ii) a perturbation of the complex permittivity inside the process vessel. We parameterize the permittivity with space dependent basis functions and determine the coefficients by means of solving an estimation problem based on an over-determined system of linear equations, where the right-hand side contains the measured deviations in the resonance frequencies as compared to the empty vessel.

MEASUREMENT SYSTEM

Process Equipment

Figure 1 shows a schematic picture of a fluidized-bed process vessel of Wurster type, which is the type of process we consider in this article. This process is used for coating and drying of pharmaceutical particles, such as pellets or granules, and the particular vessel we consider is a custom built lab-scale device for batches consisting of up to 1 kg of particles. The particles are dielectric objects of size that normally range from 250 μm to 1 mm and in this article, we consider particles that mainly consist of microcrystalline cellulose (MCC).

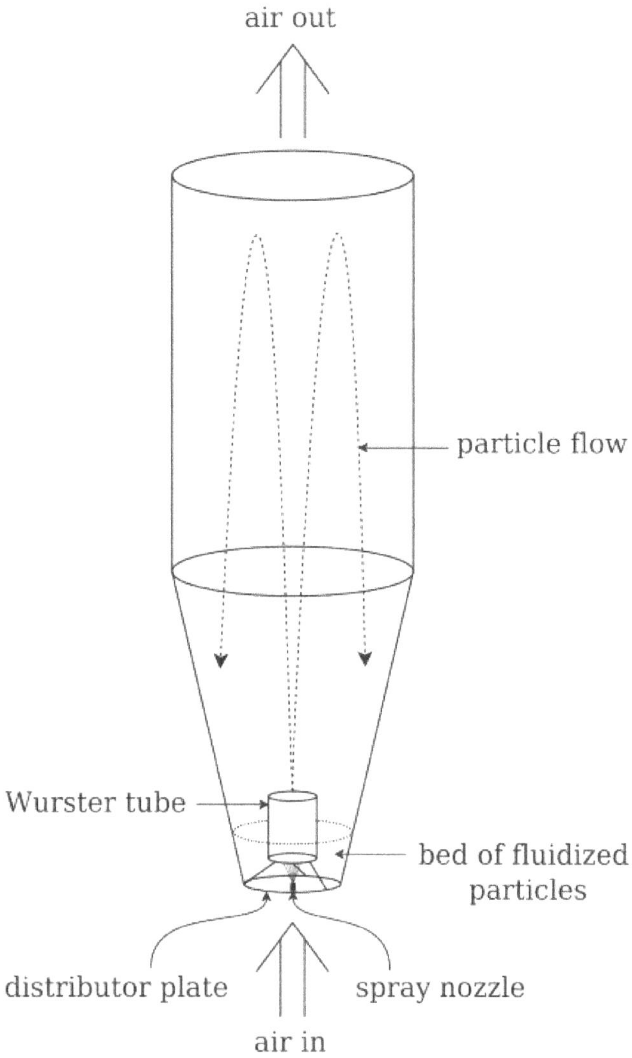

Figure 1. Schematic picture of the process vessel.

The bottom plate of the process vessel, also referred to as the distributor plate, is perforated with a large number of small holes. An airflow is applied through the distributor plate, which fluidizes the particles and creates a fluidized bed. Above the distributor plate resides a metal tube, which is referred to as the Wurster tube. At the center of the distributor plate below the Wurster tube, a pneumatic spray nozzle injects a rapidly moving mixture of air and liquid that accelerates the particles in its immediate vicinity in an upward motion. The particles move through the Wurster tube and create a fountain-like flow pattern above the Wurster tube. As the particles exit the fountain region, they fall back into the fluidized bed outside the Wurster tube.

The liquid injected through the spray nozzle is a solution that is sprayed onto the surface of the particles. The solvent evaporates mainly inside and immediately above the Wurster tube, which results in a deposited layer of the solute on the surface of the particles [7]. A common solvent is water and there is a large variety of solutes such as active pharmaceutical ingredients (API), polymers used for film coating and various filler materials. Thus, this type of process allows for the manufacturing of multi-layered particles. Here, the different layers provide designed functions such as protection from humidity during storage, delay of API release in the digestive system and the actual delivery of the API.

The water and its distribution inside the vessel for the duration of the process influence the process state and, consequently, the quality of the final product. For example, API that are sensitive to long-term exposure to moisture may be coated with a film that protects the API from ambient humidity. In such cases, it is important to establish that the particles are sufficiently dry before this film is applied. In addition, a too fast drying rate may have negative impact on the coating quality. A more extreme situation is the undesirable event of excessive amounts of water inside the vessel, which can cause extensive particle agglomeration and eventually process break-down. If such a condition can be detected, it is feasible to change the process parameters in order to counter-act agglomeration. The water inside the vessel can occur in a number of different forms: (i) water vapor; (ii) sorbed water inside the particles; (iii) free water on the surface of the particles; (iv) water droplets in the air; (v) condensed water droplets on the vessel walls; and (vi) water in agglomerates or dense gatherings of particles.

Measurement Principle

From an electromagnetic point of view, the process vessel that we consider is a rather good approximation of a microwave cavity resonator. The geometry of the vessel, which is used in the electromagnetic model described later in the article, is shown in Figure 2. We exploit two magnetic-field probes to excite eigenmodes supported by the microwave resonator and the corresponding eigenfrequencies are measured with the same set of probes. A network analyzer is connected to the two probes and the 2×2 scattering matrix is measured as a function of frequency. The measured scattering parameters are used to extract the complex resonance frequencies of a selection of the lowest eigenmodes supported by the process vessel.

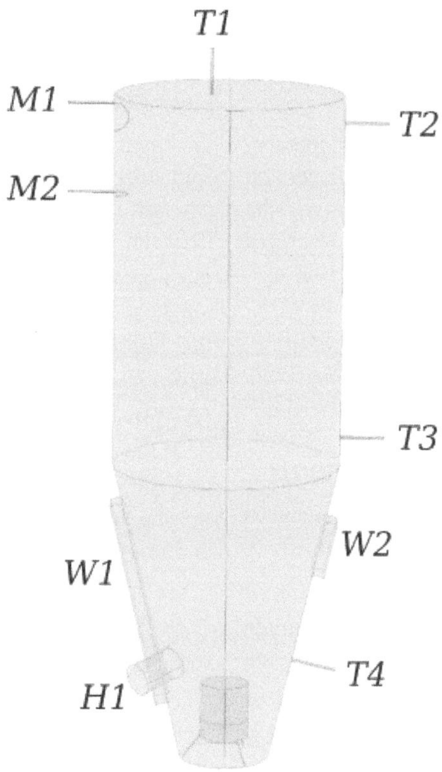

Figure 2. The process vessel including its two rectangular windows (W1 and W2) and the cylindrical tube used for loading of particles (H1). This figure also illustrates four temperature probes (T1–T4) and two microwave probes (M1 and M2) although these are not part of the electromagnetic model.

It should be emphasized that although these probes are rather small in comparison to the microwave resonator, i.e., the process vessel, the electromagnetic field for basically all eigenmodes extends throughout most of the process vessel's volume. The eigenfrequencies are influenced by the geometry of the process vessel and, more importantly for this application, the material and its distribution inside the process vessel. Consequently, important information on the distribution of particles and their moisture content can be derived from the measured eigenfrequencies.

Homogenization of a Mixture of Air and Particles

The mixture of particles and air influences the electromagnetic fields inside the process vessel. The typical length-scales associated with the lowest eigenmodes are comparable to the size of the cavity and, as a consequence, the particles are extremely small in relation to the spatial variation of the electromagnetic field. Thus, it is beneficial to represent the mixture of particles and air in terms of an effective permittivity, where the possible presence of other small dielectric objects, such as liquid droplets, contributes to the effective permittivity in the same manner as the particles. Given the material parameters and geometry of the particles together with their spatial distribution, the effective permittivity can be calculated by means of electromagnetic mixing formulas [8]. Here, we focus on the well-known Maxwell–Garnett mixing formula

$$\frac{\epsilon_{\mathrm{eff}}-\epsilon_{\mathrm{b}}}{\epsilon_{\mathrm{eff}}+2\epsilon_{\mathrm{b}}}=\frac{1}{3\epsilon_{\mathrm{b}}}\sum_{n=1}^{N}\vartheta_n\frac{\alpha_n}{V_n}, \tag{1}$$

where ϵ_{eff} is the effective permittivity and ϵ_b is the permittivity of the background medium for a mixture with N different types of inclusions indexed by n . In this article, we assume that ϵ_b is real, i.e., we have a lossless background medium. For the inclusions of type n , we have the polarizability α_n, the inclusion volume V_n and the volume fraction ϑ_n. For homogeneous spherical inclusions with permittivity ϵ_i, we have the polarizability

$$\alpha=3\epsilon_{\mathrm{b}}V\frac{\epsilon_{\mathrm{i}}-\epsilon_{\mathrm{b}}}{\epsilon_{\mathrm{i}}+2\epsilon_{\mathrm{b}}}, \tag{2}$$

where V is the volume of the particle. Layered spherical particles with a core and an outer shell with different permittivities can also be incorporated into Eq. (1) by using the associated polarizability for this type of inclusions [8].

A dispersive inclusion material yields a dispersive behavior of the effective permittivity, although the dispersion relation of the mixture may be different from that of the inclusions. For example, air with inclusions that are described as a Debye medium yields an effective permittivity that also behaves like a Debye medium, but with different values of the Debye parameters [9]. Furthermore, air and inclusion particles with non-zero conductivity yields Debye-characteristics of the effective permittivity, due to the Maxwell–Wagner effect [10].

Density Independent Moisture Measurements

A feasible way to extract information about moisture content of the particles from the estimated complex effective permittivity $\epsilon_{\mathrm{eff}}=\epsilon'_{\mathrm{eff}}-j\epsilon''_{\mathrm{eff}}$ in the process vessel, is the approach of density-independent measurement techniques [10]. The main idea of this approach is that if the material density in a given volume changes, both ϵ'_{eff} and $\epsilon''_{\mathrm{eff}}$ will be affected in a similar way, since they both depend on the total amount of polarizable material in the volume.

In a situation where a certain material is dispersed in a background medium with permittivity ϵ_b, the ratio

$$\eta=\frac{\epsilon''_{\mathrm{eff}}}{\epsilon'_{\mathrm{eff}}-\epsilon_{\mathrm{b}}}, \tag{3}$$

is an example of a density-independent function that has shown to be more or less independent on the density for many material mixtures [11] and [12]. This is not surprising, and it can be shown using the Maxwell–Garnett mixing formula applied to lossy spherical inclusions dispersed in a lossless background medium, that η is independent on the volume fraction ϑ in the low-density limit. Therefore, it is expected that measurements of $\epsilon''_{\mathrm{eff}}$ and ϵ'_{eff} taken at different material densities would fall on a straight line that intersects the origin if $\epsilon''_{\mathrm{eff}}$ is plotted versus $\epsilon'_{\mathrm{eff}}-\epsilon_{\mathrm{b}}$, given that the particles have constant permittivity. The slope η of this line usually depends on the moisture content of the inclusion material in a characteristic manner. The slope η may therefore be used as a density-independent metric of the moisture content, if it is known how η depends on the moisture content for the material considered. Such relations are usually obtained from calibration procedures which involve measurements on material samples with known moisture content. It should be emphasized that η may depend on many other factors as well, such as the physical state of the water (e.g. whether it is free or bound in the material) and the temperature. In the case of free water, its distribution in the material may also be of importance.

In practical measurement situations, the estimated effective permittivity ϵ_{eff} may be biased by, e.g., temperature variations of the metal parts of the cavity, or by changes in the permittivity of the background medium (which is usually air) caused by, e.g., a varying humidity. In such cases, there may be non-zero residual components of $\epsilon''_{\mathrm{eff}}$ and $\epsilon'_{\mathrm{eff}}-\epsilon_{\mathrm{b}}$ at zero material density due to biased estimates of ϵ_{eff}, which makes a direct use of the ratio η unreliable. We therefore make an attempt to extend the classical density-independent measurement approach to situations where an uncontrolled bias of the estimated ϵ_{eff} is present. Our approach is based on the covariance between $\epsilon''_{\mathrm{eff}}$ and $\epsilon'_{\mathrm{eff}}-\epsilon_{\mathrm{b}}$, which is computed for a fairly large number of measurement samples. The material (in our case the particles in the process) is re-distributed between different sample points which yields a variation in the material density and hence a variation in $\epsilon''_{\mathrm{eff}}$ and ϵ'_{eff}, which is used to calculate the covariance. As a motivation to this approach, we consider a situation where lossy spherical inclusions with permittivity $\epsilon_{\mathrm{i}}=\epsilon'_{\mathrm{i}}-j\epsilon''_{\mathrm{i}}$ are dispersed in a background medium with permittivity ϵ_b with volume fraction □. If the dispersion is dilute (□≪1) and if the loss factor of the inclusion material is low ($\epsilon''_{\mathrm{i}}/\epsilon'_{\mathrm{i}}\leqslant 0.1$), the Maxwell–Garnett mixing formula yields the following expressions for the real and imaginary parts of the effective permittivity

$$\epsilon'_{\mathrm{eff}}/\epsilon_{\mathrm{b}}-1 \simeq 3\frac{\epsilon'_{\mathrm{i}}/\epsilon_{\mathrm{b}}-1}{\epsilon'_{\mathrm{i}}/\epsilon_{\mathrm{b}}+2}\vartheta \tag{4}$$

$$\epsilon''_{\mathrm{eff}}/\epsilon_{\mathrm{b}} \simeq 9\frac{\epsilon''_{\mathrm{i}}/\epsilon_{\mathrm{b}}}{(\epsilon'_{\mathrm{i}}/\epsilon_{\mathrm{b}}+2)^2}\vartheta. \tag{5}$$

If we now assume that $\vartheta, \epsilon'_{\mathrm{i}}$ and ϵ''_{i} vary randomly around their respective mean values, this will obviously yield variations in ϵ_{eff}. Hence, by studying the covariance between the real and imaginary part of ϵ_{eff}, one may extract important information about the parameters $\vartheta, \epsilon'_{\mathrm{i}}$ and ϵ''_{i} and their variabilities. In particular, we notice from Eqs.(4) and (5) that both the real and imaginary part of ϵ_{eff} depend linearly on the volume fraction ϑ, which explains the density-independent nature of the ratio η and the reason why $\epsilon''_{\mathrm{eff}}$ and ϵ'_{eff} - ϵ_b obtained from different volume fractions tend to fall on a straight line. Deviations from this line can be explained by means of variations in ϵ'_i and ϵ''_i, since they influence the real and imaginary part of ϵ_{eff} in different ways, as seen from Eqs. (4) and (5).

Experimental Setup

The experimental setup is shown in Figure 3a and it includes the process vessel, the process machine and the two-port network analyzer used to perform the microwave measurements.

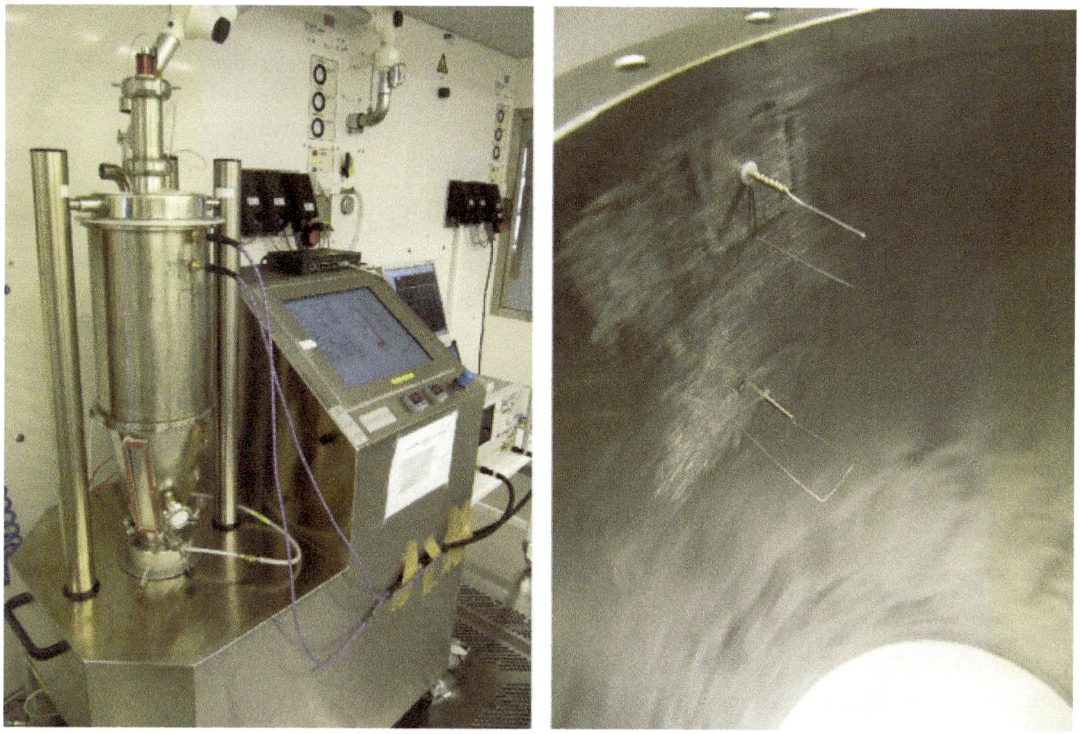

(a) Experimental setup (b) H-probes

Figure 3. Photographs of (a) the process vessel in the experimental setup and (b) the two H-probes.

Reduction of Undesired Losses

The original process vessel has a couple of apertures, such as windows that are used for visual inspection of the process, and such apertures damp the microwave cavity by means of radiation losses. We are interested in measuring the damping associated with the moisture content of the particles and, thus, we wish to reduce other contributions to the total damping of the cavity. Therefore, we shield the process vessel by covering the apertures W1, W2 and H1 in Figure 2 by aluminum foil, and use a perforated plate to shield the microwave cavity from the air outtake in the upper part of the process vessel. In addition, the shielding mitigates possible problems with external electromagnetic sources that may disturb the measurements. With these arrangements we obtain unloaded Q -values of the cavity which are in the order of 10^3–10^4 for the lowest eigenmodes.

Choice of Probes and their Design

Pharmaceutical mixtures processed in steel containers often become electrically charged due to the triboelectric effect [13] and [14]. Consequently, the statically charged particles yield a quasi-static electric field as they move inside the vessel. In order to protect the network analyzer from induced voltages of large magnitude and electric discharges that may occur in the process vessel, we exploit magnetic field probes (also referred to as H-probes) that are mainly sensitive to the time variations of the magnetic flux that passes through the loop of the H-probe.

Figure 3b shows the process vessel equipped with two H-probes located in the upper part of the cavity for measurement of the azimuthal and vertical components of the magnetic field. At these locations, we can efficiently couple to the magnetic field of the lowest eigenmodes that we use for sensing. In addition, the size of each loop is chosen such that we get sufficiently good coupling to all of these modes. For industrial usage, it may not be feasible to let the probe extend into the process vessel. A possible remedy is to create a shallow dent in the cavity wall and to cover this dent by a dielectric "radome".

Measurement Settings

The network analyzer (Agilent E8361A) is controlled from a PC via GPIB interface, in order to perform repeated scans of a frequency band that includes the resonance frequencies of the sensing modes (0.7–1.7 GHz). This frequency band takes approximately 2 s to scan (for excitation on one of the two ports), using totally 6400 frequency points and the IF bandwidth set to 5 kHz, which means that it takes 1–10 ms to scan the frequency band occupied by a single resonance. The time between two consecutive frequency sweeps is set to 10 s, where a substantial part of this time is required for data transfer over the GPIB interface. In the post-processing of the measurement data, we use a statistical approach which benefits from a large number of measurement samples. Therefore, a measurement setup that allows for faster sampling of the process would be beneficial for this measurement technique.

ELECTROMAGNETIC MODEL

For the analysis of the resonance frequencies and the corresponding eigenmodes, we use an idealized model of the process vessel: (i) all the metal parts of the process vessel are assumed to be perfect electric conductors (PEC); (ii) all apertures are completely sealed such that the interior of the process vessel is completely enclosed by a perfect electric conductor; and (iii) the probes are excluded from the model. Further, we assume that the material inside the process vessel can be described by relatively simple dispersion relations for the frequency band of interest and, as a starting point, we consider a frequency-independent permittivity that may vary with respect to space. Also, we assume that the magnetization for the media involved is negligible and use the permeability $\mu=\mu_0$.

Thus, we consider the eigenvalue problem

$$\nabla \times \boldsymbol{E} = -j\omega\boldsymbol{B} \text{ in } \Omega, \tag{6}$$

$$\nabla \times \frac{1}{\mu_0}\boldsymbol{B} = (j\omega\epsilon + \sigma)\boldsymbol{E} \text{ in } \Omega, \tag{7}$$

$$\hat{\boldsymbol{n}} \times \boldsymbol{E} = 0 \text{ on } \Gamma. \tag{8}$$

We seek the eigenmodes $(\boldsymbol{E}_m, \boldsymbol{B}_m)$ and the corresponding eigenfrequencies ω_m that satisfy the eigenvalue problem (6), (7) and (8), where the index *m* labels the resonances supported by the process vessel.

Weak Formulation

The eigenvalue problem (6), (7) and (8) is expressed in the weak form: Find$(\boldsymbol{E},\boldsymbol{B},\omega)\in H_0(\text{curl};\Omega)\times H(\text{div};\Omega)\times C$ such that

$$c_{\text{FL}}(\boldsymbol{E}, \boldsymbol{v}) = -j\omega m_1(\boldsymbol{B}, \boldsymbol{v}), \tag{9}$$

$$c_{\text{AL}}(\boldsymbol{B}, \boldsymbol{w}) = j\omega\mu_0\epsilon_0 m_{\epsilon_r}(\boldsymbol{E}, \boldsymbol{w}) + \mu_0 m_\sigma(\boldsymbol{E}, \boldsymbol{w}), \tag{10}$$

for all $(\boldsymbol{v},\boldsymbol{w})\in H(\text{div};\Omega)\times H_0(\text{curl};\Omega)$. Here, we express the real (absolute) permittivity $\epsilon=\epsilon_0\epsilon_r$ in terms of the corresponding relative permittivity. Further, we have the bilinear forms

$$c_{\text{FL}}(\boldsymbol{u}, \boldsymbol{v}) = \int_\Omega (\nabla \times \boldsymbol{u}) \cdot \boldsymbol{v} d\Omega, \tag{11}$$

$$c_{\text{AL}}(\boldsymbol{u}, \boldsymbol{v}) = \int_\Omega \boldsymbol{u} \cdot (\nabla \times \boldsymbol{v}) d\Omega, \tag{12}$$

$$m_1(\boldsymbol{u}, \boldsymbol{v}) = \int_\Omega \boldsymbol{u} \cdot \boldsymbol{v} d\Omega, \tag{13}$$

$$m_{\epsilon_r}(\boldsymbol{u}, \boldsymbol{v}) = \int_\Omega \epsilon_r \boldsymbol{u} \cdot \boldsymbol{v} d\Omega, \tag{14}$$

$$m_\sigma(\boldsymbol{u}, \boldsymbol{v}) = \int_\Omega \sigma \boldsymbol{u} \cdot \boldsymbol{v} d\Omega. \tag{15}$$

Finally, we have the spaces

$$H(\text{curl};\Omega) = \{\boldsymbol{u} : \boldsymbol{u} \in L_2(\Omega) \text{ and } \nabla \times \boldsymbol{u} \in L_2(\Omega)\}, \tag{16}$$

$$H(\text{div};\Omega) = \{\boldsymbol{u} : \boldsymbol{u} \in L_2(\Omega) \text{ and } \nabla \cdot \boldsymbol{u} \in L_2(\Omega)\}, \tag{17}$$

and $H_0(\text{curl};\Omega)$ is the subspace of $H(\text{curl};\Omega)$ that satisfies the boundary condition(8).

Finite Element Approximation

We expand the electric field in terms of curl-conforming elements and the magnetic flux density in divergence-conforming elements, see Ref. [15] for further details. This yields the discrete eigenvalue problem

$$\begin{bmatrix} \mathbf{0} & -\mathbf{C} \\ \mathbf{C}^{\mathrm{T}} & -Z_0\mathbf{M}_\sigma \end{bmatrix} \begin{bmatrix} c_0\mathbf{b} \\ \mathbf{e} \end{bmatrix} = \frac{j\omega}{c_0} \begin{bmatrix} \mathbf{M}_1 & \mathbf{0} \\ \mathbf{0} & \mathbf{M}_{\epsilon_r} \end{bmatrix} \begin{bmatrix} c_0\mathbf{b} \\ \mathbf{e} \end{bmatrix}, \tag{18}$$

where we have scaled the magnetic flux density with the speed of light and Ampère's law with the impedance of vacuum. This form is useful since it yields an eigenvalue problem that is linear in terms of the eigenvalue $j\omega/c_0$. Reference [16] contains further details on the treatment of eigenvalue problems by means of the finite element method.

Elimination of the magnetic flux density from Eq. (18) yields the corresponding eigenvalue problem

$$\left(\mathbf{C}^{\mathrm{T}}\mathbf{M}_1^{-1}\mathbf{C} + j\omega\mu_0\mathbf{M}_\sigma - \omega^2\mu_0\epsilon_0\mathbf{M}_{\epsilon_r}\right)\mathbf{e} = \mathbf{0}, \tag{19}$$

formulated in terms of only the electric field. This eigenvalue problem is nonlinear since it is quadratic in terms of the eigenfrequency ω.

For problems with conducting media, it is computationally advantageous to solve Eq.(18). Such situations may occur in pharmaceutical processes that feature solutions with ions. For dry processes, the non-conducting background medium with possibly conducting inclusions, displays a more complicated dispersion for the effective permittivity. Given the somewhat limited frequency-band used for the measurements in this article, we choose to approximate the effective permittivity by a frequency-independent function $\epsilon(\boldsymbol{r})=\epsilon'(\boldsymbol{r})-j\epsilon''(\boldsymbol{r})$. This yields the eigenvalue problem

$$\left(\mathbf{C}^{\mathrm{T}}\mathbf{M}_1^{-1}\mathbf{C} - \omega^2\mu_0(\mathbf{M}_{\epsilon'} - j\mathbf{M}_{\epsilon''})\right)\mathbf{e} = \mathbf{0}, \tag{20}$$

which is not particularly attractive for direct computation but useful for the sensitivity analysis that follows below.

Losses due to Finite Conductivity in the Cavity Walls

It is possible to approximately determine the losses due to cavity walls of finite conductivity based on the electromagnetic field solution computed for a PEC boundary. This type of approach yields the quality factor associated with a particular eigenmode and, consequently, its damping [17]. The losses are given by

$$P_{\text{wall}} = \text{Re}\left\{\frac{1}{2}\int_\Gamma Z_\Gamma |\hat{\boldsymbol{n}} \times \frac{\boldsymbol{B}_m}{\mu_0}|^2 d\Gamma\right\}, \tag{21}$$

where $Z_\Gamma=(1+j)/(\sigma\delta_m)$ is the surface impedance for the wall of finite conductivity. Here, $\delta_m=(2/(\omega_m\mu_0\sigma))^{1/2}$ is the penetration depth for the eigenfrequency ω_m and the magnetic flux density $\boldsymbol{B}_m$ of the corresponding eigenmode is computed based on the approximation that the wall conductivity is infinite. Finally, the quality factor is given by $Q_{\text{wall},m}=\omega_m W_m/P_{\text{wall},m}$ where W_m is the time-average energy stored in the vessel by the eigenmode $(\boldsymbol{E}_m,\boldsymbol{B}_m)$, i.e.,

$$W_m = \frac{1}{4}\int_\Omega \epsilon|\mathbf{E}_m|^2 d\Omega + \frac{1}{4}\int_\Omega \frac{1}{\mu_0}|\mathbf{B}_m|^2 d\Omega. \qquad (22)$$

ESTIMATION PROCEDURE

Given the measured scattering parameters, we estimate the complex resonance frequencies by means of a subspace-based multivariable system identification algorithm[18]. This measurement procedure is repeated in a regular fashion throughout the entire duration of the process time, and in the following, we refer to the estimated resonance frequencies as the measured resonance frequencies. First, the vessel is empty and the measured resonance frequencies are denoted $\bar{\omega}_m^{\text{ref}}$, where the overline denotes a measured quantity. Next, the process vessel is loaded with particles and the subsequent measured resonance frequencies $\bar{\omega}_m$ are used to form the deviations in the resonance frequencies $\Delta\bar{\omega}_m = \bar{\omega}_m - \bar{\omega}_m^{\text{ref}}$. As the process evolves, we exploit the relative deviation $\Delta\bar{\omega}_m/\bar{\omega}_m^{\text{ref}}$ to estimate the state of the process and, later in this section, we present formulas that relate a perturbation in resonance frequency to the corresponding perturbations in (i) the material parameters and (ii) the shape of the vessel due to thermal expansion.

The estimation of the material and shape parameters is based on minimization of the misfit between the measured and computed relative deviations in resonance frequencies. Thus, we wish to minimize the following expression

$$\sum_m \left|\frac{\Delta\omega_m}{\omega_m^{\text{ref}}} - \frac{\Delta\bar{\omega}_m}{\bar{\omega}_m^{\text{ref}}}\right|^2, \qquad (23)$$

where $\Delta\omega_m = \omega_m - \omega_m^{\text{ref}}$ is the deviation in the computed resonance frequencies ω_m as compared to the corresponding (computed) result ω_m^{ref} for the empty reference case. The computed resonance frequencies ω_m are functions of the material and shape parameters, and this dependency may be expressed according to

$$\omega_m = \omega_m|_{\text{LP}} + \sum_n \frac{\partial\omega_m}{\partial p_n}|_{\text{LP}}\delta p_n, \qquad (24)$$

for small deviations around a certain linearization point, which is denoted by the subscript LP. The empty cavity may be used as linearization point, but if the permittivity inside the vessel deviates substantially from vacuum, it is beneficial to use a linearization point with higher permittivity in certain regions of the vessel. The perturbations δp_n in the parameters with respect to the linearization point, are stored in a vector **x** and we compute an estimate $\hat{\mathbf{x}}$ of **x** by means of the regularized approach

$$\hat{\mathbf{x}} = \arg\min_{\mathbf{x}} ||\mathbf{A}\mathbf{x} - \mathbf{b}||^2 + \gamma||\mathbf{L}(\mathbf{x} - \tilde{\mathbf{x}})||^2. \qquad (25)$$

Here, the elements of the matrix **A** and the right-hand-side vector **b** are given by

$$A_{mn} = \frac{1}{\omega_m^{\text{ref}}}\frac{\partial\omega_m}{\partial p_n}|_{\text{LP}} \qquad (26)$$

$$b_m = \frac{\Delta\bar{\omega}_m}{\bar{\omega}_m^{\text{ref}}} - \frac{\omega_m|_{\text{LP}} - \omega_m^{\text{ref}}}{\omega_m^{\text{ref}}}. \qquad (27)$$

The first term in Eq. (25) resembles a system of linear equations which is over-determined, i.e., the number of rows in **A** exceeds the number of columns. Hence, the residual $||\mathbf{Ax}-\mathbf{b}||$ can be used as a metric of how well the column space of **A** is able to represent the data vector **b**. The second term in Eq. (25) is a regularization term, which penalizes solutions $\hat{\mathbf{x}}$ away from a given vector $\tilde{\mathbf{x}}$ using a weighted Euclidean norm. The matrix **L** in the regularization term is diagonal with all diagonal elements equal to unity, except for the elements that multiply the shape parameters, which are substantially larger. The balance between the two terms in Eq. (25), i.e., a solution which makes $\mathbf{A}\hat{\mathbf{x}} \approx \mathbf{b}$ and a solution $\hat{\mathbf{x}} \approx \tilde{\mathbf{x}}$, is controlled by the regularization parameter γ. From a Bayesian point of view we can interpret Eq. (25) as the posterior mean estimate of **x** under the assumption that the measurement errors in **b** are Gaussian distributed, have zero mean and covariance matrix $\varsigma^2\mathbf{I}$ and the parameter vector **x** has a Gaussian prior distribution with mean value $\tilde{\mathbf{x}}$ and covariance matrix$(\varsigma^2/\gamma)(\mathbf{L}^T\mathbf{L})^{-1}$[19].

For normal process conditions, the losses in the process vessel are small and, therefore, we compute the sensitivity matrix **A** for linearization points that corresponds to the vessel without losses. Consequently, the permittivity is represented by $\epsilon = \epsilon'_{\text{LP}} + \delta\epsilon' - j\delta\epsilon''$, where ϵ'_{LP} is the permittivity at the linearization point. Here, the perturbations with respect to the linearization point are denoted $\delta\epsilon'$ and $\delta\epsilon''$, which are estimated by means of Eq.(25).

Thus, we use the eigenvalue problem (20) with $\epsilon'(\boldsymbol{r})>\epsilon_0$ and $\epsilon''(\boldsymbol{r})=0$ to compute the eigenfrequencies and eigenmodes at the linearization point, which reduces to a special case of Eq. (18). The computed eigenfrequencies and eigenmodes are also used to compute the sensitivities as described later in this section. The down-bed region with the fluidized particles yields an effective permittivity that deviates substantially from vacuum. Therefore, we use a linearization point

$$\epsilon'_{\text{LP}}(\boldsymbol{r}) = \begin{cases} \epsilon'_{\text{bed}} & \text{in the bed region,} \\ \epsilon_0 & \text{outside the bed region,} \end{cases} \qquad (28)$$

where $\epsilon'_{\text{bed}} \geqslant \epsilon_0$. We solve the eigenvalue problem for a range of bed heights h_{bed} and bed permittivities ϵ'_{bed}, where the solutions from the eigenvalue problem together with the corresponding sensitivities are stored in a database. It is noticed that the resonance frequencies and their sensitivities do not change significantly as h_{bed} decreases, if ϵ'_{bed} is increased to an appropriate extent. This is reasonable since it merely corresponds to a redistribution of the bed material to a smaller region. During the evolution of the process, we continuously seek the linearization point ϵ'_{LP} that yields the best fit with the corresponding deviations in the measured resonance frequencies, where the bed height is fixed at a representative value $h_{\text{bed}} = 5$cm.

Hence, this estimation procedure can be used for a process where the particles grow significantly in size, by means of changing the linearization point ϵ'_{LP}. To summarize, Eq.(25) yields reliable estimates of the perturbations $\delta\epsilon'_{\text{eff}}$ and $\delta\epsilon''_{\text{eff}}$

with respect to the linearization point. This approach is sufficient for most normal process states.

Sensitivity Analysis

The sensitivity analysis is based on the eigensolution ($\boldsymbol{E}_m$, $\boldsymbol{B}_m$, ω_m) computed for a linearization point with $\epsilon'(\boldsymbol{r}) \geq \epsilon_0$ and no losses, i.e., $\epsilon''(\boldsymbol{r})=0$ and $\sigma(\boldsymbol{r})=0$. Given this linearization point, we formulate the sensitivities that relate a small perturbation $\delta\omega_m$ in the resonance frequency ω_m to perturbations in (i) the material parameters $\delta\epsilon'$ and $\delta\epsilon''$ and (ii) the normal displacement δn of the cavity's boundary. Figure 4a shows a subdivision of the cavity's volume that is used for the parameterization of the material and Figure 4b shows the cavity's wall with a set of basis functions used for the representation of the normal displacement, where both parameterizations are axisymmetric.

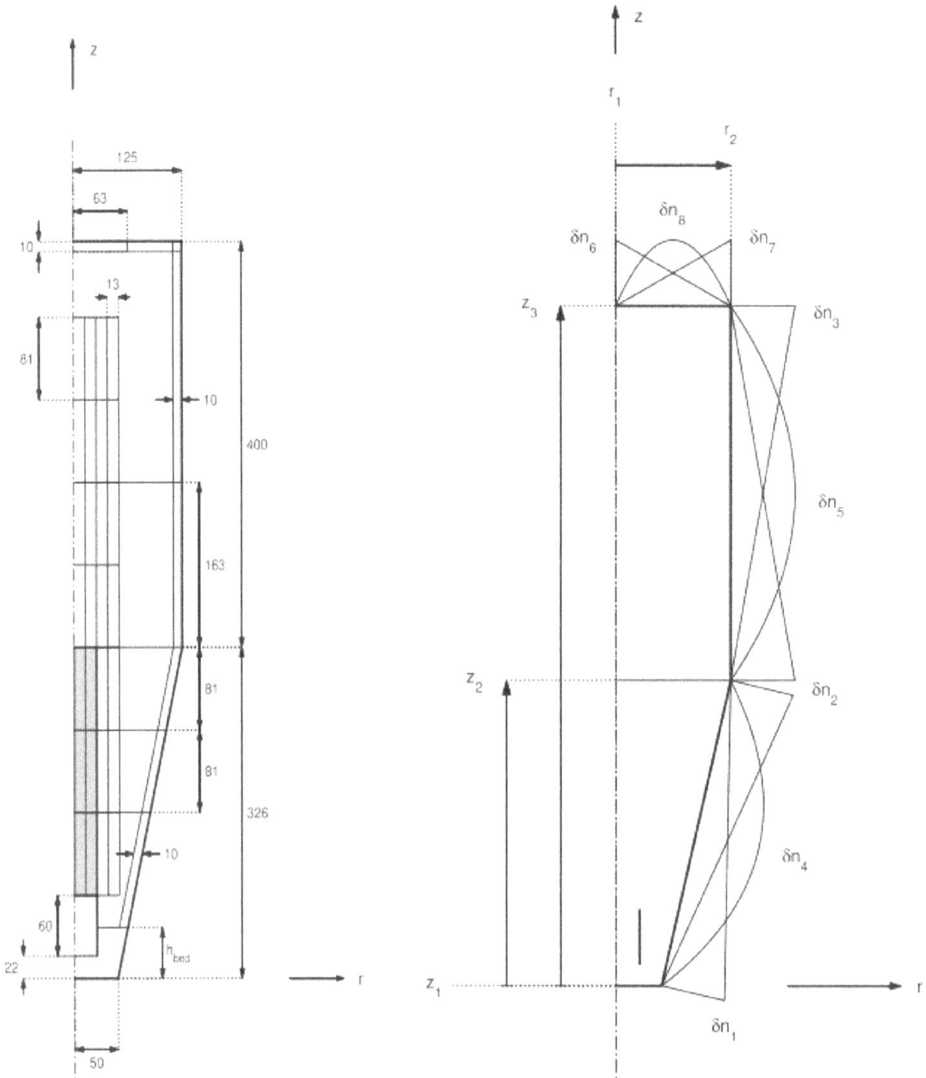

(a) Material representation (b) Shape representation

Figure 4. Two-dimensional projection of the process vessel showing (a) the basic subregions that may be combined in different ways to obtain the material basis functions φ_n and (b) basis functions used to represent shape deformations. All dimensions are in millimeters.

Material Sensitivities

Based on the eigenvalue problem (20), the sensitivity of the resonance frequency ω_m with respect to changes in the material parameters $p=\epsilon'$ and ϵ'' is expressed as

$$\frac{\delta\omega_m}{\omega_m} = -\frac{m_{\delta\epsilon'}(\boldsymbol{E}_m, \boldsymbol{E}_m) - jm_{\delta\epsilon''}(\boldsymbol{E}_m, \boldsymbol{E}_m)}{2m_{\epsilon'}(\boldsymbol{E}_m, \boldsymbol{E}_m)}, \tag{29}$$

where we have

$$m_{\delta p}(\boldsymbol{u}, \boldsymbol{v}) = \int_{\Omega} \delta p \boldsymbol{u} \cdot \boldsymbol{v} d\Omega, \tag{30}$$

for a perturbation δp with respect to the material parameter p [17] and [20]. From Eq.(29), we notice that $\delta\epsilon'$ affects only the real part of the eigenfrequencies, and that $\delta\epsilon''$ only affects the imaginary part. The discrete version of this linearization is

$$\frac{\delta\omega_m}{\omega_m} = -\frac{\mathbf{e}_m^{\mathrm{T}}[\mathbf{M}_{\delta\epsilon'} - j\mathbf{M}_{\delta\epsilon''}]\mathbf{e}_m}{2\mathbf{e}_m^{\mathrm{T}}\mathbf{M}_{\epsilon'}\mathbf{e}_m}. \tag{31}$$

Representation of Material Perturbations

The perturbation δp of the material parameter p is represented in terms of a set of N space dependent basis functions $\varphi_n(\boldsymbol{r})$, which are multiplied with the coefficients δp_n, according to

$$\delta p(\mathbf{r}) = \sum_{n=1}^{N} \delta p_n \varphi_n(\mathbf{r}). \tag{32}$$

Here, we subdivide the vessel's volume Ω into regions Ω_n such that Ω_n do not overlap and $\bigcup_{n=1}^{N} \Omega_n = \Omega$. We exploit piecewise constant basis functions $\varphi_n(\boldsymbol{r})$ that are equal to one if $\boldsymbol{r} \in \Omega_n$ and zero elsewhere. In particular, we associate the basis functions with three regions: (i) the bed region of height h_{bed} with a relatively high density of particles; (ii) a cylinder shaped volume that coincides with the core of the fountain; and (iii) the rest of the process vessel. These regions are formed by combining the basic subregions shown in Figure 4a.

The residual $||\mathbf{A}\hat{\mathbf{x}} - \mathbf{b}||$ in the solution to the estimation problem (25) is used as a metric of how well these basis functions are able to represent the actual permittivity distribution in the experiment. For instance, the height and radius of the fountain region is chosen such that the residual is minimized. Based on measurement results presented later in this article, we find that the fountain region indicated by gray color in Figure 4a yields the best agreement with the measured resonance frequencies.

Shape Sensitivities

For a normal displacement δn of the vessel's boundary Γ, the resulting perturbation in the resonance frequency is given by [17] and [20]

$$\frac{\delta\omega_m}{\omega_m} = -\frac{d_{\delta n}(\boldsymbol{B}_m, \boldsymbol{B}_m)/\mu_0 - d_{\delta n}(\epsilon' \boldsymbol{E}_m, \boldsymbol{E}_m)}{2W_m}, \tag{33}$$

where W_m is the time-average energy in the cavity given by Eq. (22), and

$$d_{\delta n}(\boldsymbol{u}, \boldsymbol{v}) = \int_{\Gamma} \boldsymbol{u} \cdot \boldsymbol{v}^* \delta n \, dS. \tag{34}$$

A positive deformation $\delta n > 0$ here corresponds to that the volume of the cavity increases. We notice that shape deformations only affect the real part of the eigenfrequencies.

The corresponding discrete formulation of Eq. (33) is

$$\frac{\delta\omega_m}{\omega_m} = -\frac{\frac{1}{\mu_0}\mathbf{b}^{\mathrm{H}}\mathbf{D}_{\delta n}\mathbf{b} - \mathbf{e}^{\mathrm{H}}\mathbf{D}_{\epsilon'\delta n}\mathbf{e}}{\frac{1}{\mu_0}\mathbf{b}^{\mathrm{H}}\mathbf{M}_1\mathbf{b} + \mathbf{e}^{\mathrm{H}}\mathbf{M}_{\epsilon'}\mathbf{e}}, \tag{35}$$

where the superscript ^{H} indicates conjugate transpose.

Representation of Shape Perturbations

We use an axisymmetric representation of the normal displacement δn by means of a quadratic hierarchical basis for three different parts of the vessel's boundary: (i) the upper lid; (ii) the cylindrical surface; and (iii) the conical surface. Figure 4b shows the basis functions in relation to the geometry of the vessel. On each of the three parts of the boundary, we exploit the (local) basis functions

$$\psi_1(u) = 1 - u \tag{36}$$

$$\psi_2(u) = u \tag{37}$$

$$\psi_3(u) = 4u(1 - u) \tag{38}$$

where the part of the boundary is parameterized by the local coordinate u such that the interval $0 < u > 1$ corresponds to the entire part of the boundary.

In order to constrain the number of degrees of freedom associated with the cavity's shape to a number that is practical, and to suppress non-physical shape deformations, we form two new basis functions by linearly combining δn_1–δn_8 according to

$$\delta n_{\text{lid}} = 0.97\delta n_6 + 0.24\delta n_8 \tag{39}$$

$$\delta n_{\text{cone/cyl}} = 0.69\delta n_2 + 0.16\delta n_3 + 0.71\delta n_4. \tag{40}$$

Thus, δn_{lid} describes a smooth deformation of the top lid and $\delta n_{cone/cyl}$ describes a joint deformation of the cylindrical and conical parts of the vessel. The coefficient values in Eqs. (39) and (40) are chosen such that the residual $||\mathbf{A}\hat{\mathbf{x}} - \mathbf{b}||$ is minimized, given only these two independent degrees of freedom for the shape.

Temperature Measurements

Before the vessel is loaded with particles, it is normally pre-heated. We exploit this part of the process to gain information that is used to compute the a priori information $\tilde{\mathbf{x}}$ for later parts of the process. During the pre-heating of the vessel, the vessel is empty and we perform simultaneous and synchronized measurements on (i) the resonance frequencies and (ii) the temperature registered by the set of temperature probes shown in Figure 2. Given the deviations $\mathbf{y}_{\delta T}$ in the temperatures (relative to the initial temperatures) and the shape perturbations $\hat{\mathbf{x}}_{\delta n}$ estimated from the resonance frequencies for the empty cavity, we use the relatively simple model $\hat{\mathbf{x}}_{\delta n} = \mathbf{B}\mathbf{y}_{\delta T}$, where the matrix $\mathbf{B}$ is determined by means of linear regression from the data acquired during the pre-heating interval. This model is used to obtain a priori estimates $\hat{\mathbf{x}}_{\delta n} = \mathbf{B}\mathbf{y}_{\delta T}$ from the measured temperatures during later parts of the process, after the particles have been loaded. In the following, we exploit the a priori information

$$\tilde{\mathbf{x}} = \begin{bmatrix} \tilde{\mathbf{x}}_{\delta\epsilon_{\text{eff}}} \\ \tilde{\mathbf{x}}_{\delta n} \end{bmatrix} = \begin{bmatrix} 0 \\ \mathbf{B}\mathbf{y}_{\delta T} \end{bmatrix}, \tag{41}$$

in the estimation problem (25), where $\tilde{\mathbf{x}}_{\delta\epsilon_{\text{eff}}}$ corresponds to all the material parameters that are incorporated in the estimation procedure.

Tests demonstrate that about 95% of the perturbations in the resonance frequencies due to thermal expansion of the vessel during heating and cooling can be accounted for by this model, when the vessel is empty. This approach is less successful once the vessel is loaded with particles, since the presence of particles yields a different temperature distribution in the vessel. However, it is presumed that it is feasible to improve the a priori information, should the metal temperature be measured at more places on the vessel's wall and top lid.

RESULTS

In the measurement results presented in the following, we perform repeated microwave measurements on the process vessel as the process is running. Each process experiment typically takes a few hours in total. The external process parameters such as the flow-rate, temperature and humidity of the fluidizing air, the pressure and flow in the pneumatic spray-nozzle and the flow-rate of the spraying liquid, are controlled from the user interface of the process machine. These external process parameters together with measured temperatures at the locations indicated in Figure 2, are logged by the process machine.

Cavity without Particles

Among the eigenmodes hosted by the process vessel we have identified eight modes that are feasible to use for sensing. These modes are listed in Table 1 together with their resonance frequencies and unloaded Q -values. All these modes have similarities with the TE and TM modes hosted by a circular cylindrical cavity. In particular, the modesm=1,2,5 and 7 resemble the modes TE_{111},TE_{112},TE_{113}, and TE_{114} of a circular cylindrical cavity, and the modes m=3,4 and 6 resemble the modesTM_{010},TM_{011} and TM_{012} respectively. The mode m=7 (TE_{114}) has a strong electric field in the lower part of the vessel, which makes it well-suited for probing the permittivity in the down-bed region. The mode m=8 resembles TE_{011} which has an electric field that is zero on all cavity walls, which implies that its resonance frequency is insensitive to dielectric material adhering to the cavity walls. This mode also has a notably high Q-value.

Table 1. Computed and measured values of the resonance frequencies and unloaded Q-values of the lowest eigenmodes in the process vessel. The variabilities (here represented by one standard deviation) in the measured quantities arise from disassembling and reassembling the complete vessel.

m	Computed		Measured	
	f_0 (MHz)	Q	f_0 (MHz)	Q
1	759	5100	761 ± 0.07	5400 ± 100
2	894	5800	894 ± 0.05	5100 ± 400
3	928	5900	930 ± 0.02	4400 ± 400
4	1008	6000	1009 ± 0.03	3200 ± 800
5	1051	6100	1051 ± 0.06	4700 ± 200
6	1149	6300	1150 ± 0.03	3400 ± 700
7	1207	5600	1206 ± 0.08	3500 ± 300
8	1495	9800	1498 ± 0.05	10,300 ± 400

The loaded Q -values in the experiment can be calculated from the real and imaginary part of the measured complex resonance frequency ω according to

$$Q_l = \frac{\mathrm{Re}(\omega)}{2 \cdot \mathrm{Im}(\omega)}, \tag{42}$$

for the weakly damped cavity modes [17]. Since the cavity is loaded by two probes with different coupling coefficients, the following expression is used to calculate the unloaded Q-value

$$Q_u = Q_l\left(\frac{2}{|S_{11}| + |S_{22}|}\right), \tag{43}$$

where the reflection coefficients S_{11} and S_{22} are evaluated at each resonance frequency [21].

We calculate the Q-value from the electromagnetic model using the method described in Section 3.3, and here we assume that the cavity walls have a conductivity $\sigma_c = 10^6$ S/m, which should be representative for the stainless steel that the vessel is made of.

As can be seen in Table 1, the measured and computed resonance frequencies agree well. There is also a generally good agreement between measured and computed Q-values although, for some modes, the measured Q-value is significantly lower than the computed. The latter may be a result of insufficient electric contact in the joint between the conical and cylindrical part of the cavity, or at the aluminum foil that covers the windows. As the process vessel is disassembled and reassembled, a certain variability in the measured resonance frequencies and Q-values is introduced, which is presented in Table 1. This variability is significant in relation to the perturbations in resonance frequencies caused by changes in the state of the process. Therefore, the particles are inserted into the process vessel through a small hole (indicated by H1 in Figure 2) after the necessary reference measurements on the empty cavity have been collected, in order to eliminate this variability. Investigations show that the variability introduced from inserting the particles in this manner is negligible.

Effects from Air Humidity and Spray

In a Wurster-process, several different components that influence the resonance frequencies are present simultaneously. The most important of these components are (i) particles, (ii) the temperature of the vessel (iii) liquid droplets from the spray nozzle and (iv) water vapor in the fluidizing air. The temperature of the vessel affects the resonance frequencies mainly via perturbations of the cavity's shape, whereas the other components are assumed to influence the resonance frequencies via the permittivity. In order to assess how each of these components contribute to the microwave response, they have been studied separately. Some of these components, such as the spraying liquid and the temperature, are inherently interconnected, i.e., the spraying rate cannot be changed without also affecting the temperature.

The humidity of the fluidizing air can be controlled independently using the process machine and humidities ranging from 0 to 20 (g H_2O)/(kg air), corresponding to dew-points in the range from −30 °C to 25 °C, can be achieved. Our measurements show that the humidity influences only the real part of the permittivity of the air and hence the resonance frequencies, to a degree that is clearly detectable.

Water sprayed from the pneumatic spray nozzle (also referred to as the atomizer) yields similar changes in the permittivity inside the cavity as the humidity does. Only ϵ'_{eff} is influenced by the spray, and if the same amount of water is added through the atomizer as via the fluidizing air, the change in ϵ'_{eff} is very similar. The contributions to the permittivity of the air from humidity in the fluidizing air and from water spray are additive, as long as the spraying rate is sufficiently low to allow for complete evaporation of the liquid droplets.

Measurements with Particles

In the following, we present measurement results for processes involving particles. The particles used are made of MCC

(Cellets® brand) and have diameters in a distribution from 500 to 700 μm. A batch consisting of 200 g of particles is used in all of the following experiments. All process settings are identical in the following experiments, except for the type of spraying solution and the spraying rate. The fluidization flow is set to 30Nm³/h (where N denotes normal temperature and pressure) with an inlet temperature of 70 °C. The atomizer pressure is 2.0bar and the associated atomizer flow is 2.0Nm³/h.

The spraying liquid used in the experiments is water containing either (i) Mannitol or (ii) NaCl. The permittivity of these spraying liquids is measured using the dielectric probe kit 85070E from Agilent Technologies, and the results are shown in Figure 5.

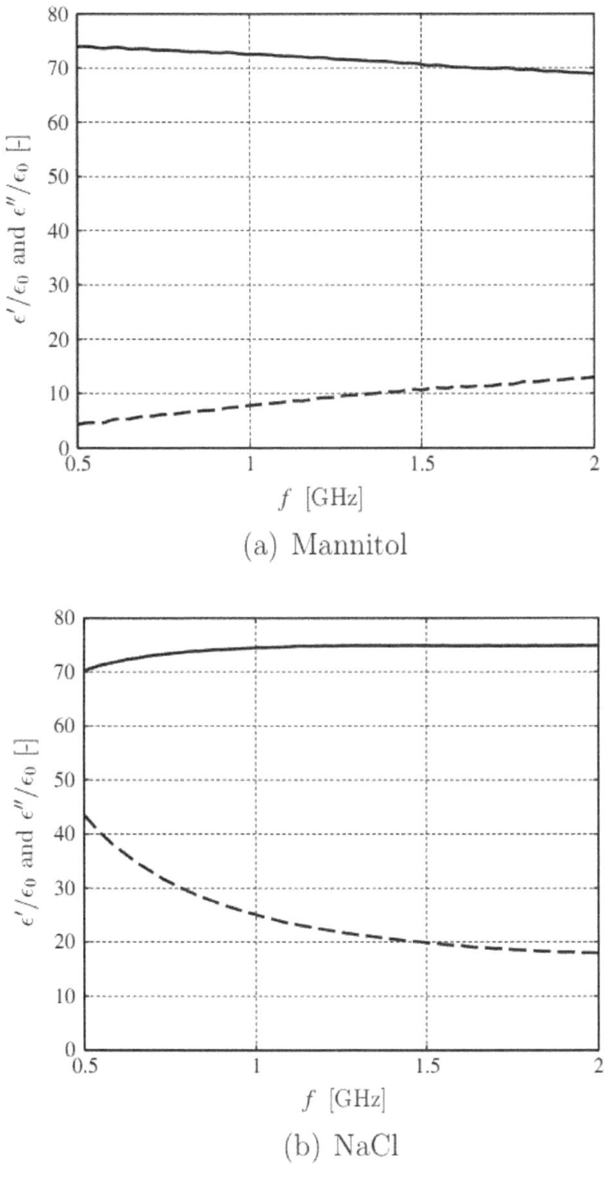

Figure 5. Relative permittivity of distilled water at room temperature that contains (a) Mannitol (13.5%) and Kollicoat IR® (1.5%), and (b) NaCl (0.5%). Solid curves represent ϵ'/ϵ_0 and dashed curves ϵ''/ϵ_0.

In the beginning of each experiment, the vessel is pre-heated and the particles are loaded, before the experiment-specific part of the process (e.g. coating) is initiated. In the results presented in the following, we omit the pre-heating interval and focus on the experiment-specific parts.

The average error between the model and measurement is computed as the root-mean-square of the difference between measurements ($\mathbf{b}$) and model fit ($\mathbf{A}\hat{\mathbf{x}}$), averaged with respect to the number of sensing modes M. With particles present in the vessel, this average error is in the order of 10^{-5} during normal process conditions (no extensive agglomeration, etc.) which should be compared with the average magnitude of the measurement response ($\|\mathbf{b}\|/M$), which is in the order of 10^{-3}.

According to the sensitivity analysis in Section 4.1, perturbations in the real part of the resonance frequencies are related to both $\delta\epsilon'_{\text{eff}}$ and δn. Similarly, a perturbation in the imaginary part of the resonance frequency is related to $\delta\epsilon''_{\text{eff}}$. As described in Section 4.2,δn is to a large extent determined based on a priori information that exploits temperature measurements. An error in the a priori information may yield a corresponding error in $\delta\epsilon'_{\text{eff}}$, which varies on a timescale that is significantly slower than the fast timescales associated with the fluid-dynamics of the process. Based on these two time-scales, we make the assumption $\epsilon_{\text{eff}}(t)=\epsilon_{\text{eff,slow}}(t)+\epsilon_{\text{eff,fast}}(t)$, where $\epsilon_{\text{eff,fast}}(t)$represents fast random fluctuations and $\epsilon_{\text{eff,slow}}(t)$ represents slow changes in the permittivity caused either by errors in the a priori information or by changes in the actual process state (such as particles growing in size or accumulation of moisture in the particles). We model $\epsilon_{\text{eff,slow}}(t)$ by a low-order polynomial with respect to time t , and this polynomial is determined by means of linear regression based on the estimated $\epsilon_{\text{eff}}(t)$. Furthermore, we compute the covariance matrix from the real and imaginary part of $\epsilon_{\text{eff,fast}}(t)$ according to $\mathbf{C}$=cov($\mathbf{z}_1$,$\mathbf{z}_2$) where $\mathbf{z}_1 = [\epsilon'_{\text{eff,fast}}(t_n)/\epsilon_0 - 1]^{\text{T}}$ and $\mathbf{z}_2 = [\epsilon''_{\text{eff,fast}}(t_n)/\epsilon_0]^{\text{T}}$ are vectors containing samples of $\epsilon_{\text{eff,fast}}$ taken at discrete time instants t_n. If the eigenvectors of $\mathbf{C}$ are scaled by the square root of its eigenvalues (i.e. the standard deviations), they define the major and minor semi-axes of an ellipse that represent the co-variation of the data. Such ellipses are included in the scatter-plots in the presentation of the following measurement results. Variations in $\epsilon_{\text{eff,fast}}(t)$ along the eigenvector associated with the largest eigenvalue of $\mathbf{C}$, are mainly attributed to fluctuations in the particle volume fraction ϑ. Therefore, we take the slope of this eigenvector (i.e. the major semi-axis of the ellipse) as a density-independent ratio which is equivalent to the ratio η in Section 2.2.2.

The relationship between this slope and underlying physical parameters such as the moisture content of the particles and the concentration of salt (NaCl) in the spraying liquid, have been studied by means of Monte-Carlo simulations. These simulations are based on measurements of the complex permittivity of MCC [11] and empirical models for the permittivity of water with and without salt [10], together with the mixing formulas described in Section 2.2.1. Thus, we generate random mixtures of MCC particles and air by using statistical normal-distributions to represent the volume fraction and water content of the particles. The water is assumed to be either absorbed uniformly in the MCC-particles or distributed in a uniform water-layer on the surface of the particles. The case with a pronounced water-layer may be representative also for situations where the surface of the particles have a substantially higher moisture content than the core. Based on these simulations, we observe the following trends:

- An increase in the moisture content of the MCC particles (absorbed water) *increases*the slope, if the initial moisture content is less than around 7%. For higher moisture contents, the slope instead *decreases* with increasing moisture content.
- Pure water distributed on the surface of the particles *decreases* the slope.
- Water containing salt distributed on the surface of the particles *increases* the slope if the salt concentration exceeds 1%.
- The slope is *not* affected by changes in the volume fraction, if the volume fraction is sufficiently low (around 5% or lower), whereas at higher volume fractions, the slope *increases* with increasing volume fraction.

Fast Coating Process

Here, we present results from a process where MCC particles are coated with a mixture of Mannitol (a sugar) and a polymer labeled Kollicoat IR® (ethylene glycol vinyl acetate grafted polymer), where the Mannitol and the Kollicoat IR are dissolved in de-ionized water. The dry content of the solution (the fraction that ideally stays on the surface of the particles) is 15 mass% (13.5% Mannitol and 1.5% Kollicoat IR). The complex relative permittivity for this solution is shown in Figure 5a, and here we notice that the response is similar to that of pure water. In this experiment, a relatively high spraying rate is used to obtain fast build-up of the coating layer. First, the spraying rate is set to 11.3 g/min and kept constant during approximately 15 min. Next, the spraying rate is increased even further to 14.4 g/min, which eventually leads to extensive agglomeration and particles adhering to the walls of the vessel.

Figure 6 shows the estimated permittivity in three different regions: (i) the down-bed region; (ii) the fountain region above the Wurster tube; and (iii) the rest of the volume above the down-bed region, as function of process time in hours. At t = 0.9 h, the particles are fluidized in the process with the fountain off, and the temperature is in equilibrium. The following changes in process settings are marked with dashed vertical lines: at t = 1.04 h the fountain is switched on; in the interval 1.2 h < t < 1.26 h the spraying rate is successively increased to 11.3 g/min; at t = 1.51 h the spraying rate is increased to 14.4 g/min. Some interesting features can be observed in these figures:

The estimated permittivities fluctuate substantially, which is reasonably caused by the continuous re-distribution of the particles in the process which occurs on a short time scale (<0.1 s).

- As the fountain is switched on, the estimated permittivity increases clearly in the fountain and remaining volume regions, whereas it decreases in the bed region. This is reasonable since as the fountain is switched on, particles start to circulate in the upper part of the vessel which reduces the amount of particles in the down-bed region.
- The estimated permittivity in the fountain decreases slightly as the spraying starts. This could be an artifact (incorrect estimation) resulting from insufficient compensation for temperature variations, but it could also be due to a lower particle density in the fountain as the spraying starts. The estimated permittivity in the remaining volume increases as the spraying starts, reasonably due to increased air humidity inside the vessel.
- During the interval 1.5 h < t < 1.75 h, extensive agglomeration occurs in the process and the particles start to build up dense layers on parts of the cavity walls, as a result of the high spraying rate (this could be observed visually after the experiment was finished). This process is clearly reflected in the estimated permittivities, which are changing dramatically during the agglomeration interval.

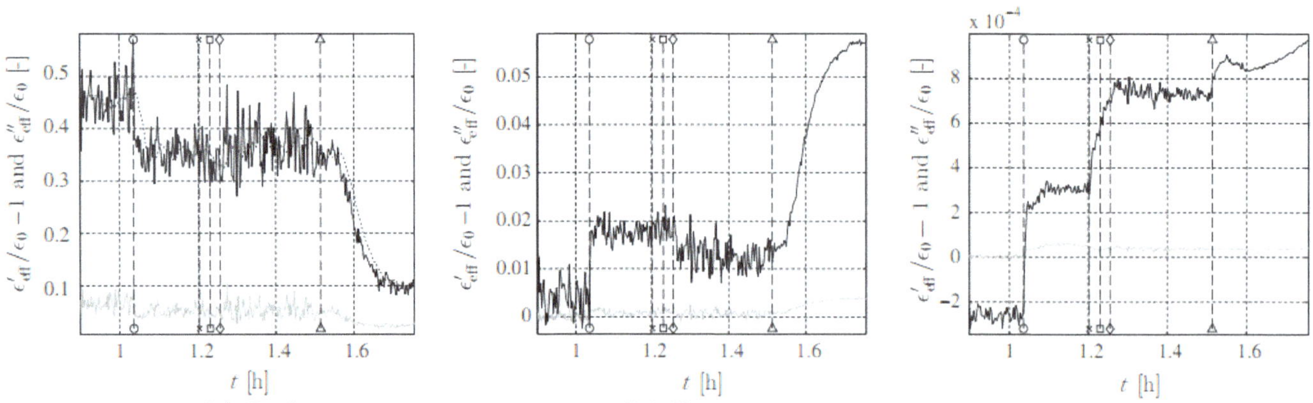

(a) Bed (b) Fountain (c) Volume

Figure 6. Estimated permittivity as a function of process time in the fast coating process. Black lines represent $\epsilon'_{\text{eff}}/\epsilon_0 - 1$ and gray lines $\epsilon''_{\text{eff}}/\epsilon_0$. The dashed black curve in (a) shows the permittivity in the bed region at the current linearization point.

In situations similar to the agglomeration interval described in item 4 above, i.e. at very moist process conditions that yields material build-up on the vessel walls, we repeatedly observe a larger discrepancy between the model and measurements, i.e. an increasing residual $||\mathbf{A}\hat{\mathbf{x}} - \mathbf{b}||$. Thus, a model that allows only for a constant permittivity in the bed, fountain and remaining volume regions, is unable to accurately represent the material distributions formed during extensive agglomeration.

This indicates the possibility of detecting anomalous process conditions based on the residual. Studies show however that it is possible to maintain a low residual also during extensive agglomeration if the cavity is divided into sub-regions that better represent the actual material distribution. The sub-regions in the direct vicinity of the cavity walls shown in Figure 4a may be well-suited for this purpose.

Figure 7 shows scatter plots of $\epsilon''_{\text{eff}}/\epsilon_0$ versus $\epsilon'_{\text{eff}}/\epsilon_0 - 1$ in the three regions before and after spraying starts. The ellipses in these scatter plots represent the co-variation of the data, where the length of each semi-axis is three standard deviations. All ellipses are highly elongated and the variations along the major semi-axis is assumed to be caused by variations in material density, according to the discussion on density-independent functions in Section 2.2.2. The slopes in the scatter plots are shown in Table 2. Here we notice that (i) the slope is higher in the bed region, as compared to the fountain and volume regions and (ii) the slope decreases in the fountain and in the volume after the spraying starts, but remain unchanged in the bed. According to the results from the Monte-Carlo simulations described previously, the decreasing slope in the fountain and volume regions could indicate that the moisture content at the surface of these particles increases during the spraying interval. The higher slope in the bed is in agreement with the Monte-Carlo simulations, since the volume fraction in the bed is substantially higher than in the other regions.

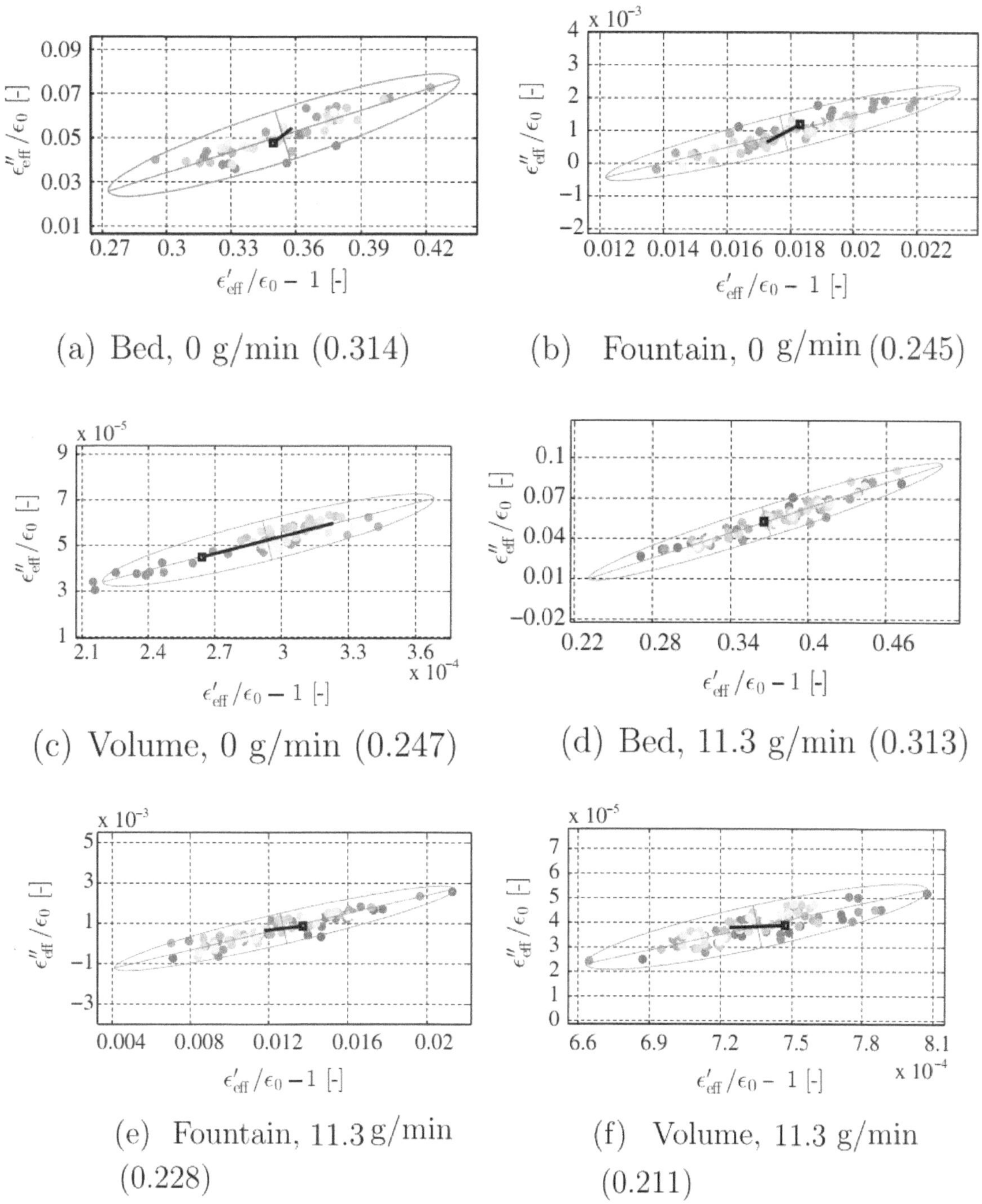

(a) Bed, 0 g/min (0.314)
(b) Fountain, 0 g/min (0.245)
(c) Volume, 0 g/min (0.247)
(d) Bed, 11.3 g/min (0.313)
(e) Fountain, 11.3 g/min (0.228)
(f) Volume, 11.3 g/min (0.211)

Figure 7. Scatter plots of $\epsilon''_{\text{eff}}/\epsilon_0$ versus $\epsilon'_{\text{eff}}/\epsilon_0 - 1$ in the fast coating process. Values inside parentheses are the slopes of the major semi-axis of the ellipses. The black lines starting with squares represent the slowly varying component $\epsilon_{\text{eff,slow}}$ of the estimated permittivity.

Table 2. Slopes of the major semi-axis in the scatter plots of $\epsilon''_{\mathrm{eff}}$ versus $\epsilon'_{\mathrm{eff}} - 1$.

	Bed	Fountain	Volume
Spraying off	0.314	0.245	0.247
Spraying 11.3 g/min	0.313	0.228	0.211

Slow Coating Process

In this experiment, MCC particles are coated with Mannitol and Kollicoat IR® with the mixing ratio described in Section 5.2.1, under relatively dry process conditions, i.e., a relatively low spraying rate of 6.6 g/min is used.

Figure 8 shows the estimated permittivity in the bed, fountain and remaining volume regions, as function of process time in hours. The time interval 0 h < t < 0.9 h, which has been omitted, includes pre-heating of the vessel and loading of particles, so at t = 0.9 h, the particle fountain is switched on and the temperature is close to equilibrium. The process conditions are changed at the time instants indicated by the dashed vertical lines: at$t$ = 0.89 h spraying (coating) starts with a spraying rate of 4.6 g/min; at t = 0.96 h the spraying rate is increased to the final value 6.6 g/min; at t = 1.13 h the temperature is considered stable and at t = 1.78 h the spraying ends. In these figures, some interesting features can be observed:

- The estimated permittivity in the bed increases slowly during the coating interval 0.90 h < t < 1.78 h, whereas the permittivities in the fountain and the remaining volume regions remain fairly constant. This may be explained by particles growing in size as a result of the coating, which results in more dielectric material that accumulates mainly in the bed region. The growth in particle size was verified by means of sieving and weighing the particles after the coating process was finished, and the mass-increase was found to be approximately 25%. A bed that grows in size in the experiment is interpreted by our model as an increase in the effective permittivity in the bed-region, which has a fixed size. This effect can however not be clearly observed in the fast coating experiment described in Section 5.2.1, which is somewhat surprising given that we use the same coating solution in both processes.
- The permittivity in the remaining volume region (Figure 8) increases as the spraying starts, it increases further as the spraying rate increases and drops as the spraying is stopped. This may be attributed to changes in the permittivity of the air, due to increased humidity caused by the evaporating water in the spray.

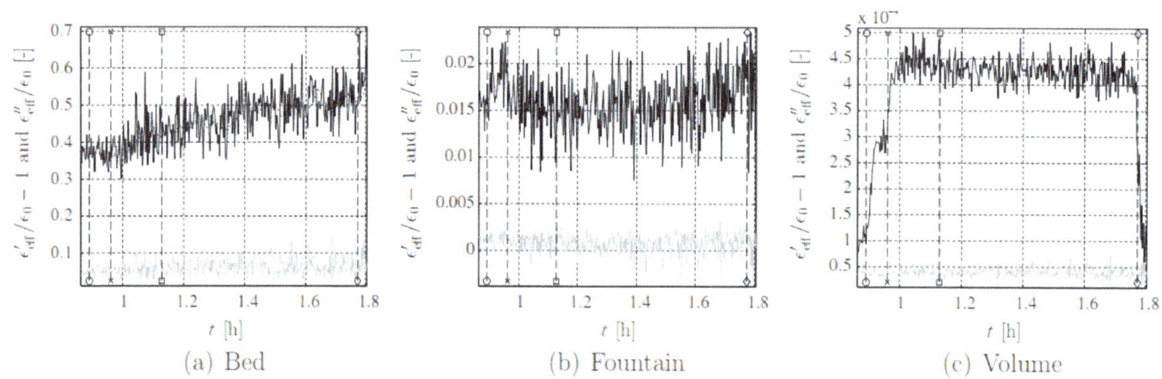

Figure 8. Estimated permittivity as a function of process time in the slow coating process. Black curves represent $\epsilon'_{\mathrm{eff}}/\epsilon_0 - 1$ and gray curves $\epsilon''_{\mathrm{eff}}/\epsilon_0$. The dashed black curve in (a) shows the permittivity in the bed region at the current linearization point.

Figure 9 shows scatter plots of $\epsilon''_{\mathrm{eff}}/\epsilon_0$ versus $\epsilon'_{\mathrm{eff}}/\epsilon_0 - 1$ for the tree regions, during the coating interval. Here, we notice that the slope is higher in the bed as compared to the other two regions, which is reasonable given the higher volume fraction in the bed. We also notice how the slowly varying component $\epsilon_{\mathrm{eff,slow}}$ shown by black lines in Figure 9, varies substantially in the bed region, reasonably due to the growth in particle size.

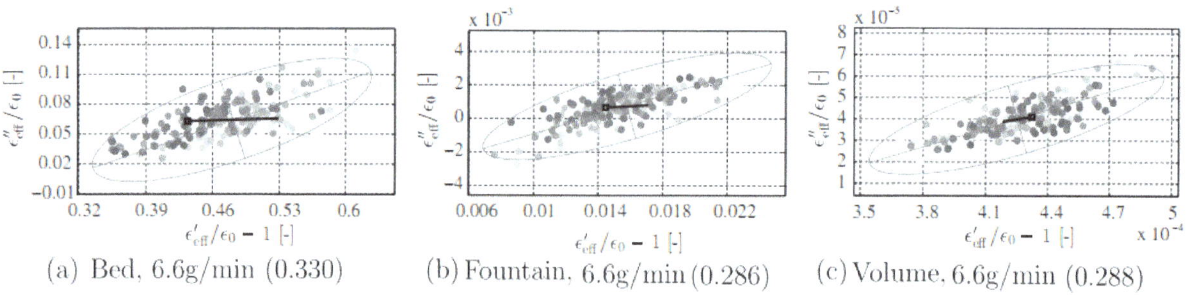

Figure 9. Scatter plots of $\epsilon''_{\mathrm{eff}}/\epsilon_0$ versus $\epsilon'_{\mathrm{eff}}/\epsilon_0 - 1$ in the slow coating process. Values inside parentheses are the slopes of the major semi-axis of the ellipses. The black lines starting with squares represent $\epsilon_{\mathrm{eff,slow}}$, i.e. the slowly varying component of the estimated permittivity.

Particles Sprayed with Salt Water

In this experiment, MCC particles are sprayed with salt water (0.5 mass% NaCl dissolved in distilled water at room temperature). The complex relative permittivity for this solution is shown in Figure 5b and here we notice a significant increase in ϵ'' at low frequencies due to the conductivity introduced by the salt. Two different spraying rates are studied, namely 11.6 g/min and 13.3 g/min.

Figure 10 shows the estimated permittivity in the bed, fountain and remaining volume regions, as function of process time in hours. At the first time instant shown in Figure 10, the particles are fluidized, the fountain is on and the temperature is stable. The following changes in process settings are marked with dashed vertical lines: at $t = 1.84$ h the spraying starts (11.6 g/min); at $t = 2$ h the temperature is considered stable; at $t = 2.24$ h the spraying rate is increased to 13.3 g/min which yields very moist process conditions and presumed initiation of extensive agglomeration. The increase in spraying rate to 13.3 g/min is followed by clear changes in the estimated permittivities, possibly due to free water (with salt) on the surface of the particles in the fountain region and/or agglomeration with particles adhering to the walls of the vessel.

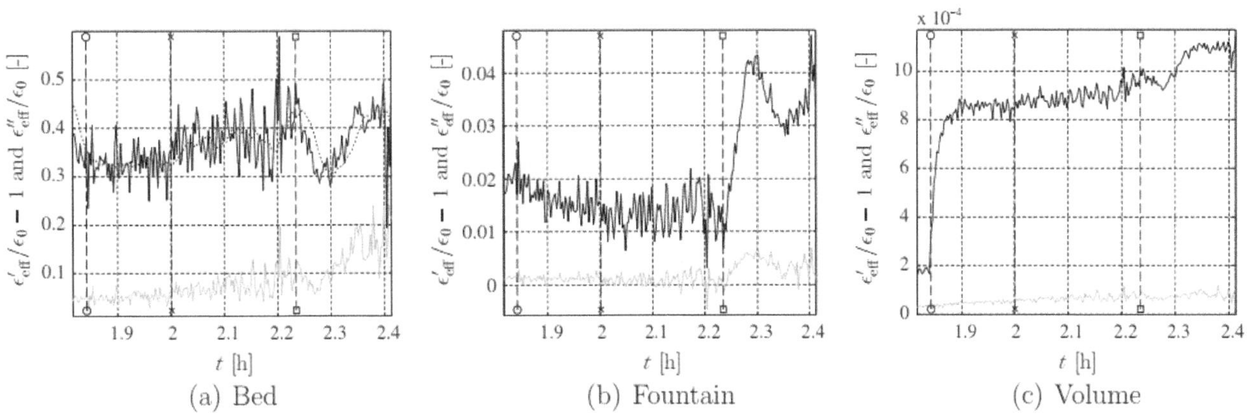

Figure 10. Estimated permittivity as a function of process time in the salt water experiment. Black curves represent $\epsilon'_{eff}/\epsilon_0 - 1$ and gray curves $\epsilon''_{eff}/\epsilon_0$. The dashed black curve in (a) shows the permittivity in the bed region at the current linearization point.

Figure 11 shows scatter plots of $\epsilon''_{eff}/\epsilon_0$ versus $\epsilon'_{eff}/\epsilon_0 - 1$ in two intervals: (i) no spray and (ii) spraying at 11.6 g/min (which does *not* yield agglomeration). The slopes of the major axis for the three regions is also presented in Table 3. Here, it is interesting to notice how the slopes in all regions clearly increase after the spraying is switched on. This is expected, given the high dielectric losses of the salt water in the frequency range of the cavity resonances, and it is also in agreement with the Monte-Carlo simulations presented previously.

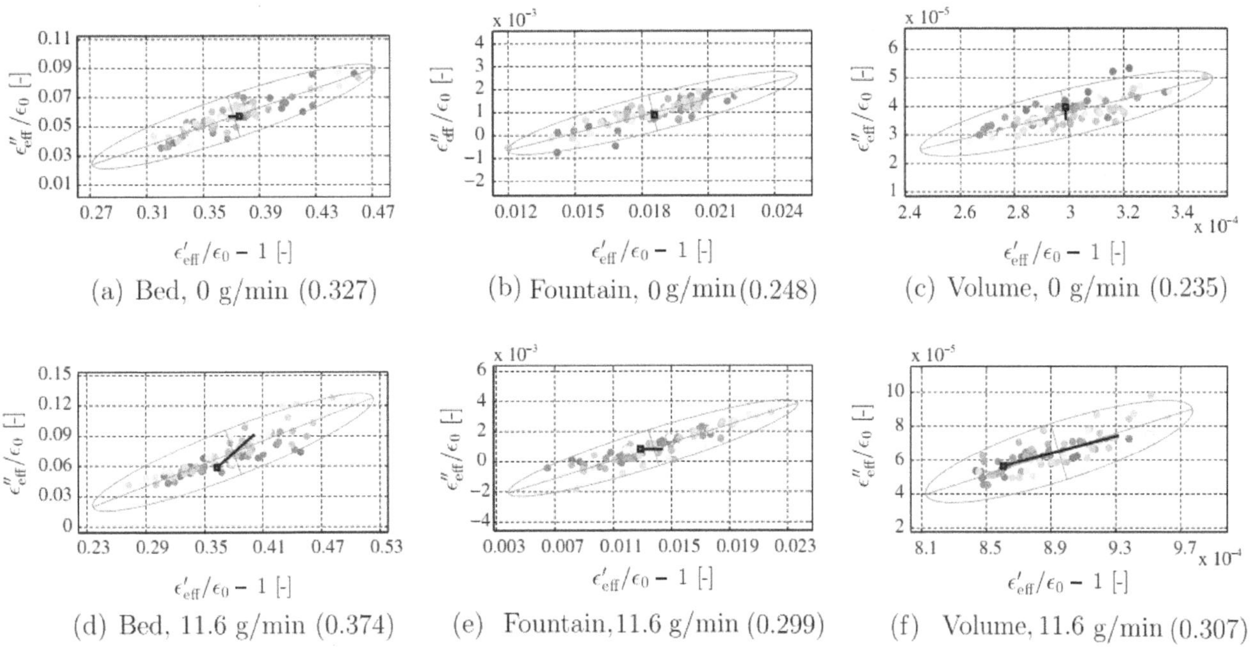

Figure 11. Scatter plots of $\epsilon''_{eff}/\epsilon_0$ versus $\epsilon'_{eff}/\epsilon_0 - 1$ in the salt water experiment. Values inside parentheses are the slopes of the major semi-axis of the ellipses. The black lines starting with squares represent the slowly varying component $\epsilon_{eff,slow}$ of the estimated permittivity.

Table 3. Slopes of the major semi-axis in the scatter plots of ϵ''_{eff} versus $\epsilon'_{\text{eff}} - 1$ for the salt water experiment.

	Bed	Fountain	Volume
Spraying off	0.327	0.248	0.235
Spraying: 11.6 g/min	0.374	0.299	0.307

CONCLUSIONS

We have presented a microwave measurement technique for industrial processes contained in closed metal vessels, which exploits a number of electromagnetic resonances of the metal vessel to estimate the process state. In particular, we estimate the effective permittivity inside a vessel that hosts a pharmaceutical fluidized-bed process, which is used for particle coating. This process involves spraying a solution onto the particles, where water acts as the solvent for different types of coating substances. For this particular application, it is important to monitor the moisture content of the particles and the particle distribution. Electromagnetic waves interact strongly with water in a non-invasive and non-destructive manner, which makes this type of measurement technology very attractive. In addition, our measurement system is sensitive to changes in permittivity throughout the process vessel and, thus, it can be considered to perform a global measurement of the process state, which is challenging for more conventional techniques.

We exploit two magnetic field probes and measure the scattering parameters as function of frequency using a network analyzer. The complex resonance frequencies are estimated from the scattering parameters and these are related to a reference measurement of the empty vessel in order to achieve relative changes in the resonance frequencies. The perturbations in the resonance frequencies are related to (i) the perturbations in the complex effective permittivity inside the vessel and (ii) the thermal expansion of the vessel caused by variations in the temperature. We exploit an estimation algorithm with regularization to estimate the effective permittivity and the shape of the vessel, where the unknown effective permittivity and shape is parameterized with the aid of spatial basis functions. The real and imaginary part of the effective permittivity are assumed to be frequency-independent in the frequency band of the used resonance frequencies. We model the vessel by means of the finite element method and the computed eigenmodes and eigenfrequencies are used to calculate sensitivities with respect to changes in the material parameters and the shape of the vessel. The estimation incorporates a priori information that is collected from probes that measure the temperature of the vessel's metal walls and the air that flows through the vessel. This approach accounts for substantial parts of the temperature variations but it is believed that it can be significantly improved.

The measurement technique performs well for the estimation of the effective permittivity that is parameterized for a small number of sub-regions. The re-distribution of material occurring when the particle fountain is switched on, is clearly reflected in the estimated permittivity in all these sub-regions. The effect of particles that grow in size during coating processes can be observed in the estimated permittivities, and we have also shown that material build-up on the cavity walls can be clearly detected. We have related the real part of the effective permittivity to its imaginary part by means of scatter plots and covariance matrices, where measurements taken at different time instants fall approximately on a straight line. This line is defined by the eigenvector associated with the largest eigenvalue of the covariance matrix, and the slope of the line is affected by the liquid sprayed onto the particles in the process. In particular, the slope of this line clearly increases if the particles are sprayed with a liquid that features substantial dielectric losses.

ACKNOWLEDGEMENTS

Livia Cerullo was financially supported by AstraZeneca. In addition, Johan Winges was financially supported by the Swedish Research Council (dnr 2010-4627) in the project "Model-based Reconstruction and Classification Based on Near-Field Microwave Measurements". The computations were performed on resources at Chalmers Centre for Computational Science and Engineering (C3SE) provided by the Swedish National Infrastructure for Computing (SNIC).

REFERENCES

1. J. Rantanen, S. Lehtola, P. Rämet, J.-P. Mannermaa, J. Yliruusi, On-line monitoring of moisture content in an instrumented fluidized bed granulator with a multi-channel NIR moisture sensor, Powder Technol. 99 (1998) 163–170.
2. T.D. Beer, A. Burggraeve, M. Fonteyne, L. Saerens, J. Remon, C. Vervaet, Near infrared and Raman spectroscopy for the in-process monitoring of pharmaceutical production processes, Int. J. Pharm. 417 (2011) 32–47.
3. C. Buschmüller, W. Wiedey, C. Döscher, J. Dressler, J. Breitkreutz, Inline monitoring of granule moisture in fluidized-bed dryers using microwave resonance technology, Euro. J. Pharm. Biopharm. 69 (2008) 380–387.
4. V. Rimpiläinen, L.M. Heikkinen, M. Vauhkonen, Moisture distribution and hydrodynamics of wet granules during fluidized-bed drying characterized with volumetric electrical capacitance tomography, Chem. Eng. Sci. 75 (2012) 220–234.
5. M. Dressel, O. Klein, S. Donovan, G. Grüner, Microwave cavity perturbation technique: Part III: applications, Int. J. Infrared Millimeter Waves 14 (1993) 2489–2517.
6. B. Meng, J. Booske, R. Cooper, Extended cavity perturbation technique to determine the complex permittivity of dielectric materials, IEEE Trans. Microwave Theory Tech. 43 (1995) 2633–2636.
7. S. Karlsson, A. Rasmuson, B. van Wachem, I.N. Björn, CFD modeling of the Wurster bed coater, AIChE J. 55 (2009) 2578–2590.
8. A. Sihvola, Electromagnetic Mixing Formulas and Applications, The Institution of Engineering and Technology (2008).
9. G. Kristensson, S. Rikte, A. Sihvola, Mixing formulas in the time domain, J. Opt. Soc. Am. 15 (1998) 1411–1422.
10. E. Nyfors, P. Vainikainen, Industrial Microwave Sensors, Artech House, 1989.

11. L. Gradinarsky, H. Brage, B. Lagerholm, I.N. Björn, S. Folestad, In situ monitoring and control of moisture content in pharmaceutical powder processes using an open-ended coaxial probe, Measure. Sci. Technol. 17 (2006) 1847–1853.
12. S. Trabelsi, S.O. Nelson, Density-independent functions for on-line microwave moisture meters: a general discussion, Measure. Sci. Technol. 9 (1998) 570–578.
13. D.A. Engers, M.N. Fricke, R.P. Storey, A.W. Newman, K.R. Morris, Triboelectrification of pharmaceutically relevant powders during low-shear tumble blending, J. Electrostat. 64 (2006) 826–835.
14. D.A. Engers, M.N. Fricke, A.W. Newman, K.R. Morris, Triboelectric charging and dielectric properties of pharmaceutically relevant mixtures, J. Electrostat. 65 (2007) 571–581.
15. J.C. Nédélec, Mixed finite elements in R3, Numer. Math. 35 (3) (1980) 315–341.
16. T. Rylander, A. Bondeson, P. Ingelström, Computational Electromagnetics, Springer, 2012.
17. D. Pozar, Microwave Engineering, second ed., Wiley, 1998.
18. T. McKelvey, H. Akçay, L. Ljung, Subspace-based multivariable system identification from frequency response data, IEEE Trans. Automatic Control 41 (1996) 960–979.
19. S.M. Kay, Fundamentals of Statistical Signal Processing: Estimation Theory, Prentice Hall, 1993.
20. I.W. Lee, D.O. Kim, G.H. Jung, Natural frequency, Natural frequency and mode shape sensitivities of damped systems: Part I, distinct natural frequencies, J. Sound Vib. 223 (1999) 399–412.
21. K. Leong, J. Mazierska, Measurements of unloaded Q-factor of transmission mode dielectric resonators, in: IEEE MTT-S International Microwave Symposium Digest 3, 1997.

CHAPTER 10

OFFSHORE WIND TURBINE BLADES MEASUREMENT USING COHERENT LASER RADAR

Jeremy Talbot [a], Qing Wang [a], Neil Brady [b], Roger Holden [b]

[a] School of Engineering and Computing Sciences, Durham University, DH1 3LE, UK
[b]Nikon Metrology UK Ltd, Staffordshire B77 5ES, UK

ABSTRACT

To maximize aerodynamic efficiency, large-scale offshore wind turbine blades require inspection during the production stage to ensure strict tolerance requirements are met. During production, the blade is fixed at the root, restricting movement in the *Z* direction. *X*,*Y*, *Rx*, *Ry* and *Rz* remain unconstrained causing blade flex due to gravity. This deforms the blade away from the theoretical CAD blade location, causing measurement results that do not accurately represent the blade profile. Measurement error can be minimized using rigorous B-spline data alignment. Such alignment compensates for blade flex by varying the constrained Degrees of Freedom (DoF), and provides manufacturers with confidence in the design process. This paper used Coherent Laser Radar and Spatial Analyzer to establish the optimal constrained DoF variation, giving the most accurate data alignment solution. Of the constraints investigated, the optimal data transformation solution was found with a double B-spline alignment method, whilst constraining movement in *Y*, *Z* and *Ry*.

INTRODUCTION

Changing climate and increasing awareness of environmental issues in recent years has resulted in a huge increase in the use of low-carbon technologies, particularly wind energy. The contribution of electricity generation from renewable energy has increased from 2.6% in 2000 to 11.5% in 2012 [1] (a target of 15% is set for 2020). Of this renewable energy, 47% is produced by wind power (29% onshore and 18% offshore). This is expected to increase further still.

To produce a large quantity of energy (usually between 1.5 and 4 MW), wind turbines must be extremely large and are therefore subject to strong wind loads. To ensure blades can survive these high wind loads whilst remaining lightweight, they are manufactured with an internal frame supporting an outer shell made from reinforced plastics [2] and [3].

To maximize aerodynamic efficiency, turbine blades must satisfy extremely tight tolerances and are therefore inspected during manufacture. However, inspection shows a divergence from the theoretical blade design due to shrinkage of the reinforced plastic during manufacture and blade flex under its own weight. Hence, minimizing the three-dimensional (3-D) measurement error during blade inspection is imperative. It

Citation: Jeremy Talbot, Qing Wang, Neil Brady, Roger Holden, Offshore wind turbine blades measurement using Coherent Laser Radar, Measurement, Volume 79, February 2016, Pages 53-65, ISSN 0263-2241, http://dx.doi.org/10.1016/j.measurement.2015.10.030.

provides manufacturers with confidence in the design process; highlighting areas of the blade that deviate away from the CAD model allows the production of more aerodynamically efficient blades.

If flexible, deformation of a turbine blade (or any structure) can consist of up to six Degrees of Freedom (DoF): translational (*X*, *Y* and *Z*), and rotational (*Rx*, *Ry* and *Rz*) directions. Figure 1 shows a typical large-scale wind turbine blade with a reference coordinate frame showing the translational *X*, *Y* and *Z* directions.

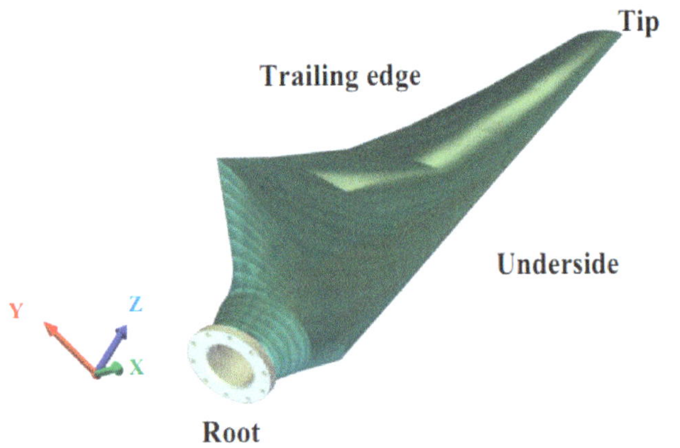

Figure 1. A typical wind turbine blade. (Blade image by M.A. Homel [4].)

The blade is also free to rotate about the *X*, *Y* and *Z* axes, relating to *Rx*, *Ry* and *Rz* rotational directions.

The inspection of turbine blades compares the measured blade data to nominal point groups on a computer-aided design (CAD) model using data alignment techniques. These techniques transform the measured data using different constrained DoF variations to account for blade flex. The measurement error is minimized by establishing the most accurate DoF constrained data alignment variation.

There are a number of software packages available that can provide a platform for such data alignment including Spatial Analyzer (SA) which was used in this investigation.

Nikon have recently developed a method [5] for inspecting turbine blades using CLR; it is currently in use by Vestas Winds Systems A/S. The method divides a blade into sections using multiple scanner locations to measure the entire blade (six locations are needed for a 60 m blade). The blade is clamped at the root, positioned horizontally with the trailing edge directed upward and supported on a 'tray' half way along the length of the blade. It can be assumed clamping the blade at the root in this manner restricts any movement in the *Z* direction, allowing inspection through five DoF.

An advantage of this method is CLR's unique ability to precisely measure with retroreflective mirrors. Using dedicated mirrors expands the range of sight, enabling the measurement of difficult-to-access areas such as the underside of the blade, allowing accurate inspection of the entire blade.

The data alignment technique currently used by Nikon implements *Z*, *Rx*, and *Ry* constraints, allowing movement in the *X*, *Y* and *Rz* directions. As this method has only recently been developed, very little experimental evaluation has been carried out; DoF constraint variations have been chosen using trial and error. This is especially true of the use of mirrors in CLR. The method therefore requires validation before it becomes common practice industrially.

This research investigates the Nikon inspection method and continues the work that demonstrated a data alignment solution based on a 'D-shaped', semi-circular blade design [6]. This research uses a similar method to that in [6] on a more complex, realistic blade profile which necessitates the use of mirrors.

Measurement error is minimized by evaluating each DoF constraint variation to determine the optimal data alignment solution. The results will propose the most accurate and time-efficient data alignment measurement solution for the large-scale metrology of wind turbine blades using CLR technology. Additionally, effects of using a mirror on measurement accuracy are investigated.

THEORY

Coherent Laser Radar technology

There are a number of metrology techniques available today [7] capable of measuring structures on a large-scale. The high accuracy of Coherent Laser Radar (CLR) along with its non-contact technology, application to large-scale structures, speed and portability are all key features shown in Figure 2 which make CLR the optimal metrology technique for turbine blade inspection.

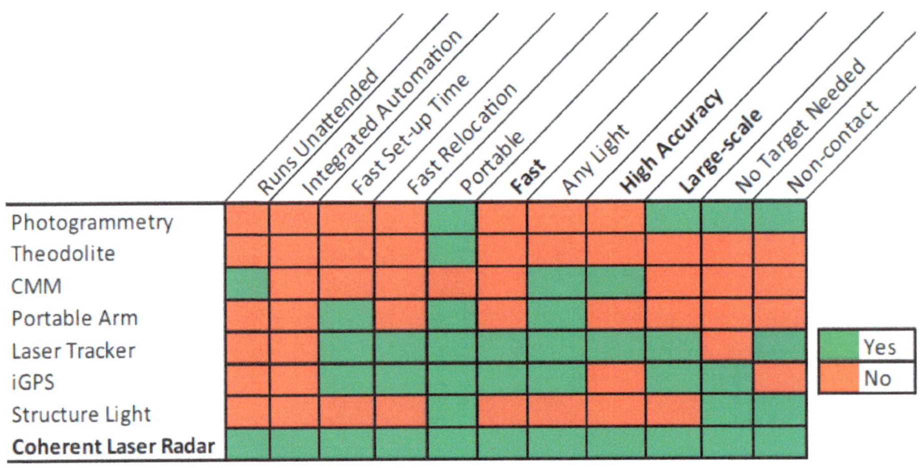

Figure 2. Comparison of various metrology techniques across wind turbine blade inspection criteria [8].

Contact metrology typically uses touch probes in contact with a surface to measure 3-D coordinates. Historically, contact devices have been able to measure surfaces to a higher accuracy than that of non-contact devices. Nikon's CMM contact devices can measure a 3-D point to a volumetric accuracy of 1.8 μm [9]. More recently, non-contact technology has advanced and is now capable of measuring to a high accuracy [10].

There are several disadvantages to using touch probes for contact methods. They are slow, require an operator, are difficult to manipulate and must be in contact with the surface which could potentially deform the measurement surface and requires the calculation of touch probe radius offsets [11]. Using touch probes for large-scale applications would therefore be extremely time-consuming; CLR can achieve a 90% inspection cycle time reduction compared with alternative contact methods [12].

Due to the large size of turbine blades, multiple CLR scanner locations are needed to inspect a complete blade. However, CLR scanners are portable and the method is fast. The CLR equipment used in this piece of research is Nikon's FM CLR Scanner (LR 200)[13], which is capable of inspecting up to 2000 points per second with a range of 50 m.

The scanner works by emitting a linear infrared laser beam onto the measurement surface and recapturing a portion of the reflected light. The laser signal's strength and ability to accurately focus at any distance from the scanner is maximized with an adjustable, large-aperture fixing [8]. Heterodyne detection [14] of the reflected beam mixed coherently with a controlled reference signal of calibrated wavelength (Figure 3) can precisely measure the change in frequency (Δf) and the change in time (Δt) of the waveform.

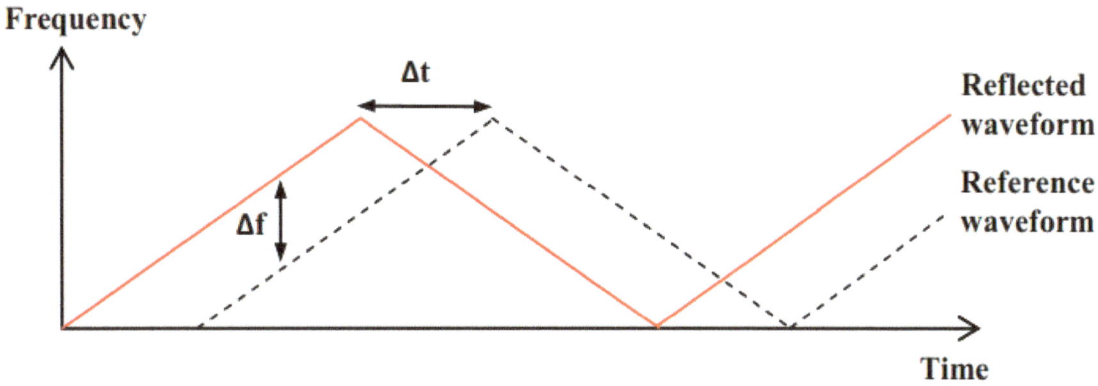

Figure 3. Comparison of the reflected laser signal with the controlled reference signal. Δf can be calculated due to a base frequency 200 THz and a fixed wavelength of 1500 nm [8].

The absolute range is determined using frequency modulation as shown in Eq. (1)[8].

$$\text{Range} = \frac{\Delta f}{0.667} \quad (\mu, \text{microns}) \tag{1}$$

Calculating the measurement points using frequency modulation produces a more accurate reading than if using light modulation shown in Eq. (2)[8]. At a distance of 2 m (similar to that used during this research), a 3-D measurement can be calculated using frequency modulation to within 24 μm [13].

$$\text{Range} = \frac{\Delta t \times c}{2} \quad (\mu\text{m}) \tag{2}$$

where c is the speed of light and 1 μm = 6.7×10^{-15} s.

Reliance on frequency shift means the scanner is largely insensitive to surface reflection and therefore only needs to recapture about 1% of the reflected beam [15]. This allows the inspection of composite materials such as turbine blade outer shells in any light. It also reduces human error when using tooling balls. Tooling balls used commonly as alignment reference points due to their geometrical precision for CAD alignment.

The laser beam is directed using a scan mirror mounted on a two-axis-gimbal shown in Figure 4, allowing movement through 360° azimuth (Az) and 120° elevation (El) to determine the point range (Rg).

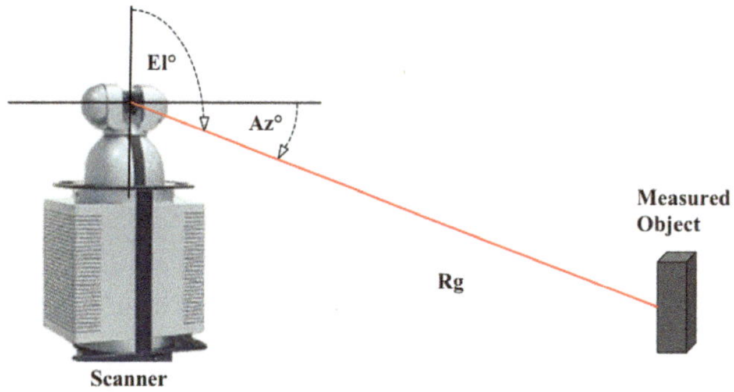

Figure 4. CLR scanner showing azimuth, elevation and range measurement capabilities.

To determine the point range, the scanner measures a distance and two angles. The scanner can measure the X and Y values (angles) precisely using probes within the two-axis movement mechanism. The Z value (range) on the other hand is determined by processing the reflected beams and is therefore less accurate [10]. Nikon [16], state that CLR has an angle uncertainty of 6.8 μm/m compared to a range uncertainty of 10 μm/m + 2.5 μm/m.

The range accuracy characteristics were also demonstrated using the flip test within SA. Ten tooling balls were measured and errors (i.e. magnitude away from the theoretical point location) for azimuth, elevation and range were recorded; an absolute average was calculated, shown in Figure 5.

	Azimuth	Elevation	Range
Average (mm)	0.0022	0.0007	0.0054

Figure 5. Absolute average (in mm) of measurement errors in azimuth, elevation and range.

Figure 5 shows that errors in range measurement are 2.45 and 7.71 times greater than azimuth and elevation measurement errors respectively.

There are various uncertainties associated with CLR because of assumptions compensating for environmental factors [15]. In this investigation, it was assumed that the laser beam travelled in a straight line [17]. This assumes that: the temperature was uniform across the measurement volume, the air refraction index was constant and effects due to air turbulence from noise vibrations and other mechanical instabilities were negligible and therefore did not affect the laser beam path.

These assumptions can be made in this investigation because measurements were taken from a 2 m distance; environmental factors would therefore be insignificant.

Mirrors

It was observed that on the underside of the blade where the laser beam was near tangential to the blade surface (i.e. an incidence angle close to 90°), CLR was unable to focus precisely on nominal points, causing measurements to fail.

CLR has a unique ability to measure precisely using mirrors, increasing the available area of vision. This allows measurement around corners and of difficult-to-access areas such as the underside of the blade. Hence, the entire blade profile perimeter can be measured. The micron-polished nickel-plated aluminium mirror used in this investigation was circular with a diameter of 15 cm. Surface nickel plating minimizes the effect of temperature change on the metal as well as improving durability and resistance to scratching, whilst the polished finish creates an extremely smooth surface, reducing beam scatter.

The CLR laser beam is transmitted at a wavelength of 1500 nm. At this wavelength, Laser Beam Products [18] have shown that protected aluminium mirrors offer close to 92% reflectivity. In this investigation, measuring using the mirror required the laser beam to be reflected twice, resulting in an overall reflectivity of ~84.64%. It is suggested that at wavelengths above 700 nm, gold or silver coatings should be used as opposed to nickel[18]. However these are not concerns as CLR needs only 1% reflectivity to operate.

Such a mirror can be challenging and time-consuming to align as there is no well-defined axis. Mirrors were traditionally aligned by measuring the angle normal to the mirrored surface with auto-collimating theodolites [19].

A simpler method of aligning the mirror plane within SA uses a tooling ball. The actual position of the tooling ball is measured normally. This fixed location allows an image of the same tooling ball to be measured using the mirror. With the two measurements, the mirror is determined as the plane halfway between the two points shown in Figure 6[20]; where Δa and Δb are the tooling ball uncertainties in the normal and perpendicular directions to the mirror plane, L is the length between the actual and apparent tooling ball positions and U is the uncertainty in the direction of the flat mirror to its normal.

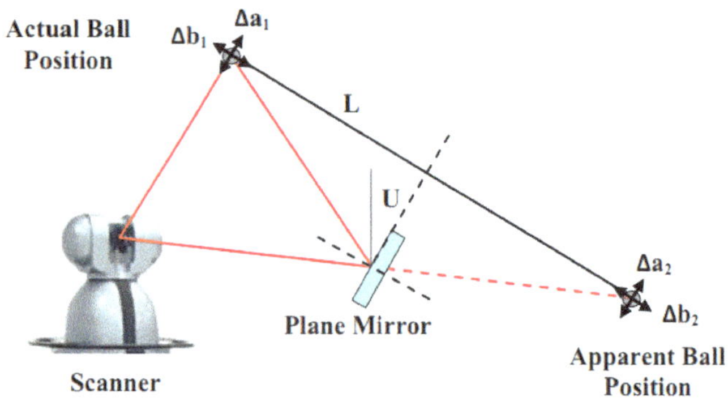

Figure 6. Tooling ball alignment, needed in order to create a mirror plane in SA [20].

The uncertainty in the direction of the mirror can be determined using Eq. (3).

$$U = \frac{\sqrt{\Delta a_1^2 + \Delta a_2^2}}{L} \tag{3}$$

Uncertainty is therefore greater when the tooling ball is closer to the mirror since L has decreased.

The complex nature of the blade profile and the small mirror size means the laser is only reflected to small sections of the blade. For each experiment configuration, multiple mirror locations were therefore required to inspect the underside of the blade.

The mirror was adjusted using trial and error (which was time-consuming) in order to maximize the number of points that could be measured. It was especially difficult to align the mirror when simulating large offsets as the reference points often fell outside the mirror's line of sight, resulting in failed measurements.

There are two main variables that can affect measurement accuracy when using mirrors: the angle of incidence between the mirror and the laser beam, and the distance away from the mirror.

It was found that a mirror can be used within an acceptable accuracy at an angle of incidence below 50° [21]. A 10° incidence angle yields an uncertainty of less than 3 arc-sec (0.0008°). Although still capable of using the mirror at an angle of incidence of 80°, at angles above 50° the uncertainty increases with incidence angle.

Similarly, at distances less than ~15 m, [21] shows a measurement uncertainty of ±5 arc-sec (0.0014°), which is an acceptable accuracy. However, beyond 15 m, the uncertainty increases with distance.

Uncertainty due to mirror distance was investigated by varying the distance of the mirror away from the blade; ten measurement points were investigated on the underside of the blade and averaged, shown in Figure 7.

	Mirror distance from blade (m)				
	No Mirror	0.25	0.5	0.75	1
Average point variation (mm)	5.59	3.93	3.93	3.92	3.26

Figure 7. Average (in mm) variation between measured points and the nominal points on the underside of the blade at different mirror distances away from the blade.

Although the 1 m mirror position appears the most accurate, it must be discounted as CLR was only capable of measuring one out of ten nominal points. Beyond 1 m, CLR was unable to measure any points.

Figure 7 shows that using a mirror increases measurement accuracy by 29.7% for points on the underside of the blade. It was found that varying the mirror distance below 1 m from the blade did not have an effect on measurement accuracy.

During this investigation, the mirror was used in positions less than 1 m away from the blade whilst the angles of incidence were kept below 50°. Measurements could therefore be made with confidence.

Spatial Analyzer

There are a number of software packages compatible with CLR, such as Spatial Analyzer and Polyworks. Both of them are highly flexible, instrument-independent, and can be used to develop and deploy automatic inspection processes or guided operator-driven workflows for effective shop floor operations. However this research was carried out using SA. SA is a traceable 3-D graphical software platform [22] that can be used in combination with CLR to manipulate and analyze the measured data in relation to a CAD model.

SA is capable of using multiple scanner locations to inspect the full span of a wind turbine blade. This is achieved using the Unified Spatial Metrology Network (USMN) [23] feature in SA which complies with the Guide to the expression of Uncertainty in Measurement (GUM) [24]. It accurately combines numerous measurement scanners (or relocates a single scanner) throughout the measurement volume. A single coordinate frame is created from the individual coordinate frames from each scanner's data set. Known points along the blade act as an alignment reference for each scanner. However, each scanner has measurement uncertainties which must be combined in order to provide a proper uncertainty statement [25].

This investigation used the best-fit transformation function of SA. The transformation reduces measurement errors by minimizing the distance between the measured blade data and a nominal point group defined in CAD. The measured data is transformed towards the nominal set, altering the measured data (including point-to-point distances, surface distances and scanner movements). The change in data is analyzed to determine the effectiveness of each transformation.

The best-fit transformation is based on the Least Squares method [17], a standard approach to the approximate solution of over-determined systems to adjust parameters of a model function to best fit a data set. However, the Least Squares method is sensitive to extreme outliers. At least three data points are required for a best-fit alignment; however using four or more data points significantly increases the transformation accuracy [17].

In addition, SA allows the raw measured data to be fabricated for other DoF constraints, reducing the time required significantly. SA works by converting the measured CLR data into *X*, *Y* and *Z* coordinates. The resulting coordinates have uncertainties relating to the original data (SA claim a 98% confidence in results) due to the conversion model applied[25]. Complex conversion models are therefore needed to accurately represent the new coordinates.

Physical measurement errors occur when algorithms are used to compensate for properties such as material expansion from temperature [17]. Furthermore, errors can arise within algorithms used to generate curve and point geometry.

Generating Curve and Point Geometry

A nominal point group must be defined on the surface of the CAD model in SA, to act as a measurement reference for CLR. There are several parametric mathematical methods available which can represent curved surfaces, including Non Uniform Rational B-Splines (NURBS) and Bézier curves [26].

NURBS surfaces are a generalization of B-spline and Bézier surfaces; however the B-spline method is a mathematically simplified version of the NURBS method [27]. B-splines have two advantages over Bézier surfaces: Bézier surfaces have a practical limit to the number of control points that can be used and Bézier point evaluation is less efficient [27].

To generate the required nominal points, first a B-spline curve is constructed on the CAD blade surface. SA uses the de Boor B-spline algorithm [28] to generate the approximate curve $F(t)$ shown in Eq. (4); where A_i are the control points and k the order of polynomial sections.

$$F(t) = \sum_{i=0}^{j} A_i N_{i,k}(t) \tag{4}$$

The curve is located by intersecting with a predefined plane and the CAD surface allowing flexible B-spline positioning. CAD surfaces are often constructed in sections, allowing more complex curved surfaces to be modelled. B-splines can be constructed on each section and combined using knot vectors [29]. Although this enhances errors, extremely complex surfaces can be inspected.

Furthermore, the number of B-splines can vary to match the level of inspection detail required. Increasing the number of B-splines however, increases the computational time.

Due to the straight line construction of B-splines, the CAD surface cannot be modelled exactly. Figure 8 shows the

Hausdorff distance [30] between the exact CAD model surface and the approximate B-spline curve.

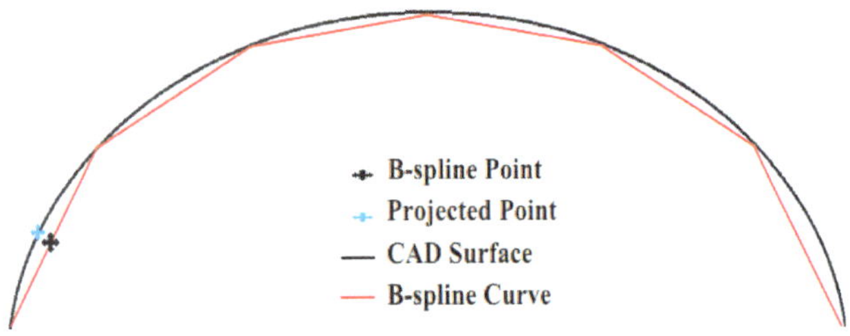

Figure 8. Difference between CAD surface to B-spline curve and point to projected point [6].

Once the B-spline is defined, points can be constructed along the curve either by defining the number of nominal points desired or defining point separation. The latter method was used in this investigation, producing a uniform distribution of points, allowing different levels of inspection for each CAD surface. Results increase in accuracy when using more data points, especially where surface gradients are more severe; however this adds to computational time.

To remove errors associated with the straight line construction of B-splines, points require projecting onto the predefined CAD surface, shown in Figure 8. The projected points can then be automatically measured using CLR.

Blade Flex

The flex (sag and twist) of a 44 m blade due to gravity was simulated using Computational Analysis [6] by applying a uniform load to the upper surface of the blade profile. The blade was constrained at the root, allowing the blade to move in all directions except *Z*.

The simulation demonstrates the worst case scenario during inspection resulting in an extreme maximum deformation of 5.605 m at the tip of the blade. This displacement relates to a 30° deviation from the *Z* axis (i.e. 30° in *Rz*).

In practice, the blade is positioned trailing edge up and supported by a 'tray'. The blade will therefore undergo less deformation. During testing a tip deviation of ~10 in. (0.254 m) is observed, relating in ~1.5° *Rz* rotation.

In this investigation, the blade test model was offset in *Rx*, *Ry* and *Rz* DoF to simulate severe flex. Optimizing measurement accuracy by determining the optimal constrained DoF variation at maximum flex provides manufacturers with confidence in the measurement capabilities of small, real-life blade deformations observed during testing.

Data Alignment

For inspection, wind turbine blades are constrained at the root, restricting any movement in the *Z* axis. This allows SA to apply best-fit transformations using 5 DoF unconstrained variables; *X*, *Y*, *Rx*, *Ry* and *Rz*. This significantly simplifies data analysis compared with that of transformations in 6 DoF. Allowing movement in 5 DoF produces 32 different constraint variations. The raw measured data is fabricated for each combination and aligned using the best-fit transformation. Each transformation is then analyzed and compared with the other 31 sets of data to determine the optimal DoF constraint variation.

METHOD

Experimental Equipment

The method used in this investigation was designed to test and evaluate the 32 DoF data alignment variations of the blade's constrained movement.

The test piece measured 1.65 m in height and 0.5 m in width. It was a 1:2 size cross-section model based on the Vestas 44 m offshore wind turbine blade section 5.5 m from the root. The test model was mounted on a turntable and a two-axis gimbal and positioned ~2.5 m from the CLR scanner within the metrology laboratory at Durham, shown in Figure 9. This allowed the blade to be rotated in the *X*, *Y* and *Z* axes.

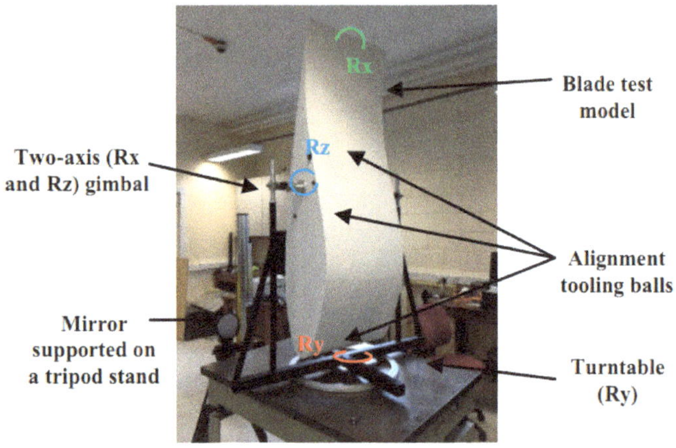

Figure 9. Blade test model setup within the laboratory. The turntable and gimbal allowed rotation in *Rx*, *Ry* and*Rz*. The mirror provided a line of sight to the underside of the blade and the tooling balls allowed quick alignment to the CAD model.

Blade flex simulated in [6] was replicated in the laboratory using a turntable allowing up to 30° rotation in *Ry* (simulating sag) and a gimbal allowing up to 15° and 20° rotation in *Rx* and *Rz* respectively (simulating twist).

The complex nature of blade profiles means that test models are difficult to manufacture accurately. The test model used in this project was manufactured from plywood and constructed using nails. This method differs significantly from how the actual blade is manufactured. The test model would therefore deviate from the theoretical CAD model in different areas and magnitude to that of a real blade. The test model used in this investigation replicated a real blade profile cross-section more truthfully than that used in[6], reducing measurement error.

For results to be reliable, the test model must accurately replicate the CAD model to avoid the generation of artificial results. The test blade was evaluated against the CAD model by measuring the point deviation from the CAD model for three full B-splines evenly separated across the blade face. A tolerance level of ±10 mm was used.

The test model had an average deviation of 4.65 mm from the CAD model, with a range of 12.31 mm. The test model deviated most from the CAD model on the underside of the blade, with an average deviation of 7.22 mm. In addition, 17 points (6.09%) measured did not meet the required tolerance limits.

This shows that the test blade accurately represents the CAD blade model. Artificial results will therefore not be created during data alignment, producing accurate and reliable results for each blade section (front, underside and back).

Experimental procedure

This investigation used an experimental procedure and data transformation method developed by Nikon. It is commonly implemented on large-scale wind turbine blades and was therefore deemed reliable. Each set of results followed the same experimental procedure, shown in Figure 10.

- *Import CAD model into SA:* The test model is imported 5.5 m from the SA coordinate origin as it was designed on a cross-section 5.5 m from the root.
- *Align CLR scanner to CAD model:* The scanner was aligned to the test model using three tooling balls placed on the root side face shown in Figure 9. The tooling balls remained in the same location throughout all experiments for quick and accurate alignment.
- *Construct B-splines and nominal point groups:* B-splines were created to construct nominal point groups, which were used as a reference for the CLR scanner to automatically measure.
- *Measure nominal point groups:* Nominal point groups were measured in three sections: front, underside and the back. The front was straight-forward to measure; however a mirror was needed for the underside of the blade. The scanner was relocated for the back of the blade and aligned, adding to the experimental time.
- *Transform measured data:* The measured data was then fabricated for all 32 DoF constraint variations.
- *Perform best-fit transformations:* The fabricated data sets were then aligned to their equivalent CLR scanners and transformed to minimize the deviation from the CAD model.
- *Export data:* All transformed data was exported into an Excel format.
- *Analyze data:* The mean and standard deviation of distance data between the transformed data sets and the CAD model were calculated within Excel.
- *Determine the optimal DoF constraint:* The analyzed data for each DoF variation was compared with each other using a ranking system to determine the most accurate transformation of the DoF variations.

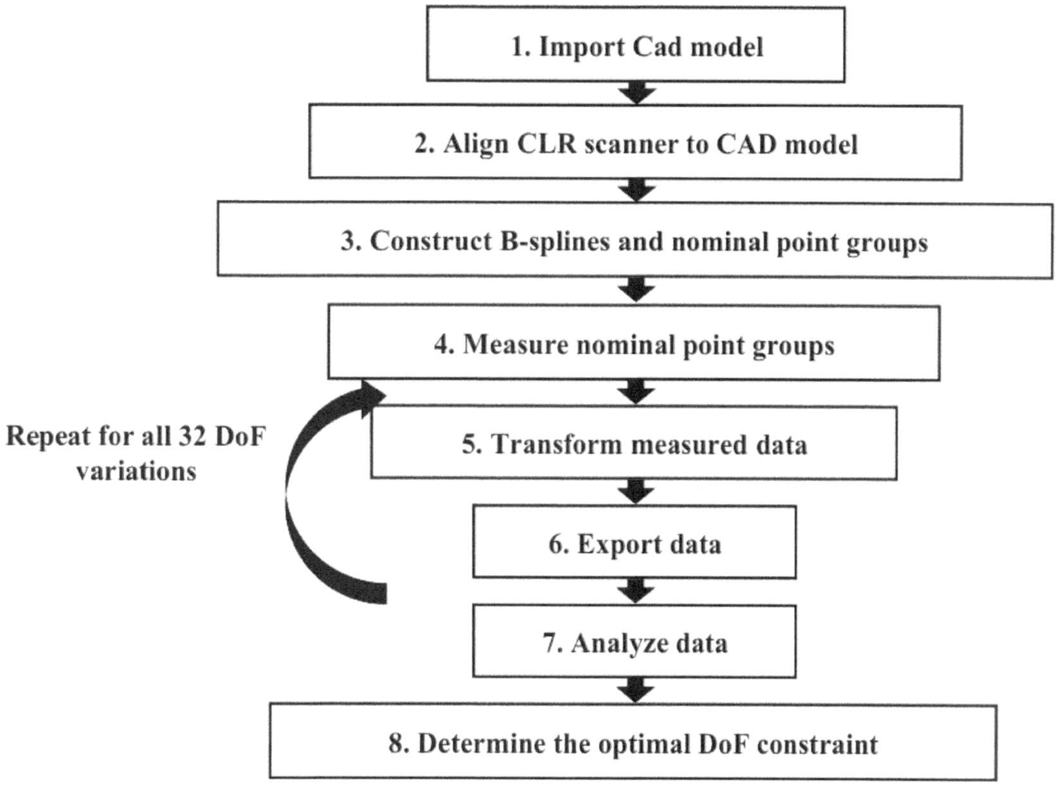

Figure 10. Experimental procedure used to measure and analyze each constrained DoF variation to establish the optimal data alignment solution.

This procedure allowed investigation of all 32 DoF variations. A main advantage of this method was that fabricated data sets were used for each DoF variation, requiring only one measurement procedure for each experiment. This saves time and eliminates human error that would be associated with taking multiple measurements. The procedure offers a flexible measurement solution as the number of B-splines and point density can be changed easily in Step 3 above.

To maximize measurement accuracy, care was taken to ensure a controlled laboratory environment was maintained. The CLR scanner and the test model were kept stationary and a clear line of sight between the scanner and blade was preserved, avoiding the need to repeat measurements.

Data Analysis

There were three components of each DoF data alignment transformation that were exported and analyzed:

Surface distance: quantifies the distance between the measured data points and the CAD surfaces. The smallest standard deviation shows the closest alignment towards the CAD surface and proves the most accurate and optimal DoF variation.

Point-to-point distance: quantifies the distance between the measured data points and the equivalent nominal points. The mean point-to-point distance shows how well the transformed data was mapped onto the nominal points; the smallest mean proves the most accurate DoF variation.

Scanner movement: quantifies the notional distance the scanner has moved away from a reference scanner location in order to transform the measured data points.

The reference scanner location is constructed during alignment when the blade is defined as being fully constrained in *X*, *Y*, *Z*, *Rx*, *Ry* and *Rz*. The fully constrained DoF variation data set will therefore always have zero scanner movement and hence the best ranked scanner movement for all experiments.

The smallest mean movement of the scanners proves the most accurate DoF variation.

Ranking

Each measurement set produced vast amounts of data. The analyzed data therefore required ranking for easy comparison between each DoF variation.

The ranking system ordered each DoF variation in terms of the three data analysis criteria discussed above. The optimal (most accurate) DoF, i.e. the smallest surface vector magnitude, point-to-point distance or scanner movement, was allocated an individual rank score of 1; the least accurate DoF variation an individual rank score of 32.

The individual rank scores were then combined for an overall total rank between 3 and 96 as shown in Figure 11. The optimal DoF constraint will achieve the lowest total rank score.

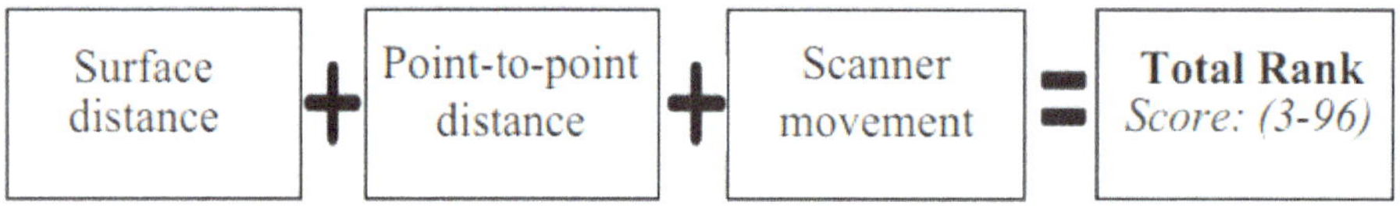

Figure 11. Summing of the individual rank scores (1–32) from each criterion to give a total rank score (3–96).

Combining the total rank scores across all experiments highlights the DoF variations regularly performing well and therefore the optimal data alignment solution.

Experimental Plan

The experimental procedure, data analysis and ranking above were all used to investigate the 32 DoF variations using a single B-spline and double B-spline arrangements for different *Rx*, *Ry* and *Rz* offsets, simulating blade flex.

A point separation of 50 mm was set for the front and back of the blade and a 12.5 mm separation was used for the underside of the blade. This resulted in 95 nominal inspection points around the blade profile for each B-spline.

The smaller point separation for the underside of the blade reduces the Hausdorff distance, offering greater accuracy than a larger separation. This was necessary as the blade surface turns through 180°.

Experiment 1 – Single B-spline: The single B-spline was defined in SA half-way (250 mm) across the test model, evenly splitting the face as shown in Figure 12. The B-spline was not constructed near the edge of the test model face as experiment offsets would rotate the blade away from the theoretical B-spline. This would result in failed measurements.

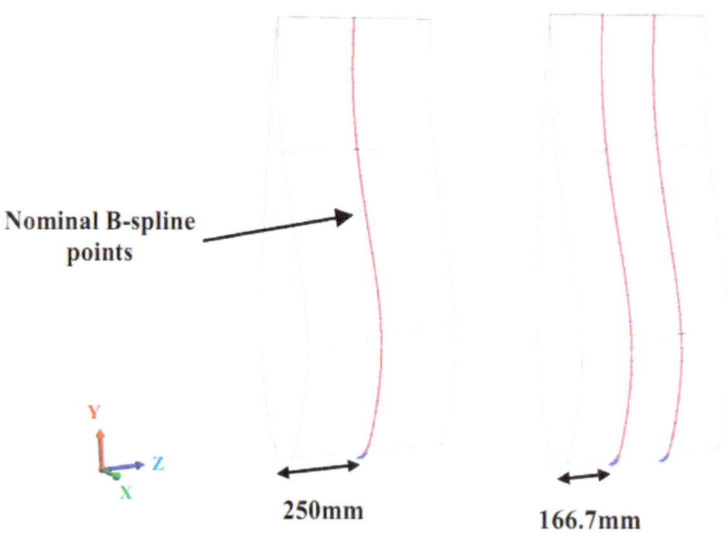

Figure 12. Left, the single B-spline arrangements (Experiment 1), 250 mm from the root end of the cross-section. Right, the double B-spline arrangement (Experiment 2), evenly distributed 166.7 mm from the root end.

Experiment 2 – Double B-spline: The two B-splines were separated evenly: 166.7 mm and 333.7 mm along the blade face, shown in Figure 12. Again, the B-spline curves were constructed away from the edge of the blade face.

Using an additional B-spline to Experiment 1 doubles the number of nominal inspection points to190.

Initially, five offsets were chosen using the extreme *Rx*, *Ry* and *Rz* flex simulations for both Experiment 1 and 2: (1) *No Offset; (2) 30° Ry; (3) 20° Rz; (4) 15° Rx; (5) 30° Ry, 20° Rz and 15° Rx.*

Change in Experimental Plan

It was found that the initial chosen offsets were not suitable for the experimental method. Implementing offset 1 (no offset) and offset 2 (30° *Ry*) yielded reliable and accurate results. However some results obtained for the 20° *Rz* offset, 15° *Rx* offset and 30° *Ry*, 20° *Rz* and 15° *Rx* offset contained huge inaccuracies shown in Figure 13.

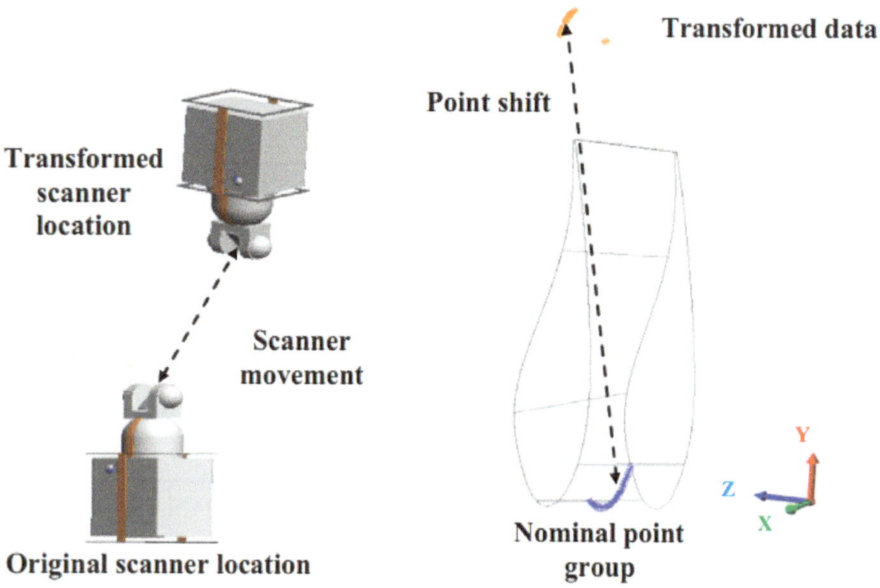

Figure 13. Huge point shifts and scanner movements were found when implementing a 20° *Rz* offset; the offset was deemed inappropriate.

Figure 13 shows a huge average point shift of 2124.85 mm and scanner movement of 639.76 mm from the nominal point group. Such a large divergence clearly shows that the CLR method used with SA is incapable of generating reliable results at large offsets simulating extreme blade flex characteristics.

Furthermore, CLR failed to measure numerous nominal points around the profile. However, this was especially true when using the mirror; only 40.7% of points on the underside were measured. This is because large offsets caused the test model to rotate outside the mirror plane.

Different offsets were therefore used simulating a less severe blade flex scenario: (1) *No Offset; (2) 10° Ry; (3) 30° Ry; (4) 5° Rz; (5) 5° Rx; (6) 10° Ry, 5° Rz and 5° Rx.*

Although these offsets are smaller than the initially planned offsets, they still represent a significantly larger blade flex than the common deformation seen during real blade inspection.

The results of this investigation will offer manufacturers confidence in the measurement process, whilst eradicating the measurement inaccuracies associated with large offsets.

The most important experiments to evaluate are Experiments 1.6 and 2.6 as these simulate the most realistic blade flex characteristics that occur during turbine blade testing.

RESULTS AND ANALYSIS

All data is presented in millimetres (mm) and rounded to the nearest 0.01 mm due to the precision of SA point measurement.

Experiment 1 – Single B-spline

Experiment 1 evaluates all 32 DoF variations for six different *Rx*, *Ry* and *Rz* offsets using a single B-spline.

Experiment 1.1: No Offset, Simulating the Measurement of the Blade in its Theoretical CAD Position with No Flex

The results for each DoF constraint variation are displayed showing: the mean point-to-point distance between the measured points and nominal points, the standard deviation of the point distance away from the surface and the mean of the scanner movements from their starting locations. The individual rank scores of each criterion have been summed and displayed as total rank score giving it a rank order compared to the other 32 DoF variations.

Table 1 shows that for Experiment 1.1, the optimal transformational solution is achieved when constraining the blade in either the *Y*, *Z* and *Ry* directions or the *X*, *Z*, *Rx* and *Rz*directions. Both DoF variations achieved the joint lowest total rank score of 26 and therefore have the best data alignment performance when measuring the blade under no deformation (see Table 2 and Table 4).

Table 1. Top 10 ranked DoF variations of Experiment 1.1

DoF constraints	Point-to-point distance	Surface distance	Scanner movement	Total rank score	Rank order
Y Z Ry	1.55	1.40	80.18*	26	1
X Z Rx Rz	1.86	1.81	8.25	26	1
Z Rx Ry	1.53	1.40	83.85*	27	3
Y Z Ry Rz	1.85	1.81	8.62	27	3
Z Rx Ry Rz	1.84	1.81	8.70	27	3
X Y Z Rz	1.88	1.82	8.13	29	6
X Y Z	1.68	1.40	80.24*	30	7
X Z Rx	1.62	1.26	91.10*	30	7
X Z Rx Ry	1.65	1.69	76.72*	31	9
X Y Z Ry Rz	2.57	2.62	6.39	32	10

* Large scanner movements above 75 mm.

Table 2. Top 10 ranked DoF variations of Experiment 1.2.

DoF constraints	Point-to-point distance	Surface distance	Scanner movement	Total rank score	Rank order
Y Z Ry	2.28	0.69	21.44	19	1
X Z Rx Ry	2.47	1.89	0.00	19	1
X Y Z Rz	2.92	0.73	11.69	20	3
Z Rx Ry	2.25	1.84	21.26	23	4
X Z Rx	2.26	1.84	21.42	25	5
X Y Z	2.29	1.84	21.57	29	6
X Z Rx Rz	2.90	2.40	11.77	36	7
X Z Rx Ry Rz	3.52	2.99	7.42	36	7
Y Z Ry Rz	2.90	2.41	12.20	38	9
X Y Z Ry Rz	3.53	2.99	7.48	38	9

Table 3. Top 10 ranked DoF variations of Experiment 1.3.

DoF constraints	Point-to-point distance	Surface distance	Scanner movement	Total rank score	Rank order
X Y Z Ry	8.04	5.01	110.00	24	1
X Z Rx	6.72	5.06	114.11	25	2
Z Rx Ry	6.68	5.08	114.01	25	2
X Z Rx Ry	8.04	4.91	116.67	27	4
Y Z Ry	6.82	5.07	114.75	28	5
Z Rx Ry Rz	11.58	10.11	32.49	30	6
X Z Rx Rz	11.63	10.11	31.68	31	7
X Y Z Rz	11.68	10.12	31.46	33	8
Z Rx	55.59	4.69	77.94	34	9
Y Z Ry Rz	11.62	10.12	32.59	34	9

Table 4. Top 10 ranked DoF variations of Experiment 1.4.

DoF constraints	Point-to-point distance	Surface distance	Scanner movement	Total rank score	Rank order
X Z Rx	7.25	6.14	161.77	24	1
Z Rx Ry	7.13	6.18	161.77	27	2
Z Rx	14.13	6.20	102.88	30	3
Y Z Ry	7.39	6.16	161.96	30	3
X Y Z	7.37	6.18	162.22	32	5
X Z Rx Ry	10.99	6.75	124.29	32	5
Y Z	15.93	6.31	105.74	33	7
X Y Z Ry	10.92	6.80	164.11	40	8
X Z Ry Rz	21.60	21.33	0.00	41	9
X Z Rx Rz	21.65	21.15	54.94	43	10

However, there is not much difference in data alignment performance compared with the three DoF variations with a total rank score of 27, which performed almost as well as the two optimal DoF variations. All 10 DoF variations shown in Table 1 have extremely low point-to-point distances and surface distances; however certain DoF variations scored poorly due to large scanner movements of above 75 mm.

Experiment 1.2: A 10° Ry Offset, Simulating Some Blade Sag

Experiment 1.2 found that the best performing DoF variation is *Y*, *Z* and *Ry* constraints once again, along with *X*, *Z*, *Rx* and *Ry* constraints. The former performed particularly well with a surface distance of 0.69 mm and the latter with no notional scanner movement.

X, *Y*, *Z* and *Rz* constraints also performed well with a total rank score of 20. It had a low surface distance of 0.73 mm.

Comparing Experiments 1.1 and 1.2, it is clear that the introduction of an *Ry* offset simulating blade sag has decreased the accuracy of the data transformation shown by larger point-to-point and surface distances.

Experiment 1.3: A 30° Ry Offset, Simulating the Maximum Blade Sag Scenario

Table 3 shows that the constrained variation of *X*, *Y*, *Z* and *Ry* is the best data alignment technique for maximum blade sag. *X*, *Z* and *Rx* as well as *Z*, *Rx* and *Ry* constrained DoF variations also performed well.

It is clear however that the data alignment transformation is not as accurate for such large sag characteristics compared with the simulation less extreme sag in Experiment 1.2, as shown by greater values for all three criteria.

Experiment 1.4: A 5° Rz Offset, Simulating Some Blade Twist

The *X*, *Z* and *Rx* constrained DoF variations is the optimal data transformation for Experiment 1.4. Similarly to Experiment 1.3, large values for all 3 criteria were obtained. This is especially true again for the scanner movement with certain movements upwards of 160 mm.

Experiment 1.5: A 5° Rx Offset, Simulating Some Blade Twist

Table 5 shows that the optimal DoF variation for a small *Rx* offset is when *Y*, *Z* and *Ry* are constrained. However there is not much difference in the three analysis criteria for the top three ranked DoF variations shown by a range of: 0.05 mm, 0.02 mm and 1.46 mm for the point-to-point distance, surface distance and scanner movement respectively.

Table 5. Top 10 ranked DoF variations of Experiment 1.5

DoF constraints	Point-to-point distance	Surface distance	Scanner movement	Total rank score	Rank order
Y Z Ry	4.93	5.04	23.51	16	1
X Y Z	4.97	5.06	22.15	18	2
X Z Rx	4.98	5.06	22.19	20	3
Y Z Rx	5.23	3.81	24.27	20	3
Z Rx Ry	4.94	5.04	27.20	23	5
X Z Rx Rz	5.84	5.87	0.00	28	6
X Y Z Rx	5.26	5.51	26.50	30	7
Y Z Rx Ry	5.26	5.50	29.22	35	8
Z Rx Ry Rz	5.82	5.86	26.75	37	9
Y Z Ry Rz	5.82	5.87	26.79	39	10

Experiment 1.6: A 10° *Ry*, 5°*Rz* and 5° *Rx* Offset, Simulating *the Most Realistic Blade Flex Scenario during Testing; Sag and Twist*

Experiment 1.6 is a combination of Experiments 1.2, 1.4 and 1.5 and best simulates the blade in a real test environment. Interestingly, Experiment 1.6 suggests that the optimal data alignment is achieved when constraining *Y* and *Z*.

This was not expected as the *Y* and *Z* constrained variation did not perform well in previous experiments.

Y, *Z* and *Ry* constrained and *Z*, *Rx* and *Ry* constrained DoF variation also performed strongly, both ranking second.

Due to the large offset however, large values for each criterion were obtained: above 20 mm, 16 mm and 100 mm for point-to-point distance, surface distance and scanner movement respectively (see Table 6).

Table 6. Top 10 ranked DoF variations of Experiment 1.6

DoF constraints	Point-to-point distance	Surface distance	Scanner movement	Total rank score	Rank order
Y Z	7.92	6.12	138.34	27	1
Y Z Ry	8.15	6.10	147.12	30	2
Z Rx Ry	21.35	5.84	111.20	30	2
X Z Rz	22.18	23.18	0.00	31	4
Z Ry Rz	22.18	23.18	0.00	31	4
X Z Rx	7.98	6.15	147.25	32	6
Y Z Rx	20.79	16.88	61.76	32	6
X Y Z Ry	11.11	7.28	132.01	35	8
X Z Rx Ry	11.24	6.90	141.44	37	9
X Y Z	8.22	6.15	251.73	42	10

Experiment 2 – Double B-Spline

Experiment 2 evaluates all 32 DoF variations for the same *Rx*, *Ry* and *Rz* offsets as in Experiment 1 using a double B-spline.

Experiment 2.1: No Offset, Simulating the Measurement of the Blade in Its Theoretical CAD Position with No Flex

Table 7 shows the strong performing data alignment DoF variations without any blade deformation. Similar to Experiment 1.1, both *Y*, *Z* and *Ry* constrained and *Z*, *Rx* and *Ry* constrained DoF variations performed very well along with *X*, *Z* and *Rx* constrained and*X*, *Y* and *Z* constrained.

Table 7. Top 10 ranked DoF variations of Experiment 2.1.

DoF constraints	Point-to-point distance	Surface distance	Scanner movement	Total rank score	Rank order
X Z Rx	1.55	1.43	15.28	18	1
Y Z Ry	1.55	1.42	15.87	18	1
Z Rx Ry	1.53	1.42	15.95	18	1
X Y Z	1.57	1.43	15.15	19	4
X Z Rx Rz	1.89	1.87	9.49	23	5
Y Z Ry Rz	1.88	1.87	9.90	23	5
Z Rx Ry Rz	1.87	1.87	9.98	23	5
X Y Z Rz	1.90	1.88	9.33	28	8
Z Rx Rz	2.39	1.87	14.11	31	9
X Z Rx Ry Rz	2.61	2.70	4.16	33	10

The lower total rank scores (18) to that of the equivalent investigation (26) in Experiment 1.1 show that the DoF variations are performing more consistently with the criteria categories. This is due to less extreme scanner movements of ~15 mm against ~80 mm in Experiment 1.1.

Experiment 2.2: A 10° Ry Offset, Simulating Some Blade Sag

Table 8 shows that *Z*, *Rx* and *Ry* constrained DoF variation achieved the highest rank for Experiment 2.2. However, there was not much difference in performance with the other high ranking DoF constraint variations.

Table 8. Top 10 ranked DoF variations of Experiment 2.2.

DoF constraints	Point-to-point distance	Surface distance	Scanner movement	Total rank score	Rank order
Z Rx Ry	23.02	3.30	39.57	27	1
Y Z	11.74	2.37	70.09	28	2
Y Z Ry	23.08	3.30	39.93	29	3
Z Rx Ry Rz	23.93	3.82	19.51	30	4
Z Rx	14.52	2.24	72.98	31	5
Y Z Ry Rz	23.94	3.80	19.77	31	5
Z Rx Rz	12.76	4.18	40.96	34	7
X Y Z Ry Rz	24.75	4.08	15.74	34	7
X Z Rx Ry Rz	24.73	4.08	15.96	34	7
Y Z Rz	13.73	4.16	44.27	36	10

The data alignment transformations performed poorly due to the point-to-point distance criterion. The *Z*, *Rx* and *Ry* constrained variation achieved a point-to-point distance of 23.02 mm for the double B-spline arrangement compared with 2.25 mm for a single B-spline in Experiment 1.2.

Experiment 2.3: A 30° Ry Offset, Simulating the Maximum Blade Sag Scenario

The larger rotational offset in *Ry* limited the ability of CLR to measure certain points in this particular configuration. Only three points from 28 (10.7%) from the underside of the test model were measured for one of the B-splines. As mentioned in 'Theory: C. Spatial Analyzer', SA requires four or more data points to significantly improve the accuracy of the best-fit transformation.

This is the reason scanner movements as large as 506.33 mm were obtained. A 30° *Ry*offset was therefore deemed inappropriate for the double B-spline arrangement.

Experiment 2.4: A 5° Rz Offset, Simulating Some Blade Twist

Table 9 shows that *Y*, *Z* and *Ry* constrained and *Z*, *Rx* and *Ry* constrained DoF variations are the best data alignment transformations. However, large scanner movements and some large point-to-point and surface distances can be seen for other DoF variations.

Table 9. Top 10 ranked DoF variations of Experiment 2.4.

DoF constraints	Point-to-point distance	Surface distance	Scanner movement	Total rank score	Rank order
Y Z Ry	6.76	4.89	125.62	27	1
Z Rx Ry	6.52	4.92	125.39	27	1
Y Z	17.72	4.75	114.35	29	3
X Z Rx	6.60	4.90	130.43	29	3
X Y Z	6.78	4.93	130.65	34	5
X Z Ry	11.10	7.08	117.41	36	6
X Z Rx Ry	9.72	5.82	134.97	38	7
X Y Z Ry	9.82	6.00	139.67	41	8
X Z Rx Rz	19.84	21.00	59.37	42	9
Z Rx Ry Rz	19.82	21.00	62.45	42	9

Experiment 2.5: A 5° Rx Offset, Simulating Some Blade Twist

Table 10 shows that fully constrained in *X*, *Y*, *Z*, *Rx*, *Ry* and *Rz* is the optimal data alignment DoF variation for a 5° *Rx* offset. No notional scanner movement was seen because the scanner reference is predefined using a fully constrained configuration. Although a low surface distance of 2.50 mm was observed, a huge point-to-point distance of 26.70 mm was obtained; it is therefore performing inconsistently across the three analysis criteria (see Table 11).

Table 10. Top 10 ranked DoF variations of Experiment 2.5.

DoF constraints	Point-to-point distance	Surface distance	Scanner movement	Total rank score	Rank order
X Y Z Rx Ry Rz	26.70	2.50	0.00	23	1
X Z Rx	4.16	4.53	15.86	24	2
X Y Z Rx	4.41	4.91	11.73	25	3
Y Z Ry	5.11	4.51	15.89	27	4
X Y Z	4.14	4.52	39.87	28	5
X Y Z Rz	4.92	5.46	5.49	28	5
X Z Rx Rz	4.92	5.46	5.49	28	5
Z Rx Ry	5.16	4.54	14.65	29	8
Y Z Rx Ry	5.43	4.93	12.55	32	9
Y Z Rx	10.89	4.50	34.97	36	10

Table 11. Top 10 ranked DoF variations of Experiment 2.6.

DoF constraints	Point-to-point distance	Surface distance	Scanner movement	Total rank score	Rank order
X Z Rx	8.20	6.20	140.97	27	1
Z Rx	13.10	6.17	120.60	28	2
Y Z Ry	8.31	6.23	140.76	28	2
Z Rx Ry	8.07	6.21	141.00	28	2
X Y Z	8.43	6.23	140.95	30	5
Y Z	12.64	6.26	139.18	33	6
Z Ry Rz	21.39	22.38	12.72	33	6
X Y Z Rz	19.27	20.17	87.95	37	8
X Z Rx Ry	11.52	7.26	145.13	37	8
X Y Z Ry	11.75	7.80	160.76	40	10

Experiment 2.6: A 10° Ry, 5° Rz and 5° Rx Offset, Simulating the Most Realistic Blade Flex Scenario during Testing; Sag And Twist

Similar to Experiment 1.6, *Y*, *Z* and *Ry* constrained and *Z*, *Rx* and *Ry* constrained DoF variations performed well, ranking in joint second along with the *Z* and *Rx* constrained variation. However, the top ranking variation was the *X*, *Z* and *Rx* constraint, slightly outperforming those ranked second.

The large values observed for all three criteria for most DoF variations are due to the large offset simulating significant blade flex.

DISCUSSION

Single B-Spline Arrangement

The optimal data alignment transformation using one B-spline was achieved when constraining in *Y*, *Z* and *Ry* DoF whilst keeping *X*, *Rx* and *Rz* unconstrained. This DoF variation was ranked first for Experiments 1.1, 1.2 and 1.5 and achieved the lowest total rank score across all six offset scenarios. Averaged across the six offsets, the constraint obtained a low point-to-point and surface distance of 5.19 mm and 4.08 mm respectively, however it achieved a large scanner movement of 91.49 mm.

The *Z*, *Rx* and *Ry* constrained DoF variation ranked second overall. Compared with the *Y*,*Z* and *Ry* constraint, the variation achieved a higher average point-to-point and surface distance of 7.31 mm and 4.23 mm respectively; however it outperformed the *Y*, *Z* and *Ry*DoF constraint with an average scanner movement of 86.55 mm.

There is no apparent relationship between the number of constrained DoF and data alignment performance; however four of the top five DoF variations imposed *Ry* constraints. Constraining in the *X* direction also appears to increase the accuracy of the alignment transformation with six of the top ten variations using *X* constraints.

CLR was capable of measuring the majority of points along the B-spline with Experiments 1.1–1.5 achieving point measurement success percentages of above 90%. Experiment 1.6 achieved 84.21% measurement success which still ensures reliable results.

The large scanner movements observed when using a single B-spline could introduce significant errors when defining a general coordinate system when combining multiple CLR scanners in the six locations needed to measure a 60 m blade.

CLR has a very short measurement time for a single B-spline arrangement, measuring 95 points in ~6 min. The entire process however, including mirror alignment and results fabrication, is time-consuming: a full set of results can take up to 6 h to complete. In reality, mirror positioning would not be done by hand and multiple B-splines would be measured for each scanner location, saving a considerable amount of time.

Therefore, for a single B-spline arrangement, the best-fit data alignment transformation should be implemented with constraints in *Y*, *Z* and *Ry*, offering a simple, quick and accurate large-scale measurement method. However, care must be taken when aligning multiple CLR scanners.

Double B-Spline Arrangement

Similarly to the single B-spline, the most accurate data alignment solution for a double B-spline arrangement was found with *Y*, *Z* and *Ry* constrained DoF. This DoF variation ranked first for Experiments 2.1 and 2.4 and ranked in the top four for all offsets. Again, it achieved the lowest combined total rank score across the different offsets. The data alignment solution had an average point-to-point distance of 8.96 mm, a surface distance average of 4.07 mm and a scanner movement of 67.61 mm.

Once again, the *Z*, *Rx* and *Ry* constraint also performed strongly; ranking second with an average point-to-point distance of 8.86 mm, an average surface distance of 4.08 mm and an average scanner movement of 67.31 mm.

No clear relationships between constraints and results can be seen. However, the variations with fewer constrained DoF generally rank poorly due to large scanner movements.

The addition of a second B-spline did not affect the measurement capability of CLR. CLR achieved similar point measurement success (excluding Experiment 2.3) to that of the single B-spline data alignment, with an average of 89.6% of nominal points measured across the different offsets. Experiment 2.3 achieved a measurement success percentage of 60% resulting in a less accurate transformation.

When using a double B-spline arrangement, it is recommended to use the best-fit data alignment transformation constrained in *Y*, *Z* and *Ry*. The addition of a second B-spline doubled the experimental time to up to 12 h; this would only increase further with more B-splines.

Comparison of Single and Double B-Spline Arrangements

The optimal data alignment transformation for both the single and double B-spline arrangements were achieved with *Y*, *Z* and *Ry* constrained DoF.

For the *Y*, *Z* and *Ry* constrained DoF, the double B-spline data alignment had no effect on the average surface distance across all flex offsets.

However, the average point-to-point distance was less accurate; 8.96 mm compared with 5.19 mm with the single B-spline. This is due to the single B-spline being constructed along the *Y* axis (situated in the center of the blade cross-section) shown in Figure 12. The blade rotates around the single B-spline when simulating sag with a rotational offset in *Ry*, as investigated in Experiments 1.2, 1.3 and 1.6 and for the equivalent in Experiment 2. Points measured along the B-spline will therefore be less affected by the offset than points away from the center of rotation (such as points along the double B-splines) because the initial measured point-to-point distances are greater.

This is also the reason why some nominal point measurements failed when using the mirror for the double B-spline. Points on the underside of the blade rotated outside of the mirror's line of sight whereas the single B-spline points did not rotate through a large angle and therefore remained in view.

Interestingly, the double B-spline achieved a smaller average scanner movement (67.61 mm) than that seen with the single B-spline (91.49 mm). This reduces the potential error that could be introduced when aligning a new scanner location in SA.

Similar comparisons, between the single and double B-spline data alignments can be made with other DoF variations. However, other DoF variations, including the *Z*, *Rx* and *Ry* constraint achieved more accurate surface distance results for the double B-spline arrangement. The surface distance for the *Z*, *Rx* and *Ry* constrained DoF decreased from 4.23 mm for the single B-spline to 4.08 mm for the double B-spline.

Although the inspection time is twice as long as a single B-spline, it is recommended that a double B-spline data alignment method should be implemented with *Y*, *Z* and *Ry* constrained DoF. The double B-spline achieved more accurate results with a slight decrease in point distance away from the CAD surface and a significant decrease in scanner movement. Although the average point-to-point distance increased, this may have been due to inaccuracies associated with the offset rotation in the *Y* axis. The large *Ry* offsets also cause some point measurement failure. In reality, blade flex is small, meaning nominal points will be measured consistently.

Experimental Evaluation and Recommendations

The accuracy of results obtained in this investigation relies heavily on the closeness of the measurement data to the real data of the product. It was found that the blade test model accurately represented the CAD model; certain parts of the blade did not meet with tolerance limits. This would have a slight negative effect on measurement accuracy. If possible, the data alignment method should be applied to a test model that more accurately represents the CAD blade profile (particularly the underside of the blade).

B-splines were defined to evenly separate the blade meaning the single B-spline was positioned halfway across (i.e. the center) of the blade face, along the *Y* axis. This meant an accurate comparison between the single and double B-splines could not be made when offset in *Ry*. The single B-spline should be tested in a location away from the *Ry* axis to enable a better comparison with double B-spline results.

This investigation showed that the double B-spline was more accurate than the single B-spline averaged across all offsets. Therefore, a triple B-spline arrangement should be investigated to examine what effect further increasing the number of B-splines has on the accuracy of results.

Due to the small mirror size, the mirror was unable to measure a proportion of points when simulating large blade flex characteristics as the nominal points were offset to beyond the mirror's line of sight. A larger mirror would increase the number of points that CLR was able to measure. However a bigger mirror may be expensive and less accurate.

In the future, similar investigations should be performed on a real large-scale wind turbine blade to validate results found in this piece of research.

CONCLUSION

This investigation found that although accurate, the data alignment technique currently used by Nikon (*Z*, *Rx* and *Ry* constraints) can be improved. The constrained DoF combination of *Y*, *Z* and *Ry* (allowing movement in *X*, *Rx* and *Rz*) for a double B-spline arrangement delivered the optimal transformation solution. It was the most accurate and consistent performing DoF variation; minimizing the measurement error introduced by blade flex characteristics.

It was found that using the double B-spline arrangement increased data post-processing accuracy. Therefore, using more B-spline curves during blade inspection reduces the overall measurement error.

Additionally, it was found that the mirror increased measurement accuracy by 29.70% for parts of the underside of the blade where CLR could not accurately focus due to a large incidence angle. However, the number of nominal points the mirror could measure decreased with greater blade flex.

The proposed data alignment solution offers a simple, quick and accurate measurement solution, minimizing measurement error during inspection. This provides manufacturers with confidence in their manufacturing procedures. The techniques described in this paper can easily be developed and used for the metrology of other large-scale structures.

ACKNOWLEDGMENTS

This research is part of the wind turbine structural condition monitoring program funded by EPSRC, United Kingdom, through the Centre for Through Life Engineering Services (EP/1033246/1, Project SC006) and EPSRC Impact Acceleration Award (EP/K50336811).

REFERENCES

1. Digest of the United Kingdom energy statistics (DUKES), Renewable sources of energy, 2013 (Chapter 6), Available: <https://www.gov.uk/government/uploads/system/uploads/attachment_data/file/65850/DUKES_2013_Chapter_6.pdf>.
2. Gurit, Wind turbine blade structural engineering, Wind Energy Handbook – 3 – Structural Design.
3. Gurit, Gurit materials for wind turbine blades, Wind Energy Handbook – 4 – Gurit Composite Materials for Wind Turbine Blades.
4. M.A. Homel, Blade Geometry and Solid Models, 2005, Available: <http://www.mech.utah.edu/senior_design/05/index.php/WindTurbine/MikeGroupReport>.
5. Quality Digest, Metris, Laser radar supports engineering of wind turbine blade aerodynamics, 2009, Available: <http://www.qualitydigest.com/inside/twitter-ed/laser-radar-supports-engineering-wind-turbine-blade-aerodynamics.html>.
6. A. Summers, Investigating the measurement of offshore wind turbine blades using coherent laser radar, MEng Technical Report, Durham University, April 2013.
7. G.N. Peggs et al., Recent developments in large-scale dimensional metrology, Proc. Inst. Mech. Eng., Part B: J. Eng. Manuf. 223 (B6) (2008) 571–595.

8. D.A. White, Quality Digest, Coherent Laser Radar: True Noncontact 3- D Measurement Has Arrived, 1999, Available: <http://www.qualitydigest.com/aug99/html/body_radar.html>.

9. Nikon, High-Accuracy Bridge CMM, 2014, Available: <http://www.nikonmetrology.com/en_EU/Products/Coordinate-Measuring-Machines/Bridge-CMM/High-Accuracy-Bridge-CMM/(specifications)>.

10. K.H. Lee, H. Woo, T. Suk, Data reduction methods for reverse engineering, Int. J. Adv. Manuf. Technol. 17 (10) (2001) 735–743.

11. B.L. Curless, New Methods for Surface Reconstruction from Range Images, PhD Dissertation, Stanford University, 1997.

12. Quality Digest, Study: Comparison Between Photogrammetry and Laser Radar, 2010, Available: <http://www.qualitydigest.com/inside/cmsc-article/study-comparison-between-photogrammetry-and-laser-radar.html#>.

13. Nikon, MV330/MV350 Laser Radar, 2014, Available: <http://www.nikonmetrology.com/en_EU/Products/Large-Volume-Applications/Laser-Radar/MV330-MV350-Laser-Radar/(key_features)>.

14. O.E. DeLange, Optical heterodyne detection, Spectrum, IEEE 5 (10) (1968) 77–85.

15. W.T. Estler, K.L. Edmundson, G.N. Peggs, D.H. Parker, Large-scale metrology – an update, CIRP Ann. – Manuf. Technol. 51 (2) (2002) 587–609.

16. Nikon, Laser Radar MV330/350 Automated Large Volume Inspection, 2010, Available: <http://www.smart-solutions.pl/download/laser-radar-mv330.pdf>.

17. Q. Wang, N. Zissler, R. Holden, Evaluate error sources and uncertainty in large scale measurement systems, Robot. Comput.- Integr. Manuf. 29 (1) (2013) 1–11.

18. Laser Beam Products, Mirror Coatings: Protected Aluminium, 2014, Available: <http://www.lbp.co.uk/pdfs/pdf_files/coatings.pdf>.

19. C.J. Brochu, J.W. Moffat, G.A. Morin, Optical Alignment of the Spherical Antenna Measurement System, Defence Research Establishment Ottawa, 1997.

20. J.H. Burge, P. Su, C. Zhao, T. Zobrist, Use of a commercial laser tracker for optical alignment, Opt. Syst. Alignment Toleranc. 6676 (2007).

21. R.P Elliott, Quality Digest, Measurement Accuracy of a Mirrored Surface Using a Laser Tracker, 2013, Available: <https://www.qualitydigest.com/inside/cmsc-article/measurement-accuracy-mirrored-surface-using-laser-tracker.html#>.

22. New River Kinematics, SpatialAnalyzer®: Features & Benefits, 2014, Available: .

23. New River Kinematics, SpatialAnalyzer®: Unified Spatial Metrology Network (USMN), 2014, Available: <http://www.kinematics.com/spatialanalyzer/usmn.php>.

24. Joint Committee for Guides in Metrology (JCGM/WG 1), Evaluation of Measurement Data – Guide to the Expression of Uncertainty in Measurement, 2008.

25. J.M. Calkins, Quantifying Coordinate Uncertainty Fields in Coupled Spatial Measurement Systems, PhD Dissertation, Virginia Polytechnic Institute and State University, 2002.

26. N.M. Patrikalakis, T. Maekawa, W. Cho, Shape Interrogation for Computer Aided Design and Manufacturing, Hyperbook Edition, 2009, Available: <http://web.mit.edu/hyperbook/Patrikalakis-Maekawa-Cho/node12.html>.

27. E. Savio, L. De Chiffre, R. Schmitt, Metrology of freeform shaped parts, Ann. CIRP 56 (2007).

28. C.D. Boor, On calculating with B-splines, J. Approx. Theory 6 (1970) 50–62.

29. J. Chen, G.-J. Wang, Approximate merging of B-spline curves and surfaces, Appl. Math. – A J. Chin. Univ. (2010).

30. X.-D. Chen, W. Ma, G. Xu, J.-C. Paul, Computing the Hausdorff distance between two B-spline curves, Comput.-Aided Des. 42 (2010) 1197–1206.

CHAPTER 11

INDOOR ROBOT POSITIONING USING AN ENHANCED TRILATERATION ALGORITHM

Pablo Cotera[1], Miguel Velazquez[1], David Cruz[1], Luis Medina[1] and Manuel Bandala[1*]

[1]Center for Engineering and Industrial Development - CIDESI, Querétaro, Querétaro, Mexico

ABSTRACT

This paper presents algorithms implemented for positioning a wheeled robot on a production floor inside a factory by means of radio-frequency distance measurement and trilateration techniques. A set of radio-frequency transceivers located on the columns of the factory (anchors) create a grid with several triangular zones capable of measuring the line-of-sight distance between each anchor and the transceiver installed in the wheeled robot. After measuring only three of these distances (radii), an enhanced trilateration algorithm is applied to obtain X and Y coordinates in a Cartesian plane, i.e., the position of the robot on the factory floor. The embedded systems developed for the anchors and the robot are robust enough to establish communication, select the closest anchors for measuring radii, and identify in which of the grid zones the robot is located.

INTRODUCTION

Precise positioning and location has long been sought for autonomous and semi-autonomous vehicles. Since the advent of low-priced GNSS technologies (mostly integrated circuits for embedded systems), these technologies have been widely implemented. Even though the accuracy of commercial systems is still only to within a couple of metres, these tools have proven useful for most applications.

A drawback of this technology is its inability to work properly in indoor environments. In such situations, wheeled robots needing to know their position (on the plane they are moving in) usually implement dead-reckoning strategies based on the information obtained from an inertial navigation system. However, most such data, e.g., from odometers, gyroscopes, magnetometers and accelerometers, are prone to have noise and generate cumulative errors; furthermore, the position of a fixed initial point always needs to be known.

As the underlying technology for the Global Navigation Satellite System (GNSS) has proven so reliable, researchers have tried to mimic its functionality for indoor uses. However, this requires wireless transceivers that are fast enough to detect very small periods of time, allowing the system to accurately

Citation: Pablo Cotera, Miguel Velazquez, David Cruz, Luis Medina and Manuel Bandala. Indoor Robot Positioning using an Enhanced Trilateration Algorithm. Int J Adv Robot Syst, 2016, 13:110. doi: 10.5772/63246.

measure distances between the transceivers in the order of metres.

Until recently, the wireless technology capable of measuring the time-of-flight of a signal between two transceivers was bulky and/or expensive, and reproducing the geo-spatial positioning systems in a research or other indoor environment was therefore very difficult.

However, miniaturization and embedding of many of the enabling technologies have finally produced inexpensive commercial systems that can be used in developing a solution with the required characteristics. Such a system is presented in this paper. The algorithms and design are explained, as well as the performed experiments. The results are discussed and further development is proposed.

RELATED WORK

Trilateration techniques for location purposes have been implemented in many different systems throughout the years. Some of them have been studied in depth, and the gained knowledge is applied directly in this work.

In 1990, Fang [1] proposed a simplification algorithm for trilateration, which makes a few interpretation assumptions and follows certain restrictions; this allowed for a simple solution and computation of the trilateration equations. This analytical solution was programmed in our system and the inaccuracies taken into account afterwards; this approach simplified the design and produced reliable results.

Two decades ago, wireless technology was neither as developed nor as cheap as it has become in recent years. This is one of the main reasons why distance measurement and location using wireless transceivers has just started to gain momentum. Furthermore, many of the first research systems required a lot of computational power; for example, in 2007, Mautz et al. [2] proposed a solution for locating nodes within a network. Five nodes were used and they communicated to a computational system in charge of minimizing errors using least-squares adjustment; however, this solution presented the inconvenience of centralization.

Towards achieving the computational requirements needed to solve the geometric algorithms, in 2008, Doukhnitch et al. [3] proposed a vector rotation method that proved more efficient than the existing methods. In 2013, Oguejiofor et al. [4] demonstrated a localization system with a wireless sensor network (WSN) that measured the distance between transceivers using the signal propagation and power law models from an RSSI (Received Signal Strength Indicator); however, this required complex computation and logarithmic functions, and resulted in a distance error higher than 70 cm.

High deviation is not uncommon for WSN systems. Therefore, several approaches have been implemented to reduce the measurement uncertainties. For instance, an improved weighted centroid algorithm was proposed by Bin et al. [5] in 2013; however, an optimized solution based on the Lemoine point that was presented in 2014 by Huang et al. [6], as well as a fuzzy optimization solution implemented by Jyoti et al. [7], gave better results.

Other AI programming techniques such as particle swarm optimization (PSO) were also studied by Ramesh et al. [8] in 2012, and many summaries and comparisons have been written regarding different localization methods. One such paper was presented in 2014 by Dyal et al. [9], for example.

More recently, the development of better hardware with high inherent accuracy (specifically the*DW1000* chip from *Decawave*) has allowed programming of simpler algorithms where it may not be necessary to minimize measurement errors. This requires less computational power and enables implementation in a cheap embedded system. This is the approach presented in this paper.

TRILATERATION ALGORITHMS

In simple terms, trilateration is a mathematical technique in which the location of a point in space is calculated using the distances from such a point to a series of known geometrical entities, e.g., a sphere or a circle.

In a 3D space, if a point of interest lies on the surface of three intersecting spheres, then, knowing the centre positions of the spheres and their radii, it is possible to narrow the possible locations of the point to no more than two. Starting with the general case, the equation for a sphere centred at (x_a, y_a, z_a) is shown in (1):

$$r^2 = (x - x_a)^2 + (y - y_a)^2 + (z - z_a)^2 \tag{1}$$

Applying the analytical approach for the general case would require the equations for the three spheres to be formulated and solved for the three unknowns (x,y,z), i.e., the location of the intersection point; however, two assumptions can be made in order to simplify this problem:

- In the case of an indoor location system, all the anchors (sphere centres) can be located at a fixed height on the same elevated plane, regarded as zero ($z_1=z_2=z_3=0$).
- The geometric algorithm can be simplified using Fang's method [1] so that one sphere centre is located at the origin of the elevated plane, another along the x-axis, and the third orthogonally to the previous two, effectively positioning three non-collinear points on a unique plane (Figure 1).Equations (2-4) depict these anchor coordinates:

$$A_1(0, 0, 0) \tag{2}$$

$$A_2(x_2, 0, 0) \tag{3}$$

$$A_3(x_3, y_3, 0) \tag{4}$$

$$r_1^2 = x^2 + y^2 + z^2 \tag{5}$$

$$r_2^2 = (x - x_2)^2 + y^2 + z^2 \tag{6}$$

$$r_3^2 = (x - x_3)^2 + (y - y_3)^2 + z^2 \tag{7}$$

$$x = \frac{r_1^2 - r_2^2 + x_2^2}{2x_2} \tag{8}$$

$$y = \frac{r_1^2 - r_3^2 + x_3^2 + y_3^2 - (2x_3x)}{2y_3} \tag{9}$$

$$z = \sqrt{r_1^2 - x^2 - y^2} \tag{10}$$

As can be seen from equation (10), depending on the values of x and y, it is possible to obtain a negative root for the value of z, which in the embedded system produces an error. This drawback and the way it is approached is explained in the following sections.

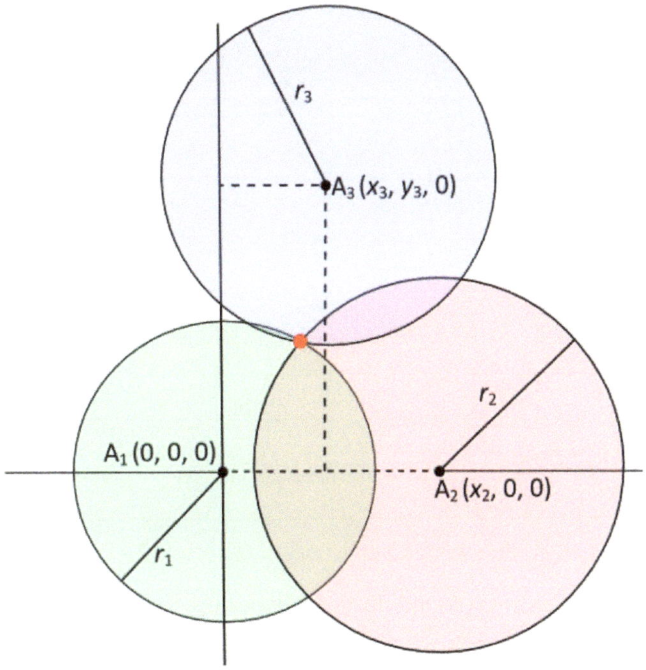

Figure 1. Simplified trilateration in 2D.

Equations (2-10) form the basis for locating a point in space, knowing the coordinates for the spheres' centres and the measured radii from each centre to the point of interest. In a 3D space, such equations would yield two possible solutions, but based on the aforementioned simplifications it is possible to work in a 2D space and analytically obtain a single solution.

Since all the anchors are positioned at the same height (plane), which is parallel to the floor where the robot moves (Figure 2), it is safe to regard the z coordinate of the robot's location as a constant; in this particular case, there is no need to calculate it.

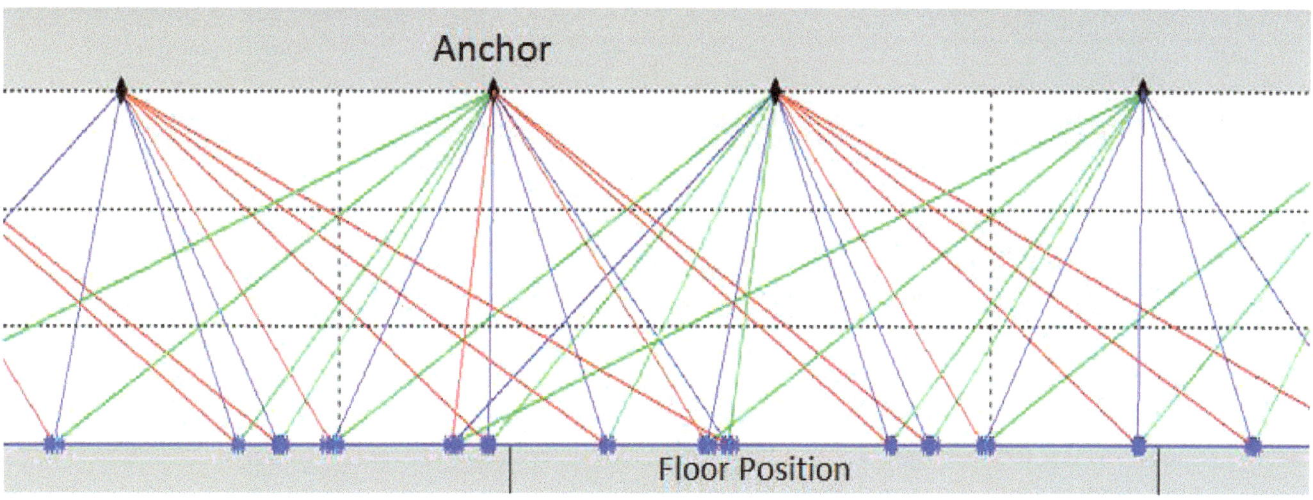

Figure 2. MATLAB® simulation of anchor positions, plane parallel to floor.

The algorithms depicted so far allow for the location of a single point within the area covered by three transceivers (anchors). However, the transceivers give a precision of about 10 cm within a line-of-sight of approximately 30 metres; therefore, in order to create a system able to detect the robot's position inside a wide area, the installation of several anchors is required, thus generating a grid of triangular zones.

Each one of these zones is formed by arranging three transceivers, whose coordinates follow equations (2-4); the whole grid is created using regular triangular patterns in which two adjacent zones share two transceivers. Each zone has a different offset with respect to a global origin point.

The position of the robot is calculated by first obtaining the local coordinates inside the zone comprising the closest anchors to it and then adding that zone's offset to get the global position within the area of the factory floor (Figure 3).

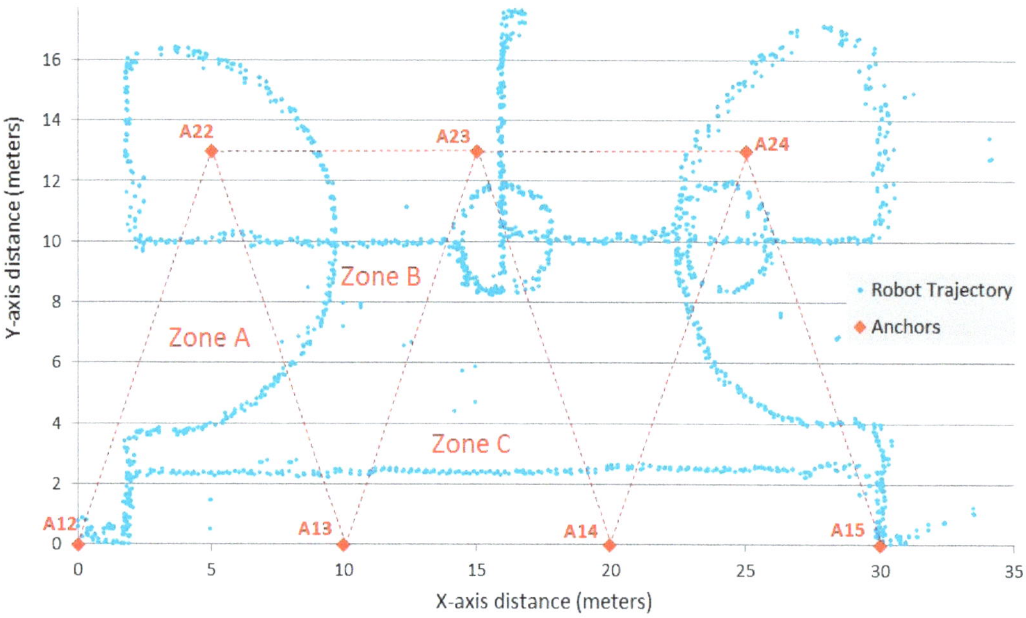

Figure 3. Anchor locations and zone delimitation for robot trajectory test.

An example based on Figure 3 can be given to aid understanding of the procedure to establish anchor location and zone delimitation:

- Zone A comprises anchors 12, 13, and 22, whose local coordinates (x_a, y_a, z_a) are A12(0,0,0), A13(10,0,0), and A22(5, 13, 0). For the depicted test, anchor 12 lies on the origin, so the global offset for that zone is Δ(0,0).
- Zone B shares anchors 13 and 22, but these anchors now have a different set of local coordinates: A22(0,0,0), A23(10,0,0), and A13(5,-13,0). The offset for this zone's origin (anchor A22) is Δ(5,13).
- For Zone C, the local origin would be at anchor A13 and the global offset Δ(10,0).

This locating approach is applied throughout the grid definition and can be reproduced for different zone sizes following the algorithm depicted in Figure 4. In the example from Figure 3, each anchor along the same row is separated by 10 metres; however, the chip range allows for separation of around 25-30 metres, depending on the environmental noise.

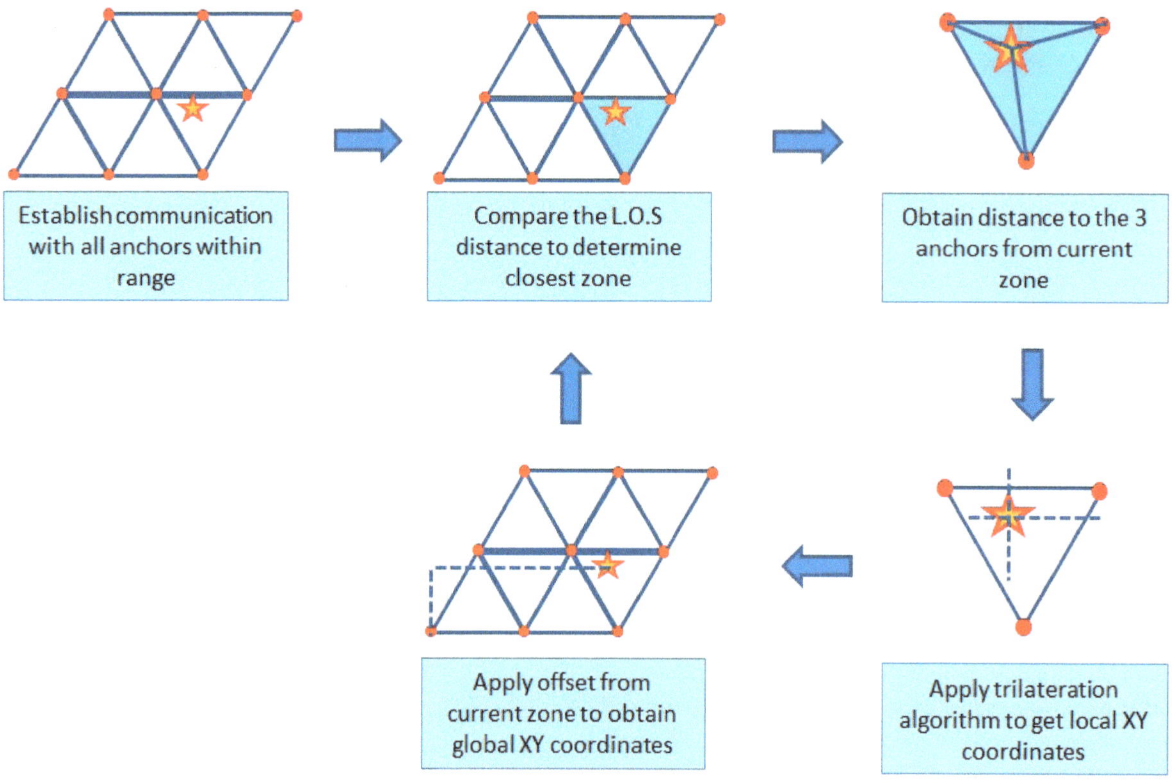

Figure 4. Zone detection and global positioning algorithm

EMBEDDED SYSTEM

As previously stated, the advantages of a system based on the GNSS technology make it desirable to implement for an indoor environment. The communication protocol can be replicated on an embedded scheme and simplified based on the following steps:

- A receiver starts listening to the satellite broadcasts, which contain information on the precise time they were sent, the location of the satellite, and a global standardized time (with atomic clock precision).
- The receiver measures the time-of-flight of the signal travelling from the satellite along the line-of-sight to the receiver, thereby calculating the distance between them. It performs the same measurement with at least three satellites.
- Knowing the distances from the receiver to three of the satellites, and their orbital position, a trilateration algorithm can be applied to obtain the geo-spatial position of the receiver.

The system presented in this paper was developed using the wireless chip *DW1000* from the Irish company *Decawave*, which allows precise distance measurement between two identical transceivers (however, each one is configured differently).

The idea behind the system was to apply the underlying principles used in GNSS technology by installing an elevated grid of anchors deliberately positioned to create triangular zones. Inside this grid, the robot's transceiver communicates with the three closest anchors. By measuring the time-of-flight of the signals and calculating the line-of-sight distances, it applies the trilateration algorithm enhanced for indoor applications in order to obtain the robot's position on a plane (factory floor).

After obtaining the local coordinates from the set of three anchors, the system installed on the robot verifies the zone and calculates the global position by analysing its offset in relation to the plane origin (Figure 4).

The hardware implemented for this system comprises two different electronic boards: one type for the anchors and another for the location and trilateration algorithms. The anchor boards consist of a microcontroller connected to the *DW1000* chip along with the power supply and indicator LEDs (Figure 5), while the embedded system for the trilateration algorithms uses a *Raspberry Pi B+ (RPi)* microcomputer with a shield for connecting the *DW1000* chip (Figure 6).

All the communication protocols and algorithms were programmed using C language in the microcontrollers and the *RPi*; the trilateration algorithms, zone lookup table for the coordinates, and control routines were saved in the *Raspberry*'s solid-state memory.

Figure 5. Anchor embedded system

Figure 6. RPi embedded system with two chips and IMU

EXPERIMENTS

The indoor positioning system described thus far was originally devised for locating a wheeled robot within the areas of a factory floor; however, the data presented were obtained through experiments in a controlled environment.

The grid of triangular zones for testing was relatively small compared to the factory dimensions; nevertheless, it was big enough for a proof of the concept of the system. The electronic boards for the anchors were enclosed for protection (Figure 5) and located over a planar surface following the pattern depicted in Figure 3.

With the seven anchors, five triangular zones were created, and the information for each one, including coordinates, global offset, and zone number, was stored in the solid-state memory of the *RPi* microcomputer.

The algorithm from Figure 4 was followed during the experiments:

- The transceiver chip connected to the *RPi* communicated initially with all the anchors.
- By identifying the three closest anchors, an algorithm selected the correct zone in which the system was located from the microcomputer memory.
- The system measured the line-of-sight distance to each of the three anchors comprising the actual zone, and obtained the local XY coordinates using the trilateration algorithm.
- Knowing the zone and its offset, the local coordinates were converted into global coordinates, taking as the global origin the position of anchor A12 (Figure 3).

Figure 7. Anchor positions for court test.

Several tests were performed in order to validate the algorithms and robustness of the system. One of these involved the positioning of three anchors inside a basketball court (Figure 7), and another involved seven anchors forming a grid with five zones, as depicted in Figure 3 and Figure 8.

A moving platform holding the transceiver shield connected to the RPi followed the paths painted on the court whilst saving the data for the X and Y coordinates as well as the zone in which it was located. Taking the position of anchor A12 as the origin, the global coordinates were calculated after measuring the local position inside each zone.

Figure 8 shows a graph of the data obtained directly from the measurements without any filtering. The colour of each point represents the zone identified directly from the system; as can be seen, the zone delimitation is acceptable and the measurement noise results are very low as well. Approximately 3% of the data was corrupted due to measurement or communication errors, but the ±10 cm accuracy remains within the specification.

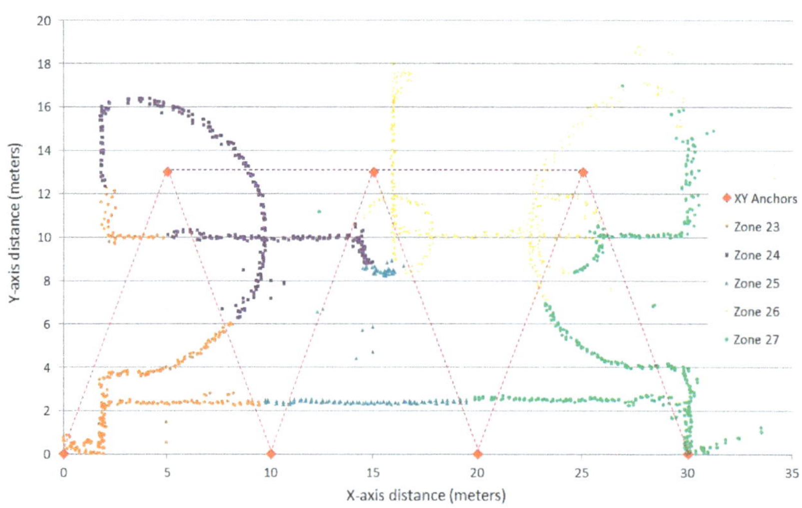

Figure 8. Zone identification from embedded system.

A subsequent test was conducted to verify the accuracy of the system, i.e., to validate that the measured distances were within the ±10 cm limits. For this test, 35 anchors were installed on columns inside a factory at 3.5 metres above the ground; the tag was then placed on a table 80 centimetres above the ground in an XY position below five of the anchors selected randomly. At each of the five positions, a series of 100 samples was taken and the deviation measured. Each sample consists of the line-of-sight distance between the tag and the three anchors, creating a triangular zone. Using the three distances, a pair of XY coordinates was obtained through the trilateration algorithms, thus obtaining 100 data pairs of XY coordinates for each sample. With these data sets, the standard deviation of each group of points was obtained.

A map of the anchors and the zones they create is depicted in Figure 9, whilst Table 1 shows the labels for each test and their selected anchors.

Table 1. Anchors table for accuracy test/

Test Number	Anchor Number	Triangular Zone
T1	D19	44
T2	E16	25
T3	E16	9
T4	E22	30
T5	F21	12

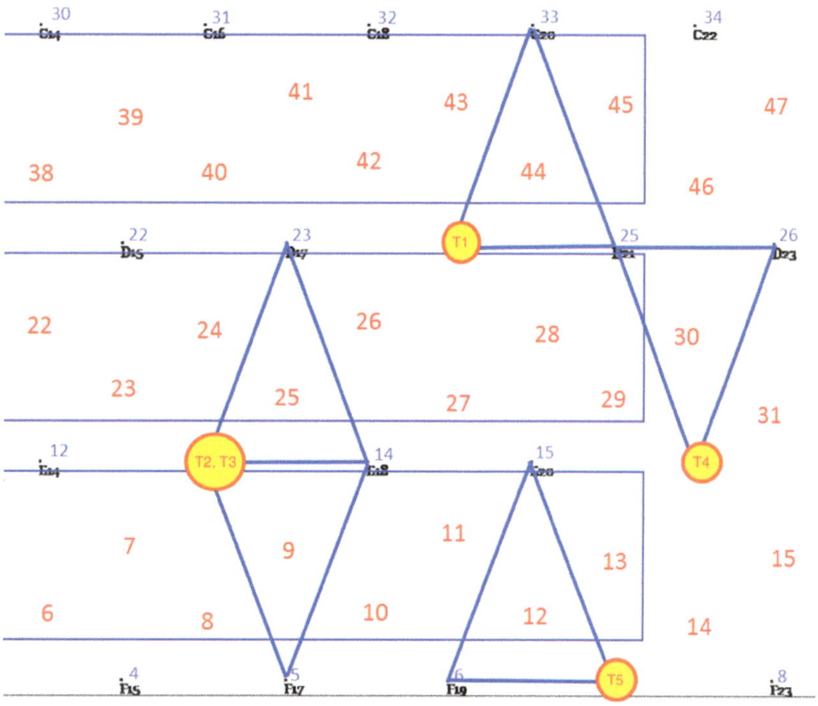

Figure 9. Map of anchors and zones inside factory.

Figures 10, 11, 12, 13 and 14 depict the accuracy tests conducted on the five selected anchors, corresponding to the previous table. As can be noted, except for test 4, all the other measurements remain within the ±10 cm specified limit, with an overall average accuracy of ±4.09 cm.

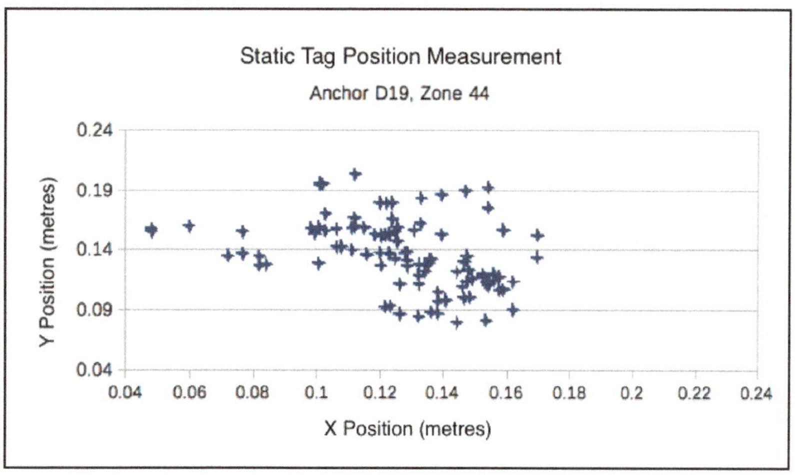

Figure 10. Test 1 XY position accuracy data. Test 1, Anchor D19, Zone 44: Standard Deviation X coordinate: 2.56 cm; Standard Deviation Y coordinate: 2.91 cm.

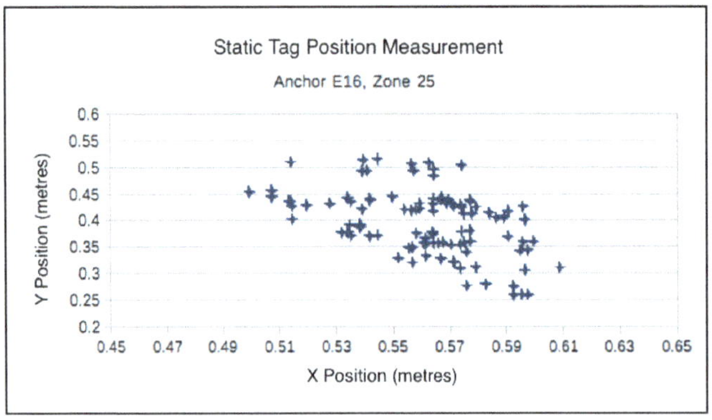

Figure 11. Test 2 XY position accuracy data. Test 2, Anchor E16, Zone 25: Standard Deviation X coordinate: 2.42 cm; Standard Deviation Y coordinate: 6.15 cm.

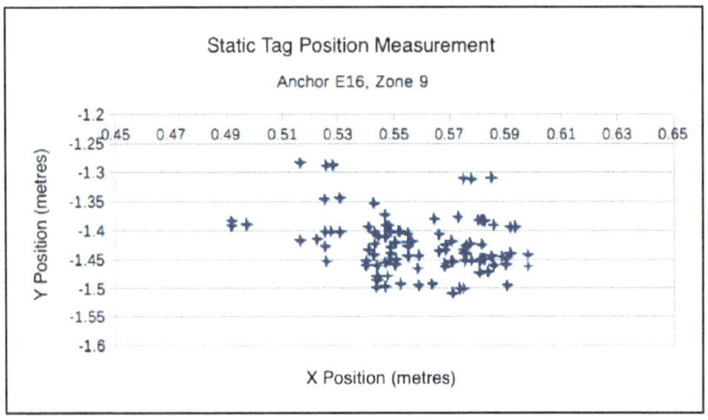

Figure 12. Test 3 XY position accuracy data. Test 3, Anchor E16, Zone 9: Standard Deviation X coordinate: 2.35 cm; Standard Deviation Y coordinate: 4.79 cm.

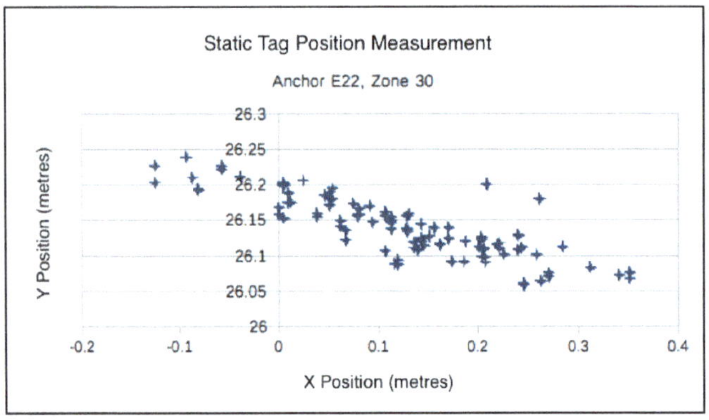

Figure 13. Test 4 XY position accuracy data. Test 4, Anchor E22, Zone 30: Standard Deviation X coordinate: 10.41 cm; Standard Deviation Y coordinate: 4.15 cm.

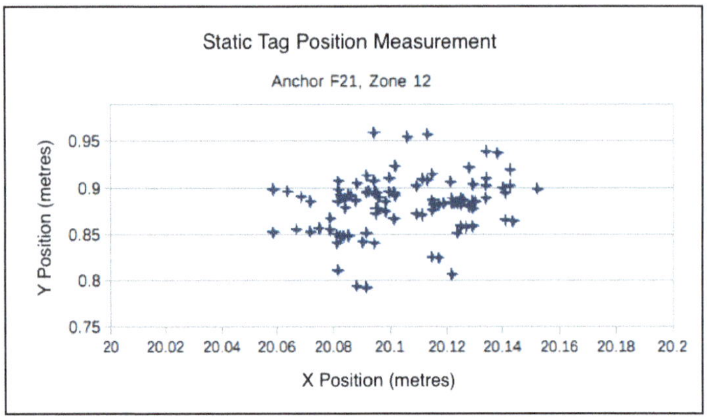

Figure 14. Test 5 XY position accuracy data. Test 5, Anchor F21, Zone 12; Standard Deviation X coordinate: 2.23 cm; Standard Deviation Y coordinate: 3.01 cm.

In 2013, Oguejiofor et al. [4] presented the experimental results of their RSSI-based location system where the position of a node was calculated from its distance to three fixed anchors (located at no more than 20 metres from the node). The estimated position was relatively accurate, remaining within 74 centimetres of the actual location of the node. But comparing that error to our average of ±4.09 cm, it is clear that the solution proposed here presents a significant accuracy increase.

Data Filtering

From the first tests, it was observed that due to electromagnetic noise and/or physical obstacles, the line-of-sight measurements varied considerably, resulting in highly deviated XY coordinates. To overcome this, a simple filter was designed. This filtering algorithm took the trilateration calculation for the Z coordinate (Equation 10) and compared it with the actual Z position of the tag, which remained a fixed value due to the level floor. Thus, if the difference between the real value and the calculated one was greater than 20%, this data set was disregarded and the previous one used instead.

CONCLUSIONS

This paper presents a robust positioning solution involving several areas of development and expertise. A lot of work was invested in the design and programming of the trilateration algorithms, as well as in the design and manufacture of the embedded systems.

The *DW1000* chips are state-of-the-art devices that present many challenges to make them work seamlessly. However, once that learning curve was overcome, their functionality and advantages were fully exploited.

Although this research was based on several examples for positioning devices using trilateration, the real contribution of the project is in revamping such techniques by applying the zone selection algorithms, effectively increasing the final reach and scope.

It is clear that precise indoor positioning has a long way to go, but further developments seem inevitable and present very interesting implications.

FUTURE WORK

In addition to the actual system presented here, several improvements are being developed to create a more robust solution. The implementation of an Inertial Measurement Unit (IMU) to obtain the orientation and direction of the movement, as well as enhancements of the algorithms will allow for more precise location and data analysis for automatic navigation control.

Other techniques, such as Kalman filtering and centroid calculation, also feature in on-going research. A second prototype implementing a DV-HOP strategy, as proposed by Xu et al. [10], will allow for a more robust system. The miniaturization and embedding of this solution for different platforms will eventually provide the desired outcome.

ACKNOWLEDGEMENTS

The authors want to thank the Engineering and Industrial Development Center (CIDESI) for their support in this project.

REFERENCES

1. Fang B.T, Simple solution for hyperbolic and related position fixes, IEEE Transactions on Aerospace and Electronic Systems, Vol. 26 (No. 5) (1990) 748-753.
2. Mautz R, Ochieng W, Brodin G, Kemp A, 3D Wireless Network Localization from Inconsistent Distance Observations, Journal of Ad Hoc and Sensor Wireless Networks, Vol. 3 (2007) 141-170.
3. Doukhnitch E, Salamah M, Ozen E, An efficient approach for trilateration in 3D positioning, Journal of Computer Communications, Vol. 31 (2008) 4124-4129.
4. Oguejiofor O.S, Aniedu A.N, Ejiofor H.C, Okolibe A.U, Trilateration based localization algorithm for wireless sensor network, International Journal of Science and Modern Engineering, Vol. 1 (No. 10) (2013) 21-27.
5. Bin L, Zheng D, Yu N, Yun L, An improved weighted centroid localization algorithm, International Journal of Future Generation Communication and Networking, Vol. 6 (No. 5) (2013) 45-52.
6. Huang M.S, Narayanan R.M, Trilateration-based localization algorithm using the Lemoine point formulation, IETE Journal of Research, Vol. 60 (No. 1) (2014) 60-73.
7. Jyoti R.K, Singh N. Localization in WSN using modified trilateration based on fuzzy optimization, International Journal of Advanced Research in Computer Science and Software Engineering, Vol. 3 (No. 7) (2013) 657-663.
8. Ramesh M, Divya P, Kulkarni R, Manoj R, A Swarm Intelligence based distributed localization technique for Wireless Sensor Network, International Conference on Advances in Computing, Communications and Informatics (2012).
9. Dyal R, Kaur V, Analysis of some localization algorithm in wireless sensor networks, International Journal of Computer and Organization Trends, Vol. 11 (No. 1) (2014).
10. Xu C-X, Chen J-Y, Research on the Improved DV-HOP Localization Algorithm in WSN, International Journal of Smart Home, Vol. 9 (No. 4) (2015) 157-162.

CHAPTER 12

GLOBAL COVERAGE MEASUREMENT PLANNING STRATEGIES FOR MOBILE ROBOTS EQUIPPED WITH A REMOTE GAS SENSOR

Muhammad Asif Arain , Marco Trincavelli, Marcello Cirillo, Erik Schaffernicht and Achim J. Lilienthal

Mobile Robotics & Olfaction Lab, Center of Applied Autonomous Sensor Systems (AASS), School of Science & Technology, Örebro University, Örebro SE-70182, Sweden

ABSTRACT

The problem of gas detection is relevant to many real-world applications, such as leak detection in industrial settings and landfill monitoring. In this paper, we address the problem of gas detection in large areas with a mobile robotic platform equipped with a remote gas sensor. We propose an algorithm that leverages a novel method based on convex relaxation for quickly solving sensor placement problems, and for generating an efficient exploration plan for the robot. To demonstrate the applicability of our method to real-world environments, we performed a large number of experimental trials, both on randomly generated maps and on the map of a real environment. Our approach proves to be highly efficient in terms of computational requirements and to provide nearly-optimal solutions.

INTRODUCTION

In a large number of situations, it is of utmost importance to assess whether a specific gas is present in an area, and establish its concentration, distribution, and the location of the source. This includes the inspection of pipelines and chemical plants to detect the presence of dangerous gas leaks, the longterm monitoring of air pollution levels in big cities or of landfill sites to identify exploitable gas sources. It is often the case that the gases of interest are dangerous for humans and, therefore, human involvement should be minimized.

In the case of pollution monitoring in cities, the deployment of a network of stationary sensors has often proven to be an effective solution [1]. Similar solutions have also been tried in more critical environments, such as coal mines [2] and landfill sites [3]. However, whenever stationary solutions are employed in large areas, two problems arise: the measurements are inherently sparse, so that there can be no guarantee for the absence of the measured gases in the environment (e.g., in a chemical plant a leak could go undetected), and it is very difficult to decide where to place the sensors in the first place, especially if the environment is subject to changes in wind direction and speed, either caused by weather conditions or moving objects.

Mobile robotic olfaction is a recent research direction which aims at overcoming the problems of fixed sensor networks. It combines gas sensors, as used in the fixed networks, with the

Citation: Arain, M.A.; Trincavelli, M.; Cirillo, M.; Schaffernicht, E.; Lilienthal, A.J. Global Coverage Measurement Planning Strategies for Mobile Robots Equipped with a Remote Gas Sensor. Sensors 2015, 15, 6845-6871. DOI: 10.3390/s150306845

flexibility of a mobile platform [4]. Different platforms have been successfully tested in recent years for mapping the gas concentration and source localization, both in indoor and outdoor environments [5–7]. The use of mobile platforms has become even more appealing since the introduction of sensors which are capable of detecting gases remotely [8–10]. In particular, sensors based on tunable diode laser absorption spectroscopy (TDLAS) [11,12] allow for ranged sensing up to considerable distances (if the field of view is unobstructed), but are expensive and relatively bulky. Therefore, they are not suitable to be distributed in large numbers in the environment, but they can be used on-board a mobile platform.

The use of ranged sensing devices on mobile platforms introduces new challenges due to the robot's limited autonomy. The largest restriction is posed by battery power and hence the limited duration of monitoring activities. Thereto, it is required to explore the environment efficiently. Here, we focus on the problem of gas detection in large environments, both outdoor and indoor. We elaborate on the problem of verifying if a specific gas is present in the environment or not. More in particular, in this paper we present a novel approach for quickly generating exploration strategies for a mobile robot equipped with a sensor for remote gas detection. The resulting strategies are efficient, in the sense that they minimize the overall time the robot will take to inspect the whole environment.

The contributions of this paper are two-fold. Our first contribution is to formalize the remote gas sensing coverage problem for mobile robotics. Our second contribution is to present a two-step approximate solution to the aforementioned problem which relies on a novel convex optimization-based approximation. Here, we test our solution on large-scale scenarios, to demonstrate its applicability in real-world situations.

The paper is organized as follows: After presenting an overview of relevant work (Section 2), we define and formalize the gas detection coverage problem and how to obtain optimum solutions in Sections 3 and 4. We then present two alternative approaches to decompose the problem, their advantages and drawbacks (Section 5). A first experimental evaluation is detailed in Section 6, where the solution quality and the scalability of the different approaches is discussed. To overcome the scalability issues common to all previous approaches, we present in Section 7 a novel solution which uses re-weighted convex relaxation. We then evaluate this novel solution in Section 8 and conclude in Section 9.

RELATED WORK

Stationary networks of gas sensors have been used over the years in different environments and in a variety of applications, such as pollution monitoring [13] and the measurement of the concentration of methane in the atmosphere [14]. The sensors commonly employed in such networks are in situ [15], which means that they have to come in direct contact with the gas they need to sense. This is obviously a strong limitation: the choice of their positioning is not trivial, new deployments could be necessary whenever the environment changes and the sensors must be re-calibrated and maintained over time. For specific gases, e.g., methane, other viable sensing technologies are available, such as infrared thermography [16], whose results, however, are highly affected by weather conditions, by the nature of the ground surface and by the distance between sensors and sources.

In recent years, many research groups have started to focus on mobile robot olfaction, a solution for gas sensing which would implicitly guarantee more flexibility than sensor networks, as it combines gas sensors with mobile platforms [17,18]. Mobile robots equipped with in situ gas sensors have been successfully used for mapping gas distributions [19,20] and leak detection [9,10]. A limitation of current approaches is that the robots generally move between pre-defined positions, or reactively follow gas plumes [21]. Finding good sensing positions for a mobile robot is still an open problem and it is closely related to the sensor placement problem for static networks.

In our work, we do not use in situ sensors, but we equip our robot with a TDLAS sensor, which can provide remote measurements up to a well defined sensing range [20]. A sensing action, that is, scanning an area of the environment with our sensor, is expensive, both in terms of time and of battery consumption. Similar costs should be accounted for whenever the robot moves from one sensing position to the next. Since here we are interested in detecting if a specific gas is at all present in a known environment, we need to scan the whole environment. Therefore, if we want to find battery- and time-efficient exploration strategies, we must take into account and minimize the costs associated both to movement and to sensing.

Neglecting the cost associated to the movements of the robot, the problem we would need to solve could be reduced to anArt Gallery Problem [22] or to a View Planning Problem [23]. The art gallery problem is NP-hard (non-deterministic polynomial-time hard [24]) in its most common variants and view planning is isomorphic to the Set Covering Problem [25], a well known NP-complete problem. This family of problems has been extensively studied over the past decades [26,27], but the algorithms proposed that solve the problems optimally are effective under restricted assumptions, such as not considering occlusions in the field of view of the sensors [28], or work only when the number of possible sensing configurations is relatively small [29,30]. There exist, however, efficient algorithms for calculating approximated solutions [31] for problems with a large number of possible sensing configurations. The solutions described above, however, present two major drawbacks in our context: first, they are mostly concerned with camera placement problems, which means that they rarely consider limited field of view; second, they inherently do not consider the cost of moving from one sensor position to the next. On the other hand, assuming known sensing positions, we would still need to solve a (Metric) Traveling Salesman Problem (TSP), that is, we should find the shortest tour that connects all the sensing positions. This is a well-known NP-hard problem, but it is possible to optimally solve very large TSP instances, with thousands of locations [32].

Solving an art gallery problem first, and then calculating the shortest tour among the selected sensing positions by solving a TSP is an approach which has proven to be effective in many applications [33–36]. However, all the algorithms proposed rely on simplifying assumptions on the field of view of the sensors (e.g., 360 ° or unrestricted). More important still, the solution of an art gallery problem for a large number of candidate sensing positions with overlapping fields of view remains computationally challenging. By contrast, we use re-weighted convex relaxation to handle a high number of candidate sensing positions.

Our problem can also be related to planning for mobile robotics inspection, although in this case sensing cost is usually neglected. For instance, in [37,38], the authors present an inspection strategy for submerged ship hull with an autonomous underwater vehicle. They address the problem by developing a probabilistic planner to maximize uncertainty reduction while providing coverage of the mesh surface. In our case, we cannot ignore the cost of sensing actions and we aim at guaranteeing that the whole environment is observed. The Generalized Covering Salesman Problem [39] can explicitly address this last issue, as a solution of one of its instances is a tour which respects given covering constraints, but still sensing costs are not taken into accounts. A similar problem is solved in [40], where an approach based on mixed integer linear programming is used for finding a surveillance route for a mobile camera. The optimal solution approach we present here is highly reminiscent of the work of Tamioka et al. [40] and could be considered an extension of their algorithm: in our case, we can solve larger instances, while also considering costs associated with sensing actions.

A combination of the View Planning Problem and the Metric TSP, the Traveling View Planning Problem has been defined and analyzed in [41,42]. It refers to finding a sequence of sensing actions while minimizing the overall cost. Here, the authors use an approximation method to find a bounded suboptimal solution which considers both traveling time and sensing time with respect to the available candidate sensing positions. However, it is not yet clear how to optimally select the candidate sensing positions and the approach presented in these papers does not scale well (in terms of computation time) when the size of the environment grows (in terms candidate sensing positions).

PROBLEM DEFINITION

In this article, we address the problem of detecting the presence of a specific gas in large environments, both indoor and outdoor. We assume a static environment with no major changes in the average concentration of the target gas due to wind or by other means. This assumption does not imply a major limitation to the gas detection problem, since it is possible to repeat an inspection tour over and over again, in case things are changing. However, these problems need to be addressed for subsequent tasks, like gas source localization and gas distribution mapping as discussed in Section 9.

Furthermore, we assume that the gas, if present, can be detected near ground level. This assumption applies to many real world applications: for instance, methane leaks in a landfill necessarily occur close to the ground. We assume that the presence of the gas in question can be observed by means of a ranged sensor, such as TDLAS. An example of such sensors is the Remote Methane Leak Detector, which can measure the integral concentration of methane over a reflected laser beam (see Figure 1a). The sensor emits light in the near-infrared (NIR) band and analyses the spectrum of the reflection. Each gas has a unique absorption spectrum and the sensor is tuned to the target gas [43]. Hence, the TDLAS technology is very selective and only responds to its target gas even in the presence of multiple airborne substances. Thus, we can safely neglect the complex issue of dealing with gas mixtures. If the utilized sensor is only partially selective, e.g., like the often used metal oxide (MOX) sensors, it would be required to introduce an additional gas discrimination step. Different studies dealing with this particular issue are for example [44–46].

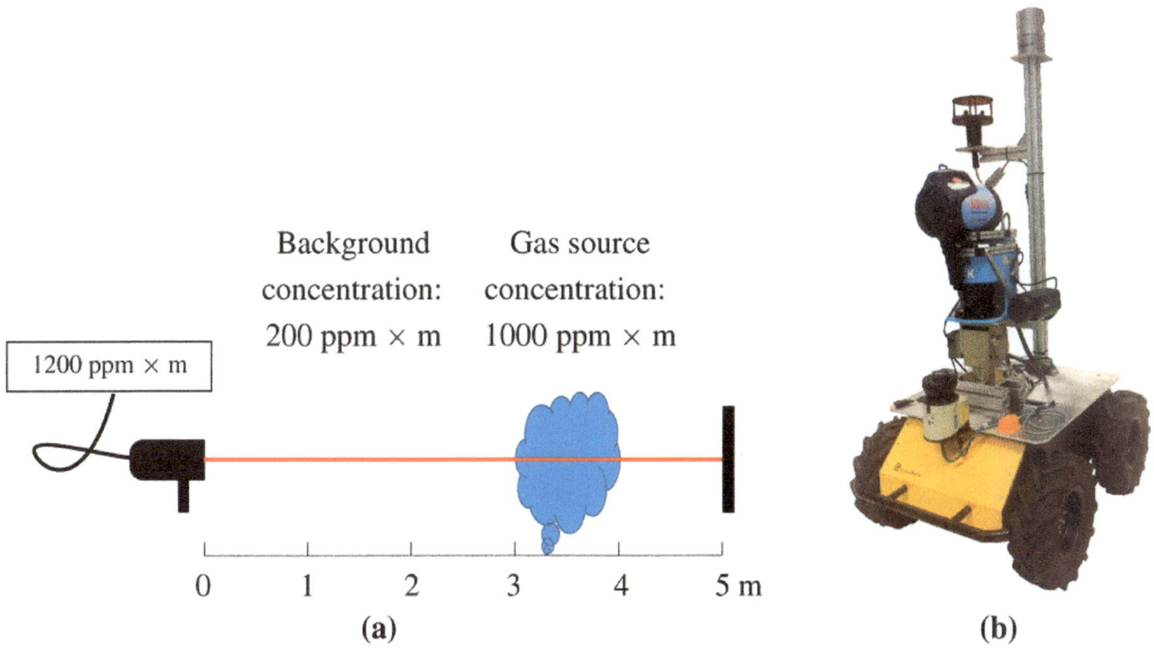

Figure 1. (**a**) the Remote Methane Leak Detector is a TDLAS sensor which can report the integral concentration of methane along its laser beam (parts per million × meter); (**b**) Gasbot, a robotic platform for gas detection. Gasbot is a research platform based on a Husky A200. It is specially equipped with a methane sensitive remote sensor (RMLD) mounted in conjunction with a laser scanner on a pan-tilt unit, a laser scanner for self-localization and mapping, an anemometer and a thermal camera.

Note that methane is not the only gas that can be detected using this technology. Finally, we assume that the ranged sensor is mounted on a mobile platform, such as the one in Figure 1b, so that its beam can be directed towards the ground at a specific range from the robot and that the sensor can be rotated around its vertical axis.

Given a map of the environment in which we need to assess the presence of a specific gas, we divide it into a Cartesian grid, thus obtaining a set $\mathcal{A}$ of n cells of identical size: $\mathcal{A} = \{a_1, \cdots, a_n\}$. The set of all cells $\mathcal{A}$ is partitioned into subsets $\mathcal{O}$ and $\mathcal{S}$, where $\mathcal{O}$ includes all the cells which contain an obstacle and therefore (1) are not traversable by the robot; and (2) can stop the beam of the gas sensor. $\mathcal{S}$ includes all the cells which do not contain obstacles and are therefore traversable by the robot. A solution to the detection problem is an obstacle free closed path, or tour, within the map that the robot can traverse and a set of sensing actions along the tour which enables the robot to sense every cell in $\mathcal{S}$.

We limit the movement of the robot between a finite set of poses $\mathcal{P}$. Each $p_j \in \mathcal{P}$ is defined by a two-tuple (a_j, θ_j), where $a_j \in \mathcal{S}$ and $\theta_j \in \Theta$. Θ is a finite set of allowed orientations, equally spaced between $[0, 2\pi)$. Note that the robot position within a cell corresponds to its center, regardless of the orientation. Here, we assume that the movements of the robot are limited to forward motions and rotations. Thus, for $\Theta = \{0, \frac{\pi}{2}, \pi, \frac{3}{2}\pi\}$, the movements of the robot can be captured by a directed graph like the one represented in Figure 2—backward motions and a higher cardinality of Θ could be obviously represented in a similar graph. In our problem definition, we use time as measure of cost. Given the times t^r, representing the time necessary to rotate $2\pi/|\Theta|$, and t^c, the cost associated to moving to the adjacent cell, the movement time $t^m_{p_i \to p_j}$ from p_i to p_j can be easily calculated by finding the shortest path on the movement graph, where each rotation edge has a weight of t^r and every edge leading to a different cell has a weight of t^c. With this decomposition, and with a knowledge of the terrain in each cell, it would be straightforward to express the cost of movement in terms of battery consumption.

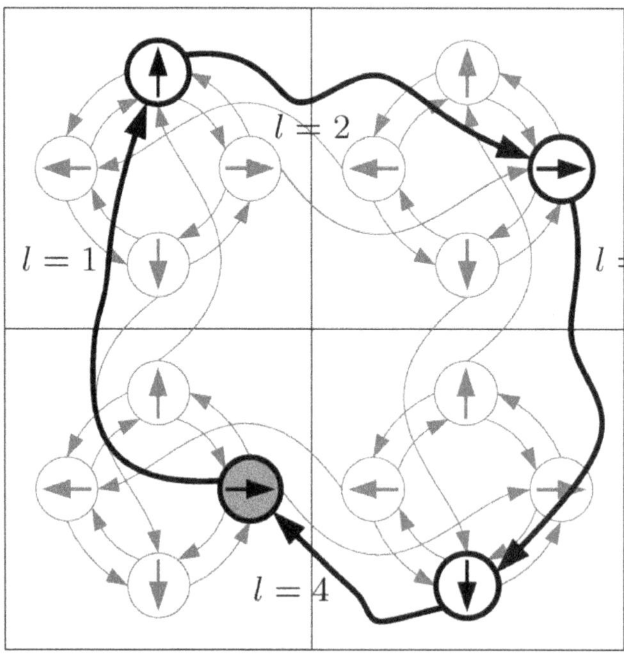

Figure 2. The graph captures the allowed movements of the robot on a grid map when $\Theta = \{0, \frac{\pi}{2}, \pi, \frac{3}{2}\pi\}$ and only forward movements are allowed. Small circles indicate the poses where the robot can stop (the internal arrow indicates the orientation of the robot) and the directed edges its allowed movements. Note that in the figure the poses do not correspond to the centers of the cells, but this is only for clarity reasons.

We can now define a candidate sensing configuration on the movement graph described above. A candidate sensing configuration corresponds to a possible sensing action of the robot in a specific pose p, that is, it represents the possibility of performing a sensing action over a well defined area. Formally:

Definition 1

A candidate sensing configuration c_i *is defined as a tuple* (p_i, ϕ_i, r_i), where $p_i \in \mathcal{P}$ is the robot's pose, ϕ_i *is the central angle of a circular sector and* r_i *its radius*.

Hence, a candidate sensing configuration c_i would allow the robot positioned in p_i to scan a circular sector of central angle ϕ_i and radius r_i, as shown in Figure 3a. Because of obstructions, not all the cells in the circular sector would be necessarily observable. Also, we need to define which portion of a cell should be swept by the laser beam to consider the cell itself as observed. Let us define $\mathcal{C}_{\mathcal{P}}$ as the set of all candidate configurations defined over a set of robot poses $\mathcal{P}$. We can then define the visibility function $\upsilon_{\mathcal{P}} : \mathcal{C}_{\mathcal{P}} \mapsto 2^{\mathbb{Z}^2}$, such that $\upsilon_{\mathcal{P}}(c)$ denotes the set of cells $\in \mathcal{S}$ visible from c. In the remainder of this paper, we define $\upsilon_{\mathcal{P}}$ so that a cell k is considered visible to a sensing configuration c_i if the line segment connecting the centers of c_i and k is in the circular sector of ϕ_i and r_i, and does not intersect any occupied cell, as shown in the example in Figure 3b.

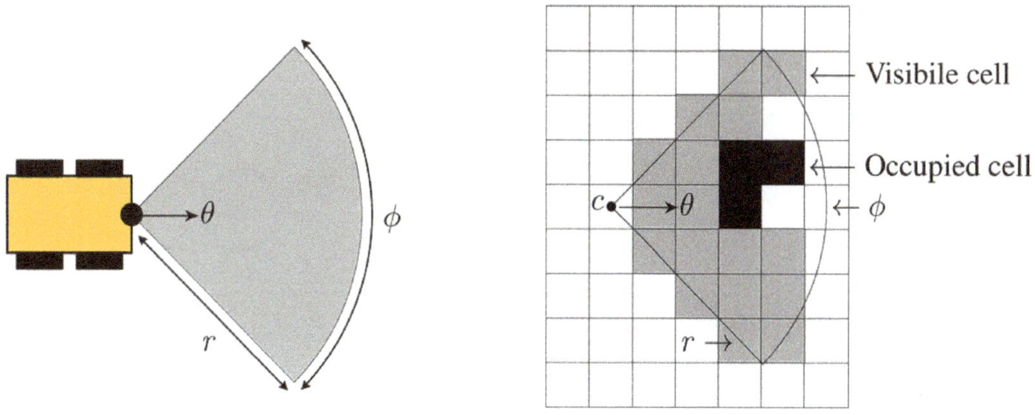

(a) (b)

Figure 3. (**a**) a candidate sensing configuration c allows the robot to scan a circular sector of central angle ϕ and radius r; (**b**) $\upsilon_{\mathcal{P}}(c)$ is the visibility function which defines which are the cell that are observable from candidate sensing configuration c.

The cost associated to perform a sensing action in a candidate sensing configuration c is tsc and it depends on the central angle ϕ associated with c.

FINDING AN OPTIMAL SOLUTION

As defined above, given a discretized map of the environment, the sets of occupied and unoccupied cells $\mathcal{O}$ and $\mathcal{S}$ (such that $\mathcal{A} = \mathcal{O} \cup \mathcal{S}$), the set of allowed poses $\mathcal{P}$ for the robot, the set of candidate sensing configurations $C_{\mathcal{P}}$ defined over $\mathcal{P}$, the visibility function $\upsilon_{\mathcal{P}}(c)$ and the cost functions $t^m_{p_i \to p_j}$ (movement time from p_i to p_j, calculated as the shortest path on the movement graph) and t^{s}_{c} (sensing time of candidate configuration c), a solution to the detection problem is an obstacle free tour defined as an ordered, finite set of sensing configurations $\pi = \{c_1, \cdots, c_k\}$ such that $\bigcup_{c_i \in \pi} \upsilon_{\mathcal{P}}(c_i) = \mathcal{S}$. The cost associated to a solution π is equal to the sum of the traveling costs (including the traveling cost to go from the last sensing configuration back to the first one) and sensing costs:

$$cost(\pi) = \sum_{i=1}^{k-1} t^m_{p_i \to p_{i+1}} + t^m_{p_k \to p_1} + \sum_{c_i \in \pi} t^s_{c_i} \qquad (1)$$

Given the set Π of all valid solutions to a given problem instance, an optimal solution π_{opt} is the one with minimum cost:

$$\pi_{opt} = \operatorname*{argmin}_{\pi_i \in \Pi} cost(\pi_i)$$

Problem Formulation

Practically, we can calculate optimal solutions for the problem defined above by casting it as an optimization problem. Here, we use a formulation inspired by the work of Tomioka et al. [40]. More specifically, we define flow variables and label variables over the movement graph of the robot. Let us consider a problem instance with a set of candidate sensing configurations $C_{\mathcal{P}}$. Flow variables are binary variables associated with all pairs of candidate sensing configurations $C_{\mathcal{P}}$. A solution $\pi = \{c_1, \cdots, c_k\}$ to a problem instance represents a closed walk on the selected candidate sensing configurations. We can then refer to the configurations in a solution π by means of their indexes 1, ⋯, k and, as we are considering a tour, we can define the configuration of index k + 1 = 1, that is, $c_{k+1} = c_1$ in π. Given a solution π, flow variable $f_{ci,cj}$ represents whether c_i and c_j are consecutively traversed in π:

$$\forall c_i \in C_{\mathcal{P}}, \forall c_j \in C_{\mathcal{P}}, c_i \neq c_j \quad f_{c_i,c_j} = \begin{cases} 1 \text{ if } & c_i = c_h \in \pi, c_j = c_{h+1} \in \pi. \\ 0 & \text{otherwise.} \end{cases} \qquad (2)$$

Label variables $l_{ci,cj}$ are non-negative, real variables. They are assigned to the shortest path on the movement graph between any two candidate sensing configurations $C_{\mathcal{P}}$. As detailed in [40], label variables enforce flow conservation constraints, that is, constraints which guarantee that a valid solution to a problem instance is composed by closed walks in the movement graph. Given a solution $\pi = \{c_1, \cdots, c_k\}$, label variables are assigned in such a way that, given $c_i \in \pi$, the label variable assigned to the (shortest) path between the predecessor in π of c_i has a lower value than the one assigned to the (shortest) path between c_i and its successor in π. This rule has a single exception in the special configuration, or special vertex, which is unique for each solution π. All other label variables are set to 0. Formally:

$$\forall c_i \in C_{\mathcal{P}} \quad \begin{cases} l_{c_h,c_i} < l_{c_i,c_j} & \text{if } c_i = c_g \in \pi, c_h = c_{g-1} \in \pi, c_j = c_{g+1} \in \pi \\ l_{c_h,c_i} > l_{c_i,c_j} & \text{if } c_i = c_g \in \pi, c_h = c_{g-1} \in \pi, c_j = c_{g+1} \in \pi, c_i \text{ special vertex of } \pi \\ l_{c_h,c_i} = l_{c_i,c_j} = 0 & \text{otherwise} \end{cases} \qquad (3)$$

Finally, special vertex variables υ_{ci} are binary variables defined over every candidate sensing configuration in $C_{\mathcal{P}}$, where $\upsilon_{ci} = 1$ if configuration c_i is the unique special vertex in a given solution, $\upsilon_{ci} = 0$ otherwise. An example of a solution π where the shortest paths between configurations are annotated with label variables is shown in Figure 4.

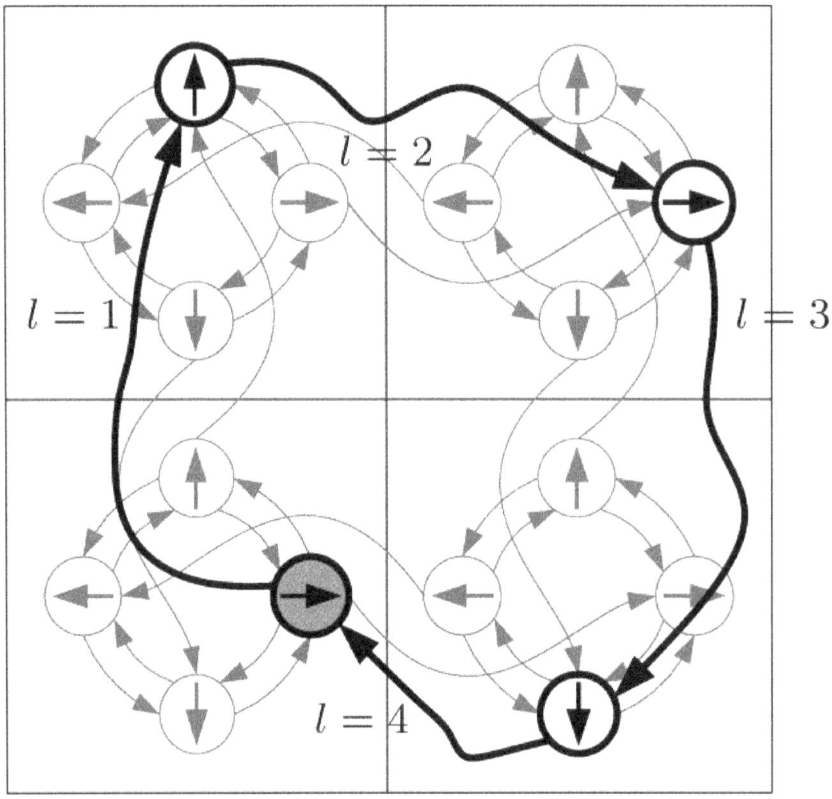

Figure 4. An example of a solution to a problem instance. Configurations in the solution are represented by thick circles and the shortest paths between them by thick lines. Each path between two consecutive candidate sensing configurations in the solution is annotated with a label variable l, and the marked configuration corresponds to the special vertex.

We can find an optimal solution to a problem instance by casting it as a mixed integer linear programming problem, which we define as described in Equation (4). Given the set of candidate sensing configurations $\mathcal{C}_\mathcal{P}$, flow and label variables are represented in matrices F and L, both of size $|\mathcal{C}_\mathcal{P}| \times |\mathcal{C}_\mathcal{P}|$, where F[i, j] and L[i, j] represent flow variable $f_{ci,cj}$ and label variable $l_{ci,cj}$, respectively. Please note that |A| denotes the cardinality of A. Special vertex vector S is a boolean vector of size $|\mathcal{C}_\mathcal{P}|$ with each element corresponding to a configuration in $\mathcal{C}_\mathcal{P}$.

minimize $\mathbf{1}^T(F \circ T_m)\mathbf{1} + \mathbf{1}^T(FT_s)$

subject to (4a)

$(VF)1 \geqslant 1$ (4b)

$F1 = (1^T F)^T$ (4c)

$F \leqslant L \leqslant uF$ (4d)

$-uS \leqslant ((L1-(1^T L)^T)-F1) \leqslant -S$ (4e)

$1^T S = 1$ (4f)

The objective function (Equation (4a)) minimizes total cost, expressed as the sum of traveling cost and sensing cost. In this function, matrix T_m of size $|\mathcal{C}_\mathcal{P}| \times |\mathcal{C}_\mathcal{P}|$ captures the traveling costs $(T_m[i,j] = t^m_{p_i \to p_j})$ and T_s is a column vector of size $|\mathcal{C}_\mathcal{P}|$ represents the sensing cost associated to each candidate sensing configuration $(T_s[i] = t^s_{c_i})$.

The optimization problem is subject to the following constraints:

- **Coverage constraints** (Equation (4b)) require that each cell $a_i \in \mathcal{S}$ is visible from at least one of the candidate sensing configurations selected in the solution π. Given the problem definition, we can calculate V as a binary matrix of size $n \times |\mathcal{C}_\mathcal{P}|$, where n is the number of cells in the problem. In particular:

$$V[a,c] = \begin{cases} 1 & \text{if } a \in \upsilon_\mathcal{P}(c) \\ 0 & \text{otherwise} \end{cases}$$

- **Flow conservation constraints** (Equation (4c)) ensure that the solution will consist of one or more closed paths among the selected candidate sensing configurations.
- **Traveling route constraints** (Equations (4d) and (4e)) ensure that each candidate sensing configuration is visited only once in a solution, and, therefore, a solution consists of a single closed path. u is an upper limit constant for label variables.
- **Special vertex constraint** (Equation (4f)) restricts the number of special vertices to one and only one for each solution.

To reduce the size of the problem, without affecting the quality of the solution, it is possible to identify and disregard all those candidate sensing configurations which would not, in any case, belong to an optimal solution. For instance, all those sensing configurations c_i that do not observe any other cells $\upsilon\,\mathcal{P}(c_i) = \emptyset$ (for instance, the ones placed in the cells at the border of the map; can be safely disregarded. Figure 5a is an example test map whose optimal solution is given in Figure 5b.

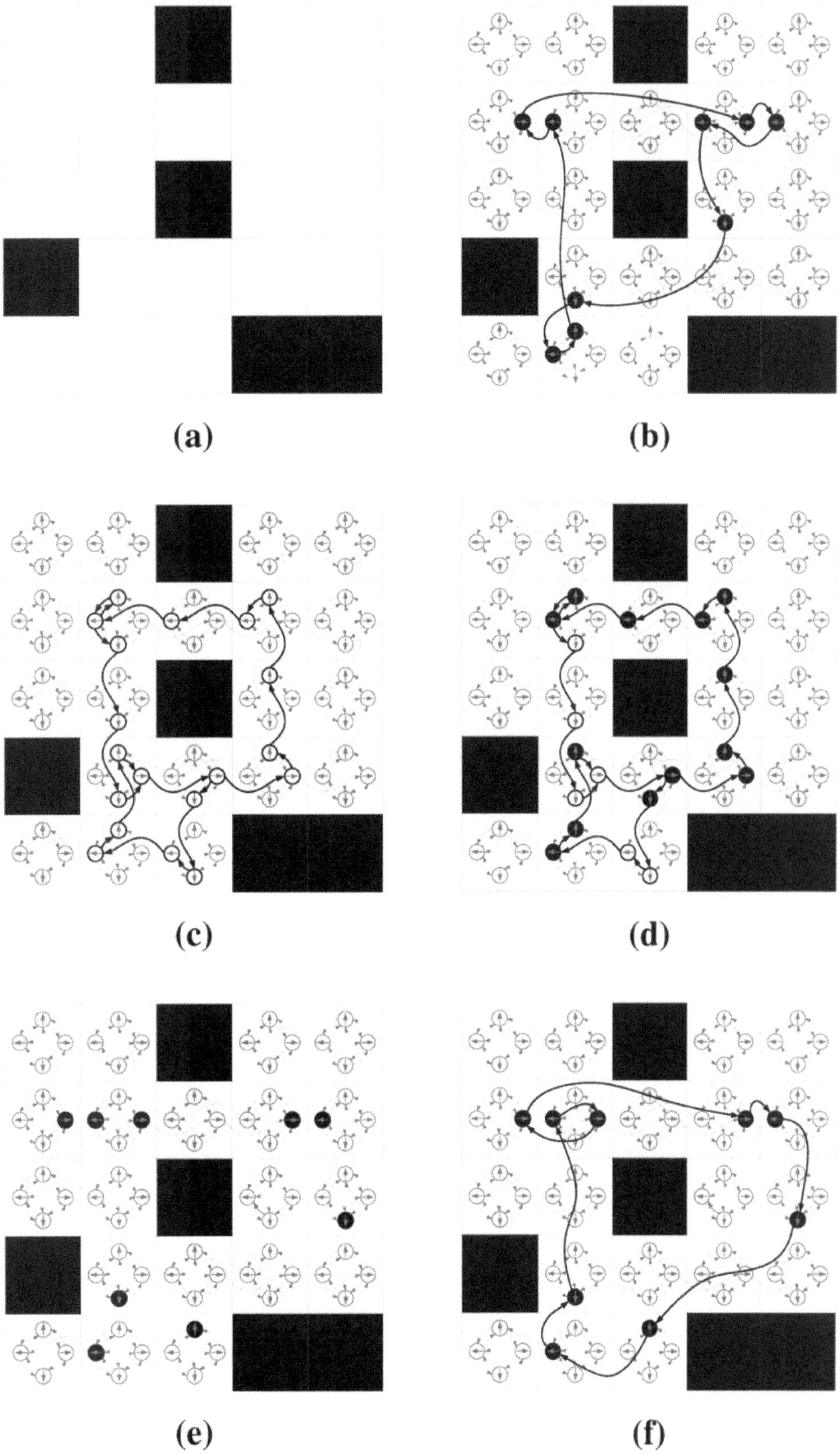

Figure 5. (**a**) a simple test map, where obstructed cells are represented in black and traversable ones in white. Candidate sensing configurations are defined over poses in the cells such that $\Theta = \{0, \frac{\pi}{2}, \pi, \frac{3}{2}\pi\}$ and have identical ϕ and r (r = 2 cells, ϕ=π/2). In this example setup, the movement from one cell to the next requires 1 s, the rotation of π2 requires 0.5 s, and a sensing action takes 4 s; (**b**) the optimal solution, when traveling and sensing costs are considered at the same time. Here, the curved arrows represent the minimum distance from a sensing configuration to the next on the underlying graph. In this case, the total exploration time is 52.5 s (16.5 s for traveling and 36 s for sensing). (**c,d**): Here, traveling time is minimized first (see Section 5.1). (c) the minimum cost closed path from which all cells can be observed is calculated; (d) and then the minimum set of sensing configurations is selected, yielding an overall exploration time of 66.5 s. (**e,f**): Here, sensing time is minimized first (see Section 5.2). (e) the set of minimum cost sensing configurations is selected from which all cells are observable; (f) and then connecting them with the shortest closed path. This approach yields to an overall exploration time of 55 s.

Solving the mixed integer linear programming problem as described above yields an optimal solution to a given problem instance, with respect to the total operation time of the robot (traveling and sensing time). However, intuitively speaking, this corresponds to solving a combination of a Watchman Route Problem (that is, the problem of computing the shortest route to guard a known area) and an Art Gallery Problem (that is, selecting the minimum number of observation points to completely observe a known area). Therefore, the combined problem is NP-hard and, as shown in Section 6, this approach becomes practically unfeasible as the number of candidate sensing configurations grows.

DECOMPOSING THE PROBLEM

Solving the problem so that the overall exploration time is minimized is of no practical applicability. However, a viable solution would be to sacrifice optimality by solving the problem in a two-step approach. This can be done in two different ways: (1) minimizing the overall traveling time (while guaranteeing complete coverage) and then selecting the sensing configurations along the route for minimum sensing time; or, (2) minimizing the sensing time, by selecting the sensing configurations necessary to cover the map and then finding the shortest path to connect them. In this section, we formally define such approaches.

Minimizing the Traveling Time

The core idea behind this decoupled approach is to find a minimum cost closed path for the robot which connects adjacent candidate sensing configurations. This path is subject to the requirement that all cells in $\mathcal{S}$ are visible by at least one candidate sensing configuration included in the path. This first step roughly corresponds to solving a Watchman Route Problem. The second step of this approach consists of selecting the minimum cost sensing configurations which would allow the coverage of the whole environment among the ones included in the path.

Intuitively, this approach should yield good results whenever the time required for sensing actions is significantly less than the one for movement. Figure 5c,d show an example of this two phase approach, where first a tour is calculated, such that it connects contiguous candidate sensing configurations from which the whole environment is observable (Figure 5c). Then, in the second phase, the algorithm selects the set of minimum cost candidate sensing configurations on the path necessary to observe the whole environment (Figure 5d). The robot will follow the tour as generated in the first phase, but will only stop for sensing according on those sensing configurations selected in the second phase.

We use the mixed integer linear programming problem as defined by Tomioka et al. [40] to find a traveling path that provides full coverage without considering sensing cost. In order to then select the set of minimum cost candidate sensing configurations, we solve a sensor placement problem where the only variables considered are the candidate sensing configurations included in the traveling route. The formulation we use to solve the sensor placement problem is presented in the next section.

Minimizing the Sensing Time

The core idea behind this approach is dual with respect to the one presented above and consists in first selecting a set of minimum cost candidate sensing configurations from which all cells in $\mathcal{S}$ are visible, and then solving a traveling salesman problem over them to find the shortest tour. This approach should therefore yield to better solutions than the previous one whenever the time necessary to perform sensing actions is much higher compared to the time required for the robot to move.

Figure 5e,f show how this approach would work compared to the ones described above: First the set of minimum cost candidate sensing configurations is selected (Figure 5e), and then a shortest closed path is calculated so that the robot will visit all the configurations (Figure 5f).

We find the set of minimum cost candidate sensing configurations by solving the following integer linear programming problem:

$$\underset{C}{\text{minimize}}\ C^T T_s$$

$$\text{subject to} \qquad (5a)$$

$$VC \geqslant 1 \qquad (5b)$$

$$C \in \{0,1\} \qquad (5c)$$

where C is a column vector of cardinality $|\mathcal{C_P}|$ whose elements are binary variables representing if a given candidate sensing configuration is selected or not. Equation (5b) captures the necessary coverage constraints. For calculating the minimum traveling route, we use the Repetitive Nearest-Neighbor Algorithm [47] for calculating a Hamiltonian cycle.

EXPERIMENTAL EVALUATION

We present an evaluation of the three approaches described above: the one in which traveling time and sensing time are minimized at the same time (Section 4), and the decoupled ones, in which either traveling time or sensing time are minimized first (Section 5). The evaluation is conducted in sets of randomly generated maps of varying size, whose cells have a fixed side length of 1 m. Candidate sensing configurations are defined over poses in the cells such that $\Theta = \{0, \frac{\pi}{2}, \pi, \frac{3}{2}\pi\}$ and have identical ϕ and r (r = 15 meters, π/2). Sensing and movement times are set based on our previous experience with our mobile platform [20]: Each sensing action takes 4 s, movement from one cell to the adjacent towards which the robot is oriented requires 1 s, while a rotation of the mobile platform of π/2 requires 0.5 s.

For the first part of the experimental evaluation, we prepared 6 sets of 10 randomly generated maps and one truncated set. Each set contains maps of identical size, starting from small 4 × 4 cell maps, up to 9 × 9 cell maps of which 10% cells contain an obstacle. For 3 × 3 cells there are only three unique maps with a single occupied cell (approx. 10%), all other maps are just rotations. Thereto, the results for 3 × 3 maps are averaged using the unique 3 maps only. Furthermore the maps are checked, whether the observable area is a single connected component. This means, there are no enclosed spaces, which cannot be reached by the robot.

The evaluation focuses on two aspects: solution quality (in terms of overall exploration time) and time required to compute the solutions. However, finding the optimal solution for maps as small as 6 × 6 becomes already an unpractical task, as the computation of 12 h on an Intel i7 Quad Core 2.60 GHz computer with 8 GB RAM was not enough to find the solution. The aggregated results for the comparison of all three methods up to maps of size 5 × 5 are presented in Figure 6. Here, Opt indicates the optimal solution, D_T the decoupled approach in which traveling time is minimized first, and D_S the decoupled approach in which sensing time is minimized first. In particular, Figure 6a illustrates the aggregate quality of the solutions provided by the two decoupled approaches and the optimal one. In this limited set of examples, the disjoint approach where traveling time is minimized first shows better solution quality whenever it is possible to discard high number of redundant sensing configurations along the traveling route, whose field of view overlap. On the other hand, the disjoint approach where sensing time is minimized first shows better solution quality whenever the traveling path along the sensing configurations in the solution is small. When we consider the time necessary to compute a solution, the disjoint approach where the sensing times are minimized first is the most efficient, while the optimal solution becomes soon too expensive to be of any practical use (Figure 6b).

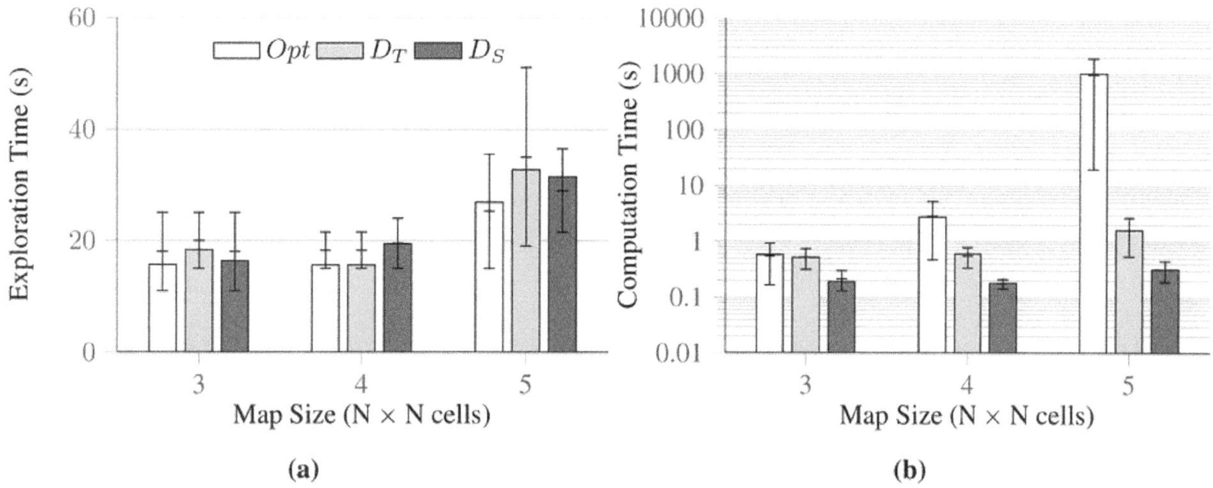

Figure 6. Comparison of all approaches, optimal and disjoint (Opt, D_T, D_S), on three sets of maps. Each set contains maps of varying size, from 3 × 3 to 5 × 5. The solid bars indicate the average values over the 3 maps for 3 × 3 and 10 maps for the rest, and error bars show the minimum and maximum values observed during the trial. (**a**) shows the solution quality; and (**b**) shows the computation time taken by the all three approaches on a logarithmic scale.

We tested the decoupled approaches on the remaining 4 sets of maps (of size 6 × 6 to 9 × 9), and the results are presented in Figure 7. Since we could not calculate the optimal solutions for those runs, we directly compare the quality of the solutions found by the two disjoint approaches. In this second set of tests, D_T, in average, generates solutions of slightly better quality (Figure 7a), that is, solutions where the overall exploration times are lower. On the other hand, the computation times for this decoupled approach grow quickly as the size of the maps increase, as it is clearly shown inFigure 7b. It is very important to note that 9 × 9 maps are still not large enough for representing real-world problems, and that D_T requires a time in the order of hours to solve such simple instances. On the other hand, D_S generates solutions for the same instances in less than two seconds, with a reasonable loss of quality (in this set of experimental runs, between 0.69% and 13.81% to the D_T).

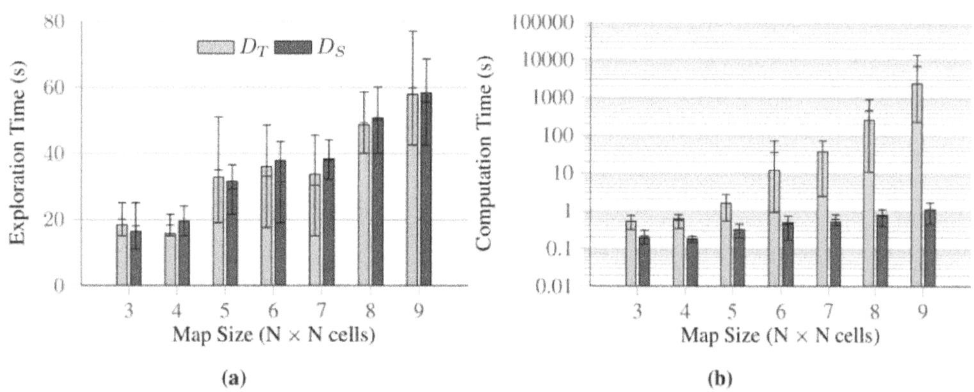

Figure 7. Comparison of two disjoint approaches up to maps of size 9 × 9. Notations are similar as in the previous figure, i.e., bars indicate the average values over the 3 maps for 3 × 3 and 10 maps for the rest, and error bars indicate the minimum and maximum values observed during the trial. (**a**) shows the solution quality of two disjoint approaches (D_T, D_S); and (**b**) shows the average computation time on a logarithmic scale for both approaches.

We continued our tests focusing on D_S up to maps of size 26 × 26 when also this decoupled approach begins to require computation times in the order of hours. The bottleneck of the computation lies in finding a set of minimum cost candidate sensing configurations which can cover the whole map. This is a Sensor Placement Problem (SPP), which has been studied extensively in the past decades [27]. In the following, we propose a new convex relaxation method which allows us to drastically curtail the computation times.

A NEW RE-WEIGHTED CONVEX RELAXATION FOR SOLVING SENSOR PLACEMENT PROBLEMS

To solve the problem of finding the set of sensing configurations that covers the whole environment at minimum cost (as we do in the decoupled approach described in Section 5.2) for instances with a high number of variables, we propose an iterative convex relaxation method which introduces intense sparsity, thus drastically reducing the number of variables to consider. The reduced problem can then be cast into an integer linear programming problem as formulated in Equation (5).

We call this algorithm SPP with Re-weighted Convex Relaxation, or in short conv-SPP. The SPP described in Equation (5) is NP-hard and corresponds to ℓ_0-minimization whose solutions are binary variables, i.e., sensing configurations. The combinatorial search soon becomes impracticable as the size of the problem grows. It is because ℓ_0-minimization penalizes non-zero elements identical and the overall solution cost is based on the cardinality of the set. ℓ_1-minimization on the other hand is much faster than ℓ_0-minimization, and therefore, can handle relatively high number of variables. However, it penalizes each element proportional to the magnitude and thus the solution is no longer a binary vector (see Figure 8).

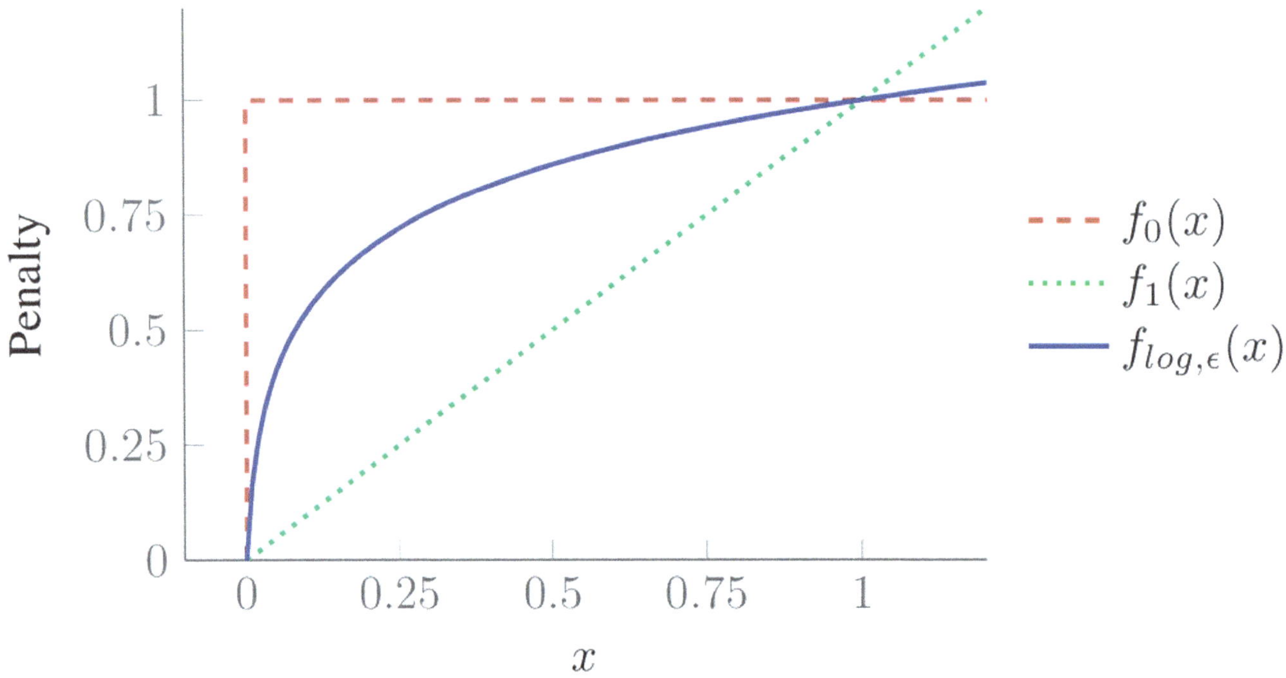

Figure 8. The concave loss function $f_{log,\epsilon}(x)$ approximates better the ℓ_0 sparsity count$f_0(x)$ than by the traditional convex ℓ_1 relaxation f_1 (x) [48].

Candès et al. [48] proposed a re-weighted (iterative) formulation of ℓ_1-minimization to handle large instances of a discrete problem in continuous domains. The approach effectively uses a concave loss function ($f_{log,\epsilon}(x)$), which is much closer to the penalty function of ℓ_0-minimization (see $f_0(x)$ in Figure 8). The essential part of the approach is to solve an overall non-convex optimization problem through iterations of convex optimization. To sparsify the solution vector along the curve$f_{log,\epsilon}(x)$ in Figure 8, a weight vector is introduced which evolves iteratively. The resultant vector is much sparsifer than the result of simple ℓ_1-minimization with a linear penalty function. Afterwards, the reduced set of non-zero elements forms a reduced search space in which the combinatorial problem in its integer linear formulation (Equation (5)) can be solved in a feasible time.

Formally, our conv-SPP formulation, inspired by [48] is:

$$\underset{C}{\text{minimize}}\ (W \circ C)^T T_s$$

subject to (6a)

$$VC \succcurlyeq 1 \tag{6b}$$

$$0 \preccurlyeq C \preccurlyeq 1 \tag{6c}$$

W is a weight vector of cardinality $|\mathcal{C_P}|$. At the beginning, all the elements of W are equal to 1, i.e., W = 1. In subsequent iterations, the weights are updated according to Equation (7).

$$W_{(i)} = \frac{\epsilon}{C_{(i)} + \epsilon} \quad i = 1, ..., |W| \tag{7}$$

The parameter ϵ is strictly positive and determines the rate of convergence. Smaller values of ϵ allow for faster convergence (see Figure 9a) but at an increased risk of getting trapped in a local-minimum.

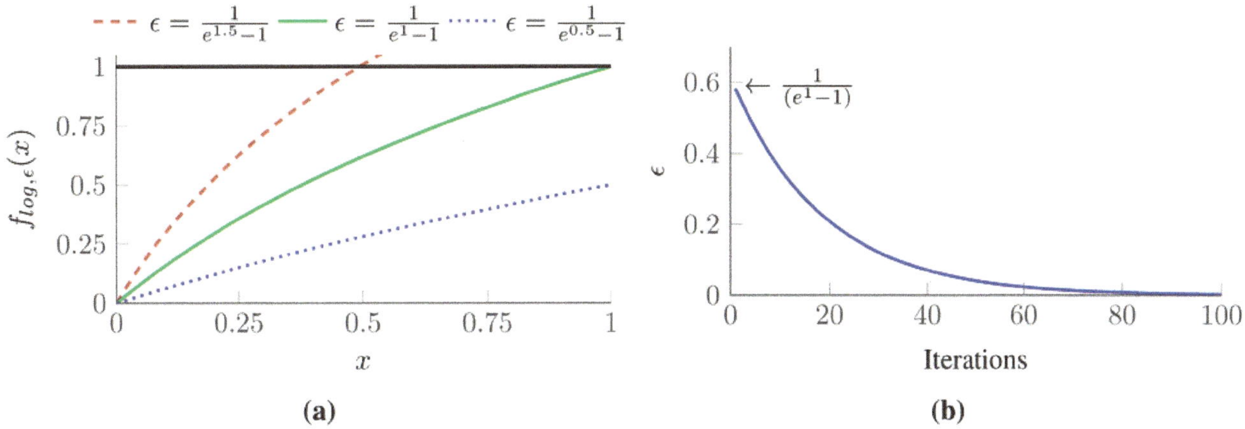

Figure 9. (**a**) Smaller value of ϵ defines steeper penalty functions; (**b**) The exponential decay of ϵ during the iterative procedure.

A local-minimum may occur whenever two or more variables have identical values and none of them is converging to the maximum or minimum limit. This could happen, for example, when two or more candidate sensing configurations have overlapping field of view and share a common subset of coverage. In Figure 10, c_1 and c_2 are two sensing configurations facing each other. The set $\mathbb{O}$ contains elements (cells) under the overlapping field of view, i.e. $\mathbb{O} \subset \upsilon_{\mathcal{P}}(c_1)$ and $\mathbb{O} \subset \upsilon_{\mathcal{P}}(c_2)$. If $\upsilon_{\mathcal{P}}(c_1) \setminus \mathbb{O}$ and $\upsilon_{\mathcal{P}}(c_2) \setminus \mathbb{O}$ arepartiallyobservedbyanotherselectedsensingconfiguration(s) in the solution. The configurations c_1 and c_2 will hold half of the magnitude to minimize the overall solution cost in continuous domain, and any further iteration will not improve the situation, since the algorithm can not decide which of the two sensing configurations it prefers. This problem has then to be resolved in the subsequent combinatorial optimization step.

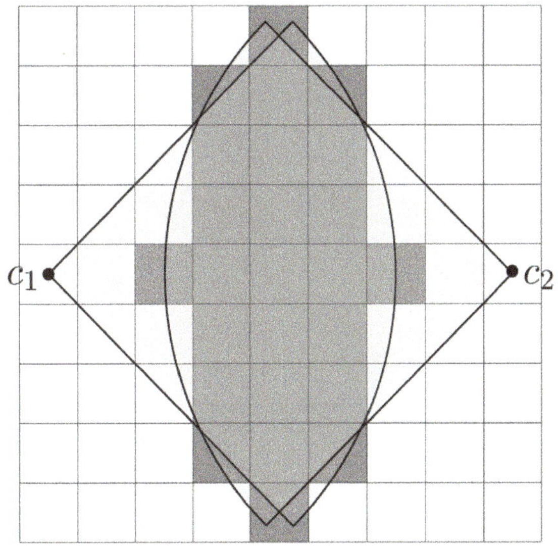

Figure 10. The sensing configurations c_1 and c_2 are trapped in a local minimum. The cells under the overlapping field of view are indicated in gray. Assuming that the remaining visible cells of both configurations, indicated in white under the coverage window, are partially (50%) observed by another configuration(s) in the solution. Thec_1 and c_2 are set to 0.5 in the solution vector and any subsequent iteration does not improve the situation.

In our implementation we adapt the value of ϵ in each step i according to Equation (8), so that it sets relatively high value at beginning to avoid a local minimum in early iterations to explore globally better solution, and then deceases to exponential decay with each iteration to speed up the convergence (see Figure 9b).

$$\epsilon_{(i)} = \left(\frac{1}{e^1 - 1}\right)^{1+\left((i-1)\times 10^{-1}\right)} \quad (8)$$

The procedure stops if no improvement in sparsity for the last n (n = 5 in our implementation) iterations is observed, or if a predefined number of iterations (150 in our case) is reached. We use l0τ sparsity measure with τ equals 10^{-2}. In addition, we stop if the number of non-zero elements is low enough so that the combinatorial search is feasible (the value was empirically set to 80). Finally, the reduced vector of candidate sensing configurations in C is solved with the combinatorial method according to Equation (5). The overall procedure is summarized in Algorithm 1.

Algorithm 1 conv-SPP	
1:	Set W = 1;
2:	Solve Equation (6);
3:	Update W as in Equation (7) and ϵ as in Equation (8);
4:	Go back to step 2, if none of the stopping criteria is true;
5:	Discard zero elements of C;
6:	Solve Equation (5) with updated C;

In addition to the solution itself, our algorithm provides an interval in which the optimal solution for the sensing time minimization lies. The upper bound for this interval is the found solution of Algorithm 1, since the optimal solution is either equal to our solution or faster. The lower bound is obtained by considering the continuous problem instead of the combinatorial one. Simple ℓ_1-minimization with linear penalty function $f_1(x)$; which is the first iteration of step 2 in Algorithm 1, will provide a solution, which is always less than or equal to the optimal solution of combinatorial search in terms of sensing time.

The limits can be used to assess the quality of the solution when the optimal solution is not available. However, in practice, the optimal sensing time is very close to the upper bound of conv-SPP, as shown in the following section (Section 8).

EVALUATION OF THE NEW METHOD

We evaluated conv-SPP on sets of randomly generated grid maps of varying size, from 3 × 3 to 90 × 90 cells. In each map, 10% of the cells are occupied by obstacles, and we assume that cells have an identical side length of 1 m. Each set is composed by 10 different maps of identical size, with the only exception of the set of maps of size 3×3, where, considering all symmetries, only 3 unique instances are possible.

In each test run, the candidate sensing configurations are defined such that $\Theta = \{0, \frac{\pi}{2}, \pi, \frac{3}{2}\pi\}$, and ϕ and r are fixed. Therefore, the sensing time for each candidate sensing configuration is set to 1 and the algorithm minimizes the number of sensing configurations. To evaluate the dependency of the algorithm with respect to the sensing parameters, we consider two different sensing radii (r = {15, 30} meters) and two central angles (ϕ={π/2,π}).

To solve the optimization problem, we used the Gurobi Solver [49], with CVX, a package for specifying and solving convex programs [50,51] in Matlab. For these runs, we used a computer with an Intel i7 Quad Core 2.60 GHz and 8 GB RAM.

We compared the optimal solutions of the sensor placement problem instances with the solutions provided by conv-SPP on maps up to the size of 26 × 26 cells. Larger maps required computation times in the order of hours to solve a single instance optimally.

The results for a single map randomly chosen from each set are shown in Figure 11. The thick black lines represent the number of sensing configurations selected in the optimal solutions, while the upper bounds of the green-colored intervals represent the number of sensing configurations selected by conv-SPP. The lower bounds of the intervals represent the number of configurations after the first iteration of the ℓ_1-minimization. The inset graphs show the results up to maps of size 26 × 26, for which the optimal solutions are available. conv-SPP obtains results which are very close to the optimal solutions in terms of the number of sensing configurations selected, as our method does not add more than two sensing configurations for any instance. Figure 12 shows the average difference of solution quality, which, for the available results, is always less than one sensing configuration.

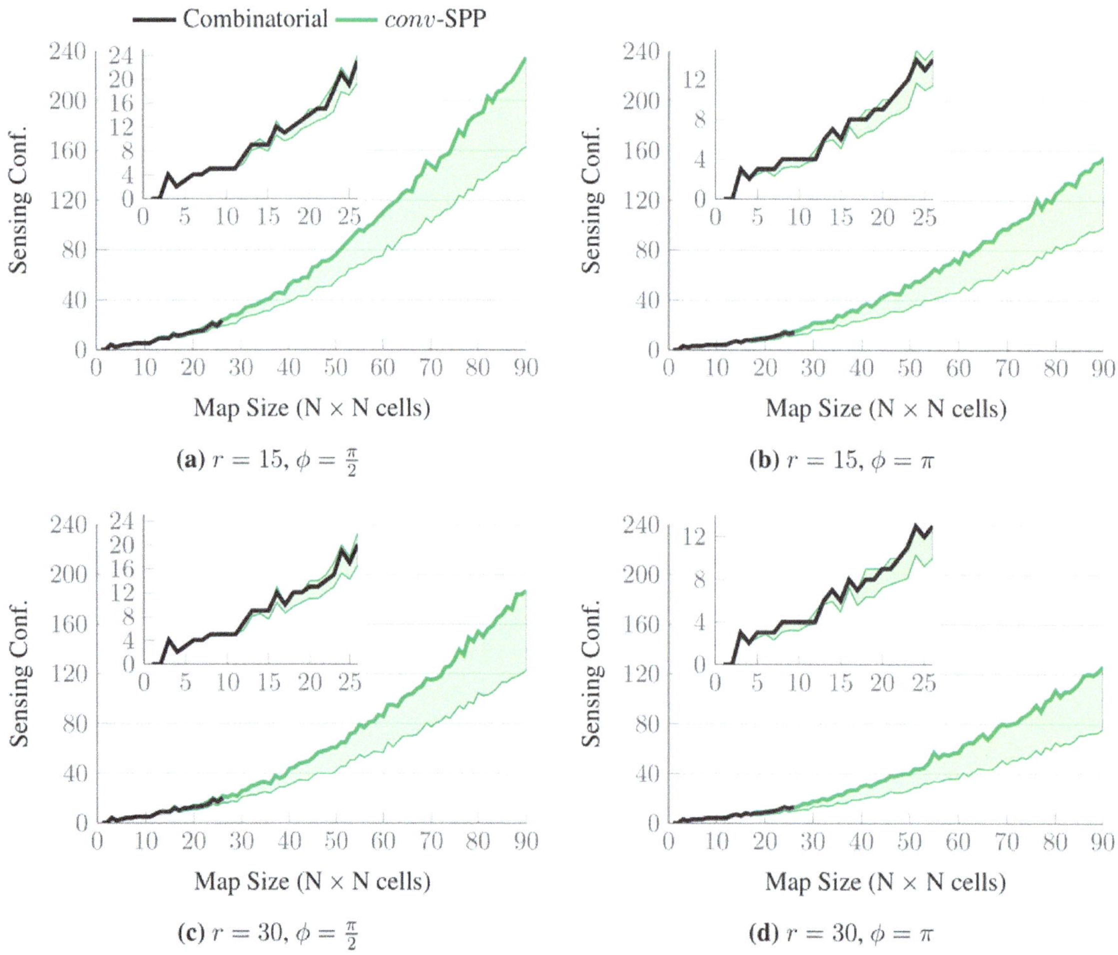

Figure 11. Comparison of the quality of the solutions obtained with conv-SPP and the combinatorial method on randomly generated maps. Sensing parameters are fixed to $\Theta = \{0, \frac{\pi}{2}, \pi, \frac{3}{2}\pi\}$, $r = \{15, 30\}$, and $\phi = \{\pi, \pi/2\}$. The optimal solutions (thick black lines) are close to the solutions provided by conv-SPP (thick green lines) with respect to the number of sensing configurations selected. The lower bounds of the green intervals represent the result of a single ℓ_1-minimization step.

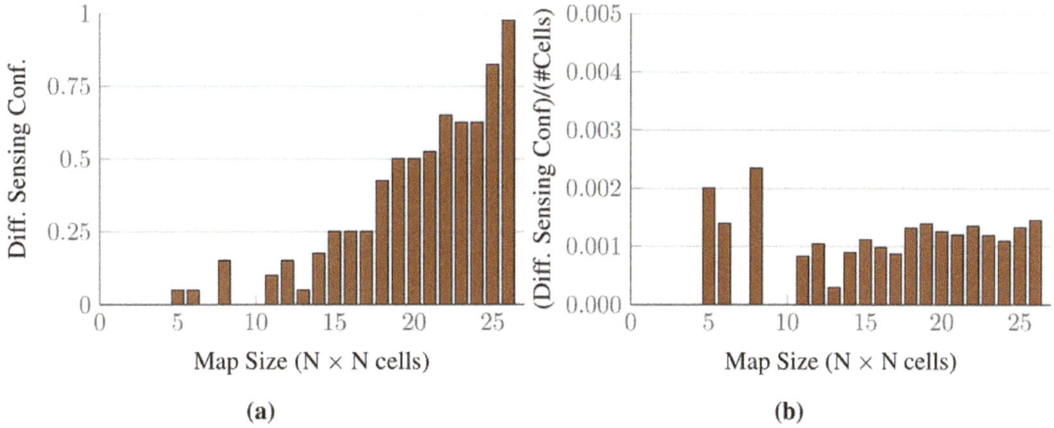

Figure 12. The solution quality gap between conv-SPP and combinatorial optimization. (**a**) shows the average difference for all possible combinations of the sensing parameters r = {15, 30} and ϕ = {π, π/2}; (**b**) shows the average difference divided by the number of cells in the maps.

Figure 13 shows the aggregated computation times for both approaches: The thick lines represent the average computation times for each set of maps, while the colored intervals are bounded by the minimum and maximum times it took to solve all the instances in one set. The inset graphs show the computation times on a logarithmic scale: As it can be seen, the time required to calculate the optimal solution of the problem increases steeply with the size of the maps considered. It is worth noticing that the calculation of the optimal solutions for maps up to 13 × 13 cells takes less time than when conv-SPP is employed. This is because the number of variables is small and the combinatorial method can find a solution quickly. On the other hand, conv-SPP runs at least one iteration of convex relaxation and then performs combinatorial optimization for the remaining variables. Therefore, the iterations of convex relaxation simply require additional time. We stopped computing the optimal solutions for a set of maps whenever at least one instance of the set required more than one hour to be solved optimally. By contrast, our method could solve all the instances in less than 800 s.

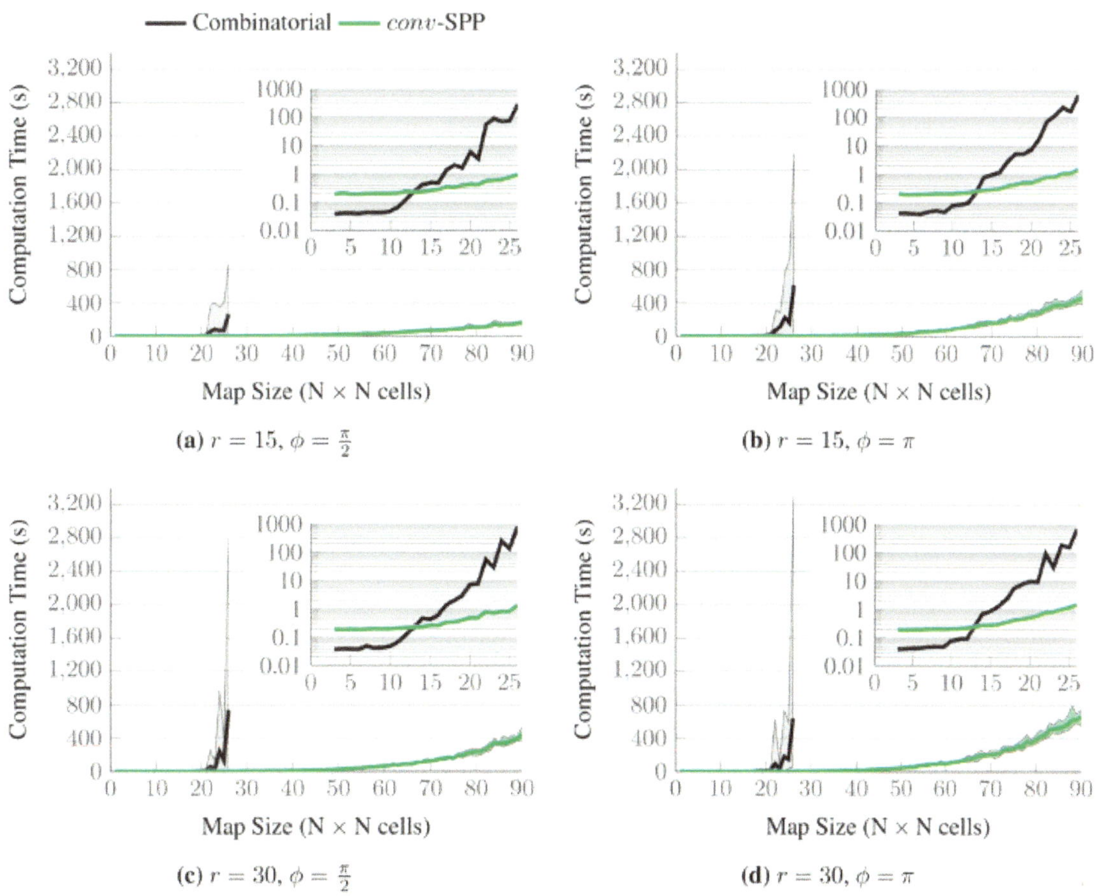

Figure 13. Computation times to calculate the optimal solutions and to solve the instances with conv-SPP. The thick lines represent the average computation times for each set of maps, while the colored intervals are bounded by the minimum and maximum times it took to solve all the instances in one set. The inset graphs show the computation times on a logarithmic scale.

Within the same set of maps of identical size, different instances can require very different computation times for computing the optimal solution. Conversely, this is not the case when using conv-SPP. This is because few changes in the map layout can completely change the discrete optimization problem, while, on the other hand, the continuous problem is much less affected, and in conv-SPP the combinatorial solver is used only on a reduced set of variables. Hence, another advantage of conv-SPP is that the computation time is largely independent of the particular problem instance within a given problem size.

We also used the Freiburg University campus map [54] to demonstrate our algorithm on a real map. We discretized the map into a Cartesian grid where each cell covers 1 m^2. The map size is 120 × 136 cells, 6113 of which are unoccupied (seeFigure 14). For the sensing parameters $\Theta = \{0, \frac{\pi}{2}, \pi, \frac{3}{2}\pi\}$, r = 15 m, and $\phi = \pi$, our algorithm generated a solution of 68 sensing configurations in 90.87 s with a lower bound of 53.03 configurations. This scenario is evaluated in based on real map data (but without actual gas sources) to verify that our approach is able to efficiently cover a large real environment. In our ongoing work, we are performing actual gas detection tasks using the described algorithm on a robot in different environments (see Figure 15a for an example). Initial results are quite promising, as shown in Figure 15b. For more details on the experiments and the evaluation, we refer the reader to our forthcoming publication [52].

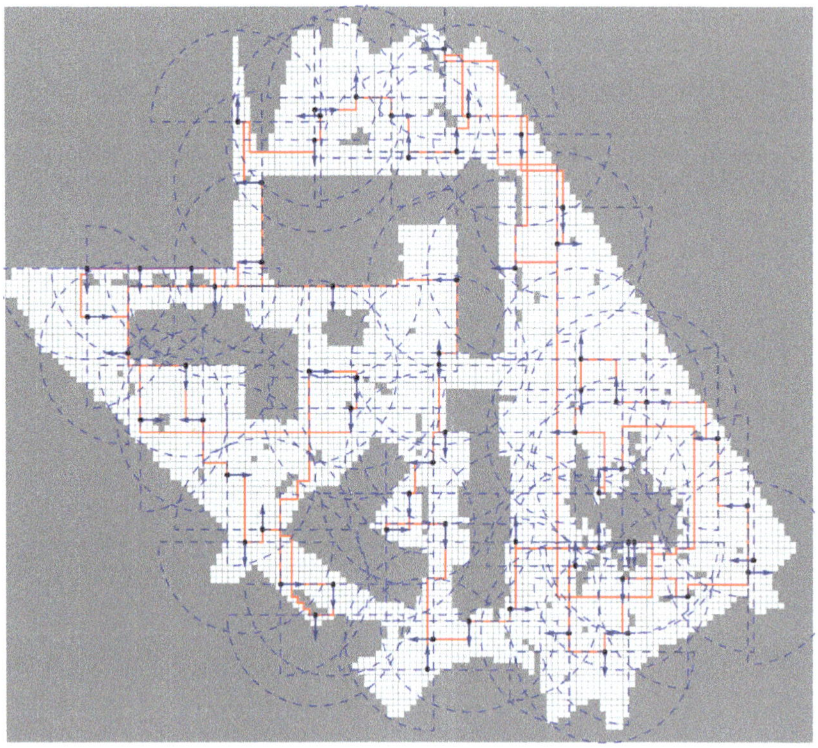

Figure 14. The exploration plan generated for the map of the Freiburg University campus (obstacles are represented in gray). The positions of the sensing configurations are represented with blue dots, blue arrows indicate their orientation and dashed-lines their field of view. A closed path through the configurations is shown in red.

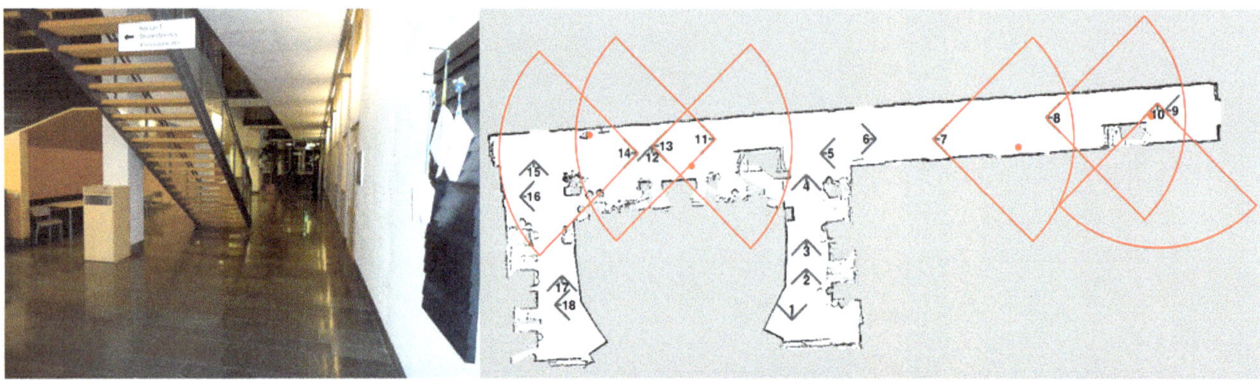

(a) **(b)**

Figure 15. (**a**) an indoor setting for the gas detection task; (**b**) graphical results: the gas sources are indicated by red dots, the planned sensing configuration by truncated cones in gray, and the small number associated to each configuration indicates the order in which they are visited. The configurations plotted in red indicate the positive readings and the actual area they cover.

CONCLUSIONS

In this paper, we discussed the problem of gas detection in large environments with a mobile robot equipped with a remote gas sensor. Generally speaking, given an environment where a specific gas may be present in detectable concentrations, a solution to the problem would be a cost-effective exploration plan for the robot so that it could ascertain whether this is the case or not. Here, we consider cost to be directly proportional to the time required to complete the exploration plan.

We provided a formal definition of the problem and suggested several possible methods for solving it. The main contribution of this paper is conv-SPP, an algorithm for quickly finding an exploration plan which guarantees a complete coverage of the environment. Our algorithm leverages a novel method based on convex relaxation that drastically reduces the number of variables and it is therefore effective for solving large problem instances. Furthermore, conv-SPP offers guarantees on the quality of the solution, as it is always possible to asses the theoretical maximum distance of the solution provided to the optimal one. In practice, we show that in all the cases for which we could experimentally solve the problems optimally, our algorithm always generated results very close to the optimal ones.

We performed extensive validation of our approach in simulation, both on randomly generated maps of various sizes and on the map of the Freiburg University campus. We also compared our methods with other approaches, both in terms of solution quality and of computational requirements. In our future work, we want to test our algorithm in real world scenarios with a mobile platform. We also intend to extend our problem definition to take further advantage of remote sensing and include non-traversable but observable cells.

In this article, we are addressing the planning problem for gas detection only. For addressing this problem a pregenerated plan, like the approach we presented here, is sufficient. Future work will address how to deal with actual gas detections. Possible questions of interest include the problem of gas source localization [17,53], which often requires an active strategy to follow the trail to the source. Robot assisted gas tomography for gas distribution mapping with remote sensors [19,46] requires the consideration of different sensing positions in order to meet constraints regarding the overlap of the sensing fields and intersection angles that are necessary to reconstruct the gas distribution. Optimizing the sensing geometry for tomographic measurements [55] will be one focus of our future work. Extending the presented work to multi-robot scenarios, in which a fleet of robots cooperatively monitors a target area [56–60] is another interesting direction for future work.

AUTHOR CONTRIBUTIONS

- M.A. Arain: Problem formalization, algorithm design, experimental design, evaluation and manuscript writing.
- M. Trincavelli: Problem formalization, algorithm design and experimental design.
- M. Cirillo: Problem formalization, algorithm design, experimental design, evaluation and manuscript writing.
- E. Schaffernicht: Problem formalization, experimental design, evaluation and manuscript writing.
- A.J. Lilienthal: Problem formalization, evaluation and manuscript writing.

Conflicts of Interest

The authors declare no conflict of interest.

REFERENCES

1. Asadi, S.; Badica, C.; Comes, T.; Conrado, C.; Evers, V.; Ilie, S.; Jensen, J.S.; Lilienthal, A.; Milan, B.; Neidhart, T.; et al. ICT solutions supporting collaborative information acquisition, situation assessment and decision making in contemporary environmental management problems: the DIADEM approach. Proceedings of the International Conference on Innovations in Sharing Environmental Observation and Information (EnviroInfo), Ispra, Italy, 5–7 October 2011; pp. 920–931.
2. Zhou, G.; Zhu, Z.; Chen, G.; Hu, N. Energy-Efficient Chain-Type Wireless Sensor Network for Gas Monitoring. Proceedings of the Second International Conference on Information and Computing Science (ICIC), Manchester, UK, 21–22 May 2009; pp. 125–128.
3. Kiernan, B.M.; Beirne, S.; Fay, C.; Diamond, D. Landfill Gas Monitoring at Borehole Wells using an Autonomous Environmental Monitoring System. Proceedings of the 5th International Conference on Climate Change and Global Warming, Heidelberg, Germany, 24–26 September 2008; pp. 166–171.
4. Trincavelli, M.; Reggente, M.; Coradeschi, S.; Loutfi, A.; Ishida, H.; Lilienthal, A.J. Towards Environmental Monitoring with Mobile Robots. Proceedings of the IEEE/RSJ International Conference on Intelligent Robots and Systems (IROS), Nice, France, 22–26 September 2008; pp. 2210–2215.
5. Ishida, H. Blimp Robot for Three-Dimensional Gas Distribution Mapping in Indoor Environment. AIP Conf. Proc. **2009**,1137, 61–64.
6. Hernandez Bennetts, V.; Lilienthal, A.J.; Neumann, P.P.; Trincavelli, M. Mobile robots for localizing gas emission sources on landfill sites: is bio-inspiration the way to go? Front. Neuroeng. **2012**, 4, 1–20.
7. Neumann, P.; Hernandez Bennetts, V.; Bartholmai, M. Adaptive Gas Source Localization Strategies and Gas Distribution Mapping using a Gas-sensitive Micro-Drone. Proceedings of the 16th GMA/ITG-Fachtagung Sensoren und Messsysteme, Nuremberg, Germany, 22–23 May 2012; pp. 800–809.
8. Baetz, W.; Kroll, A.; Bonow, G. Mobile Robots with Active IR-Optical Sensing for Remote Gas Detection and Source Localization. Proceedings of the IEEE International Conference on Robotics and Automation (ICRA), Kobe, Japan, 12–17 May 2009; pp. 2773–2778.
9. Soldan, S.; Welle, J.; Barz, T.; Kroll, A.; Schulz, D. Towards Autonomous Robotic Systems for Remote Gas Leak Detection and Localization in Industrial Environments. In Field and Service Robotics; Yoshida,

K., Tadokoro, S., Eds.; Springer: Berlin/Heidelberg, Germany, 2014; Volume 92, pp. 233–247.

10. Bonow, G.; Kroll, A. Gas leak localization in industrial environments using a TDLAS-based remote gas sensor and autonomous mobile robot with the Tri-Max method. Proceedings of the IEEE International Conference on Robotics and Automation (ICRA), Karlsruhe, Germany, 6–10 May 2013; pp. 987–992.
11. Lackner, M. Tunable Diode Laser Absorption Spectroscopy (TDLAS) in the Process Industries- A Review. Rev. Chem. Eng. **2007**, 23, 65–147.
12. Frish, M.B.; Wainner, R.T.; Green, B.D.; Laderer, M.C.; Allen, M.G. Standoff gas leak detectors based on tunable diode laser absorption spectroscopy. Proceedings of the SPIE Optics East, Boston, MA, USA, 23–26 October 2005.
13. Tsujita, W.; Yoshino, A.; Ishida, H.; Moriizumi, T. Gas sensor network for air-pollution monitoring. Sens. Actuators B: Chem. **2005**, 110, 304–311.
14. Somov, A.; Baranov, A.; Savkin, A.; Spirjakin, D.; Spirjakin, A.; Passerone, R. Development of wireless sensor network for combustible gas monitoring. Sens. Actuators A: Phys. **2011**, 171, 398–405.
15. Liu, X.; Cheng, S.; Liu, H.; Hu, S.; Zhang, D.; Ning, H. A Survey on Gas Sensing Technology. Sensors **2012**, 12, 9635–9665.
16. Lewis, A.W.; Yuen, S.T.S.; Smith, A.J.R. Detection of gas leakage from landfills using infrared thermography - applicability and limitations. Waste Manag. Res. **2003**, 21, 436–447.
17. Kowadlo, G.; Russell, R.A. Robot Odor Localization: A Taxonomy and Survey. Int. J. Robot. Res. **2008**, 27, 869–894.
18. Ishida, H.; Wada, Y.; Matsukura, H. Chemical Sensing in Robotic Applications: A Review. IEEE Sens. J. **2012**, 12, 3163–3173.
19. Hernandez Bennetts, V.M.; Lilienthal, A.J.; Khaliq, A.A.; Sesé, V.P.; Trincavelli, M. Towards Real-World Gas Distribution Mapping and Leak Localization Using a Mobile Robot with 3D and Remote Gas Sensing Capabilities. Proceedings of the IEEE International Conference on Robotics and Automation (ICRA), Karlsruhe, Germany, 6–10 May 2013; pp. 2335–2340.
20. Hernandez Bennetts, V.; Schaffernicht, E.; Stoyanov, T.; Lilienthal, A.J.; Trincavelli, M. Robot Assisted Gas Tomography—Localizing Methane Leaks in Outdoor Environments. Proceedings of the IEEE International Conference on Robotics and Automation (ICRA), Hong Kong, China, 31 May–7 June 2014; pp. 6362–6367.
21. Lilienthal, A.; Streichert, F.; Zell, A. Model-based Shape Analysis of Gas Concentration Gridmaps for Improved Gas Source Localisation. Proceedings of the IEEE International Conference on Robotics and Automation (ICRA), Barcelona, Spain, 18–22 April 2005; pp. 3564–3569.
22. Lee, D.T.; Lin, A.K. Computational Complexity of Art Gallery Problems. IEEE Trans. Inf. Theory **1986**, 32, 276–282.
23. Scott, W.R. Model-based view planning. Mach. Vis. Appl. **2009**, 20, 47–69.
24. Korte, B.; Vygen, J. Combinatorial Optimization; Theory and Algorithms, 2nd ed.; Springer-Verlag: Berlin/Heidelberg, Germany, 2002; pp. 327–259.
25. Erdem, U.M.; Sclaroff, S. Automated camera layout to satisfy task-specific and floor plan-specific coverage requirements. Comput. Vis. Image Underst. **2006**, 103, 156–169.
26. Tarabanis, K.A.; Allen, P.K.; Tsai, R.Y. A Survey of Sensor Planning in Computer Vision. IEEE Trans. Robot. Autom.**1995**, 11, 86–104.
27. Ramsden, D. Optimization Approaches to Sensor Placement Problems. Ph.D. Thesis, Rensselaer Polytechnic Institute, Troy, NY, USA, December 2009.
28. An, W.; Shao, F.M.; Meng, H. The coverage-control optimization in sensor network subject to sensing area. Comput. Math. Appl. **2009**, 57, 529–539.
29. Angella, F.; Reithler, L.; Gallesio, F. Optimal Development of Cameras for Video Surveillance Systems. Proceedings of the IEEE Conference on Advanced Video and Signal Based Surveillance (AVSS), London, UK, 5–7 September 2007; pp. 388–392.
30. Chen, X.; Davis, J. An occlusion metric for selecting robust camera configurations. Mach. Vis. Appl. **2008**, 19, 217–222.
31. Chiu, P.L.; Lin, F.Y.S. A Simulated Annealing Algorithm to Support the Sensor Placement for Target Location. Proceedings of the Canadian Conference on Electrical and Computer Engineering, Ontario, Canada, 2–5 May 2004; pp. 867–870.
32. Applegate, D.L.; Bixby, R.E.; Chvátal, V.; Cook, W.J. The Traveling Salesman Problem: A Computational Study; Princeton University Press: Princeton, NJ, USA, 2007; pp. 1–58.
33. Faigl, J.; Kulich, M.; Přeučil, L. A Sensor Placement Algorithm for a Mobile Robot Inspection Planning. J. Intell. Robot. Syst. **2011**, 62, 329–353.
34. Danner, T.; Kavraki, L.E. Randomized Planning for Short Inspection Paths. Proceedings of the IEEE International Conference on Robotics and Automation (ICRA), San Francisco, CA, USA, 24–28 April 2000; pp. 971–976.
35. Kazazakis, G.D.; Argyros, A.A. Fast Positioning of Limited-Visibility Guards for the Inspection of 2D Workspaces. Proceedings of the IEEE/RSJ International Conference on Intelligent Robots and Systems (IROS), Lausanne, Switzerland, 30 September–4 October 2002; pp. 2843–2848.
36. Vieira Neto, H.; Nehmzow, U. Real-time Automated Visual Inspection using Mobile Robots. J. Intell. Robot. Syst. **2007**,49, 293–307.
37. Hollinger, G.A.; Englot, B.; Hover, F.; Mitra, U.; Sukhatme, G.S. Uncertainty-Driven View Planning for Underwater Inspection. Proceedings of the IEEE International Conference on Robotics and Automation (ICRA), Saint Paul, MN, USA, 14–18 May 2012; pp. 4884–4891.

38. Englot, B.; Hover, F.S. Three-dimensional coverage planning for an underwater inspection robot. Int. J. Robot. Res.**2013**, 32, 1048–1073.
39. Golden, B.; Naji-Azimi, Z.; Raghavan, S.; Salari, M.; Toth, P. The Generalized Covering Salesman Problem. INFORMS J. Comput. **2012**, 24, 534–553.
40. Tomioka, Y.; Takara, A.; Kitazawa, H. Generation of an Optimum Patrol Course for Mobile Surveillance Camera. IEEE Trans. Circuits Syst. Video Technol. **2012**, 22, 216–224.
41. Wang, P.; Krishnamurti, R.; Gupta, K. View Planning Problem with Travel Costs: Program Formulation, Hardness of Approximation, and Approximation Algorithms; Simon Fraser University: Burnaby, BC, Canada, 2006.
42. Wang, P.; Krishnamurti, R.; Gupta, K. View Planning Problem with Combined View and Traveling Cost. Proceedings of the IEEE International Conference on Robotics and Automation (ICRA), Roma, Italy, 10–14 April 2007; pp. 711–716.
43. Adam, H.; Bauer, J.; Thornton, E.; Tulip, J. Greenhouse Gas Emissions and Reduction Studies Using a Portable TDL Gas Monitor. Proceedings of the climate change and GHG technology conference and tradeshow, Calgary, AB, Canada, 22–24 May 2002.
44. Trincavelli, M. Gas Discrimination for Mobile Robots. Künstliche Intell. **2011**, 25, 351–354.
45. Fonollosa, J.; Rodríguez-Luján, I.; Trincavelli, M.; Vergara, A.; Huerta, R. Chemical Discrimination in Turbulent Gas Mixtures with MOX Sensors Validated by Gas Chromatography-Mass Spectrometry. Sensors **2014**, 14, 19336–19353.
46. Hernandez Bennetts, V.; Schaffernicht, E.; Pomareda, V.; Lilienthal, A.J.; Marco, S.; Trincavelli, M. Combining Non Selective Gas Sensors on a Mobile Robot for Identification and Mapping of Multiple Chemical Compounds. Sensors**2014**, 14, 17331–17352.
47. Kizilateş, G.; Nuriyeva, F. On the Nearest Neighbor Algorithms for the Traveling Salesman Problem. In Advances in Computational Science, Engineering and Information Technology; Nagamalai, D., Kumar, A., Annamalai, A., Eds.; Springer International Publishing: Cham, Switzerland, 2013; Volume 225, pp. 111–118.
48. Candès, E.J.; Wakin, M.B.; Boyd, S.P. Enhancing Sparsity by Reweighted ℓ_1 Minimization. J. Fourier Anal. Appl. **2008**,14, 877–905.
49. Gurobi Optimizer, Version 5.6. Available online: http://www.gurobi.com (accessed on 10 December 2014).
50. Grant, M.; Boyd, S. CVX: Matlab Software for Disciplined Convex Programming, Version 2.1. Available online: http://cvxr.com/cvx (accessed on 10 December 2014).
51. Grant, M.C.; Boyd, S.P. Graph Implementations for Nonsmooth Convex Programs. In Recent Advances in Learning and Control; Blondel, V., Boyd, S.P., Kimura, H., Eds.; Springer: London, UK, 2008; Volume 371, pp. 95–110.
52. Arain, M.A.; Cirillo, M.; Hernandez Bennetts, V.; Schaffernicht, E.; Trincavelli, M.; Lilienthal, A.J. Efficient Measurement Planning for Remote Gas Sensing with Mobile Robots. Proceedings of the IEEE International Conference on Robotics and Automation (ICRA), Seattle, WA, USA, 26–30 May 2015.
53. Fukazawa, Y.; Ishida, H. Estimating Gas-Source Location in Outdoor Environment Using Mobile Robot Equipped with Gas Sensors and Anemometer. Proceedings of the IEEE Sensors Conference, Christchurch, New Zealand, 25–28 October 2009; pp. 1721–1724.
54. OpenSLAM GMapping. Available online: https://www.openslam.org/gmapping.html (accessed on 10 December 2014).
55. Hartl, A.; Song, B.C.; Pundt, I. 2-D reconstruction of atmospheric concentration peaks from horizontal long path DOAS tomographic measurements: parametrisation and geometry within a discrete approach. Atmos. Chem. Phys.**2006**, 6, 847–861.
56. Lochmatter, T. Bio-Inspired and Probabilistic Algorithms for Distributed Odor Source Localization using Mobile Robots. PhD Thesis, École Polytechnique Fédérale de Lausanne (EPFL), Lausanne, Switzerland, February 2010.
57. Rocco, M.D.; Reggente, M.; Saffiotti, A. Gas Source Localization in Indoor Environments using Multiple Inexpensive Robots and Stigmergy. Proceedings of the IEEE/RSJ International Conference on Intelligent Robots and Systems (IROS), San Francisco, CA, USA, 25–30 September 2011; pp. 5007–5014.
58. Li, F.; Meng, Q.H.; Bai, S.; Li, J.G.; Popescu, D. Probability-PSO Algorithm for Multi-robot Based Odor Source Localization in Ventilated Indoor Environments. In Intelligent Robotics and Applications; Xiong, C., Huang, Y., Xiong, Y., Liu, H., Eds.; Springer: Berlin/Heidelberg, Germany, 2008; Volume 5314, pp. 1206–1215.
59. Zou, Y.; Luo, D. A Modified Ant Colony Algorithm Used for Multi-robot Odor Source Localization. In Advanced Intelligent Computing Theories and Applications. With Aspects of Artificial Intelligence; Huang, D.S., Wunsch II, D.C., Levine, D.S., Jo, K.H., Eds.; Springer: Berlin/Heidelberg, Germany, 2008; Volume 5227, pp. 502–509.
60. Liu, Z.; Lu, T.F. Multiple Robots Plume-Tracing in Open Space Obstructed Environments. Proceedings of the IEEE International Conference on Robotics and Biomimetics (ROBIO), Guilin, China, 19–23 December 2009; pp. 2433–2439.

CHAPTER 13

REDUCTION OF LIQUID BRIDGE FORCE FOR 3D MICROSTRUCTURE MEASUREMENTS

Hiroshi Murakami [1], Akio Katsuki [2], Takao Sajima [2] and Mitsuyoshi Fukuda [1]

[1]Department of Mechanical Systems Engineering, Faculty of Environmental Engineering, The University of Kitakyushu, 1-1 Hibikino, Wakamatsu-ku, Kitakyushu, Fukuoka 808-0135, Japan

[2]Department of Mechanical Engineering, Faculty of Engineering, Kyushu University, 744, Motooka, Nishi-ku, Fukuoka 819-0395, Japan

ABSTRACT

Recent years have witnessed an increased demand for a method for precise measurement of the microstructures of mechanical microparts, microelectromechanical systems, micromolds, optical devices, microholes,*etc.* This paper presents a measurement system for three-dimensional (3D) microstructures that use an optical fiber probe. This probe consists of a stylus shaft with a diameter of 2.5 μm and a glass ball with a diameter of 5 μm attached to the stylus tip. In this study, the measurement system, placed in a vacuum vessel, is constructed suitably to prevent adhesion of the stylus tip to the measured surface caused by the surface force resulting from the van der Waals force, electrostatic force, and liquid bridge force. First, these surface forces are analyzed with the aim of investigating the causes of adhesion. Subsequently, the effects of pressure inside the vacuum vessel on surface forces are evaluated. As a result, it is found that the surface force is 0.13 μN when the pressure inside the vacuum vessel is 350 Pa. This effect is equivalent to a 60% reduction in the surface force in the atmosphere.

INTRODUCTION

Recent years have witnessed an increased demand for a method for precise measurement of the microstructures of mechanical microparts, microelectromechanical systems, micromolds, optical devices, microholes, *etc.* However, precise measurement of the shape of a microstructure with a large length-to-diameter (L/D) ratio is rather difficult because of the difficulty in probe fabrication and sensing methods where the measuring force is very small. Previous works have reported microstructure measurement techniques that employ a variety of probes such as optical probes, vibroscanning probes, vibrating probes, tunneling effect probes, opto-tactile probes, fiber deflection probes, optical trapping probes, and diaphragm probes [1,2,3,4,5,6,7,8,9].

In a previous study, we developed a system for the measurement of three-dimensional (3D) microstructures using an optical fiber probe that functions as a kind of displacement measuring probe with a small contact force and wide measurement range [10]. In this system, the shaft of the stylus does not need to be rigid for the detection of the measuring force, because the deflection of the stylus is measured by a non-contact method. In general, when the particle size is less

Citation: Murakami H, Katsuki A, Sajima T, Fukuda M. Reduction of Liquid Bridge Force for 3D Microstructure Measurements. Applied Sciences. 2016; 6(5):153. DOI: 10.3390/app6050153 y than several tens of micrometers, the effects of surface force generated from the van der Waals force, electrostatic force, and liquid bridge force are strengthened and this surface force becomes greater than the force of gravity [11]. We used a fiber stylus with a 5-μm-diameter sphere on its tip, and as a result, we found that the measurement surface draws the stylus tip closer when the tip approaches it and the distance between the stylus tip and the measured surface is less than the regular displacement. When the stylus tip comes into contact with the measured surface, it adheres to the surface and cannot be separated from it. When the measured surface is scanned point by point in the touch-trigger mode (Figure 1a), the measurement time increases because the stylus tip is required to be separated from the measured surface to overcome the surface force. When the measured surface is scanned continuously in the scanning mode (Figure 1b), the measurement accuracy reduces because of the bending of the stylus shaft due to the adhesion. The observed adhesion, which is influenced by environmental factors (e.g., humidity) and the roughness of the measurement surface, is not reproducible. Therefore, in another previous work, we developed a measurement system for 3D microstructures that uses a vibrating fiber probe (Figure 1c) to prevent adhesion of the stylus tip to the surface being measured [12]. In this system, the stylus tip is set to vibrate in a circular motion, where it traces a circle of diameter 0.4 μm in the *X*-*Y* plane. However, the stylus tip actually traces an elliptical path. The difference between the profile of the perfect circle and that of the actual elliptical path of the stylus tip leads to measurement error. Moreover, the surface roughness cannot be measured in the scanning mode when using the vibrating fiber probe.

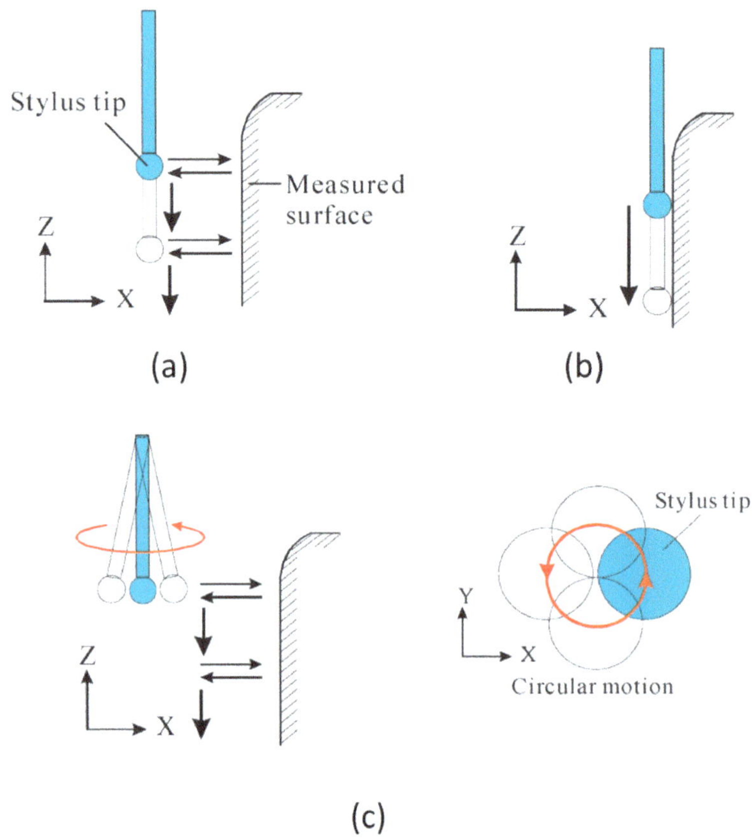

Figure 1. Various types of probes. (**a**) Touch-trigger probe; (**b**) scanning probe; and (**c**) vibrating probe (touch-trigger mode).

In the present study, the stylus characteristics are examined. Then, the effects of the surface force on the adhesion are analyzed. The results confirm that the primary cause of adhesion is the liquid bridge force. Therefore, the measurement system, placed in a vacuum vessel, is constructed suitably to prevent the adhesion caused by the surface force between the stylus tip and the measured surface. The effects of pressure inside the vacuum vessel on the surface force are evaluated experimentally. The surface force is calculated by assuming that the optical fiber probe is equal to the cantilever of the fixed support. In other words, the surface force is calculated by the deflection of the stylus shaft. There are many techniques for measuring surface forces, such as those involving the surface forces apparatus (SFA), atomic force microscopy (AFM), micro cantilever (MC), optical trapping (OT), *etc.* [13,14,15]. The measuring method that uses the optical fiber probe is, in principle, similar to the SFA and AFM. As a result, the surface force is found to be 0.13 μN when the pressure inside the vacuum vessel is 350 Pa. This effect is equivalent to a 60% reduction in the surface force in the atmosphere.

MEASUREMENT PRINCIPLE

Figure 2 shows a diagram of the developed optical measurement system. The stylus consists of a 2.5-μm-diameter optical fiber to which a 5-μm-diameter glass stylus tip is attached. The total length of the stylus is 0.38 mm. The probing system consists of the fiber stylus, two laser diodes with a 650 nm wavelength (LDX and LDY in the *X*- and *Y*-directions, respectively), and two dual-element photodiodes (PX and PY in the *X*- and *Y*-directions, respectively). The stylus shaft is fixed to a tube-type piezo driver element in order to perform attitude adjustment of the stylus shaft; the stylus shaft is installed between the laser diodes and the dual-element photodiodes, which are oriented orthogonally. The laser diodes are mounted above the stylus tip, and the focused laser beams irradiate along the *X*- and *Y*-directions onto the stylus shaft. The two dual-element photodiodes are located opposite the laser diodes beyond the stylus. The laser beams that pass through the stylus shaft are received by these dual-element photodiodes. The beam intensities detected by the photodiodes are converted into voltages; hereafter, these intensities are denoted as I_{PX1}, I_{PX2}, I_{PY1}, and I_{PY2} (V). The output signal I_X in the*X*-direction obtained using I_{PY1} and I_{PY2} and the output signal I_Y in the *Y*-direction obtained using I_{PX1} and I_{PX2} are defined as given in Equations (1) and (2), respectively. A charge-coupled device is employed to monitor the positions of the stylus and test piece during the setting up of the equipment and the measurement.

$$I_X = I_{PY1} - I_{PY2} \tag{1}$$

$$I_Y = I_{PX1} - I_{PX2} \tag{2}$$

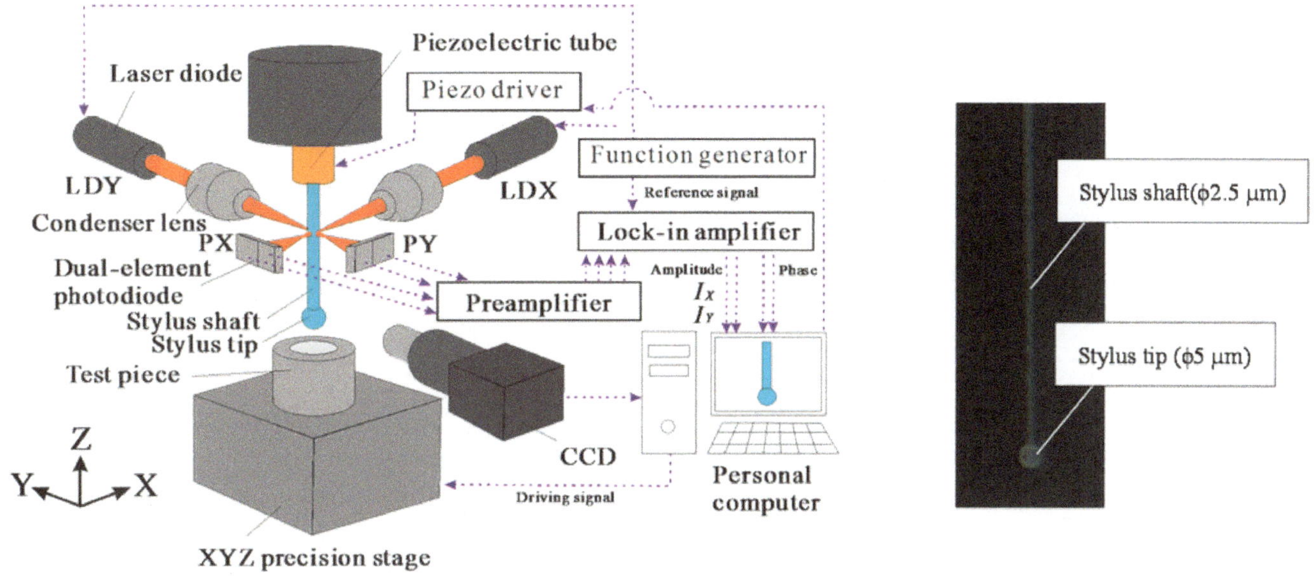

Figure 2. Measurement system using optical fiber probe.

To illustrate the measurement principle of the optical fiber probe, Figure 3 shows a cross-sectional diagram of the *X*-*Y*plane of the optical system shown in Figure 2. Before the stylus tip comes into contact with the measured surface, the light intensities measured by each element of the dual-element photodiode are equal (*i.e.*, $I_{PX1} = I_{PX2}$ and $I_{PY1} = I_{PY2}$), as shown inFigure 3a. When the stylus tip comes into contact with the measured surface in the *X*-direction, the laser-irradiated part of the stylus shaft is displaced, and the light intensities measured by each element of the dual-element photodiode are no longer equal to each other (*i.e.*, $I_{PX1} = I_{PX2}$ and $I_{PY1} > I_{PY2}$), as shown in Figure 3b. When the stylus shaft is displaced in the +*X*-direction, the angle of refraction of the laser beam passing through the stylus shaft in the *Y*-direction changes owing to a shift in the part of the stylus shaft being irradiated. Additionally, when the stylus tip comes into contact with the measured surface in the *Y*-direction, the light intensities measured by each element of the dual-element photodiode are no longer equal to each other (*i.e.*, $I_{PX1} < I_{PX2}$ and $I_{PY1} = I_{PY2}$), as shown in Figure 3c. As a result, the contact direction and the magnitude of displacement of the stylus tip can be obtained from the output signals I_X and I_Y. When the stylus tip comes into contact with the measured surface in the *Z*-direction, the stylus shaft is buckled and deflected. This deflection is also measured using the above-mentioned method.

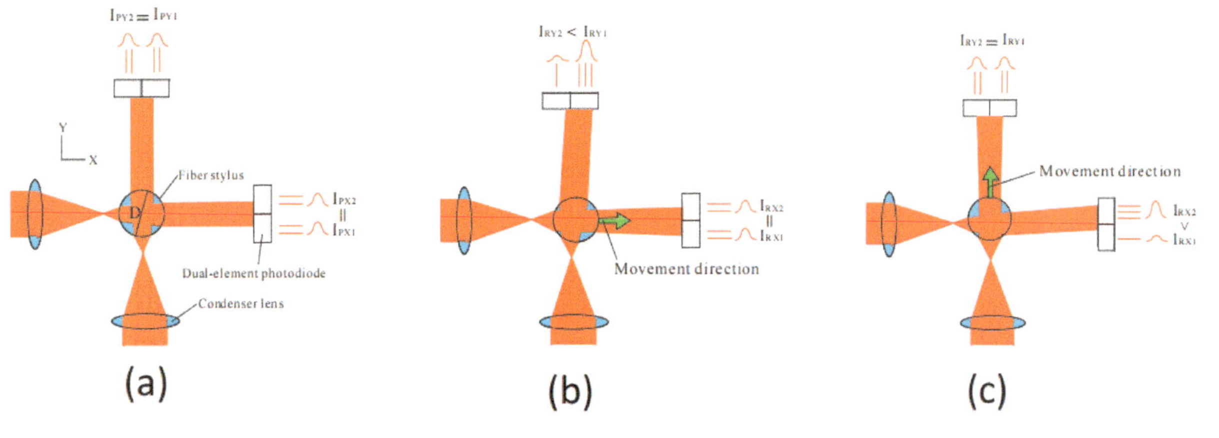

Figure 3. Principle of measurement. (**a**) initial stage; (**b**) displacement in *X*-direction; and (**c**) displacement in *Y*-direction.

Because the proposed probe measures the deflection amplitude of the stylus shaft by using a laser-based non-contact method, the stylus shaft does not need to be rigid; this principle also applies to a stylus with a much smaller diameter. This measurement system measures the deflection of the stylus shaft; however, it does not directly measure the displacement of the stylus tip. The noise present in I_X and I_Y is removed via synchronous detection using a lock-in amplifier. The displacement of the stylus is magnified by using it as a rod lens. The surface of the microstructure is measured by recording the displacement of the stylus shaft as well as the coordinates at which the stylus comes into contact with the measured surface.

STYLUS CHARACTERISTICS

Figure 4 shows the changes in outputs I_X and I_Y when the stylus tip is displaced in the ±*X*-direction. Here, the horizontal axis represents the displacement of the stylus tip and the vertical axis represents the changes in I_X and I_Y. It can be seen that when the stylus tip is displaced in the *X*-direction, barely any change occurs in the output I_Y in the *Y*-direction and the fiber probe can function as a displacement sensor because the rate of change in I_X can approximate a straight line within a ±3 μm range in the *X*-direction.

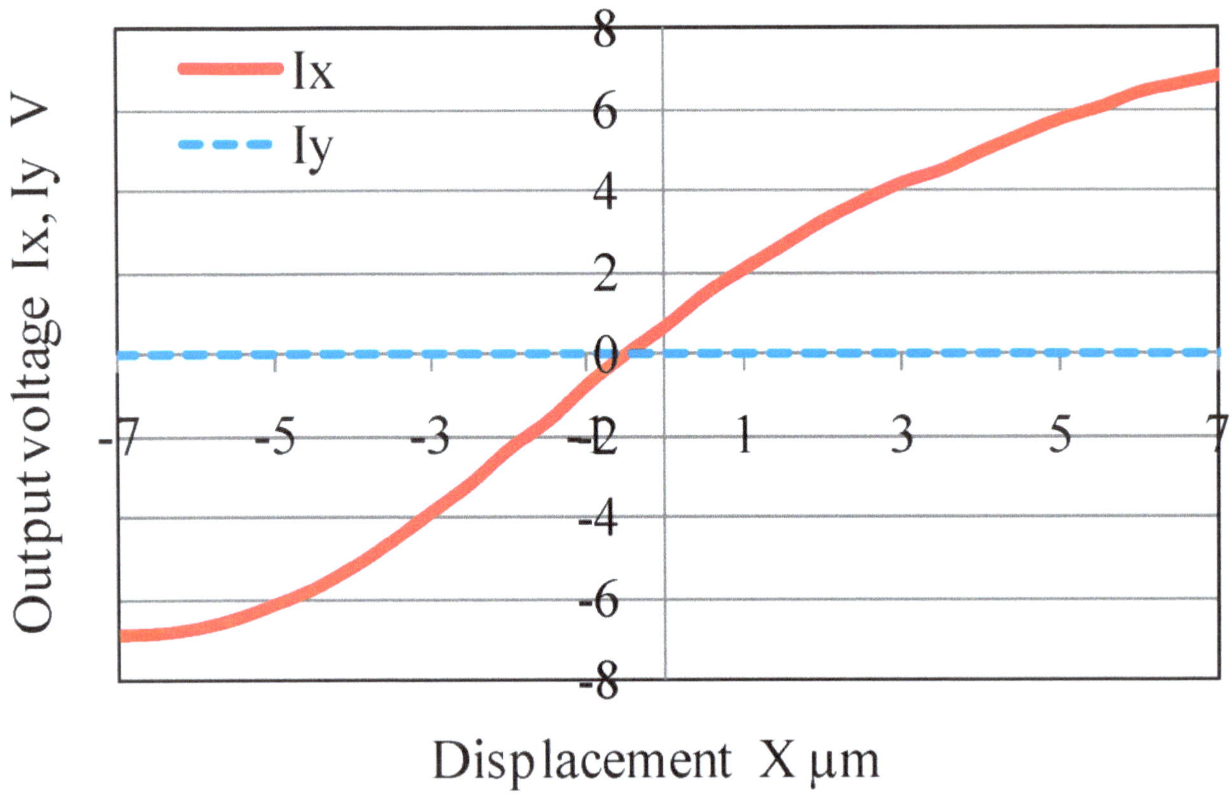

Figure 4. Changes in output voltages I_X and I_Y induced by displacement in ±*X*-direction.

EFFECTS OF SURFACE FORCE

Analyses of Liquid Bridge Force, Van der Waals force, and Electrostatic Force

In general, when the particle size is less than several tens of micrometers, the effects of the surface force generated from the van der Waals force, electrostatic force, and liquid bridge force are strengthened and this surface force becomes greater than the force of gravity [11]. Figure 5 shows a schematic diagram of the surface force between the stylus tip and the measurement surface. Because the fiber stylus has a 5-μm-diameter sphere on its tip, the measurement surface draws the stylus tip closer when the stylus tip approaches it, and the distance between the stylus tip and the measured surface is less than the regular displacement. When the stylus tip comes into contact with the measured surface, it adheres to the surface and cannot be separated from it. When the measured surface is scanned point by point in the touch-trigger mode, the measurement time increases because the stylus tip is required to be separated from the measured surface on account of the surface force. When the measured surface is scanned continuously in the scanning mode, measurement accuracy reduced because of the bending of the stylus shaft due to the adhesion. The observed adhesion, which is influenced by environmental factors (e.g., humidity) and the roughness of the measured surface, is not reproducible.

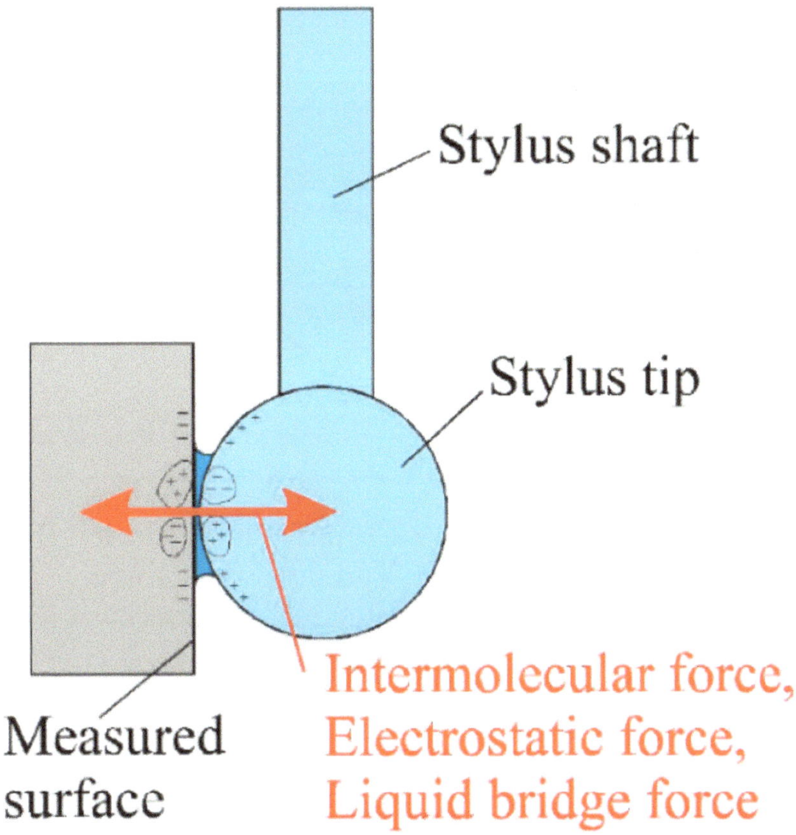

Figure 5. Schematic diagram of surface force between stylus tip and measured surface.

These surface forces are analyzed with the aim of investigating the causes of adhesion. First, we examine the liquid bridge force. This force has a large effect on adhesion in a humid environment; it is generated by the capillary action of condensed water near the contact zone of the stylus tip and measured surface. This force increases with increasing relative humidity. The liquid bridge force F_l experienced in the case of a stylus tip of diameter d and under a relative humidity of %RH can be calculated using the empirical formula in Equation (3) [11]. This is the empirical formula of the liquid bridge force between a particle and a plate with a hard and clean surface. As indicated by this formula, the liquid bridge force increases with increasing relative humidity.

$$F_l = 0.15d\{0.5 + 0.0045 \times (\%RH)\} \quad (3)$$

Next, we examine the van der Waals force. This force acts between molecules, and it is a general force of interaction that acts between objects at all times. The van der Waals force F_V can be calculated using the Hamaker constant A and the distance z between the stylus tip and the measured surface as given in Equation (4) [11]. The Hamaker constant A is a constant specific to materials, and it is expressed as in Equation (5), where n and Λ are the number of atoms per unit volume and the constant of proportionality of London forces, respectively. When the molecules of the stylus tip and the measured surface come close to each other by the van der Waals force, intermolecular repulsive force is generated between the molecules of the stylus tip and the measured surface. The molecules are stabilized at the most stable position, which corresponds to the position with the smallest potential energy. At this instant, the separation distance z is about 0.4 nm.

$$F_V = Ad/(12z^2) \quad (4)$$

$$A = n^2 \pi^2 \Lambda \quad (5)$$

Finally, the electrostatic force generated by an interaction between a charged particle and an uncharged surface is examined. When a charged particle comes into contact with an uncharged surface, the generated electrostatic force F_e can be calculated using Equation (6) in the case that the distance between the stylus tip and the measured surface is negligible because it is remarkably small compared with the diameter of the stylus tip [11].

$$F_e = \frac{\pi}{4\varepsilon_0} \cdot \frac{\varepsilon - \varepsilon_0}{\varepsilon + \varepsilon_0} \cdot d^2 \sigma^2 \quad (6)$$

where ε_0, ε, and σ are the dielectric constant of the vacuum, the dielectric constant of the measured surface, and the marginal surface density of the charge of the stylus tip, respectively.

Figure 6 shows the relationship between the stylus tip diameter and the surface forces calculated using Equations (3), (4), and (6). Here, the horizontal axis represents the stylus tip diameter and the vertical axis represents the van der Waals force, electrostatic force, liquid bridge force, and gravity. It can be seen from this figure that when the stylus tip diameter is less than about 1 mm, the effects of surface force generated from the van der Waals force, electrostatic force, and liquid bridge force are strengthened and the surface force becomes greater than the force of gravity.

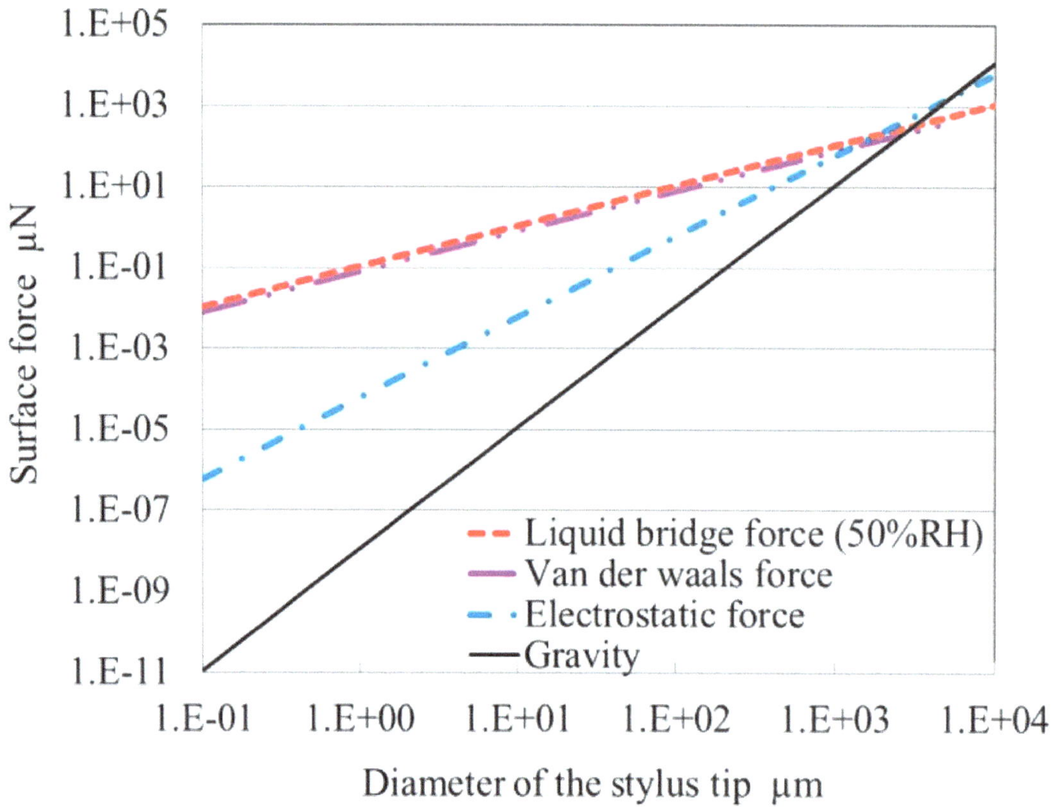

Figure 6. Relationship between stylus tip diameter and surface forces: A (SiO_2) = 1.5 × 10^{-20} J, z = 0.4 nm, ε_0 = 8.85 × 10^{-12} F·m^{-1}, $\varepsilon = \infty$, σ = 26.5 μC·m^{-2}.

This confirms that the primary cause of adhesion is the liquid bridge force. The van der Waals force and liquid bridge force are approximately equal. However, the electrostatic force is much smaller than the van der Waals force and liquid bridge force. When the surface forces are calculated using Equations (3), (4), and (6) for a stylus tip 5 μm in diameter, the liquid bridge force (50% *RH*), van der Waals force, and electrostatic force are 0.54 μN, 0.36 μN, and 0.0016 μN, respectively. In this case, the sum of the surface forces is approximately 0.9 μN. Figure 7 shows the relationship between the relative humidity and the sum of surface forces ($F_l + F_v + F_e$) calculated using Equations (3), (4), and (6). It can be seen from this figure that the liquid bridge force increases with increasing relative humidity.

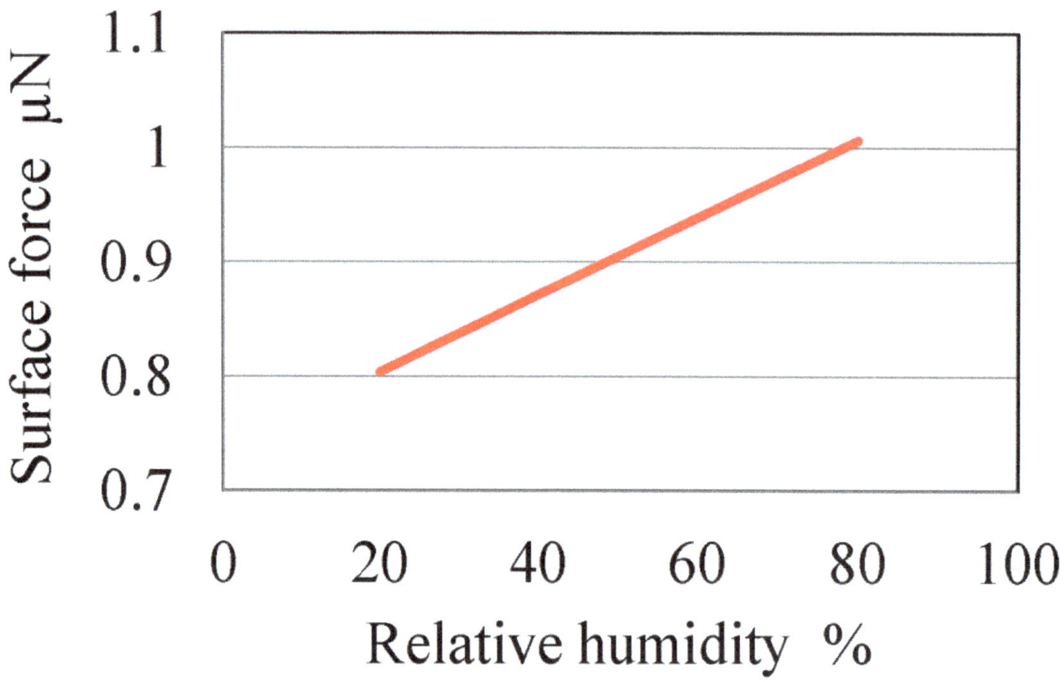

Figure 7. Relationship between relative humidity and surface force (stylus tip diameter: 5 μm).

Effect of Relative Humidity on the Liquid Bridge Force

As stated in the discussion above, the primary cause of adhesion is the liquid bridge force, which increases with increasing relative humidity. When the measured surface is scanned point by point in the touch-trigger mode, the measurement time increases because the stylus tip is required to be separated from the measured surface on account of the surface force. When the measured surface is scanned continuously in the scanning mode, the measurement accuracy reduces owing to the bending of the stylus shaft due to the adhesion. Therefore, in our previous work, a vibration mechanism was introduced to prevent the adhesion of the stylus tip to the measurement surface caused by the surface force [12]. By this mechanism, the stylus tip is set to vibrate in a circular motion, where it traces a circle of diameter 0.4 μm in the *X-Y* plane. However, the stylus shaft actually traces an elliptical path. The difference between the profile of the perfect circle and that of the actual elliptical path of the stylus tip leads to measurement errors. Moreover, the surface roughness cannot be measured in the scanning mode when using the vibrating fiber probe.

Therefore, the measurement system, placed in a vacuum vessel, is constructed suitably to prevent adhesion of the stylus tip to the measured surface caused by the surface force. Figure 8 shows a photograph of the measurement system in the vacuum vessel. The effects of pressure inside the vacuum vessel on the surface force are evaluated experimentally. In the experiment, the relative humidity is 61% and the temperature is 25.1 °C.

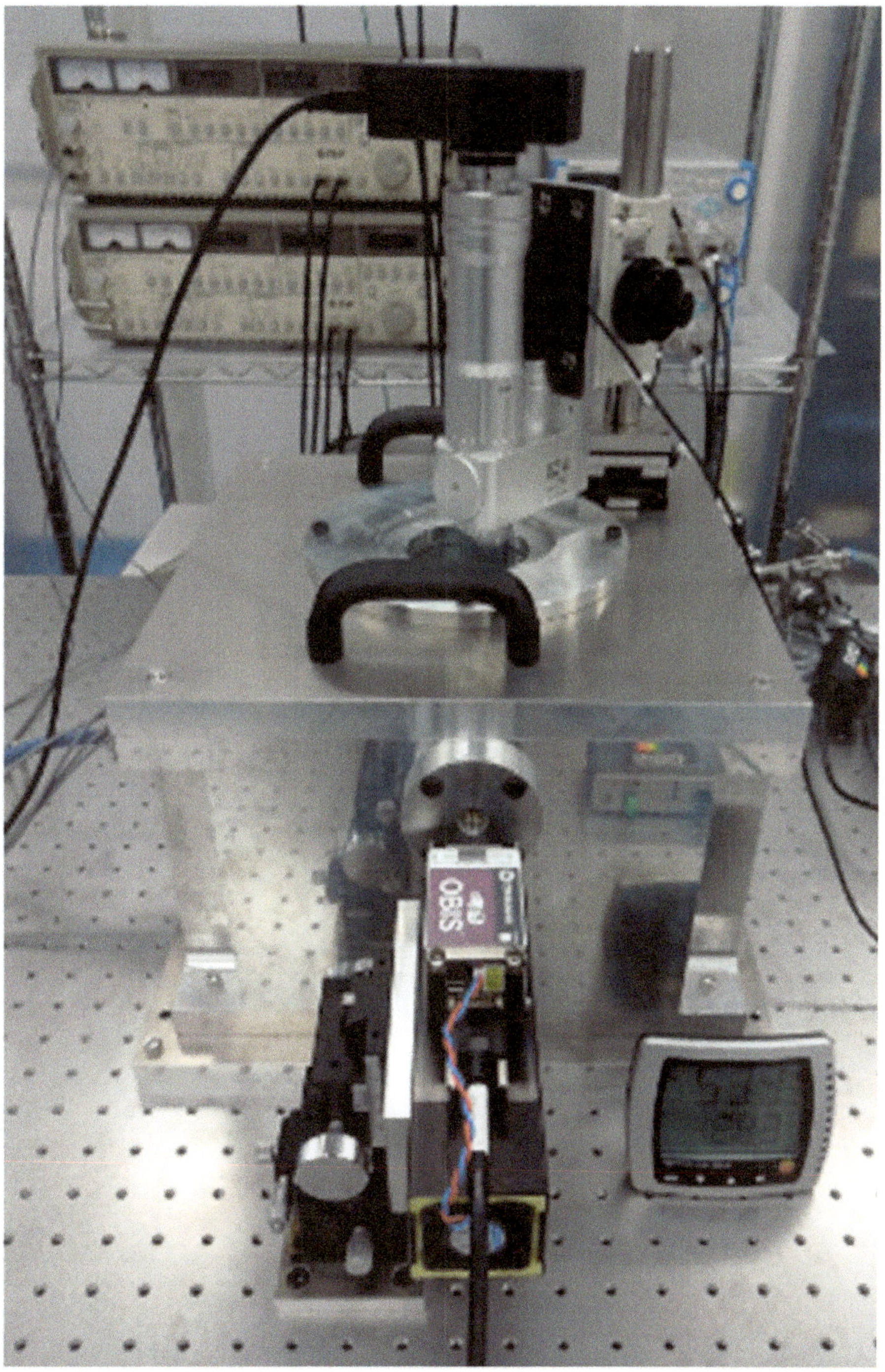

Figure 8. Photograph of measurement system in vacuum vessel.

First, the measurement method of the surface force is explained as follows. There are many techniques for measuring surface forces, such as those involving the surface forces apparatus (SFA), atomic force microscopy (AFM), micro cantilever (MC), optical trapping (OT), *etc.* [13,14,15]. The SFA can directly measure the force between two surfaces in controlled vapors or immersed in liquids. The force sensitivity of the SFA is about 10 nN. The SFA contains two curved molecularly smooth surfaces of mica between which the interaction forces are measured using a variety of force-measuring springs. AFM is, in principle, similar to the SFA except that forces are measured not between two macroscopic surfaces but between a fine tip and a surface. The force sensitivity of AFM is 1–10 pN. The measuring method using the optical fiber probe is, in principle, similar to the SFA and AFM. As shown in Figure 9, after the stylus tip comes into contact with the measured surface, the stylus tip is displaced in the backward direction by using the piezoelectric stage, due to which it separates from the measured surface. Because the stylus tip is displaced while maintaining its contact with the measured surface, the stylus shaft is deflected and the force required to separate the stylus tip from the measured surface is generated by the deflection of the stylus shaft. With an increase in the displacement of the stylus tip, this required force also increases. Therefore, the stylus tip can be separated from the measured surface when the displacement D of the stylus tip exceeds a certain value.

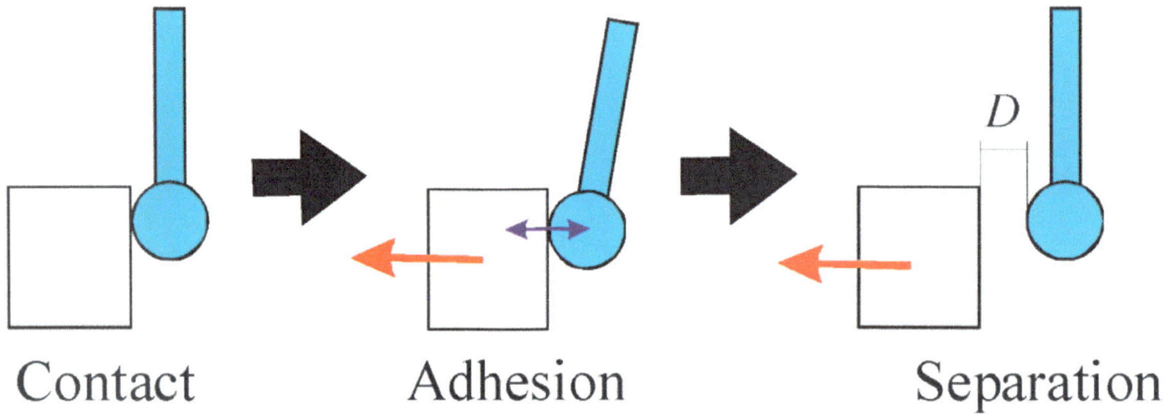

Figure 9. Experimental apparatus for measurement of surface force.

The load acting on the stylus tip, *i.e.*, the surface force, is calculated by assuming that the optical fiber probe, 2.5 µm in diameter and 0.38 mm in length, is equal to the cantilever of the fixed support. In other words, the surface force is calculated by the deflection of the stylus shaft. The surface force F is calculated using Equation (7).

$$F = \frac{3EI}{L^3} \cdot D \tag{7}$$

where E, I, and L are the Young's modulus, the geometrical moment of inertia, and the length of the stylus shaft, respectively. Figure 10 shows the load acting on the stylus tip under the assumption that the stylus shaft (Young's modulus$E$ = 72 GPa) is equivalent to the cantilever of the fixed support. In this figure, the horizontal axis represents displacement of the stylus tip, and the vertical axis represents the load acting on the stylus tip. For example, when the displacement Drequired to separate the stylus tip from the measured surface is about 20 µm, the load acting on the stylus tip is about 0.15 µN immediately before the stylus tip is separated from the measured surface as shown in Figure 10. Therefore, this value of 0.15 µN is regarded as the surface force. In this experiment, the spring constant of the stylus shaft is not calibrated. However, to accurately measure the surface force, it is necessary to calibrate it [16,17,18,19].

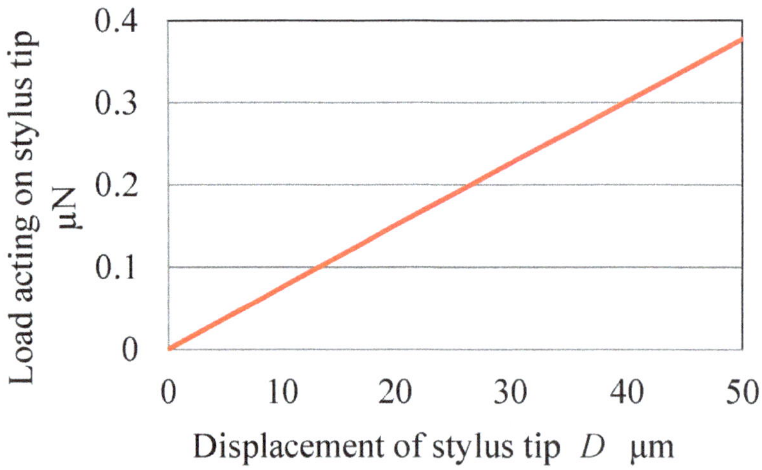

Figure 10. Load acting on the stylus tip (stem length: 0.38 mm, stem diameter: 2.5 µm, and ball diameter: 5 µm).

Figure 11 shows the relationship between the pressure inside the vacuum vessel and the relative humidity. Further, Figure 12 shows the relationship between the pressure inside the vacuum vessel and the surface forces. Finally, Figure 13 shows the relationship between the relative humidity and the surface force. From Figure 13, it is confirmed that surface force decreases with decreasing relative humidity. It is found that the surface force is 0.13 μN when pressure inside the vacuum vessel and relative humidity are 350 Pa and 7% *RH*, respectively. This effect is equivalent to a 60% reduction in the surface force in the atmosphere. However, a large residual surface force still acts between the stylus tip and the measured surface. This is because the amount of van der Waals force is almost the same as that of the liquid bridge force. Therefore, the surface force is thought to reduce by slightly a little more than 60%. When the stylus tip diameter is 5 μm as shown in Figure 6, the theoretical van der Waals force in a vacuum environment is about 0.36 μN. However, according to the obtained experimental results, the van der Waals force in a vacuum environment is about 0.13 μN. Thus, the theoretical and experimental results are different. Because the amount of van der Waals force is affected by the surface roughness of the stylus tip and that of the measurement surface [11], the surface roughness of the stylus tip and that of the measurement surface are, in turn, thought to be affected by the amount of van der Waals force.

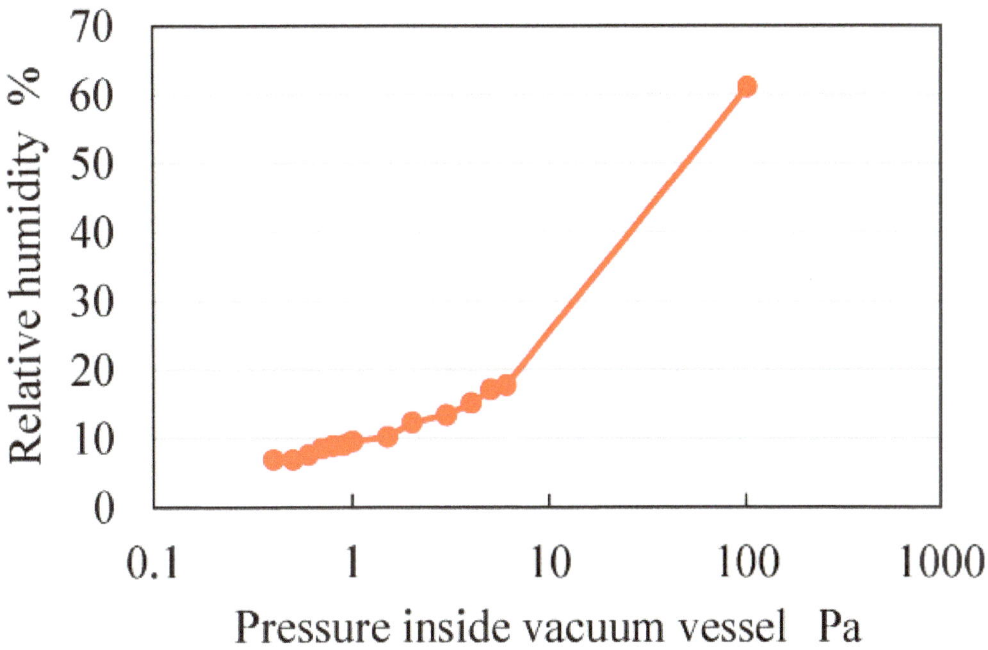

Figure 11. Relationship between pressure inside vacuum vessel and relative humidity.

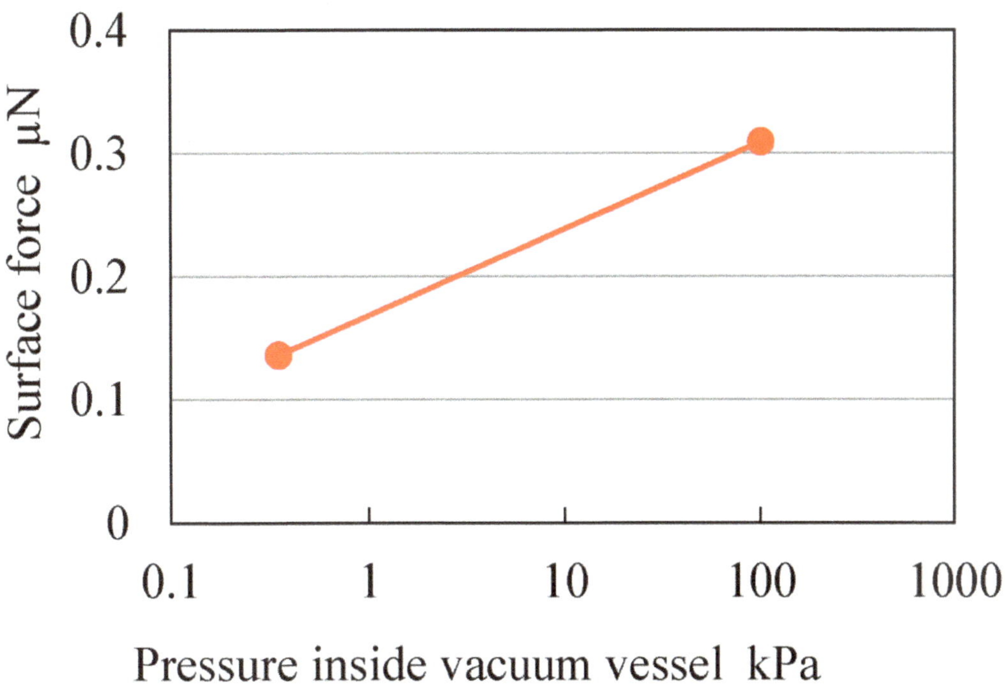

Figure 12. Relationship between pressure inside vacuum vessel and surface forces.

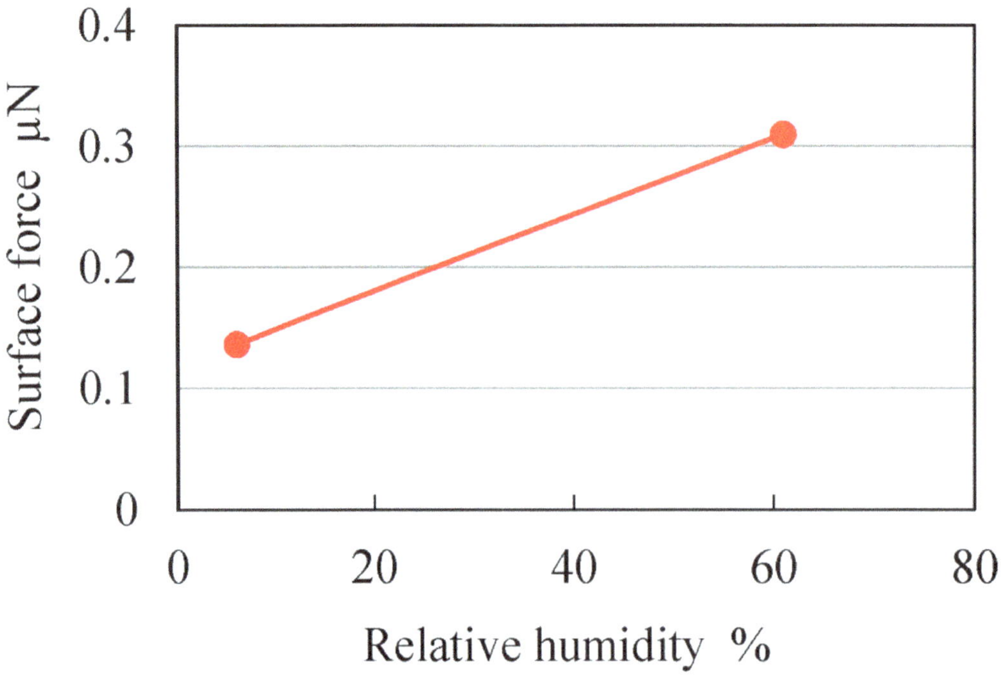

Figure 13. Relationship between relative humidity and surface forces.

Alternative Reduction Method for Surface Force

According to the above-mentioned experimental results, it is possible to reduce the liquid bridge force considerably by using a vacuum vessel. However, in scenarios where the use of a vacuum vessel is difficult, the stylus tip coated with a water-repellent coating has the potential to reduce the liquid bridge force. In order to reduce the van der Waals force, a stylus tip made of a material with a small Hamarker constant may be used. For example, the Hamaker constants of SiO_2, Fe, Si, Cu, and Au are 8.55, 21.2, 25.6, 28.4, and 45.5 × 10^{-20} J, respectively. However, in practice, the use of a material with a small low Hamarker constant can be difficult. In such cases, a stylus tip coated with a material with a small Hamarker constant may be used. The effect of the material of the stylus tip can be made negligible by using a coating that is several nanometers thick. The influence of the electrostatic force on the surface force is small; however, it is believed that various methods of removal of electricity or antistatic agents could be useful in reducing the electrostatic force.

CONCLUSIONS

In this study, the stylus characteristics are examined, and then, surface forces are analyzed in order to investigate the causes of adhesion, which reduces the measurement accuracy. The analysis results of surface forces reveal the liquid bridge force to be the primary cause of adhesion. Therefore, the measurement system placed in the vacuum vessel is constructed suitably to prevent adhesion of the stylus tip to the measured surface because of the surface forces, mainly by reducing the liquid bridge force. As a result, it is found that the surface force is 0.13 μN when the pressure inside the vacuum vessel is 350 Pa. This effect is equivalent to a 60% reduction in the surface force in the atmosphere.

ACKNOWLEDGMENTS

This study was partly supported by a research grant from the Mitutoyo Association for Science and Technology and by JSPS KAKENHI Grant Number 26420392.

AUTHOR CONTRIBUTIONS

Hiroshi Murakami conceived and wrote the paper; Akio Katsuki and Takao Sajima manufactured the vacuum vessel and analyzed the data; Hiroshi Murakami and Mitsuyoshi Fukuda performed the experiments and analyzed the surface forces.

CONFLICTS OF INTEREST

The authors declare no conflict of interest.

REFERENCES

1. Masuzawa, T.; Hamasaki, Y.; Fujino, M. Vibroscanning method for nondestructive measurement of small holes. *CIRP Ann.* **1993**, *42*, 589–592.
2. Hidaka, K.; Saito, A.; Koga, S. Study of a micro-roughness probe with ultrasonic sensor. *CIRP Ann.* **2008**, *57*, 489–492.
3. Bauza, M.B.; Hocken, R.J.; Smith, S.T.; Woody, S.C. Development of a virtual probe tip with an application to high aspect ratio microscale features. *Rev. Sci. Instrum.* **2005**, *76*.
4. Claverley, J.D.; Leach, R.K. A vibrating micro-scale CMM probe for measuring high aspect ratio structures. *Microsyst. Technol.* **2010**, *16*, 1507–1512.
5. Shiramatsu, T.; Kitano, K.; Kawata, M.; Mitsui, K. Development of a measuring method for shape and dimension of micro-components. Modification to the original measuring system, calibration of the probes and

the results of dimensional measurements. *JSME Int. J. Ser. C Mech. Syst.* **2002**, *68*, 267–274.

6. Schwenke, H.; Wäldele, F.; Weiskrich, C.; Kunzmann, H. Opto-Tactile sensor for 2D and 3D measurement of small structures on coordinate measuring machines. *CIRP Ann.* **2001**, *50*, 361–364.
7. Muralikrishnan, B.; Stone, J.A.; Stoup, J.R. Fiber deflection probe for small hole metrology. *Precis. Eng.* **2006**, *30*, 154–164.
8. Michihata, M.; Takaya, Y.; Hayashi, T. Development of the nano-probe system based on the laser-trapping technique.*CIRP Ann.* **2008**, *57*, 493–496.
9. Liebrich, T.; Knapp, W. New concept of a 3D-probing system for micro-components. *CIRP Ann.* **2010**, *59*, 513–516.
10. Murakami, H.; Katsuki, A.; Onikura, H.; Sajima, T.; Kawagoishi, N.; Kondo, E. Development of a system for measuring micro hole accuracy using an optical fiber probe. *J. Adv. Mech. Des. Syst. Manuf.* **2010**, *5*, 995–1004.
11. Okuyama, K.; Masuda, H.; Morooka, S. *Biryushi Kogaku*; Ohmsha: Tokyo, Japan, 1992.
12. Murakami, H.; Katsuki, A.; Sajima, T.; Suematsu, T. Study of a vibrating fiber probing system for 3-D micro-structures: Performance improvement. *Meas. Sci. Technol.* **2014**, *25*.
13. Israelachvili, J.N. *Intermolecular and Surface Forces: Revised Third Edition*; Academic Press: Waltham, MA, USA, 2011; pp. 227–247.
14. Lee, M.; Kim, B.; Kim, J.; Jhe, W. Noncontact friction via capillary shear interaction at nanoscale. *Nat. Commun.* **2015**, *6*.
15. Yongho, S.; Wonho, J. Atomic force microscopy and spectroscopy. *Rep. Prog. Phys.* **2008**, *71*.
16. Guebum, H.; Ahn, H.S. Calibration of effective spring constants of colloidal probes using reference cantilever method.*Coll. Surf. A Physicochem. Eng. Asp.* **2015**, *489*, 86–94.
17. Song, Y.P.; Wu, S.; Xu, L.Y.; Zhang, J.M.; Dorantes-Gonzalez, D.J.; Fu, X.; Hu, X.D. Calibration of the effective spring constant of ultra-short cantilevers for a high-speed atomic force microscope. *Meas. Sci. Technol.* **2015**, *26*.
18. Slattery, A.D.; Blanch, A.J.; Ejov, V.; Quinton, J.S.; Gibson, C.T. Spring constant calibration techniques for next-generation fast-scanning atomic force microscope cantilevers. *Nanotechnology* **2014**, *25*.
19. Slattery, A.D.; Blanch, A.J.; Quinton, J.S.; Gibson, C.T. Calibration of atomic force microscope cantilevers using standard and inverted static methods assisted by FIB-milled spatial markers. *Nanotechnology* **2013**, *24*.

CHAPTER 14 A MEMS RESONANT SENSOR TO MEASURE FLUID DENSITY AND VISCOSITY UNDER FLEXURAL AND TORSIONAL VIBRATING MODES

Libo Zhao, Yingjie Hu, Tongdong Wang, Jianjun Ding, Xixiang Liu, Yulong Zhao and Zhuangde Jiang

State Key Laboratory for Manufacturing Systems Engineering, Collaborative Innovation Center of Suzhou Nano

ABSTRACT

Methods to calculate fluid density and viscosity using a micro-cantilever and based on the resonance principle were put forward. Their measuring mechanisms were analyzed and the theoretical equations to calculate the density and viscosity were deduced. The fluid-solid coupling simulations were completed for the micro-cantilevers with different shapes. The sensing chips with micro-cantilevers were designed based on the simulation results and fabricated using the micro electromechanical systems (MEMS) technology. Finally, the MEMS resonant sensor was packaged with the sensing chip to measure the densities and viscosities of eight different fluids under the flexural and torsional vibrating modes separately. The relative errors of the measured densities from 600 kg/m^3 to 900 kg/m^3 and viscosities from 200 μPa·s to 1000 μPa·s were calculated and analyzed with different microcantilevers under various vibrating modes. The experimental results showed that the effects of the shape and vibrating mode of micro-cantilever on the measurement accuracies of fluid density and viscosity were analyzed in detail.

INTRODUCTION

The density and viscosity of fluid are the most important parameters in the oil [1,2], chemical [3] and medical [4] industries and so on. The density can be measured based on the resonant frequency shift of resonant devices [5]. Reference [6] showed that a microresonator had different resonant frequencies and quality factors due to different densities and viscosities of the loaded liquids. Generally, the resonant frequency of the microresonator decreased with increasing liquid density, and the full width at half maximum (FWHM) of the microresonator increased with increasing liquid viscosity [7,8]. With the development of micro electromechanical systems (MEMS) technology, the MEMS resonator has higher resonant frequency and quality factor, and presents many advantages to implement online measurements of fluid parameters, such as fast response times [9,10].

On the basis of the above phenomena, many MEMS resonant sensors have been studied for the measurement of fluid density and viscosity. Oden [11], Ahmed [12], and Papi [13] studied the measurement of fluid viscosity using the resonant frequency shift of an Atomic Force Microscope (AFM)

Citation: Libo Zhao, Yingjie Hu, Tongdong Wang, Jianjun Ding, Xixiang Liu, Yulong Zhao and Zhuangde Jiang, A MEMS Resonant Sensor to Measure Fluid Density and Viscosity under Flexural and Torsional Vibrating Modes, doi:10.3390/s16060830.

microcantilever before and after being submerged in the fluid, but the high cost and special installation requirements of the AFM make this method difficult to use in industry. Goodwin [14,15] developed a rectangular silicon microcantilever with a silicon on insulator (SOI) wafer to measure fluid density and viscosity through a vibrating method, and a large number of experimental data of fluid density and viscosity were given in those papers. The measuring accuracies of this sensor for viscosity and density were 10% and 1%, respectively, when the fluid viscosity was less than 1 mPa·s. The sensor also had lower measurement sensitivity because the piezoresistors were fabricated using polysilicon with a low piezoresistive coefficient. The quality factor of a rectangular microcantilever was significantly reduced with the increase of fluid viscosity due to the increase of resonator energy dissipation, so this type of sensor could not be used to measure high viscosity fluids. Boudjiet [16] studied the effects of microcantilever shape and geometrical dimensions on the density sensitivity. This study showed that a wide and short cantilever was more sensitive to the density variation, the highest sensitivity was 228 Hz/(kg·m^{-3}) and the measurement accuracy ranged from 0.4% to 0.6%. To improve the sensitivity of microcantilever sensors, Ansari [17] analyzed and compared the deflections and vibration characteristics of rectangular and trapezoidal profile microcantilevers. The results showed that the trapezoidal microcantilever had better sensitivity. To improve the quality factor of the resonator, Lucklum [18] designed a density and viscosity sensor, where a vibration plate was supported by four elastic beams and driven by Lorentz force undergo a reciprocating in-plane motion. The sensor could measure the square root of the product of density and viscosity with an accuracy of 1% in the range of 1~500 mPa·s, but the density and viscosity were not measured separately. When the microcantilever was used to detect ultrasensitive mass, an optimized electromagnetic excitation method specifically for the second resonant mode was proposed and developed for further improving the resolution [19]. In conclusion, the resonant sensors reported in the literature above mainly operate under the first order resonant mode, which is also called the flexural resonant mode. Recently, Manzaneque [20] designed a piezoelectric MEMS resonator to measure fluid viscosity and density based on the second order bending mode, the measurement ranges of viscosity and density were 0.4–7.3 mPa·s and 680–905 kg·m^{-3}, and the measuring accuracies of viscosity and density were 8% and 0.4%, respectively. All this suggests that the performance of resonant sensors under higher resonant modes should be researched further.

In this paper, the theories of fluid density and viscosity measurement using the microcantilever resonator were analyzed in detail. Sensing chips with micro-cantilevers in two different shapes were designed via numerical analysis. The MEMS resonant sensor was fabricated with the sensing chips by the packaging process. The first two order resonant modes of micro-cantilevers were both used to measure densities and viscosities of different fluids. The experimental results were analyzed to discuss the effects of the microcantilever structure and the resonant modes on the accuracies of the density and viscosity measurements. Compared with the reference values, the sensor had higher accuracy to measure the fluid viscosity under the higher order resonant mode, but had approximate accuracies to measure the fluid density under the first two resonant modes.

THEORY AND SIMULATION

The schematic diagram of the microcantilever resonant sensor is shown in Figure 1. The micro-cantilever was designed with a length l, width w and thickness d. The metal coil was positioned on the surface of the micro-cantilever. The Wheatstone bridge consisted of four piezoresistors designed to be located at the positions of stress concentration. Welding pads were used to connect with the input and output signals. B denotes the constant magnetic intensity provided by the external magnet. When the metal coil was powered with the alternating current I in a certain frequency, an alternating Lorentz force F was generated to drive the microcantilever to vibrate at the same frequency. Then, the resistance values of four piezoresistors were changed based on the piezoresistive effect because the stress conditions of the four piezoresistors varied. The Wheatstone bridge would output the corresponding signal which was proportional to the magnitude of the microcantilever vibration. Therefore, a dynamic resonant curve was obtained from the output of the Wheatstone bridge. When the frequency of the alternating current through the metal coil was close to the natural frequency of the microcantilever, then the microcantilever would vibrate in resonance. A peak value of the Wheatstone bridge output signal appeared in the resonant curve, then the resonant frequency and quality factor of the micro-cantilever could be calculated.

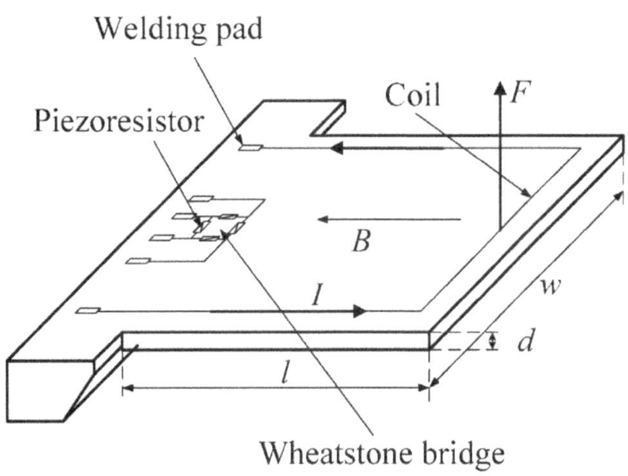

Figure 1. Schematic diagram of the micro-cantilever resonant sensor.

The fundamental frequencies f_1 and f_2 of the microcantilever under the first order flexural mode and the first order torsional mode can be expressed by Equations (1) and (2), respectively [21].

$$f_1 = \frac{d}{4\pi}\frac{\varphi_1^2}{l^2}\sqrt{\frac{E}{3\rho_c}} \tag{1}$$

$$f_2 = \frac{1}{4l}\sqrt{\frac{G\xi}{\rho_c J_P}} \tag{2}$$

where φ_1 is the first order positive root of equation $1 + \cosh(\varphi_n)\cos(\varphi_n) = 0$, $\varphi_1 = 1.8751$. ρ_c and E are the density and elasticity modulus of micro-cantilever, respectively. G is the shear modulus, and $G = E/(2 + 2v)$, where v is the Poisson ratio. J_p is polar moment of inertia and $J_p = (wd^3 + w^3d)/12$. ξ can be expressed as follows [22], where n is the order number:

$$\xi = \frac{1}{3}d^4\left(\frac{w}{d} - \frac{192}{\pi^5}\sum_{n=1}^{\infty}\frac{1}{n^5}\tanh\frac{n\pi w}{2d}\right) \quad (3)$$

Measurement Theory

The different measurement equations for the fluid density and viscosity were analyzed under different conditions as follows: some assumptions should be discussed first. In general, the microcantilever was assumed as an isotropic linearly elastic solid. The internal frictional effect was negligible and the vibration amplitude was far smaller than the length scale of the micro-cantilever in geometry. The fluid can be considered to be inviscid in practical application when the Reynolds number (Re) >> 1 [23].

The resonant frequency and half peak width of the microcantilever have close relationships with fluid density and viscosity when the microcantilever vibrates resonantly in the measured fluids. The relationship of the resonant frequency and fluid density can be described as follows under the flexural vibration [24]. This approximation is implemented with good accuracy in the derivation of the well-known inviscid result for a rectangular cantilever [23].

$$\frac{f_{\text{fluid1}}}{f_{\text{vac1}}} = \left(1 + \frac{\pi\rho_f w}{4\rho_c d}\right)^{-1/2} \quad (4)$$

where f_{fluid1} and f_{vac1} are the flexural resonant frequencies of the microcantilever in the fluid and vacuum, respectively. ρ_f is the density of the fluid.

The relationship between the resonant frequency and fluid density under the torsional vibration can be described as follows [25]:

$$\frac{f_{\text{fluid2}}}{f_{\text{vac2}}} = \left(1 + \frac{3\pi\rho_f w}{32\rho_c d}\right)^{-1/2} \quad (5)$$

where f_{fluid2} and f_{vac2} are the torsional resonant frequencies of micro-cantilever in the fluid and vacuum, respectively.

Equation (4) can be expressed in another form:

$$\rho_f = \frac{4\rho_c d}{\pi w}\left(\frac{f_{\text{vac1}}^2}{f_{\text{fluid1}}^2} - 1\right) \quad (6)$$

In addition, Equation (5) can be also expressed as follows:

$$\rho_f = \frac{32\rho_c d}{3\pi w}\left(\frac{f_{\text{vac2}}^2}{f_{\text{fluid2}}^2} - 1\right) \quad (7)$$

In Equations (6) and (7), f_{vac1}, f_{vac2}, ρ_c, d and w all need to be determined before the fluid density is measured via f_{fluid1} and f_{fluid2}. However, they are all constants for a specific microcantilever.

Equations (6) and (7) were deduced for a microcantilever with a length much larger than its width ($l/w >> 1$). In addition, [14] has discussed the fluid density and viscosity measurement method with a cantilever plate structure ($l/w \approx 1$), where the relationship of the resonant frequency and fluid density could be expressed as follows:

$$\rho_f = \frac{Ev_n^5 d^3}{24\{1-\sigma^2\}l^5(2\pi f_{\text{fluid}})^2} - \frac{\rho_c d v_n}{2l} \quad (8)$$

where σ is Poisson's ratio, v_n is the eigen value of the fluid velocity potential function. In Equation (8), all variables except ρ_f and f_{fluid} are constants for a specific micro-cantilever.

For Re >> 1, the frequency response of a rectangular cantilever in a viscous fluid is given by [26]:

$$\frac{f_{\text{fluid}}}{f_{\text{vac}}} = \left(1 + \frac{\pi\rho_f w}{4\rho_c d}\text{Real}\left[\Gamma(f_{\text{fluid}})\right]\right)^{-1/2} \quad (9)$$

where Real $[\Gamma(f_{\text{fluid}})]$ is the real part of the hydrodynamic function evaluated at the frequency in the liquid. However, the dimensions of the resonator in [26] are 2.8 mm × 2.6 mm, indicating that Equation (9) can be rearranged as Equation (10) due to the Real $[\Gamma(f_{\text{fluid}})]$ approaches 1 at Re >> 1:

$$\rho_f = \frac{4\rho_c d}{\pi w}\left(\frac{f_{\text{vac}}^2}{f_{\text{fluid}}^2} - 1\right) \quad (10)$$

Therefore, Equations (6)–(8) and (10) can be all simplified as the following Equation (11) since the parameters of cantilever itself can be all confirmed:

$$\rho_f = \frac{k_1}{f_{\text{fluid}}^2} + k_2 \quad (11)$$

where k_1 and k_2 are constants and calculated by experimental calibrations. However, both k_1 and k_2 are different under flexural and torsional resonant modes. Therefore, this paper used Equation (11) as working equation to measure the fluid density under different conditions such as the flexural and torsional resonant modes.

The resonance quality factor Q is an important parameter for fluid viscosity measurement. When a resonator with random shape oscillates in a liquid, the inertial and viscous forces apply to the motion [27], and Q is given by:

$$Q^2 \propto (2\pi f_{\text{fluid}})^3 \eta_f \rho_f \quad (12)$$

where η_f is the viscosity of the liquid.

Based on the above viewpoint, the relationship of the resonant frequency and fluid viscosity can be described as [14]:

$$\eta_f = \frac{k_3}{\rho_f f_{\text{fluid}}}\left(\frac{2g_{\text{fluid}}}{f_{\text{fluid}}} - \frac{2g_{\text{vac}}}{f_{\text{vac}}}\right)^2 \quad (13)$$

where k_3 is a constant and obtained by the experimental calibration, g_{fluid} and g_{vac} are half peak widths of the microcantilever in the fluid and vacuum, respectively. For a specific micro-cantilever, f_{vac} and g_{vac} are also constants and can be determined via calibration, so the ratio of $g_{\text{vac}}/f_{\text{vac}}$ can be replaced by k_4. Then Equation (13) becomes:

$$\eta_f = \frac{k_3}{\rho_f f_{\text{fluid}}^3}\left(\frac{2g_{\text{fluid}}}{f_{\text{fluid}}} - k_4\right)^2 \quad (14)$$

Fluid-Structure Interaction Simulation

Microcantilevers with rectangular and trapezoidal shapes were proposed. The design dimensions of the two different microcantilevers were a length of 1500 μm, width of 2500 μm, thickness of 20 μm, and the free end width of the trapezoidal micro-cantilever was 1000 μm. It is proven that a microcantilever with a larger width could help to improve the sensitivity [16] and quality factor [28] of a microcantilever, and in particular it could improve the Re (Re = $\rho_f f w/4\eta_f$), where f is the resonant frequency of the microcantilever) [23]. If The Re is much larger than 1, it means the viscosity equation and the density equation could be decoupled when the fluid density and viscosity are measured simultaneously. In this study, the Re of the proposed micro-cantilever was about 100, so the resonant frequency of microcantilever was only affected by the fluid density as shown in Equation (11), and the viscosity was calculated by the measured density, quality factor and resonant frequency of the micro-cantilever in the fluid, as shown in Equation (14).

Fluid-solid coupling simulations were carried out by the finite element method (FEM) as shown in Figure 2. The silicon microcantilever was modeled using the SOLID45 solid element (the red part), its elastic modulus was 169 GPa, its density was 2330 kg/m^3, and the Poisson's ratio was 0.064. The fluid was described by the FLUID30 acoustic fluid element (the green part), and the fluid elements were coupled with solid elements. The fluid domain covered the microcantilever except for the fixed end, and the dimension of the fluid domain was increased by 5000 μm over the dimensions of the microcantilever. The mapping method was used to mesh the structure with the number of solid elements, which was 300, and the number of fluid elements which was 2000. The resonant frequency f was obtained by modal analyses in the fluid-solid coupling simulation. However, the quality factor Q has not been obtained from the current simulation because the peak width at half height cannot be calculated by the simulation results.

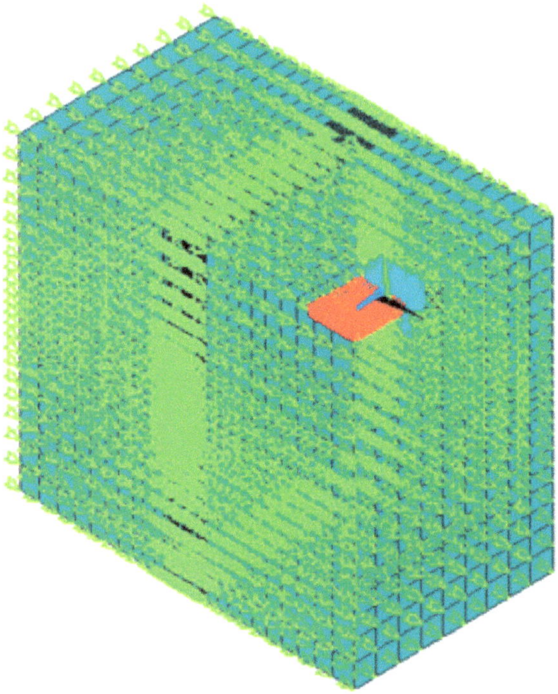

Figure 2. Fluid-solid coupling model.

The resonant modes of different micro-cantilevers were obtained by fluid-solid coupling simulation in *n*-pentane, as shown in Figure 3, the resonant modes in other fluids were also obtained using the same method. The first order mode was the first order flexural vibration, and the second order mode was the first order torsional vibration in all different fluids.Figure 3a,b showed that the vibrating shapes of the rectangular micro-cantilever under the first and second order modes, respectively. Figure 3c,d reveal the vibrating shapes of the trapezoidal microcantilever under the first and second order modes, respectively. Thus, it could be seen that the vibrating shapes of the microcantilever obviously has nothing to do with the fluid environment, but are mainly related to the microcantilever structure (the length-width ratio).

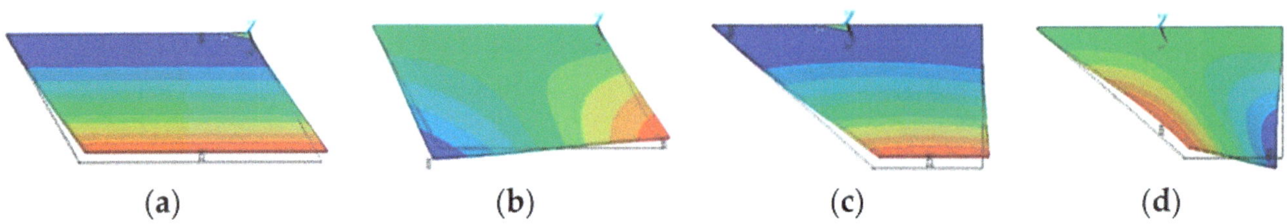

Figure 3. Vibration shapes of two micro-cantilevers in deflection under the first two order modes. (**a**) Rec.1st; (**b**) Rec.2nd; (**c**) Tra.1st; (**d**) Tra.2nd.

The rectangular micro-cantilever was taken as an example to analyze the relationship of its resonant frequency and the measured density, as shown in Figure 4. The curves were obtained from the results of coupling simulations and experiments when the microcantilever was vibrating in fluids with different densities under the first two resonant modes.

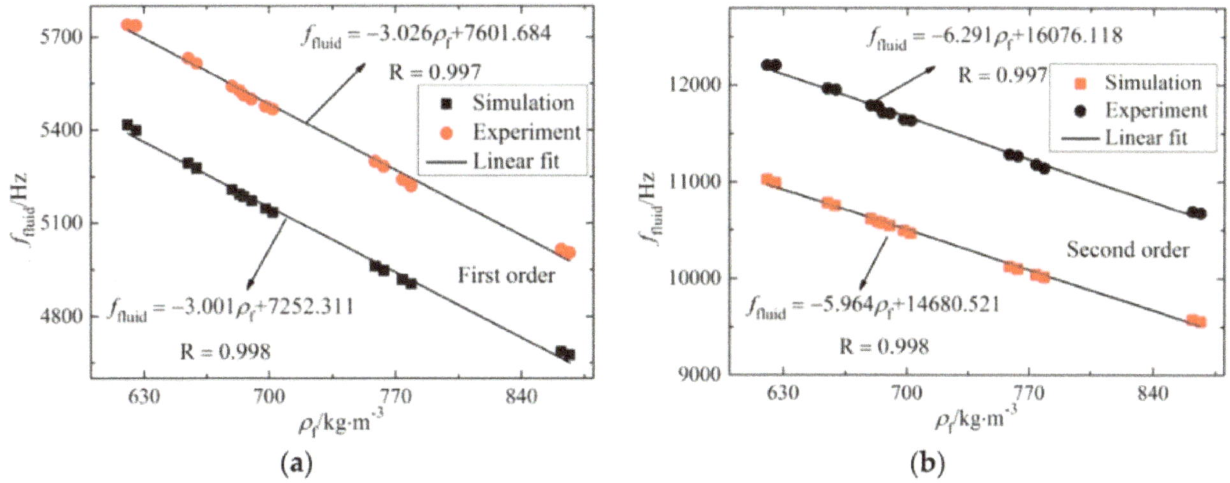

Figure 4. The relationship of resonant frequency and density under different order modes. (**a**) Under the first order mode (flexure); (**b**) under the second order mode (torsion).

In Figure 4, the resonance frequency was varied linearly with the fluid density, but we know that the fluid density is indeed inversely proportional to the square of resonant frequency according to Equation (11). The Taylor expansion at $f_{fluid} = a$ ($a > 0$) of Equation (11) can be written as:

$$\rho_f = \frac{\rho_f\big|_{f_{fluid}=a}}{0!} + \frac{\rho_f'\big|_{f_{fluid}=a}}{1!}(f-a) + \frac{\rho_f''\big|_{f_{fluid}=a}}{2!}(f-a)^2 + \dots + \frac{\rho_f^{(n)}\big|_{f_{fluid}=a}}{n!}(f-a)^n + R_n(f) \quad (15)$$

where $R_n(f)$ is the high order infinitesimal of $(f-a)^n$, and $R_n(f)$ approaching zero, thus Equation (15) can be simplified as follows:

$$\rho_f = k_1a^{-2} + k_2 + (-2k_1a^{-3})(f-a) + 3k_1a^{-4}(f-a)^2 + \dots \quad (16)$$

The resonant frequency of sensor in this paper are much larger than 4000 Hz, so the quadratic coefficient is very small due to the negative fourth power of a, and the coefficients of higher order term are all close to zero. Then, Equation (16) is changed as follows:

$$\rho_f = -2k_1a^{-3}f + 3k_1a^{-2} + k_2 \quad (17)$$

Figure 4 showed that the slopes and linearities of the two kinds of results were fairly consistent. When the density of the fluid decreased, the resonant frequency increased with a linearity larger than 0.99. Also, the relative deviations of the resonant frequency in the simulation and experiment under the same vibration were smaller than 10%, which validated the correctness of the fluid-solid coupling simulation results. Therefore, the fluid-solid coupling simulation was very useful to guide the design of the microcantilever.

In addition, it's shown that the slopes of the fitting straight lines under the second order mode were larger than those under the first order mode. This means the variation of resonant frequency under higher order mode was larger with the same fluid density fluctuation. Thus, the sensitivity of fluid density measurement was higher under the second order mode than that under the first order mode. The density sensitivity under the second order mode in this paper was about 2 times larger than that with the value of −2.6 Hz·(kg·m^{-3})$^{-1}$ in [29].

FABRICATION AND EXPERIMENT

Microcantilever Design

Two different micro-cantilever chips with rectangular and trapezoidal structures were designed according to the simulation results as shown in Figure 5 and Figure 6.

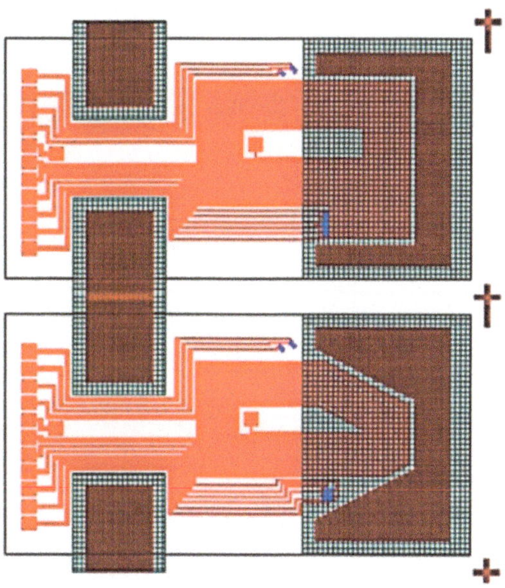

Figure 5. Two differentmicro-cantilever chips.

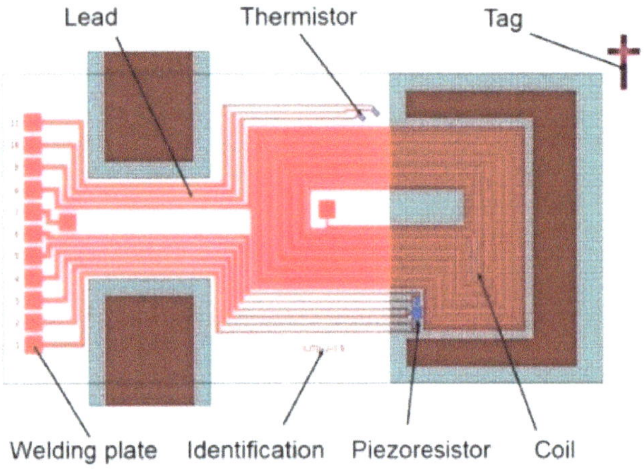

Figure 6. Layout of the rectangular microcantilever chip.

The sensing chip with the rectangular microcantilever was taken as an example to demonstrate the layout in Figure 6, where its dimensions were the same as in the simulation model. In the layout, there is a metal coil, four piezoresistors constituting the Wheatstone bridge, thermistor, identification tags, inner leads and welding pads. The sensing chip was fabricated using MEMS technology [30]. The microcantilever chip was excited to vibrate by the Lorentz force which was generated by the alternating current through the metal coil in the magnetic field. When the frequency of the alternating current was equal to the resonant frequency of the microcantilever chip, resonance would happen and the vibrating amplitude achieved its peak value. The amplitude data were obtained by the output of the Wheatstone bridge for the resistance values of piezoresistors were changed correspondingly. These data were fitted by the curves to calculate the resonant frequency and quality factor of the microcantilever chip. A thermistor was used to provide a reference temperature.

The Experimental System

The schematic diagram and object diagram of the experimental system are shown in Figure 7 and Figure 8. The resonant sensor was packaged with the developed sensing chip. The external magnetic field was provided by a samarium cobalt permanent magnet with a strength of 0.28T. The magnetic field direction was parallel to length direction of the microcantilever chip. A signal generator (33220 A Agilent, Santa Clara, CA, USA) provided a sinusoidal AC voltage with suitable amplitude and frequency to power the metal coil of the sensing chip. At the same time, it also provided a sync reference voltage signal to the phase-locked amplifier (SR830, Stanford, Sunnyvale, CA, USA). The first and second order modes were both realized by increasing the frequency of exciting signal (AC voltage through the coil) to be the first and second order resonant frequencies of the micro-cantilever. The Wheatstone bridge was powered by a constant current source with a value of 2 mA, and its output signal was detected by the Stanford SR830 to determine the vibration mode. The coupling of the AC voltage to the resistors of the Wheatstone bridge could be ignored because the resistors were located near the fixed end of the micro-cantilever and their displacements were very small. The signal generator and phase-locked amplifier were both controlled by a computer through the general purpose interface bus (GPIB) lines. The environment temperature was kept with different constant values by a thermostatic (7008, Fluke, Everett, WA, USA).

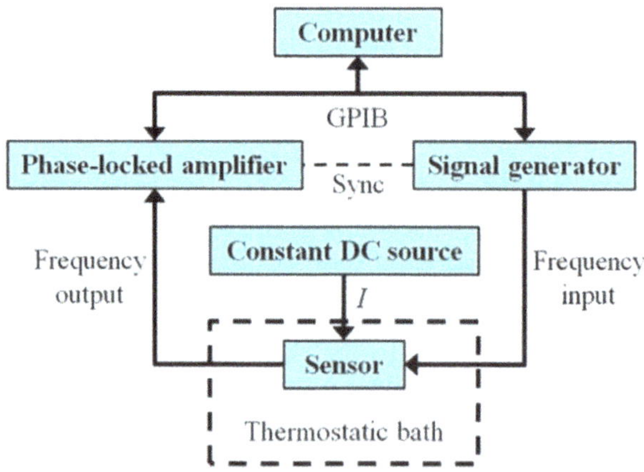

Figure 7. Block diagram of experimental system.]

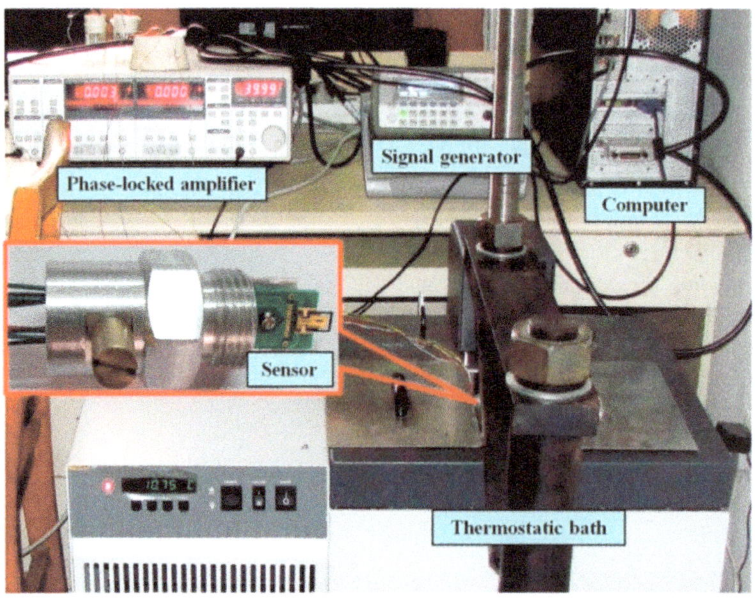

Figure 8. Experimental system.

The resonant frequency and half peak width should be obtained to calculate the fluid density and viscosity using Equations (11) and (14). The Stanford S830 detected the output signals of the Wheatstone bridge when the sinusoidal AC voltage excited the metal coil from low frequency to high frequency. The signal amplitude of the Wheatstone bridge could be intensified when the input frequency of AC voltage was close to the resonant frequency of the micro-cantilever chip in the fluid. The obtained data of the amplitude *versus* the sweeping frequency are shown in Figure 9. The data in blue * were the in-phase component of voltage amplitude, and the data in blue ○ were the quadrature component of voltage amplitude. All data were fitted by the MATLAB software to obtain the fitting equation. Then the resonant frequency, amplitude and half peak width of the resonant sensor could be calculated based on this equation.

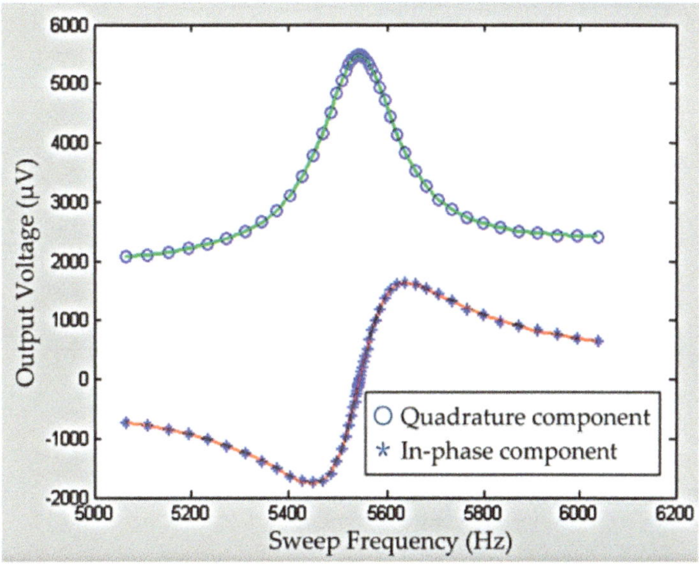

Figure 9. The resonant fitted curves.

RESULTS AND DISCUSSION

In order to test the sensor's performance, different fluids with various densities and viscosities should be measured by the proposed sensor. The measured fluids included *n*-pentane, *n*-hexane, *n*-heptane, isooctane, *n*-octane, 0.65 cs silicone oil, cyclohexane and methylbenzene. A specific fluids has different densities and viscosities under different temperatures. For the alkane fluids, their physical properties are the most stable from 20 °C to 25 °C. Their reference values of density, viscosity and sonic speed under standard atmospheric pressure and different temperature (293.15 K and 298.15 K) were calculated using the Reference Fluid Properties (REFPROP) software. In the field of thermophysical properties, the REFPROP software [31,32] is often used to provide the reference values of fluid density, viscosity, conductivity, and so on. Therefore, we used the reference values from the software to estimate the accuracy of each measurement.

For the rectangular microcantilever chip, a frequency swept curve was obtained in 0.65 cs silicone oil under different resonant modes, as shown in Figure 10. There were many irregular peaks in the curve below 2 kHz as shown in Figure 10Ba because the microcantilever chip was easily disturbed by external vibration and noise.

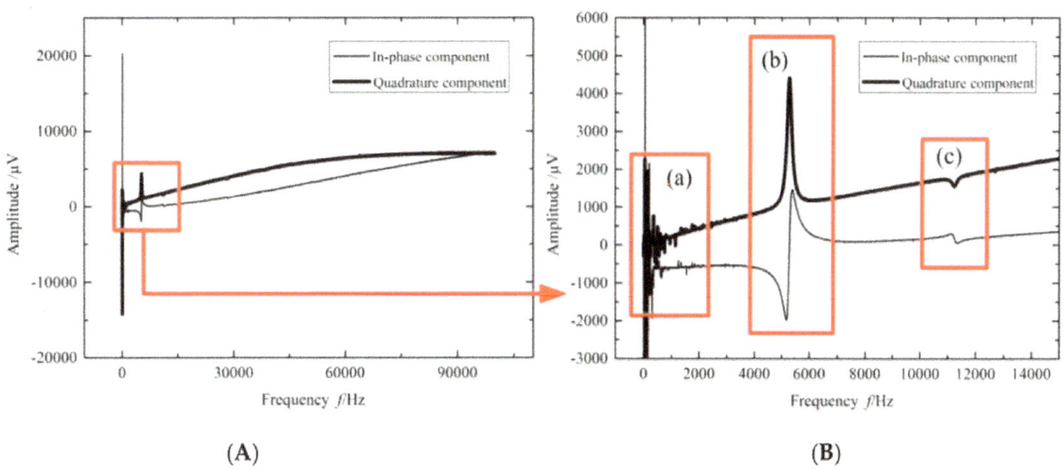

Figure 10. Sweep frequency results of rectangular micro-cantilever in 0.65 cs silicone oil. (**A**) The curve of sweep frequency; (**B**) The detail of red box in picture (**A**): The red box (a) is the noise below 2 kHz; (b) is the first order resonance (fFlexure); (c) is the second order resonance (tTorsion).

It was easy to make sure that this curve had two resonant peaks in the frequency range from 1 to 100 kHz. In theory, another resonant peaks should also appear in the frequency range from 1 to 100 kHz, but the strains of the piezoresistors in the microcantilever chip were so small under higher order resonant modes, so the Wheatstone bridge's outputs were also small and difficult to detect. The vibration modes could be estimated by the coupled fluid-solid simulation results. For example, Figure 10Bb,c show the resonant curves under first order flexural vibration and first order torsional vibration, respectively. Therefore, the fluid viscosity and density could be measured under the first two order modes, and the experimental results were discussed as follows: the measurement data of the rectangular microcantilever chip under flexural resonant mode and torsional resonant mode are shown in Table 1 and Table 2, respectively. The values of Q were calculated by $Q = f_{fluid}/2g_{fluid}$ using the MATLAB fitting method. The reference values of ρ_{ref} and η_{ref} for the fluid density and viscosity were obtained from the REFPROP software.

Table 1. Experimental results of rectangular micro-cantilever chip under flexural resonant mode.

Fluid	T (K)	f_{fluid} (Hz)	Q	ρ_{ref} (kg·m^{-3})	ρ_f (kg·m^{-3})	η_{ref} (μPa·s)	η_f (μPa·s)
n-pentane	293.15	5737.16	38.18	625.75	621.32	227.5	175.7
	298.15 [a]	5738.98	34.75	620.83	620.83	217.9	217.9
n-hexane	293.15	5615.55	32.68	659.36	655.55	312.3	253
	298.15	5632.09	30.95	654.78	650.76	296.3	285.4
n-heptane	293.15	5527.5	29.01	683.82	681.75	411.4	333.1
	298.15	5541.15	29.17	679.6	677.6	388.5	328.5
isooctane	293.15	5500.02	25.63	690.6	690.18	506.1	438.6
	298.15	5512.56	25.25	686.3	686.32	478.6	452.8
n-octane	293.15	5466.83	24.53	702.29	700.54	542	484.6
	298.15	5476.98	25.74	698.27	697.35	509.7	435.7
silicone oil	293.15	5283.39	27.41	763.61	761.36	672.3	387.1
(0.65 cs)	298.15	5299.43	28.39	758.87	755.79	650	357.5
cyclohexane	293.15	5220.12	24.24	778.63	783.84	961.8	510.5
	298.15	5240.91	25.99	773.89	776.36	884.7	437.3
methylbenzene	293.15	5004.75	30.48	866.87	866.85	588	316
	298.15 [b]	5016.13	32.48	862.2	862.2	556	273.6

Note: The groups of data which have superscripts of a and b are used to calibrate the constants in the Equations (11) and (14). k_1, k_2, k_3 and k_4 were $2.5729154926 \times 10^{10}$, -160.36, $3.9768556044 \times 10^{13}$ and 3.777×10^{-3}, respectively.

Table 2. Experimental results of rectangular micro-cantilever chip under torsional resonant mode.

Fluid	T (K)	f_{fluid} (Hz)	Q	ρ_{ref} (kg·m^{-3})	ρ_f (kg·m^{-3})	η_{ref} (μPa·s)	η_f (μPa·s)
n-pentane	293.15	12,210.45	75.63	625.75	620.32	227.5	218.9
	298.15 [a]	12,206.51	75.79	620.83	620.83	217.9	217.9
n-hexane	293.15	11,951.80	66.11	659.36	655.08	312.3	293.1
	298.15	11,964.81	67.54	654.78	653.28	296.3	280.1
n-heptane	293.15	11,777.05	57.71	683.82	679.87	411.4	391.9
	298.15	11,790.67	59.15	679.6	677.89	388.5	372.0
isooctane	293.15	11,706.48	52.73	690.6	690.19	506.1	473.9
	298.15	11,717.74	54.24	686.3	688.53	478.6	446.8
n-octane	293.15	11,633.45	51.28	702.29	701.08	542	503.7
	298.15	11,643.80	51.99	698.27	699.52	509.7	489.4
silicone oil	293.15	11,264.99	54.17	763.61	759.26	672.3	609.2
(0.65 cs)	298.15	11,282.02	54.63	758.87	756.45	650	587.6
cyclohexane	293.15	11,141.85	39.38	778.63	780.01	961.8	888.3
	298.15	11,178.47	41.91	773.89	773.77	884.7	779.9
methylbenzene	293.15	10,675.39	50.59	866.87	865.21	588	543.2
	298.15 [b]	10,690.88	51.33	862.2	862.20	556	526.5

Note: The groups of data which have superscripts of a and b are used to calibrate the constants in the Equations (11) and (14). k_1, k_2, k_3 and k_4 were $1.1844404355 \times 10^{11}$, -174.10, $1.567181537564 \times 10^{15}$ and 6.649×10^{-4}, respectively.

Here only experimental results and data acquisition method of the rectangular microcantilever chip were discussed. The deviations between the experimental values and reference values of the measured densities and viscosities under different resonant modes are shown in Figure 11 and Figure 12. In addition, the measurement results of the trapezoidal micro-cantilever chip were obtained similarly, and for brevity are not repeated in this paper.

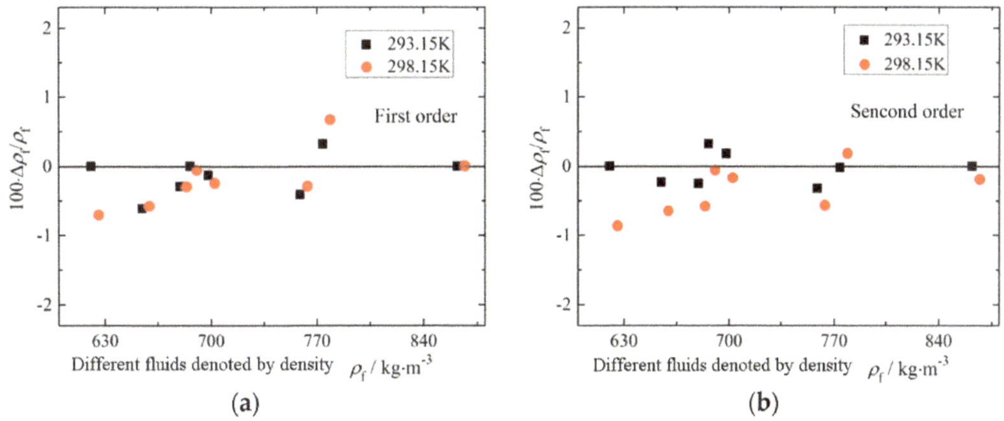

Figure 11. Relative errors of density under different order modes. (**a**) Under first order mode (flexure); (**b**) under second order mode (torsion).

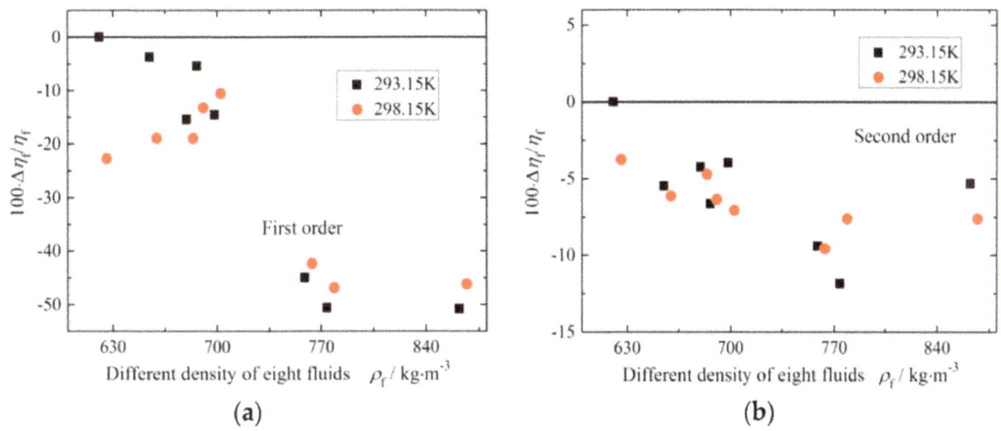

Figure 12. Relative errors of viscosity under different order modes. (**a**) Under first order mode (flexure); (**b**) under second order mode (torsion).

In order to describe and evaluate the measurement results of fluid density and viscosity, the average relative deviation (ARD) and maximum relative deviation (MRD) were introduced to analyze the measurement accuracy. The ARD is the average relative errors of all corresponding measurement results of the eight fluids at two temperature points. The MRD is the maximum relative deviation of all corresponding measurement results of eight fluids as well. Figure 13 and Figure 14showed measurement accuracies of density and viscosity in ARD and MRD under different resonant modes.

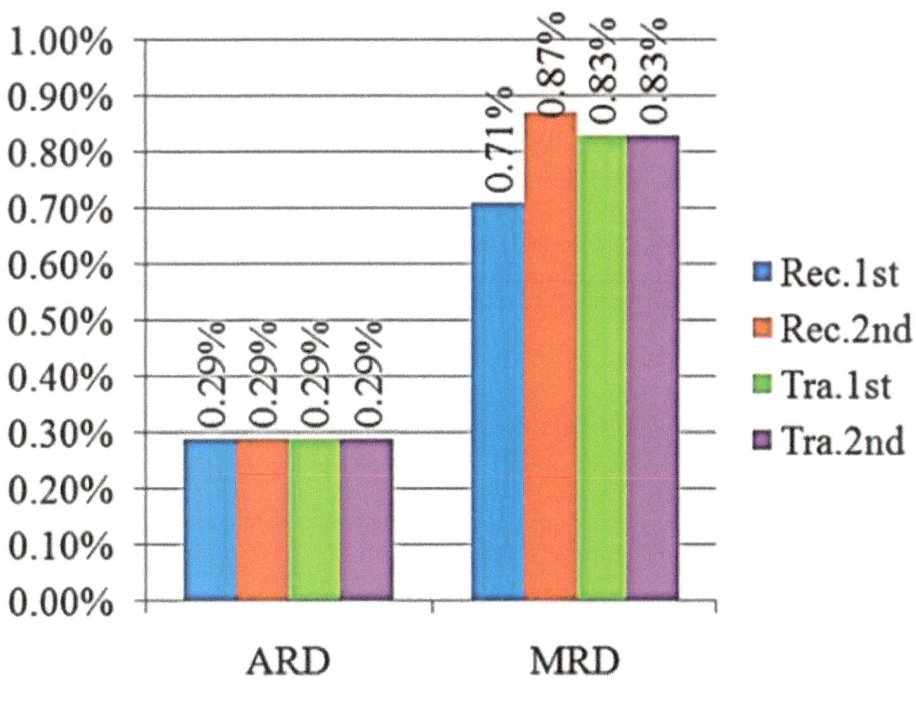

Figure 13. Deviations of density values.

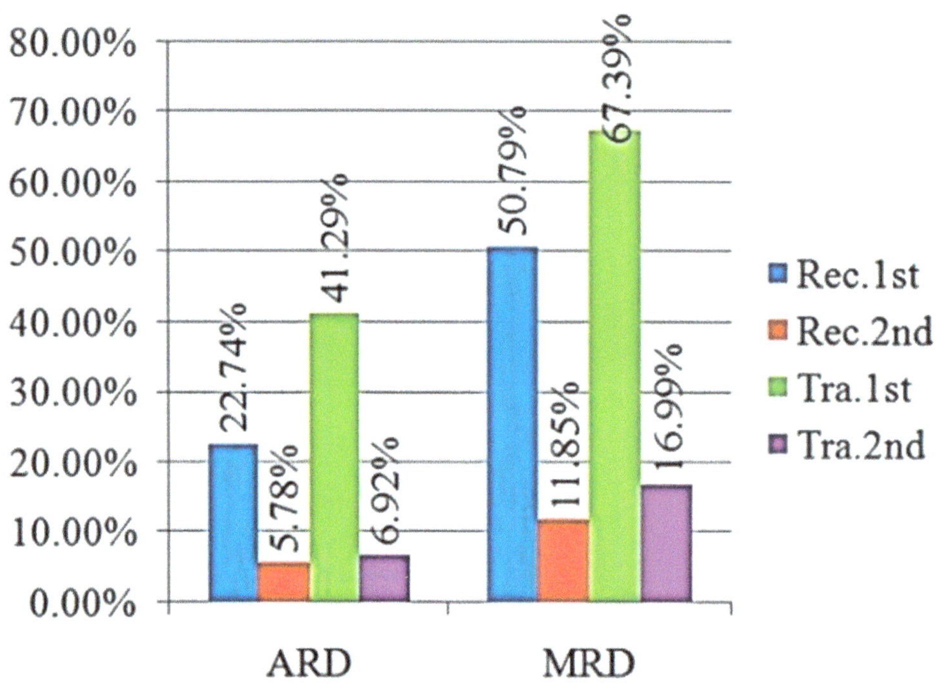

Figure 14. Deviations of viscosity values.

As shown in Figure 11 and Figure 13, the fluid density range was from 620.83 kg/m^3 to 866.87 kg/m^3. In Figure 13, all ARDs in the density were approximate and all less than 0.3%, even under different resonant modes. In addition, the MRDs in the density were all less than 1% and also did not changed too much under different resonant modes. Therefore, the chip shape and the resonant modes had only slight influences on the fluid density measurement accuracy.

In Figure 12 and Figure 14, the fluid viscosity range was from 217.9 μPa·s to 961.8 μPa·s. The ARDs in the viscosity between experimental values and reference values were 22.74% and 41.29% under first order resonant mode for the rectangular and trapezoidal micro-cantilevers, respectively. However, the ARDs in the viscosity between experiment values and reference values under the second order resonant mode were 5.78% and 6.92% for them, respectively. The ARDs of the fluid viscosity measurement were larger under the flexural mode because the quality factors of the microcantilever chip were fairly small under the first resonant mode. The ARDs of viscosity measurement under the torsional mode were about 10% and could be accepted in comparison with other viscosity measuring methods [14,15,33,34]. It meant that the accuracy of fluid viscosity could be enhanced using the torsional resonant mode due to the higher quality factors as shown in Table 1and Table 2.

CONCLUSIONS

In fluid parameter measurement, the main factor affecting the resonance frequency of a resonant sensor was the fluid density, and the main factor affecting the resonance amplitude was the fluid viscosity. In this paper, a fluid density measuring equation had been deduced from the resonant frequency of a microcantilever, and the fluid viscosity measuring equation had been deduced from the quality factor. In this research a simulation method was built to guide the design of a microcantilever resonant chip. An experimental system had been set up for the fabricated MEMS resonant sensor to measure the densities and viscosities of different fluids. The experimental results showed that the fluid density measurement accuracy could not be improved by using microcantilever chips of different shapes or high order resonant mode, but the fluid viscosity measurement accuracy under the torsional vibration mode was obviously better than that under the flexural vibration mode due to the higher quality factors under the former mode. Therefore, it's necessary to investigate methods to improve the quality factor such as optimizing coil pattern and piezoresistors location or selecting a high order resonant mode.

ACKNOWLEDGMENTS

This work was supported in part by the National Natural Science Foundation of China (51375378, 51205306, 91323303, 51421004), National Key Scientific Instrument and Equipment Development Projects of China (2012YQ03026101), the Fundamental Research Funds for the Central Universities (2012jdgz08) and the 111 Program (B12016).

AUTHOR CONTRIBUTIONS

Libo Zhao and Zhuangde Jiang conceived and designed the experiments; Yingjie Hu and Tongdong Wang performed the experiments; Jianjun Ding and Xixiang Liu. analyzed the data; Yulong Zhao contributed reagents/materials/analysis tools; Yingjie Hu wrote the paper.

CONFLICTS OF INTEREST

The authors declare no conflict of interest.

REFERENCES

1. Harrison, C.; Ryu, S.; Goodwin, A.; Hsu, K.; Donzier, E.; Marty, F.; Mercier, B. A MEMS sensor for the measurement of density-viscosity for oilfield applications. In Proceedings of the Reliability, Packaging, Testing,

and Characterization of MEMS/MOEMS V, San Jose, CA, USA, 25–26 January 2006.

2. Markova, L.V.; Makarenko, V.M.; Semenyuk, M.S.; Zozulya, A.P.; Kong, H.; Han, H.-G. A magnetoelastic viscometer for on-line monitoring of viscosity of lubricating oils. *J. Frict. Wear* **2011**, *32*, 41–48.
3. Jakoby, B.; Vellekoop, M.J. Physical sensors for liquid properties. *IEEE Sens. J.* **2011**, *11*, 3076–3085.
4. Smith, P.D.; Young, R.C.D.; Chatwin, C.R. A MEMS viscometer for unadulterated human blood. *Measurement* **2010**, *43*, 144–151.
5. Sparks, D.; Smith, R.; Straayer, M.; Cripe, J.; Schneider, R.; Chimbayo, A.; Anasari, S.; Najafi, N. Measurement of density and chemical concentration using a microfluidic chip. *Lab Chip* **2003**, *3*, 19–21.
6. Kim, H.J.; Kim, J.; Zandieh, O.; Chae1, M.-S.; Kim, T.S.; Lee, J.H.; Park, J.H.; Kim, S.; Hwang, K.S. Piezoelectric layer embedded-microdiaphragm sensors for the determination of blood viscosity and density. *Appl. Phys. Lett.* **2014**, *105*.
7. Fedorchenko, A.I.; Stachiv, I.; Wang, W.C. Method of the viscosity measurement by means of the vibrating micro-/nano-mechanical resonators. *Flow Meas. Instrum.* **2013**, *32*, 84–89.
8. Hur, D.; Lee, J.H. Determination of Liquid Density and Viscosity Using a Self-Actuating Microcantilever. *Jpn. J. Appl. Phys.* **2013**, *52*.
9. Etchart, I.; Chen, H.; Dryden, P.; Jundt, J.; Harrison, C.; Hsu, K.; Marty, F.; Mercier, B. MEMS sensors for density-viscosity sensing in a low-flow microfluidic environment. *Sens. Actuators A Phys.* **2008**, *141*, 266–275.
10. Toledo, J.; Manzaneque, T.; Ruiz-Diez, V.; Pfusterschmied, G.; Wistrela, E.; Steindl, W.; Schmid, U.; Sánchez-Rojas, J.L. Piezoelectric MEMS resonators for density and viscosity sensing in engine oil with diesel fuel. In Proceedings of the Transducers—18th International Conference on Solid-State Sensors, Actuators and Microsystems, Anchorage, AK, USA, 21–25 June 2015; pp. 436–439.
11. Oden, P.I.; Chen, G.Y.; Steele, R.A.; Warmack, R.J.; Thundat, T. Viscous drag measurements utilizing microfabricated cantilevers. *Appl. Phys. Lett.* **1996**, *68*.
12. Ahmed, N.; Nino, D.F.; Moy, V.T. Measurement of solution viscosity by atomic force microscopy. *Rev. Sci. Instrum.* **2001**, *72*.
13. Papi, M.; Arcovito, G.; De Spirito, M.; Vassalli, M.; Tiribilli, B. Fluid viscosity determination by means of uncalibrated atomic force microscopy cantilevers. *Appl. Phys. Lett.* **2006**, *88*.
14. Goodwin, A.R.H.; Donzier, E.P.; Vancauwenberghe, O.; Fitt, A.D.; Ronaldson, K.A.; Wakeham, W.A.; de Lara, M.M.; Marty, F.; Mercier, B. A Vibrating Edge Supported Plate, Fabricated by the Methods of Micro Electro Mechanical System for the Simultaneous Measurement of Density and Viscosity: Results for Methylbenzene and Octane at Temperatures between (323 and 423) K and Pressures in the range (0.1–68) MPa. *J. Chem. Eng. Data* **2006**, *51*, 190–208.
15. Goodwin, A.R.H.; Jakeways, C.V.; De Lara, M.M. A MEMS Vibrating Edge Supported Plate for the Simultaneous Measurement of Density and Viscosity: Results for Nitrogen, Methylbenzene, Water, 1-Propene, 1,1,2,3,3,3-hexafluoro-oxidized-polymd, and Polydimethylsiloxane and Four Certified Reference Materials with Viscosities in the Range (0.038 to 275) mPa·s and Densities between (408 to 1834) kg·m^{-3} at Temperatures from (313 to 373) K and Pressures up to 60 MPa. *J. Chem. Eng. Data* **2008**, *53*, 1436–1443.
16. Boudjiet, M.T.; Bertrand, J.; Mathieu, F.; Nicu, L.; Mazenq, L.; Leïchlé, T.; Heinrich, S.M.; Pellet, C.; Dufour, I. Geometry optimization of uncoated silicon microcantilever-based gas density sensors. *Sens. Actuators B Chem.* **2015**, *208*, 600–607.
17. Ansari, M.Z.; Cho, C.; Kim, J.; Bang, B. Comparison between deflection and vibration characteristics of rectangular and trapezoidal profile microcantilevers. *Sensors* **2009**, *9*, 2706–2718.
18. Luckluma, F.; Reichel, E.K.; Jakoby, B. Miniature density-viscosity measurement cell utilizing electrodynamic-acoustic resonator sensors. *Sens. Actuators A Phys.* **2011**, *172*, 75–81.
19. Jin, D.; Li, X.; Liu, J.; Zuo, G.; Wang, Y.; Liu, M.; Yu, H. High-mode resonant piezoresistive cantilever sensors for tens-femtogram resoluble mass sensing in air. *J. Micromech. Microeng.* **2006**, *16*, 1017–1023.
20. Manzaneque, T.; Ruiz-Díez, V.; Hernando-García, J.; Wistrela, E.; Kucera, M.; Schmid, U.; Sánchez-Rojas, J.L. Piezoelectric MEMS resonator-based oscillator for density and viscosity sensing. *Sens. Actuators A Phys.* **2014**, *220*, 305–315.
21. McFarland, A.W.; Poggi, M.A.; Bottomley, L.A.; Colton, J.S. Characterization of microcantilevers solely by frequency response acquisition. *J. Micromech. Microeng.* **2005**, *15*, 785–791.
22. Timoshenko, S.P.; Goodier, J.N.; Abramson, H.N. *Theory of Elasticity*; McGraw-Hill: New York, NY, USA, 1970.
23. Sader, J.E. Frequency response of cantilever beams immersed in viscous fluids with applications to the atomic force microscope. *J. Appl. Phys.* **1998**, *84*, 64–76.
24. Chu, W.H. *Tech Rep No. 2, DTMB, Contract NObs-86396 (X)*; Southwest Research Institute: San Antonio, TX, USA, 1963.
25. Green, C.P.; Sader, J.E. Torsional frequency response of cantilever beams immersed in viscous fluids with applications to the atomic force microscope. *J. Appl. Phys.* **2002**, *92*, 6262–6274.
26. Wilson, T.L.; Campbell, G.A.; Mutharasan, R. Viscosity and density values from excitation level response of piezoelectric-excited cantilever sensors. *Sens. Actuators A Phys.* **2007**, *138*, 44–51.
27. Ikehara, T.; Lu, J.; Konno, M.; Maeda, R.; Mihara, T. A high quality-factor silicon cantilever for a low detection-limit resonant mass sensor operated in air. *J. Micromech. Microeng.* **2007**, *17*.

28. Zhang, G.; Zhao, L.; Jiang, Z.; Yang, S.; Zhao, Y.; Huang, E.; Hebibul, R.; Liu, Z. Surface stress-induced deflection of a microcantilever with various widths and overall microcantilever sensitivity enhancement via geometry modification.*J. Phys. D Appl. Phys.* **2011**, *44*.

29. Corman, T.; Enoksson, P. A low-pressure encapsulated resonant fluid density sensor with feedback control electronics. *Meas. Sci. Technol.* **2000**, *11*.

30. Xu, L.; Zhang, G.; Zhao, L.; Jiang, Z.; Wang, H.; Liu, Z. A fluid viscosity sensor with resonant trapezoidal micro cantilever. In Proceedings of the IEEE International Symposium on Assembly and Manufacturing (ISAM), Xi'an, China, 30 July–2 August 2013; pp. 131–134.

31. Stanwix, P.L.; Locke, C.R.; Hughes, T.J.; Johns, M.L.; Goodwin, A.R.H.; Marsh, K.N.; May, E.F. Viscosity of $\{xCH_4 + (1 - x)C_3H_8\}$ with $x = 0.949$ for Temperatures between (200 and 423) K and Pressures between (10 and 31) MPa. *J. Chem. Eng. Data* **2014**, *60*, 118–123.

32. Locke, C.R.; Stanwix, P.L.; Hughes, T.J.; May, E.F. Viscosity of $\{xCO_2 + (1 - x)CH_4\}$ with $x = 0.5174$ for temperatures between (229 and 348) K and pressures between (1 and 32) MPa. *J. Chem. Thermodyn.* **2015**, *87*, 162–167.

33. Riesch, C.; Reichel, E.K.; Jachimowicz, A.; Schalko, J.; Jakoby, B.; Keplinger, F. A suspended plate viscosity sensor featuring in-plane vibration and piezoresistive readout. *J. Micromech. Microeng.* **2009**, *19*.

34. Cerimovic, S.; Beigelbeck, R.; Antlinger, H.; Schalko, J.; Jakoby, B.; Keplinger, F. Sensing viscosity and density of glycerol-water mixtures utilizing a suspended plate mems resonator. *Microsyst. Technol.* **2012**, *18*, 1045–1056.

CHAPTER 15

MEASUREMENT SENSITIVITY IMPROVEMENT OF ALL-OPTICAL ATOMIC SPIN MAGNETOMETER BY SUPPRESSING NOISES

Xiyuan Chen [1,2], Hong Zhang [1,2] and Sheng Zou [1,2]

[1]School of Instrument Science and Engineering, Southeast University, Nanjing 210096, China

[2]Key Laboratory of Micro-Inertial Instrument and Advanced Navigation Technology Ministry of Education, Nanjing 210096, China

ABSTRACT

Quantum manipulation technology and photoelectric detection technology have jointly facilitated the rapid development of ultra-sensitive atomic spin magnetometers. To improve the output signal and sensitivity of the spin-exchange-relaxation-free (SERF) atomic spin magnetometer, the noises influencing on the output signal and the sensitivity were analyzed, and the corresponding noise suppression methods were presented. The magnetic field noises, including the residual magnetic field noise and the light shift noise, were reduced to approximately zero by employing the magnetic field compensation method and by adjusting the frequency of the pump beam, respectively. With respect to the operation temperature, the simulation results showed that the temperature of the potassium atomic spin magnetometer realizing the spin-exchange relaxation-free regime was 180 °C. Moreover, the fluctuation noises of the frequency and the power were suppressed by using the frequency and the power stable systems. The experimental power stability results showed that the light intensity stability was enhanced 10%. Contrast experiments on the sensitivity were carried out to demonstrate the validity of the suppression methods. Finally, a sensitivity of 13 fT/Hz$^{1/2}$was successfully achieved by suppressing noises and optimizing parameters.

INTRODUCTION

In recent years, sensitive magnetometers have been used in many fields, such as biomedicine [1,2], Magnetic Resonance Imaging (MRI) [3,4], material science [5,6], geography [7,8], basic physics [9,10,11,12] and magnetic induction tomography [13]. With the rapid developments of quantum manipulation technology and photoelectric detection technology, the spin-exchange relaxation-free (SERF) atomic magnetometer [14] with ultrahigh theoretical sensitivity has received considerable attention, and it has been a hot research area [15,16,17,18].

Romalis group of Princeton University was the first to realize a SERF atomic spin magnetometer [19]. This group successfully achieved a sensitivity of 0.16 fT/Hz$^{1/2}$ in 2010, which currently still is the highest sensitivity achieved in magnetic field measurement [20]. Kamada *et al.* have used atomic magnetometers with optical gradiometer configurations to improve the sensitivity [21]. Atomic spin magnetometers are expected to be a new developmental direction in the future,

Citation: Chen X, Zhang H, Zou S. Measurement Sensitivity Improvement of All-Optical Atomic Spin Magnetometer by Suppressing Noises. Sensors. 2016; 16(6):896. DOI: 10.3390/s16060896

however, although the SERF atomic spin magnetometer has been successfully implemented, research on noise suppression methods based on constructing an output signal model and analyzing the noise sources and noises are still vitally important.

A SERF atomic spin magnetometer mainly consists of an optical pumping system, optical detection system, alkali-metal cell, magnetic shielding and magnetic compensation system, and non-magnetic electric heating system. All of these parts can produce noises. In this paper, according to the Bloch equation, the output signal was established. The factors influencing the sensitivity, such as magnetic field, temperature and light field were analyzed; meanwhile, corresponding suppression methods and the optimal operation temperature were proposed. Finally, the magnetic field measurement sensitivities were analyzed by contrast experiments.

PRINCIPLE OF ATOMIC SPIN MAGNETOMETER

The principle of the SERF atomic spin magnetometer is shown in Figure 1. Optical pumping (OP) [22,23] technology is used to polarize electron spin, and the spin angular momentum can be transferred to the nuclei via collisions. A circularly-polarized light, whose wavelength is tuned to the D1 resonance of the alkali metal atoms, can be used to polarize the atomic spin. A magnetic field perpendicular to the pump beam can induce a Larmor procession. At this time, a linearly-polarized light perpendicular to the plane of the magnetic field and pump beam can detect optical rotation which reflects such a Larmor procession. Realizing SERF regime thus needs three essential conditions, including a weak magnetic field, high heating temperature and optical pumping.

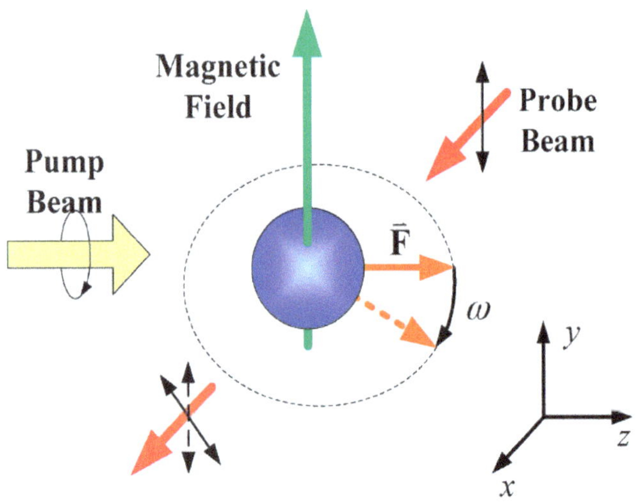

Figure 1. Principle of the SERF atomic spin magnetometer.

The arrangement of the SERF atomic spin magnetometer is shown in Figure 2. A cubic glass cell (25 mm length) contains a droplet of K atoms, 2 atm of ^{4}He to reduce the wall collision relaxation, and 30 Torr of N_2 to quench the alkali-metal excited state. The cell is heated to more than one hundred of degrees by a non-magnetic electric heating device to obtain a high atomic number density. The circularly-polarized beam is tuned to approximately 770.108 nm. The center wavelength of the linearly-polarized probe beam is about 766.2 nm. A Faraday modulator is placed in the probe optical path. Optical rotation can be detected by this linearly-polarized probe beam.

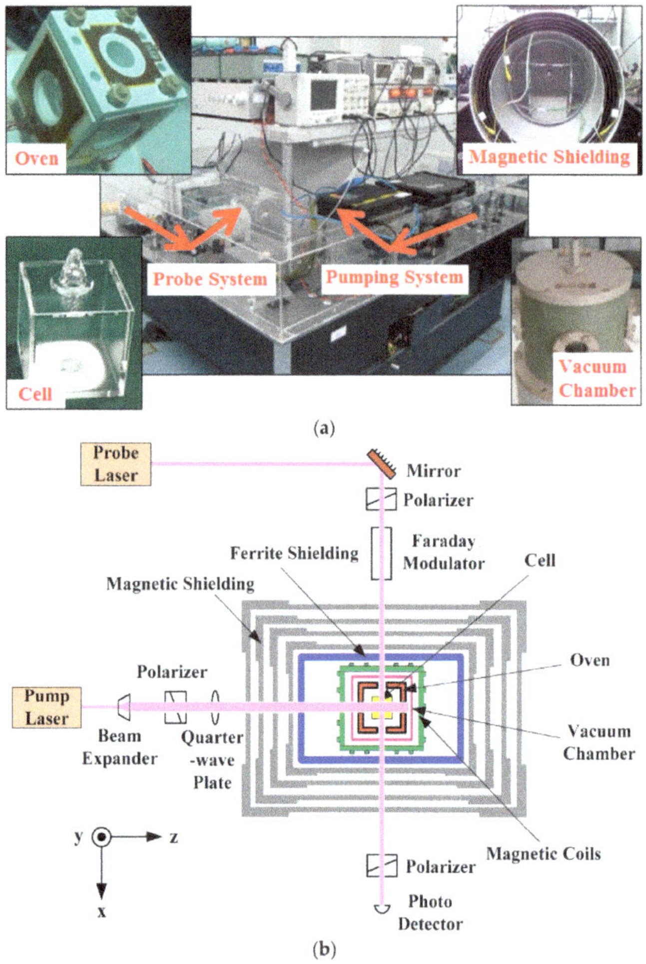

Figure 2. Arrangement of the SERF atomic spin magnetometer. (**a**) The photograph of the experimental device; (**b**) The schematic diagram of the light path. The pump beam propagates along the *z* direction. The probe beam is perpendicular to the pump beam and propagates along the *x* direction. The ferrite shielding is used to reduce the magnetic noise.

A four-layer magnetic shielding with high permeability is employed to reduce the ambient magnetic field. A tri-axial magnetic coils can produce a uniform magnetic field along the *x*, *y* and *z* directions, respectively. A vacuum chamber is used in our experiments to isolate the heat conduction from the oven due to the limitation on performance of the magnetic shielding caused by the temperature.

OUTPUT SIGNAL OF SERF ATOMIC SPIN MAGNETOMETER

Output Signal of SERF Atomic Spin Magnetometer

The evolution of the electron spin polarization can be well expressed by the Bloch equation [24]:

$$\frac{\partial \vec{P}^e}{\partial t} = \frac{1}{q}\left(\gamma^e \vec{B} \times \vec{P}^e + R_p(\hat{z} - \vec{P}^e) + R_m(\hat{x} - \vec{P}^e) - R_{sd}\vec{P}^e\right) \tag{1}$$

$$\frac{\partial \vec{P}^{e}}{\partial t} = \frac{1}{q}(\gamma^{e}\vec{B} \times \vec{P}^{e} + R_{p}(\hat{z} - \vec{P}^{e}) + R_{m}(\hat{x} - \vec{P}^{e}) - R_{sd}\vec{P}^{e}) \quad (2)$$

where $\vec{P}^{e}$ is the electron spin polarization, q is the slowing-down factor, γ^{e} is the electron gyromagnetic ratio, $\vec{B}$ is the external magnetic field in the magnetic shielding, R_p is the pumping rate of the pump beam defined as the average rate at which an unpolarized atom absorbs a photon from the pump beam, R_m is the pumping rate of the probe beam caused by the small circular component in the probe beam, R_{sd} is the spin-destruction relaxation rate, and $\hat{z}$ and $\hat{x}$ are the unit vectors of the z direction and the x direction, respectively. R_{sd}^{a-a} is the spin-destruction relaxation rate between the alkali metal atoms, R_{sd}^{a-g} is the spin-destruction relaxation rate between the alkali metal atoms and the gases including the noble and quenching gases, and R_D is the spin diffusion relaxation caused by the collision with the wall of the cell. Under the conditions of the magnetic field existing and changing slowly, by setting the left side of Equation (1) to zero, the steady-state solution in the $x$direction can be considered as [24]:

$$P_{x}^{e} = \frac{R_{P}(\beta_{y} + \beta_{x}\beta_{z}) + R_{m}(1 + \beta_{x}^{2})}{R_{tot}(1 + \beta_{x}^{2} + \beta_{y}^{2} + \beta_{z}^{2})} \quad (3)$$

where $\vec{\beta} = \gamma^{e}\vec{B}/R_{tot}$ is the magnetic field parameter, $\left|\vec{\beta}\right| \ll 1$, and R_{tot} is the total relaxation rate. The output signal of the SERF atomic spin magnetometer can be obtained by applying the lock-in amplifier to extract the first harmonic, and the signal is proportional to the optical rotation angle θ [25]:

$$I = I_0 \sin^2[\theta + \alpha\sin(\omega_{mod}t)] \quad (4)$$

$$S_{out} = 2I_0\theta\alpha \quad (5)$$

$$\theta = \frac{\pi}{2} lnr_{e}cfP_{x}^{e}\{-\mathrm{Im}[V(\nu - \nu_{D1})] + \mathrm{Im}[V(\nu - \nu_{D2})]\} \quad (6)$$

where I_0 is the light intensity, which is detected by the photoelectric detector placed in the x direction, α is the modulation amplitude of the Faraday modulator, ω_{mod} is the modulation frequency, l is the distance of the light propagating through the alkali-metal cell, n is the density number of the alkali-metal atoms, which is closely related to the temperature T in the cell, r_e is the electron classical radius, c is the speed of light, f is the oscillator strength, V is the Voigt profile, and ν_{D1} and $\nu_{D2}$are the resonance frequencies of the D_1 and the D_2 transitions, respectively. Equations (3) and (6) are inserted into Equation (5), it can be obtained:

$$S_{out} = \alpha I_0 nK \frac{R_{P}(\beta_{y} + \beta_{x}\beta_{z}) + R_{m}(1 + \beta}{R_{tot}(1 + \beta_{x}^{2} + \beta_{y}^{2} + \beta_{z}^{2})} \quad (7)$$

$$K = \pi lr_{e}cf\{-\mathrm{Im}[V(\nu - \nu_{D1})] + \mathrm{Im}[V(\nu - \nu_{D2})]\} \quad (8)$$

where K is defined as the calibration coefficient. Equation (7) is described as the output signal.

Noises Influencing on Performance of SERF Atomic Spin Magnetometer

The above output signal indicates that the magnetic field included in the magnetic field parameter $\vec{\beta} = \gamma^{e}\vec{B}/R_{tot}$, the temperature which directly impacts the density number of the alkali metal atoms and the relaxation rate, and the light which can affect the pumping rate and the light shift, are the main noise sources influencing on the output signal and the sensitivity of the SERF atomic spin magnetometer. Thus, analyzing and suppressing the noises and optimizing the operation temperature are very necessary.

Magnetic Field Noise

Light shift referring to the AC Stark shift is equivalent to a virtual magnetic field. The magnetic field including the residual magnetic field in the magnetic shielding and the light shift is one of the noise sources restricting the sensitivity. The SERF magnetometer works at a weak magnetic environment, so a larger residual magnetic field may cause a low sensitivity, and even make the magnetometer not work normally. Fortunately, the residual magnetic field can be compensated by the magnetic field produced by the magnetic coils. According to the steady state Equation (3), the simulation of the magnetic field compensation process is shown in Figure 3.

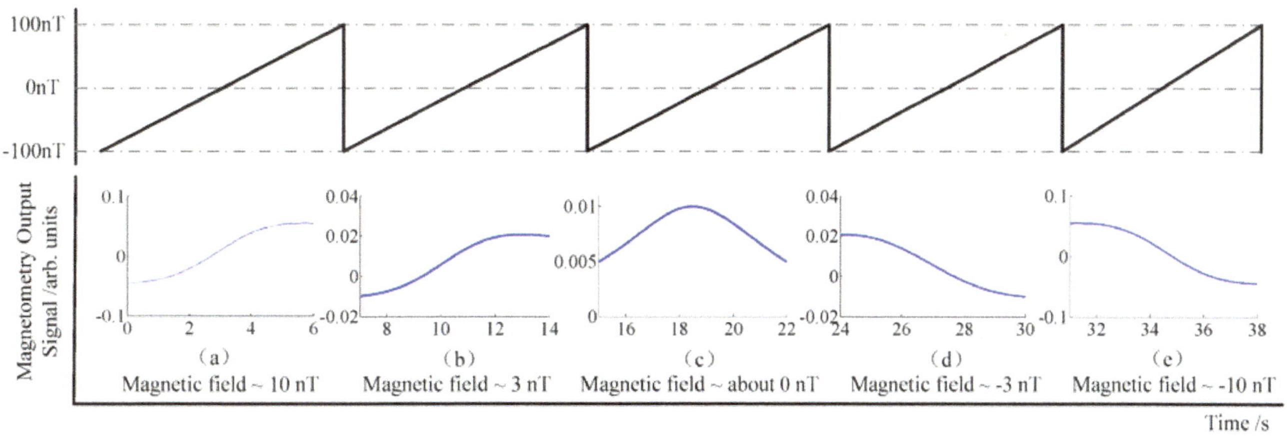

Figure 3. Simulation of the magnetic field compensation process. The magnetic field is changing from −100 nT to 100 nT by scanning the sawtooth wave.

As Figure 3 shown, the method and the steps of the residual magnetic field compensation are described as follows:

- During the simulation process, the residual magnetic field in the x direction is assumed as 10 nT, as shown in Figure 3a. A changing magnetic field from −100 nT to 100 nT is scanning in the z direction, meanwhile, the offset of the magnetic field in the x direction needs to be adjusted synchronously. By observing the output signal, it can be known that if the residual magnetic field in the shielding is counteracted gradually, the output signal will reduce, as shown in Figure 3b; otherwise, the output signal will increase. Continue to adjust the offset of the magnetic field in the x direction until the output signal reaches the minimum value, as shown in Figure 3c. After passing through this minimum value, the pattern of the output signal overturns, presently, the signal's amplitude increases gradually, as shown in Figure 3d,e, respectively. The minimum value of the output signal is considered as the compensation point in the x direction.
- In a similar way, the scanning field condition of the z direction is kept invariant and the offset of the magnetic field in the y direction adjusted to achieve the minimum output value in this direction.
- After compensating the magnetic field in the x and y directions by using steps 1 and 2, a scanning field sawtooth wave from −100 nT to 100 nT is applied to the x direction. The offsets of the magnetic field in the z and y directions are adjusted until the minimum output values are achieved, respectively.
- Steps 1 to 3 are repeated until the offset values of three directions are approximately constant. By using this compensation method, the residual magnetic field in the three directions can be counteracted, and the residual magnetic field noise can also be suppressed.

The energy levels can be shifted by the light field via the AC Stark shift, and this is known as the light shift. The atomic energy level shift, which originates from the interaction between the alkali metal spins and the applied AC electric field from the circularly-polarized light, is equivalent to a Zeeman energy shift due to a fictitious magnetic field in the direction of the photon spin, and it is impacted by the power and the frequency of the pump laser [24,25],

$$L = \frac{-\pi r_e c f \Phi}{\gamma^e} \mathrm{Im}[V(\nu - \nu_0)] s \tag{9}$$

where L is the virtual magnetic field, $\Phi = \frac{P}{h\nu}$ is the luminous flux, which is closely related with the intensity P and the frequency v of the pump laser, h is the Planck constant, v_0 is the frequency at the resonant peak of the alkali-metal atom, and s is the photon spin.

Figure 4 shows that the simulation of the relation between the light shift and the frequency of the pump beam. With changing frequency, the light shift can pass through a zero point, which is at the resonant peak of 389,286.25 GHz.

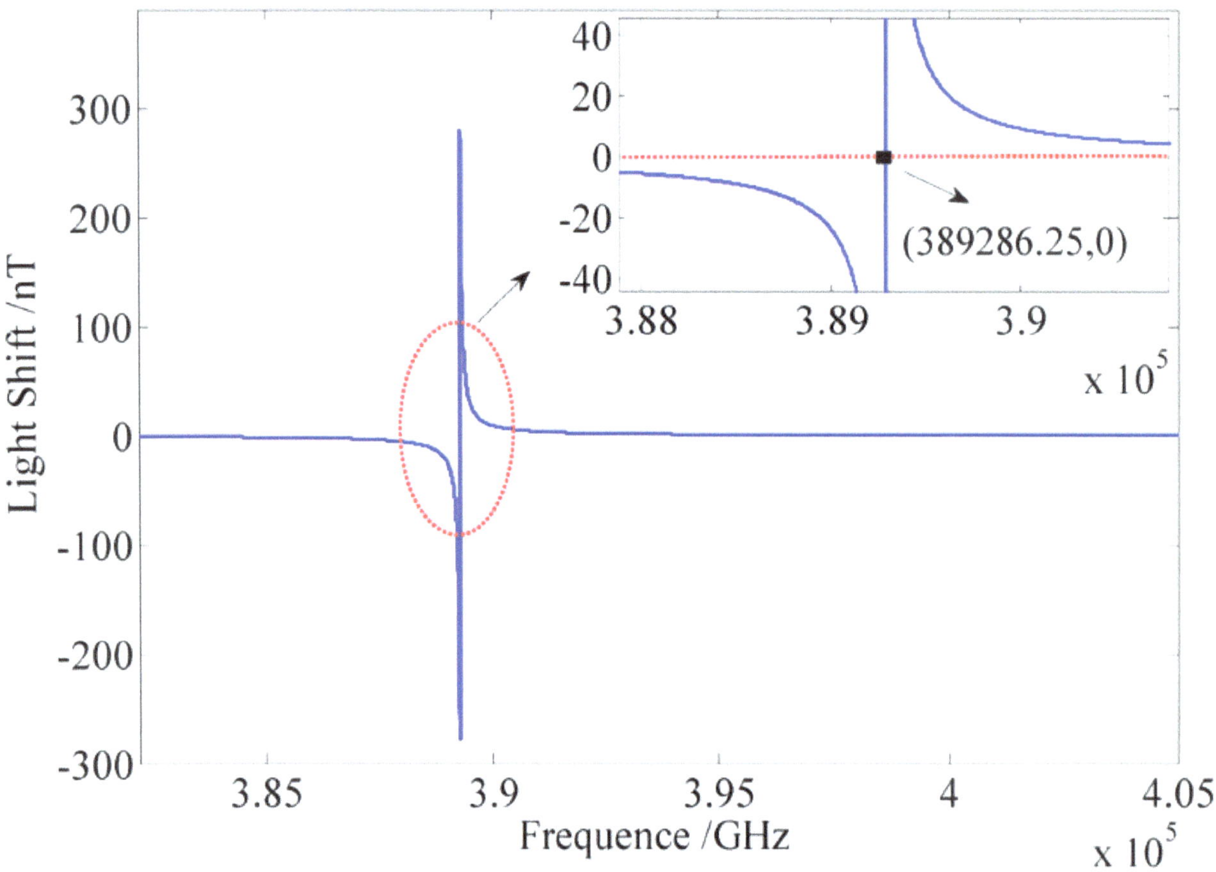

Figure 4. Simulation of the relation between the light shift and the frequency of the pump laser. The power of the pump laser is 100 mW, and the frequency at the resonant peak of potassium atom is 389,286.25 GHz.

Temperature Optimization

The alkali-metal atomic density number and the total relaxation depend directly on the temperature in the cell. On the one hand, with increasing temperature, the atomic number density increases rapidly; on the other hand, the spin-destruction relaxation rate and the diffusion relaxation rate to walls can be directly increased due to the faster thermal velocity at a high temperature. Thus, the total relaxation rate increases. The greater the total relaxation rate is, the lower the atomic spin magnetometer's sensitivity will be. Therefore, optimizing the heating temperature is particularly important.

Take the potassium atomic magnetometers as examples. Generally, the electron spin relaxation time includes the transverse relaxation time T_2 and the longitudinal relaxation time T_1. The slowing-down factor q affects the relaxation. The relationships among these relaxation times are shown as follows [19]:

$$\frac{1}{T_1} = \frac{1}{q}(R_{sd} + R_P + R_m) \tag{10}$$

$$\frac{1}{T_2} = \frac{1}{T_1} + \frac{1}{q_{se}} R_{se} + R_{gr} \tag{11}$$

$$\frac{1}{q_{se}} = \frac{2I(2I-1)}{3(2I+1)^2} \tag{12}$$

$$R_{tot} = \frac{1}{T} 10^{21.866 + A_K - B_K/T} \cdot \sigma_{K-K} \cdot \sqrt{\frac{8k_B T}{\pi M_{K-K}}} + \frac{P_{He}}{k_B T} \cdot \sigma_{K-He} \cdot \sqrt{\frac{8k_B T}{\pi M_{K-He}}} + \frac{P_{N_2}}{k_B T} \cdot \sigma_{K-N_2} \cdot \sqrt{\frac{8k_B T}{\pi M_{K-N_2}}} + R_P + R_m \tag{13}$$

$$R_{se} = \frac{(\gamma^e B)^2 [q^2 - (2I+1)^2]}{2 n v_e \sigma_{se}^e q^2} \tag{14}$$

where $\frac{1}{q_{se}}$ is the spin-exchange broadening factor, R_{se} is the spin-exchange relaxation rate, R_{gr} is the relaxation due to the magnetic field gradient across the vapor cell, which is a tiny amount that is neglected due to the cell placement in the center of a uniform magnetic field, I is the nuclear spin quantum number, T is the temperature in cell, A_k = 4.402 and B_k = 4453 are the parameters, $\sigma_{species}$ is the effective collisional cross-section, K_B is the Boltzmann constant, M is the reduced mass, $P_{species}$ is the gas pressure, v_e is the relative thermal velocity between the alkali metals atoms, and σ_{se}^e is the spin exchange collision cross section. One of the most striking features of the electron spin in the SERF regime is that the spin exchange is suppressed. In other words, if the spin-exchange relaxation rate R_{se} is suppressed, the transverse relaxation time T_2 will be approximately equal to the longitudinal relaxation time T_1, and the atomic spin SERF regime will be realized. According to Equations (10)–(14), we know that both the transverse relaxation time T_1 and the longitudinal relaxation time T_2 are related to the temperature T. The simulation of the atomic magnetometer realizing the SERF regime is shown in Figure 5.

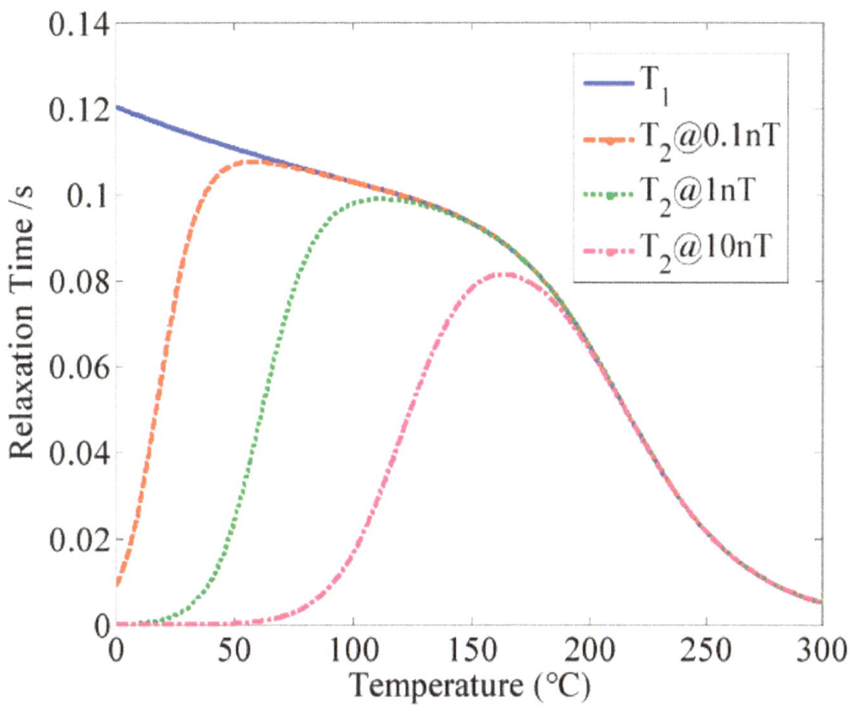

Figure 5. Simulation of the SERF atomic spin magnetometer realizing the SERF regime.

Under the different magnetic field environment, with increasing temperature, the transverse relaxation time T_2 and the longitudinal relaxation time T_1 both overlap, which means that the magnetometer realizes the SERF regime. Figure 5indicates that under the magnetic field of 1 nT, the temperature realizing the SERF regime is approximately 180 °C.

Light Field Noise

In this paper, we mainly consider the fluctuations of the frequency and the power from the pump beam, which can cause the pumping rate, the polarization rate projection and the relaxation rate to be output by the forms of noises. In addition, the laser frequency fluctuation also can cause the light shift, as shown in Figure 6, where when the pumping power fluctuates from 20 mW to 25 mW, the polarization transforms from 28% to 33%.

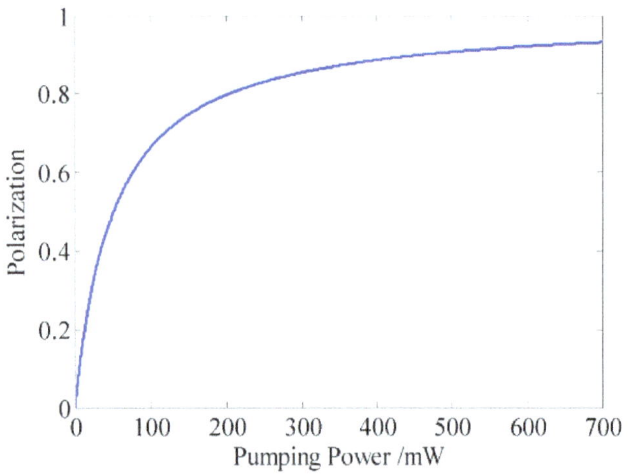

Figure 6. Relationship between the pumping power and the polarization.

If the pumping power fluctuates from 95 mW to 100 mW, the polarization will change from 65% to 66%. It is very important to stabilize the pumping power, especially in the low pumping power condition. According to Equation (9) and Figure 4, we can know that if the frequency fluctuation transforms to 472 GHz, the light shift changes about 4 nT. However, these two noises can be suppressed by employing the frequency and the power stabilization systems, as shown in Figure 7and Figure 8, respectively.

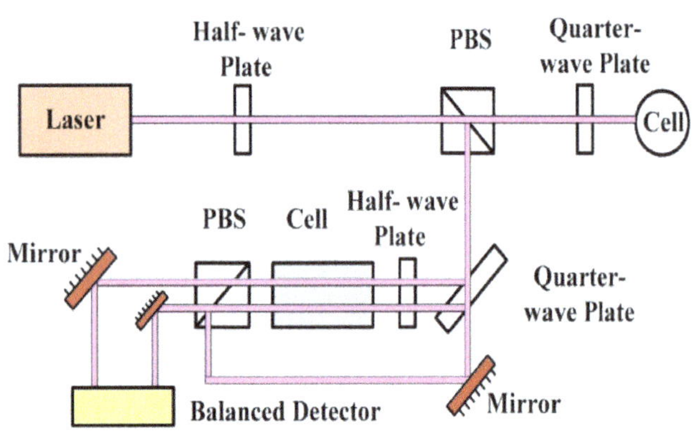

Figure 7. Schematic diagram of the frequency stabilization based on the saturation absorption method.

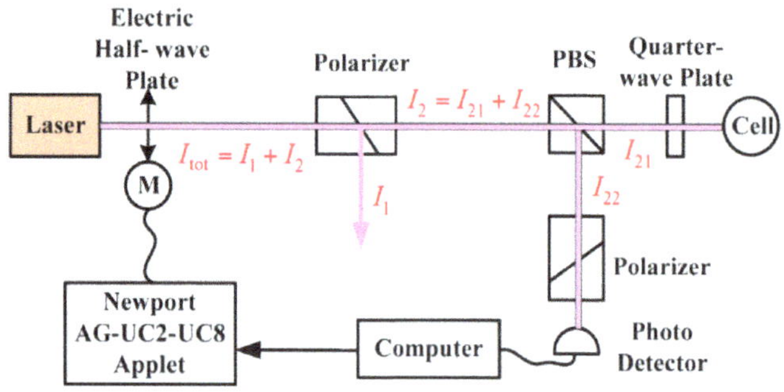

Figure 8. Schematic diagram of the power stabilization system. Newport AG-UC2-UC8 Applet is the piezo motor driven components.

$$I_{tot} = I_1 + I_2, \quad I_2 = I_{21} + I_{22} \tag{15}$$

where I_{tot} is the total light intensity, which is the sum of light intensities I_1 and I_2. I_1 is the light intensity with fluctuation, and I_2 is the stable light intensity. The stable light intensity is divided into two beams I_{21} and I_{22}, which are used to pump the atomic spin and monitor and stabilize the light intensity, respectively. The light intensity is transmitted to the computer by the photoelectric detector. The stable I_2 can be stabilized through controlling the electric wave plate by the PID algorithm, while, the fluctuation noise is allotted to I_1. Hence, the stable light intensity is successfully achieved. The experimental result of the light intensity stability is shown in Figure 9.

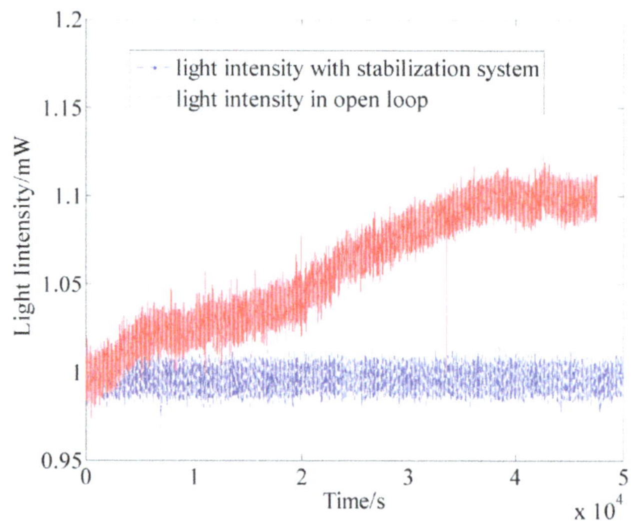

Figure 9. Experiment results of the light intensity stability.

Figure 9 shows that the power stabilization system could effectively suppress the power fluctuation. Compared with these two test results, the light intensity had a 10% fluctuation without the power stabilization system.

EXPERIMENTAL AND ANALYSIS

The experimental apparatus is described as Figure 2. In order to analyze the effects of the noises on the sensitivity, relevant experiments were implemented. A 22.5 pT calibration signal at 25 Hz was applied to the y direction in these experiments. After FFT, the output signals were processed by divided a normalized frequency-response function with a fit to $A/(f_0^2 + B_{bandwidth}^2)^{1/2}$, where A is the coefficient, f_0 is the frequency, and $B_{bandwidth}$ is the bandwidth [17]. By analyzing the signal to noise ratio, the sensitivities of the magnetometer under the different situations were achieved. The experimental results are shown in Figure 10. Figure 10a indicates that under the condition of the residual magnetic field coarse compensation, the sensitivity was about 20 fT/Hz$^{1/2}$. However, when the residual magnetic field is compensated accurately, the sensitivity can be promoted efficiently, as shown by the dotted line in Figure 10a.

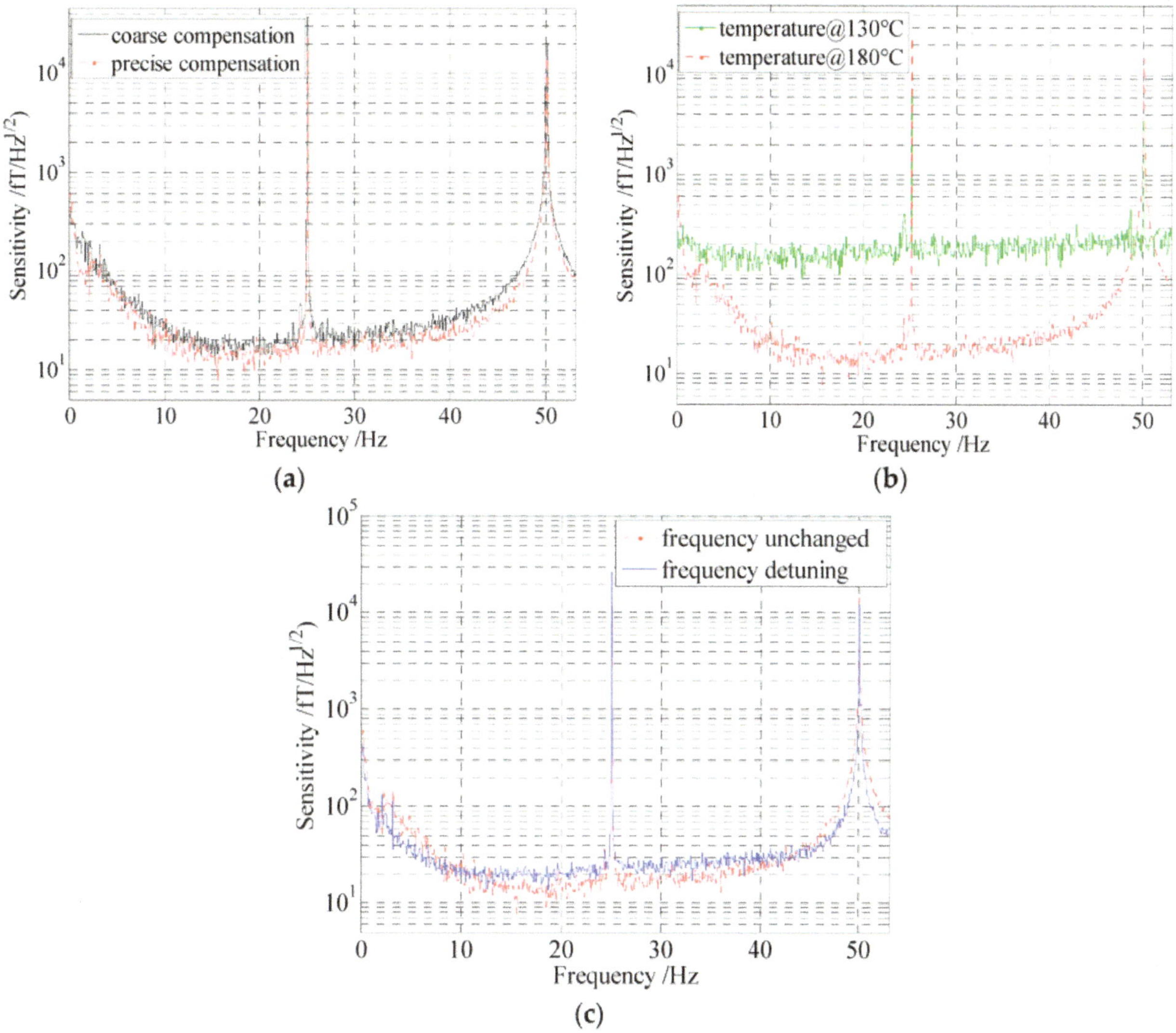

Figure 10. Experimental results of sensitivity comparison under the different situations. For each test, only one condition was changed. The dotted line acted as a comparison benchmark. (**a**) Optimization results of the residual magnetic field compensation in the magnetic shielding; (**b**) Optimization results of the temperature in cell; (**c**) Comparison results of the frequency detuning. On the basis of the un-optimized frequency, the frequency detuning was approximately 40 GHz.

Figure 10b shows that at 130 °C, the sensitivity was about 150 fT/Hz$^{1/2}$, it was much lower than that at the temperature in the SERF regime (180 °C).

Figure 10c illustrates that at the unoptimized frequency, which is detuned approximately 40 GHz, the sensitivity was about 22 fT/Hz$^{1/2}$. This is mainly attributed to the light shift, which was caused by the frequency detuning. Equation (3) is simplified and rewritten as:

$$P_x^e = \frac{R_P\left[(B_z - L_z)B_x + \varepsilon B_y\right]}{R_{tot}\left[(B_z - L_z)^2 + \varepsilon^2 + B_x^2 + B_y^2\right]} \quad (16)$$

where $\varepsilon = R_{tot}/\gamma^e$, B_x, B_y and B_z are the total magnetic fields in the x, y and z directions, respectively, B_{resz} is the residual magnetic field, $B_{resz} = B_z - L_z$, and L_z is the light shift in the z direction. According to the following fitting formula, the light shift can be obtained:

$$y = a\frac{(x-b)e + cB_y}{(x-b)^2 + c^2 + e^2 + B_y^2} + d \quad (17)$$

where b is the light shift, a, c, d and e are the fitting coefficients. After compensating the residual magnetic field in all three directions, a 15 nT bias magnetic field was utilized in the y direction and the magnetometer's output signal was measured by applying B_z values from −1000 nT to 1000 nT. According to the fitting formula (16), the light shift L_z value was approximately 2 nT, when the pumping power was 50 mW. The test result of the light shift L_z is shown in Figure 11. The light shift was equivalent to a virtual magnetic field, which caused the residual magnetic field to increase while the sensitivity was reduced.

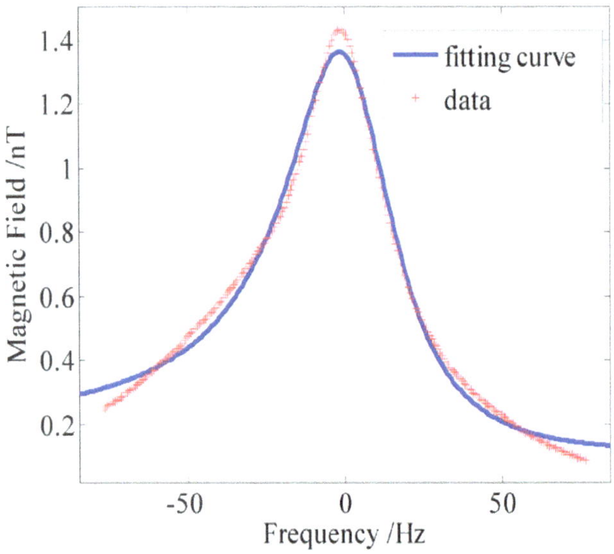

Figure 11. Test result of the light shift L_z. The pumping power was 50 mW.

After suppressing the noises by the methods presented in this paper, the optimal sensitivity of 13 fT/Hz$^{1/2}$, which was mainly limited by the Johnson noise, was achieved, as shown in Figure 12. The bandwidth of our SERF atomic magnetometer was approximately 40 Hz. In addition, there is a significant noise increase observed below 10 Hz. Unfortunately, at present, we don't know the cause of this problem. In future, we will attempt to clarify this phenomenon through further study.

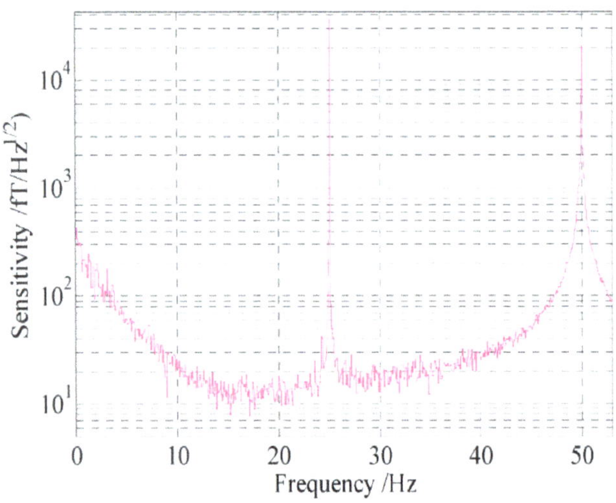

Figure 12. Optimal sensitivity of the atomic magnetometer based on suppressing the noises.

CONCLUSIONS

The principle of the SERF atomic spin magnetometer was introduced in this paper. According to the established output signal, the factors that influence he output signal and sensitivity were analyzed. The magnetic field, including the residual magnetic field and the light shift, the temperature and the light field were the main factors influencing the sensitivity. The residual magnetic field could be compensated and the light shift can be suppressed by adjusting the laser frequency to the light shift zero point. The operational temperature, which is related to the atomic density number, the polarization rate projection and the relaxation rate, should be 180 °C. The stabilization systems of the frequency and the power were presented, and the power stabilization experimental results showed that the power fluctuations could be controlled in 10%. Finally, contrast experiments on the sensitivity were implemented, and the results indicated that after suppressing the noises, an optimal sensitivity of 13 fT/Hz$^{1/2}$ was successfully achieved.

ACKNOWLEDGMENTS

This study was supported by the National Natural Science Foundation of China (Grant No. 51375087), the Fundamental Research Funds for the Central Universities of Ministry of Education of China (Grant No. CXLX13_083), and the Scientific Research Foundation of Graduate School of Southeast University (Grant No. YBJJ1573). The authors thank the helpful suggestions of experts from Inertial national lab in Beihang University, China.

AUTHOR CONTRIBUTIONS

Xiyuan Chen and Hong Zhang contributed equally in the work of the experiments, data acquisition and analysis, and manuscript writing. Sheng Zou contributed to the design of the experiments, analysis of the results, the writing of the manuscript, and modification the manuscript.

CONFLICTS OF INTEREST

The authors declare no conflict of interest.

REFERENCES

1. Ito, Y.; Ohnishi, H.; Kamada, K.; Kobayashi, T. Development of an optically pumped atomic magnetometer using a K-Rb hybrid cell and its application to magnetocardiography. *AIP Adv.* **2012**, *2*, 1–5.
2. Sander, T.H.; Preusser, J.; Mhaskar, R.; Kitching, J.; Trahms, L.; Knappe, S. Magnetoencephalography with a chip-scale atomic magnetometer. *Biomed. Opt. Express* **2012**, *3*, 981–990.
3. Savukov, I.M.; Zotev, V.S.; Volegov, P.L.; Espy, M.A.; Matlashov, A.N.; Gomez, J.J.; Kraus, R.H. MRI with an atomic magnetometer suitable for practical imaging applications. *J. Magn. Reson.* **2009**, *199*, 188–191.
4. Savukov, I.; Karaulanov, T. Anatomical MRI with an atomic magnetometer. *J. Magn. Reson.* **2013**, *231*, 9–45.
5. Tiporlini, V.; Alameh, K. High sensitivity optically pumped quantum magnetometer. *Sci. World J.* **2013**, *2013*, 1–8.
6. Jiménez-Martínez, R.; Knappe, S.; Kitching, J. An optically modulated zero-field atomic magnetometer with suppressed spin-exchange broadening. *Rev. Sci. Instrum.* **2014**, *85*, 1–5.
7. Baranga, A.B.A.; Hoffman, D.; Xia, H.; Romalis, M.V. An atomic magnetometer for brain activity imaging. *IEEE* **2005**,*5*, 417–718.
8. Robert, W.I. *The Development of a Multichannel Atomic Magnetometer Array for Fetal Magnetocardiography*; University of Wisconsin: Madison, WI, USA, 2012.

9. Brown, J.M. *A New Limit on Lorentz- and CPT-Violating Neutron Spin Interactions Using a K-3He Comagnetometer*; University of Princeton: Princeton, NJ, USA, 2011.
10. Moi, L.; Cartaleva, S. Sensitive magnetometers based on dark states. *Europhys. News* **2012**, *43*, 24–27.
11. Zhang, H.; Zou, S.; Chen, X.; Li, Y. Research on the influence of alkali polarization on an all-optical atomic spin magnetometer based on hybrid pumping potassium and rubidium. *J. Korean Phys. Soc.* **2015**, *66*, 1212–1217.
12. Lu, C.C.; Huang, J.; Chiu, P.K.; Chiu, S.L.; Jeng, J.T. High-sensitivity low-noise miniature fluxgate magnetometers using a flip chip conceptual design. *Sensors* **2014**, *14*, 13815–13829.
13. Wickenbrock, A.; Jurgilas, S.; Dow, A.; Marmugi, L.; Renzoni, F. Magnetic induction tomography using an all-optical^{87}Rb atomic magnetometer. *Opt. Lett.* **2014**, *39*, 6367–6370.
14. Fang, J.C.; Wang, T.; Quan, W.; Yuan, H.; Zhang, H.; Li, Y.; Zou, S. *In situ* magnetic compensation for potassium spin-exchange relaxation-free magnetometer considering probe beam pumping effect. *Rev. Sci. Instrum.* **2014**, *85*, 1–7.
15. Affolderbach, C.; Stähler, M.; Knappe, S.; Wynands, R. An all-optical, high-sensitivity magnetic gradiometer. *Appl. Phys. B Lasers Opt.* **2002**, *75*, 605–612.
16. Zou, S.; Zhang, H.; Chen, X.; Quan, W. Magnetization produced by spin-polarized xenon-129 gas detected by using all-optical atomic magnetometer. *J. Korean Phys. Soc.* **2015**, *66*, 887–893.
17. Kominis, I.K.; Kornack, T.W.; Allred, J.C.; Romalis, M.V. A subfemtotesla multichannel atomic magnetometer. *Nature***2003**, *422*, 596–599.
18. Ito, Y.; Ohnishi, H.; Kamada, K.; Kobayashi, T. Sensitivity improvement of spin-exchange relaxation free atomic magnetometers by hybrid optical pumping of potassium and rubidium. *IEEE Trans. Magn.* **2011**, *47*, 3550–3553.
19. Allred, J.; Lyman, R.; Kornack, T.; Romalis, M.V. High-sensitivity atomic magnetometer unaffected by spin-exchange relaxation. *Phys. Rev. Lett.* **2002**, *89*, 1–4.
20. Dang, H.B.; Maloof, A.C.; Romalis, M.V. Ultrahigh sensitivity magnetic field and magnetization measurements with an atomic magnetometer. *Phys. Rev. Lett.* **2010**, *97*, 1–3.
21. Kamada, K.; Ito, Y.; Ichihara, S.; Mizutani, N.; Kobayashi, T. Noise reduction and signal-to-noise ratio improvement of atomic magnetometers with optical gradiometer configurations. *Opt. Express* **2015**, *23*, 6976–6987.
22. Lee, S.; Kim, J.; Moon, M.K.; Lee, K.H.; Lee, S.W.; Ino, T.; Vadim, R.S.; Lee, M.; Kim, G. A compact spin-exchange optical pumping system for 3He polarization based on a solenoid coil, a VBG laser diode, and a cosine theta RF coil. *J. Korean Phys. Soc.* **2013**, *62*, 419–423.
23. Kim, K.; Begus, S.; Xia, H.; Lee, S.-K.; Jazbinsek, V.; Trontelj, Z.; Romalis, M.V. Multi-channel atomic magnetometer for magnetoencephalography: A configuration study. *NeuroImage* **2014**, *89*, 143–151.
24. Seltzer, S.J. *Developments in Alkali-Metal Atomic Magnetometry*; University of Princeton: Princeton, NJ, USA, 2008.
25. Kornack, T.W. *A Test of CPT and Lorentz Symmetry Using a $K-^3$ He Co-Magnetometer*; University of Princeton: Princeton, NJ, USA, 2005.

CHAPTER 16

MICROCANTILEVER DISPLACEMENT MEASUREMENT USING A MECHANICALLY MODULATED OPTICAL FEEDBACK INTERFEROMETER

Francisco J. Azcona, Ajit Jha, Carlos Yáñez, Reza Atashkhooei and Santiago Royo

Centre for Sensors, Instruments and Systems Development (CD6)-UPC BarcelonaTech, Rambla St. Nebridi 10, Terrassa E-08222, Spain

ABSTRACT

Microcantilever motion detection is a useful tool for the characterization of the physical, chemical and biological properties of materials. In the past, different approaches have been proposed and tested to enhance the behavior, size and simplicity of microcantilever motion detectors. In this paper, a new approach to measure microcantilever motion with nanometric resolution is presented. The proposed approach is based on the concept of mechanically-modulated optical feedback interferometry, a technique that has shown displacement measurement capabilities well within the nanometric scale and that, due to its size, compactness and low cost, may be a suitable choice for measuring nanometric motions in cantilever-like sensors. It will be shown that the sensor, in its current state of development, is capable of following a cantilever sinusoidal trajectory at different sets of frequencies ranging up to 200 Hz and peak to peak amplitudes up to λ/2 with experimental resolutions in the λ/100 range.

INTRODUCTION

The measurement of physical, chemical and biological properties, especially in the microscopic and submicroscopic scales, typically relies on the use of indirect estimation methods. In the last three decades, cantilevers [1,2,3] and transducers with a mass-spring-like behavior [4] have been used, e.g., for the characterization of micron and sub-micron sample profiles, surface force measurements (e.g., van der Waals, ionic, electrostatic) [5,6], the detection of protein interactions [7] or measuring adhesion forces between cells [8,9]. In all of these cases, the transducer motion is used for retrieving a desired property by relating its behavior to a defined mechanical model with well-known parameters.

In particular, cantilever and microcantilever transducers are interesting indirect measurement tools, since it is possible to attach a different type of tip at the free end of the cantilever, allowing versatility in the materials and analysis to be performed. Furthermore, the mechanical properties of cantilevers are simple enough to be easily estimated using well-known classical methods, such as Euler–Bernoulli [10] or Timoshenko beam theory [11]. Besides theoretical methods, other approaches have been proposed [12,13,14] for the characterization of the cantilever dynamic properties, e.g., the

Citation: Azcona FJ, Jha A, Yáñez C, Atashkhooei R, Royo S. Microcantilever Displacement Measurement Using a Mechanically Modulated Optical Feedback Interferometer. Sensors. 2016; 16(7):997. DOI: 10.3390/s16070997

quality factor Q and the resonance frequency fr, resulting in the possibility of a dynamic analysis of the cantilever motion, such as the ones described in non-contact and tapping mode atomic force microscopy (*AFM*) techniques.

Provided that the cantilever is properly characterized using one of the methods mentioned above, it is only necessary to measure the cantilever motion with enough resolution to provide an estimation of the force applied at the tip of the cantilever and, furthermore, to relate the obtained force to a desired property. To measure cantilever motion, different types of sensors and technologies have been proposed [15]. Typically, optical sensors are preferred over other technologies, since they do not exert an external loading over the cantilever, thus preventing possible bias on the measurement.

The principle of the optical lever technique [16] is one of the most used methods for measuring microcantilever deflection. This technique allows a high sensitivity of the cantilever deflection angle by sensing the position of the laser spot relative to the central position of a four quadrant photodiode. The main problems it presents are the requirement of a metal coated cantilever (which may induce deformations due to the bimetallic effect [17,18]) and a delicate alignment of the three elements involved (laser, cantilever and photodiode).

Classical interferometric methods have also been suggested for measuring cantilever motion. They provide high resolution, usually below 1 Å, at the expense of requiring an expensive and bulky setup, which might prove difficult to align. To partially solve this issue, particularly in the microscale, optical fiber interferometers have been proposed. Nevertheless, the use of optical fibers may induce additional errors in the shape of thermal drifts in low time scales [19], and it will also require special handling to prevent stress during the positioning.

In order to simplify the alignment constraints in cantilever deflection sensing, we propose the use of a sensor based on optical feedback interferometry (*OFI*) [20,21,22]. Sensors based on the OFI technique can be described as compact and self-aligned, thus minimizing alignment issues at least in one degree of freedom while the sensor is kept in a small encapsulation.

Optical feedback interferometry has been previously explored as a method for estimating cantilever deflection. In [23], Ovryn and Andrews measured micrometric scale displacements on a millimetric length cantilever using a phase shifting approach. In this method, a calibration for different amplitudes and frequencies is performed using an electro-optic modulator. After a calibration and a phase unwrapping process in the cases where amplitudes are larger than half-wavelength (λ/2), the net peak amplitude of the deformation is obtained. According to their results, it is possible to achieve similar results to those obtained with holographic interferometry in terms of peak to peak amplitude and main resonance frequency. In [24], Larsson et al. used a vertical cavity surface emitting laser (*VCSEL*) and an external photodiode (*PD*) to characterize the vibration of a polymer micrometric-sized cantilever. In their experiment, the VCSEL is placed at approximately 35 μm from the cantilever. The cantilever is then oscillated by a piezoelectric actuator in a range of 1.2 μm. The OFI signal is then recovered using a voltage junction approach [22], and it is also recovered using an external PD placed above the cantilever. As a result of the motion, a cyclic sinusoidal-like behavior correspondent to OFI is observed. In some sections of the displacement they observed linearity regions with an equivalent slope of 45 mV/nm for the PD measurements. This ratio, however, is only found in small intervals of the measurement (±25 nm), thus requiring locking the measurement region around this point to provide a linear response of the cantilever deflection. In this work, we propose the use of a method based on the differential optical feedback interferometry concept [25], further on referred to as mechanically-modulated OFI. Through this approach, it is possible to obtain resolutions in the order of λ/100. In comparison to the method presented in [23], mechanically-modulated OFI presents a smaller setup, which, due to the decrease in the number of optical elements, reduces the alignment constraints. Furthermore, the complexity of the method is reduced, as the technique depicted in [23] requires at least five intensity measurements. Comparing mechanically-modulated OFI to the method described in [24], the technique presented in this paper removes the need of an external photodiode and increases the measurement range, as the method allows estimating displacements up to an amplitude of λ/2, while the linearity region shown in [24] is limited to λ/8 approximately. The inclusion of a focusing lens also improves the system, as it allows changes of the distance between the cantilever and the sensor, as well as a degree of tuning for the amount of feedback.

In the next section, a brief review of OFI and the mechanically-modulated OFI is presented to enhance the readability of the paper. In Section 2, a complete description of the sensor and the setup used for this work is presented. Furthermore, the calculation methods and the steps followed during the measurement are described. Next, in Section 3, the results obtained are shown. A brief discussion of the results obtained and the strengths and drawbacks of the method proposed are commented on in Section 4.

MATERIALS AND METHODS

Mechanically-Modulated Optical Feedback Interferometry

Optical feedback interferometry, also known as self-mixing interferometry (SMI), is a well-known technique capable of measuring changes in the optical path of a laser beam with a typical resolution of (λ/2) for displacement measurements using a fringe counting method (FCM). OFI has been successfully implemented in diverse fields, which range from motion in solid and liquid interfaces [26,27,28] to biological applications [29].

The OFI effect is produced whenever a laser electric field is partially backscattered from an object or particle affected by an optical path change (e.g., displacement). After re-entering the laser cavity, the electric stationary wave and the backscattered wave interact, producing interference phenomena, which is observed as a beat of the laser optical output power (OOP). The OOP can then be retrieved using a photodiode placed in the same package of the laser, or, alternatively, using the voltage variations appearing in the laser junction [22,30].

The OFI process can be modeled using the excess phase equation derived in [20,22,31]:

$$\Delta\phi = (\phi_f - \phi_0)\tau_e + C\sin(\phi_f\tau_e + \arctan\alpha) \quad (1)$$

$$P = P_0[1 + m\cos(\phi_f \tau_e)] \quad (2)$$

where ϕ_0 is the phase of the free running laser, ϕ_f is the phase of the laser under feedback, τ_e is the time of flight of the external cavity, α the linewidth enhancement factor and C the feedback factor [32]. The optical power under feedback P can then be represented as the sum of the laser power before feedback P0 and the back-reflected optical power, which is described as a proportional part ($m \ll 1$) of P_0, which depends on ϕ_f and τ_e.

The feedback factor C is used to discriminate between different working regimes and consequently gives information on the behavior of the OOP AC modulation [22,33]. As the value of C increases, a deformation from the original sinusoidal signal is observed until discontinuities appear, generating the well-known sawtooth-like interferometric pattern (C=1). Beyond this point, other effects, such as hysteresis, fringe loss and even chaos, may appear on the signal [22].

For the case of displacement measurements and specially for the implementation of the techniques to be discussed further on, it is generally of interest to work with a $C \simeq 1$, in order to obtain a high signal to noise ratio (*SNR*) while avoiding potential fringe loss, typical of higher feedback regimes [26]. Working within this regime also allows detecting the target motion direction without ambiguity.

Basic OFI processing methods, such as FCM, are typically limited to a $\lambda/2$ resolution. In order to increase the resolution of the technique in displacement measurements, different methods have been reported [22,34,35,36], allowing one to obtain resolutions down to $\lambda/100$. In the case of vibrations with amplitudes smaller than $\lambda/2$, it has also been shown that it is possible to obtain a responsivity in the range of 4.25 mV/µm and theoretical noise figures in the $pm/\sqrt{Hz}$ scale by using the lock-in and phase nulling method described in [37] at the cost of increasing the complexity of the electronic card readout and requiring large distances with respect to the measurement object. Other methods [22] based on an output power calibration have also shown vibration measurement capabilities within the nanometric scale. These methods, however, require the use of other elements, such as electro-optical modulators.

In order to measure within the nanometric scale, we recently proposed the use of differential OFI [25], which has shown practical resolutions in the order of $\lambda/100$ in laboratory conditions. The concept of differential OFI, however, can also be applied to single laser diode (*LD*) setups, such as the one shown in Figure 1. For the purpose of clarity, we will refer to this method as mechanically-modulated OFI. In this type of configuration, an LD is pointed at the target and subjected to a reference motion during the whole measurement process. The measurement is divided into two steps. First, as shown inFigure 1a, an OFI signal with $C \simeq 1$ is acquired for a known piezoelectric (PZT) motion. Next, in a second step, shown inFigure 1b, the measurement signal is acquired while the target undergoes a sub-$\lambda/2$ motion, while the PZT keeps the motion used during the reference acquisition.

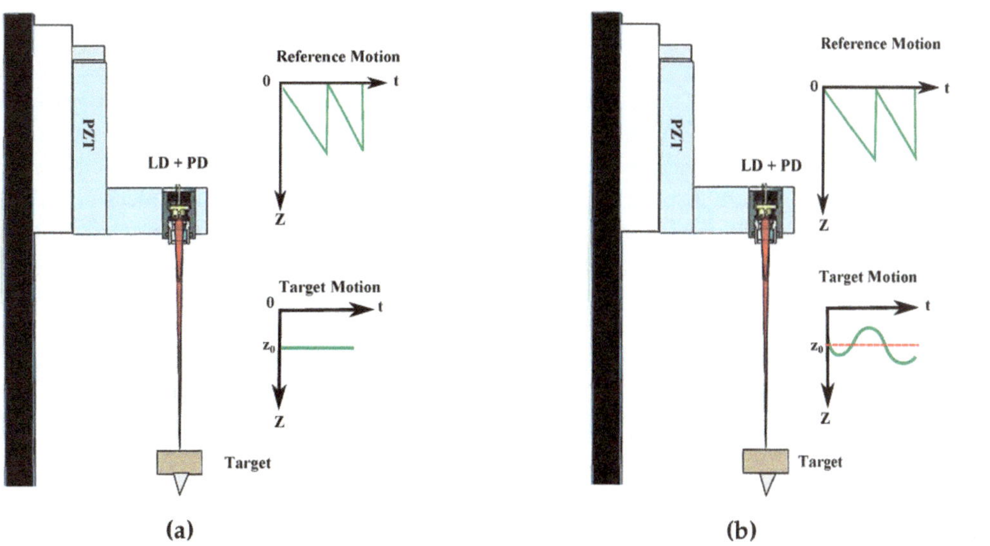

Figure 1. Example of a mechanically-modulated optical feedback interferometry (OFI) setup. (**a**) Reference signal acquired while the target is in a static position with respect to the Y axis; (**b**) The measurement signal is acquired while both the target and the PZT are in motion.

To understand the concept of the technique, consider the effects of setting into motion the displacement stage (e.g., a piezoelectric stage (PZT)) with a peak to peak amplitude $A_{pp} > k\lambda$, where $k \in Z^+ > 2$, in order to create at least 3 measurement points, while the target is kept at rest. As a result of the PZT displacement and given a $C \simeq 1$, a sawtooth-like pattern will be generated on the OOP with sharp transitions (fringes) generated at each $\lambda/2$ displacement. Furthermore, if the reference displacement is linear, the time intervals between two consecutive fringes (Δt_r) should be equal. Assuming no other perturbation is affecting the OFI signal, it is possible to consider the captured data as a reference signal for further measurements. Now, consider that, in addition to the reference displacement, the target starts moving with a peak to peak amplitude $A_t \ll \lambda/2$, hence generating the measurement signal. As a result of the target displacement, the time intervals between consecutive fringes in the OOP are no longer constant. An example of this effect is shown in Figure 2, where the dashed line represents the measurement signal. In Figure 2, as well as in all of the following formulas, the sub-index r is used to define values related to the reference signal, while the sub-index m marks those values related to the measurement signal.

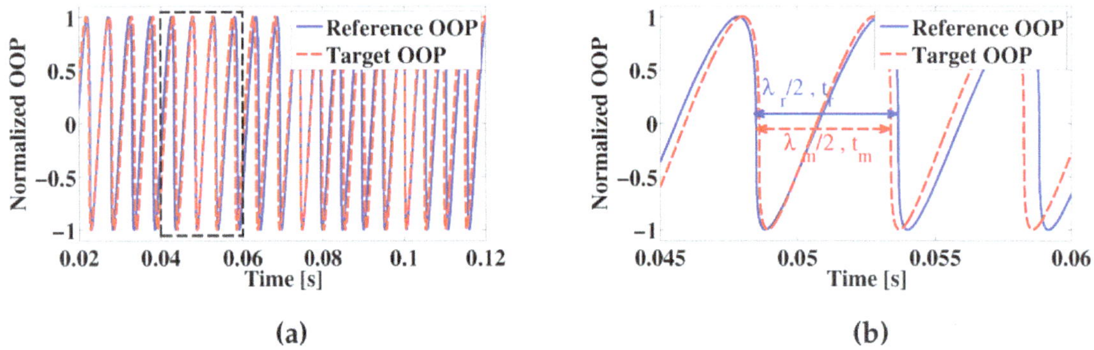

Figure 2. Simulated fringe position change on a measurement signal as a result of a target displacement. The subindex *r* is used for the reference signal and *m* for the measurement signal. (**a**) Comparison between a simulated reference OOP (solid line) and a simulated OOP during target displacement(dashed line); (**b**) Zoom around the region inside of the dashed rectangle in (**a**).

As is shown in [25], it can be proven that for a small displacement Δd, the resolution of the technique can be expressed as:

$$\Delta d = \mp\frac{\lambda_r}{2}\frac{\Delta t}{\Delta t_r} \quad (3)$$

where $\Delta t = \Delta t_m - \Delta t_r$ is the difference between the fringe time intervals in the reference and measurement signals and is limited by the sampling rate of the acquisition device and the relative separation of the fringes. In theory, resolutions better than λ/1000 might be achievable, while in practice [25], resolutions in the order of λ/100 have been obtained.

The sampling frequency of the sensor can be modeled as:

$$f_s = 2\frac{v_r}{\lambda_r} = 4\frac{A_{pp}f_r}{\lambda_r} \quad (4)$$

where vr is the reference velocity of the linear reference displacement. In order to limit the sensor size and to minimize the speckle effect [38] present on large displacements, it is preferred to use a triangular reference displacement motion with amplitudes in the order of 40 μm or lower. Thus, for a triangular displacement, the sampling frequency is proportional to App and the frequency f_r of the motion. Due to limitations in mechanical displacement with high linearity, mechanically-modulated OFI is usually limited to work in a low frequency band reaching up to a few kHz.

Although the above described setup could be bulky compared to the original OFI setup, the sensor can be built in a reduced area. Moreover, the sensor's size may be reduced even further by using an electrical modulation version, such as the one suggested in [39]. For the current implementation, however, the mechanical modulation might present an advantage over the electronic modulation due to its lower dependence on the distance to generate sampling fringes. A more detailed description of the specific methods used in this work to produce the reference signal can be found in Section 2.2.2.

Materials

To test the capabilities and limitations of the mechanically-modulated OFI sensor, the custom setup shown in Figure 3b was built. As is observed in the schematic shown in Figure 3a, the sensor performs the measurement along the Z axis, while it is shinned at the tip of the cantilever. The description of the setup is divided into three main groups of components: positioning elements, the sensor itself and the digital acquisition equipment.

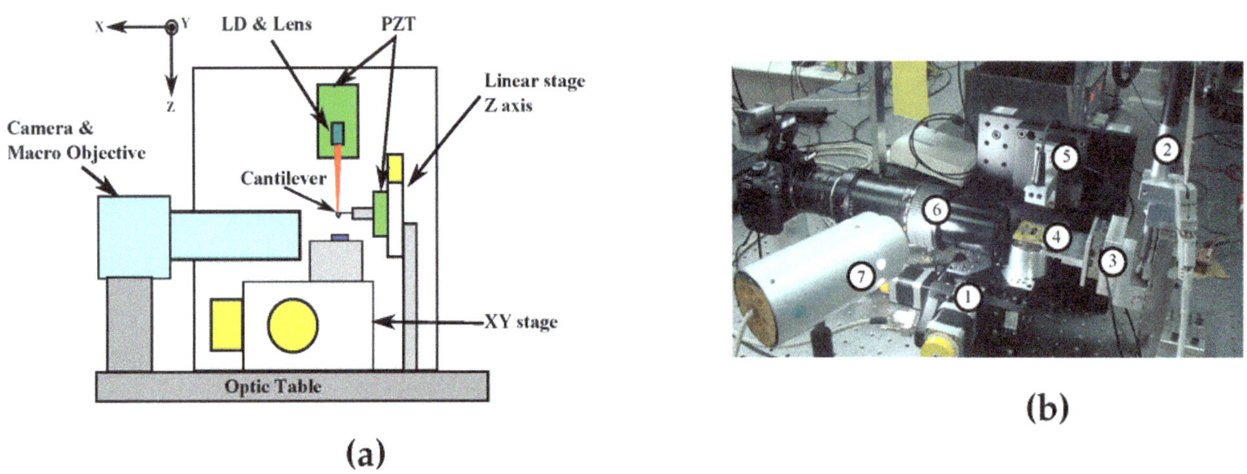

Figure 3. (**a**) Schematic of the mechanically-modulated OFI AFM setup; (**b**) Photograph of the implemented experimental setup: (1) XY translation stage; (2) Z-axis translation stage; (3) piezoelectric vibration displacement stage; (4) cantilever holder; (5) mechanically-modulated OFI sensor; (6) camera with macro objective; (7) white light source for imaging.

The mechanical setup consists of 2 PI XY LS-120 stepper motor tables (labeled as (1) in Figure 3b) with a step resolution of 0.2 μm and a maximum travel length of 40 mm, which will be used in later stages to change the relative position between a sample surface and the cantilever probe. A PI M-227.50 translation stage (labeled as (2)) is used to control the Z-axis positioning of the cantilever with a resolution of 50 nm. A PI-LISA piezoelectric (*PZT*) stage (3) with a maximum travel length of 25 μm is used to generate a sinusoidal motion for the cantilever holder (4) and the cantilever. The PZT includes an internal capacitive sensor with a maximum resolution of 2 nm, which will be used in the measurement for comparison purposes. An *AppNano SHOCONA* cantilever with aluminum coating was selected as the target and experimentally characterized yielding an experimental resonance frequency of 45.29 kHz, a quality factor on air of Q=108.70 and a length of L=225 μm.

The sensor (5) is composed of a *Hitachi HL7851G* multi-quantum well Fabry–Perot laser diode with λ=783.5 nm measured using *Instrument System's SPECTRO 320(D)R5*. The LD is mounted on a Thorlabs *LT230220P-B* collimating tube equipped with a *C230220-B* aspheric lens pair with numerical aperture NA=0.55/0.25. The tube is then attached to another PI-LISA PZT with a maximum travel length of 38 μm using a custom aluminum support, which will be used to oscillate the laser and create the interference fringes. The sensor is placed at an approximate distance of 100 mm from the cantilever. The LD is then connected to a custom-made electronic card, which collects the photodiode current (I_{PD}) and sends it to a current mirror (I mirror in Figure 4) allowing us to obtain two duplicates of IPD. One of the signals is sent to the LD driver and the second one to a passive transimpedance amplifier (*TIA*). After the conversion to voltage, the signal is subjected to a high pass filter with a cutoff frequency to remove the DC part of the signal. Finally, the signal is subjected to a gain stage (GA), and it is digitized. A block representation of the connection scheme is shown in Figure 4.

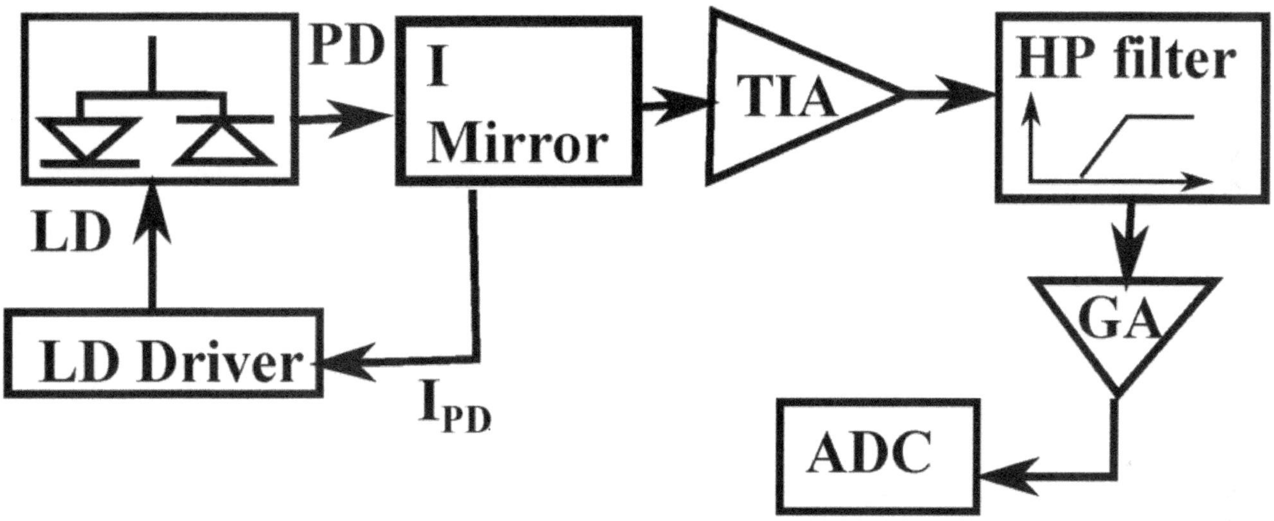

Figure 4. Electronic scheme of the mechanically-modulated OFI analogue acquisition and preconditioning stage. The PD current is passed through a current mirror (I mirror) to prevent loading effects on the transimpedance amplifier stage (TIA).

The acquisition stage uses a *Tektronix DPO2024B* oscilloscope using one channel to recover the OFI signal and a second channel to retrieve the displacement of the 25 μm LISA PZT. The oscilloscope signal is directly captured in the PC using a custom-made interface developed on VC++ 2010.net. Besides the oscilloscope acquisition, the application controls the parameters of a *Tektronix AFG3022B* signal generator, which controls the PZT drivers and commands the XYZ stage positioning. A *Canon EOS1100D* camera equipped with a *MP-E 65mm f/2.8* macro objective (6) and a white light source (7) is used to provide visual feedback regarding the vertical position of the cantilever.

Scanning and Signal Acquisition

Previous to the start of the measurement, a SHOCONA cantilever is placed on a cantilever holder, which is then fixed on a custom mechanical support connected to the PZT, as shown in Figure 3b (3) and (4), then, the LD is focused at the cantilever. The alignment between the cantilever and the laser beam is confirmed by looking at the reflection of the laser spot on the cantilever through the camera, as shown in Figure 5. Once the alignment is confirmed, the XY table is calibrated and set at its origin until a sample is placed on the stage for measurement. After displacing the sample holder and the sample to the measurement coordinates, the cantilever is moved to the required working distance from the sample, and an offline test is performed to check that the OFI signal is working under the desired feedback regime ($C \simeq 1$). After moving up the cantilever to prevent a collision with the sample, the XY table is moved to the initial scanning position. At this point, it is possible to change the starting scanning position relative to the initial preset coordinate. Next, the cantilever is approached as close as possible to the sample surface. The approach is controlled manually using custom software with the visual aid of the camera.

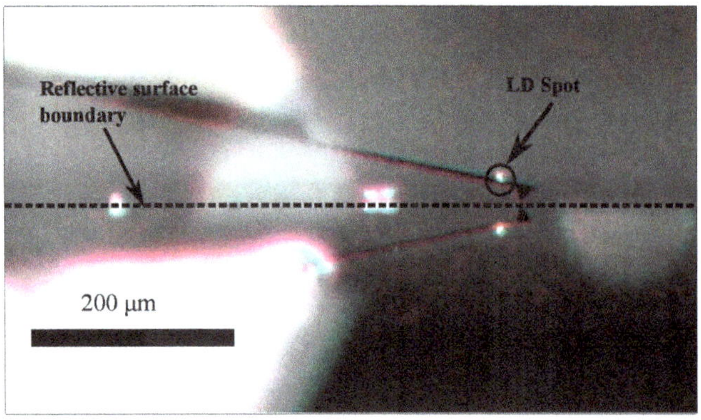

Figure 5. Microcantilever SHOCONA with the reflecting laser spot from the OFI sensor.

Once the cantilever and sample are at the initial measurement position, a reference measurement is taken by inducing a triangular mechanical modulation on the OFI sensor. The captured signal corresponds only to the positive ramp of the PZT displacement. After storing the reference measurement, a sinusoidal vibration in the range of 10 to 200 Hz is induced on the cantilever by means of the PZT shown in Figure 3b. The start of the acquisition for the measurement signals is launched using the rising edge of a square transistor-transistor logic (*TTL*) signal that is matched with the start of the positive ramp translation of the PZT, and it is stopped on the falling edge of the TTL signal. After each acquisition, the signals are automatically saved in a PC for further offline signal processing.

After each measurement, the sample is displaced either in the X- or Y-axis in order to test the effects of the raster scanning functionality to be used in further developments. Once the sample is set in the new position, the vibration process is restarted using the parameters previously selected by the user. The process is repeated for the number of points (distance) selected in the software interface.

Signal Processing

As discussed in Section 2.1, the implemented method uses a mechanically-modulated OFI sensor. The reference OOP signal can be obtained in four possible ways: (1) acquiring a reference OFI signal previous to the measurement with an static target; (2) using the piezoelectric internal capacitive sensor to estimate the reference velocity; (3) measuring the mean velocity by detecting the main frequency of the OFI signal spectra given that the measurement process is dynamic; or finally, (4) using the PZT (Figure 3b element (5) capacitive sensor signal to generate a simulated OFI signal at a suitable *C* value. For the results presented below, a reference signal was acquired for each sample set with a static target. In the case of simulations, however, the frequency spectra of the mean velocity method are evaluated, as this is the most suitable implementation in the case that no other type of reference signal is available.

Each method presents different advantages and disadvantages. The first method is preferred over the other noted methods, since it allows one to obtain a better notion of the OFI signal behavior of the laser. This type of acquisition will also reduce the effects of non-linearities, which might occur along the reference motion. Yet, this method requires the use of memory space to store the reference signal, as well as the extra computing cost used for the FCM, which is required to be performed at least once for a set of measurements. The second method reduces the amount of computing resources, as the reference velocity can be calculated directly from the ramp acquired in the PZT sensor. The use of this method may result in a bias, since the velocity is estimated using only the maximum and minimum points of the ramp. The third method reduces the use of memory resources, as the reference velocity is estimated directly from the mean velocity measured with each OFI signal. This method requires the use of an FFT algorithm and also might cause a bias in the measurement similar to the one of the second method. The fourth method is similar to the second one, as it also requires capturing the signal of the PZT capacitive sensor; nonetheless, the computing cost increases as an OFI signal simulation is required. If the laser is well characterized, the resulting bias should be lower than in the second method, as the complete displacement, as well as some of the disturbances suffered by the PZT can be considered.

After the acquisition, the measurement sample (e.g., Figure 6) is band-pass filtered using an FFT-based filter with cutoff frequencies at $f_{OFI}/5$ and $10\times f_{OFI}$, where f_{OFI} is the main frequency of the OFI signal. Next, the filtered OFI signal is treated with a fringe detection algorithm. For the case of signals presented in this paper, the fringe detection algorithm consists of the differentiation of the OFI signal, after which an initial fringe detection is performed using a threshold comparison method. The threshold is calculated as the RMS of the differentiated signal. After the initial fringe detection, a zero crossing algorithm is applied on the filtered OFI signal, and the detected positions are compared to those in the differentiated signal. This step allows one to remove the possible false positives of the first detection.

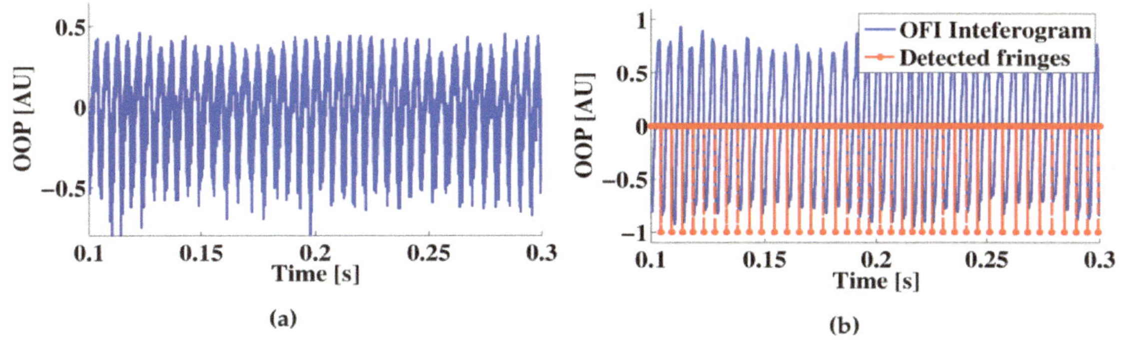

Figure 6. OFI signal fringe detection. (**a**) An example of an experimentally-acquired OFI signal of a ramp-like motion with v_r=76 μm/s; (**b**) The OFI signal shown in (**a**) after filtering and fringe detection.

After the fringe detection process, the time intervals between consecutive fringes in the measurement and reference signals are calculated, as well as the corresponding velocity given that the distance between two fringes is fixed as λ/2. After obtaining the velocity waveform, the signal is integrated using the cumulative trapezoidal method as described in [25]. In the case of AFM measurement, this process is repeated for each of the acquired signals.

RESULTS

Simulation

A simulation of a sinusoidal cantilever motion detection using the mechanically-modulated OFI method was performed in order to detect possible limitations on the detection of the sensor. The OFI algorithm solves the excess phase Equation (1) using the Schröeder root finding method and bracketed with the intervals described in [33]. The parameters of the simulation are presented in Table 1.

Table 1. Parameters for cantilever displacement simulation.

Parameter	Value	Units
Wavelength (λ)	783.50	nm
Reference velocity (v_r)	380.00	μm/s
Feedback attenuation (ϵ)	0.06	
Feedback factor (C)	0.98	
Distance to cantilever (D)	100.00	mm
Cantilever displacement (ΔD)	100.00	nm
Cantilever frequency (f)	100.00	Hz
Cantilever reflectance (R_e)	0.90	
Distance sample to cantilever 1 (d_{z1})	100.00	nm
Distance sample to cantilever 2 (d_{z2})	0.50	nm
Cantilever stiffness (k_c)	0.95	N/m
Cantilever quality factor (Q_c)	108.70	
Cantilever resonance frequency (f_{rc})	45.29	kHz

The chosen parameters were selected to provide a C≃1 (C=0.976) to comply with the expected requirements of the method. A positive ramp-like displacement with a constant velocity v_r=380 μm/s is chosen as a reference for the OFI signal. All simulated displacement vectors, as well as the corresponding OFI signals are composed by 125 kS to match the length of acquisition vectors in the experimental measurements.

For the cantilever motion, a harmonic modulation is calculated using as a reference the mass-spring resonator equation:

$$F_{ext} + F_d \times sin(2\pi f t) = m_{eff}\frac{d^2x}{dt^2} + f_v\frac{dx}{dt} + k_c x \quad (5)$$

where the effective mass m_{eff}, damping factor f_v and stiffness k_c are mechanical cantilever parameters relative to the displacement x and its time derivatives, F_d is the driving force amplitude and F_{ext} accounts for the van der Waals (VdW) force interaction using as reference the Derjaguin-Muller-Toporov (DMT) model. The simulation is tested for two cantilever tips to sample minimum distances d_{z1}=100 nm and d_{z2}=0.50 nm. The resulting cantilever displacement profiles behave following a harmonic motion with a variation in amplitude when the cantilever tip is closer to the sample, as illustrated in Figure 7. The cantilever displacement is only considered for the cantilever tip in a motion parallel to the beam axis.

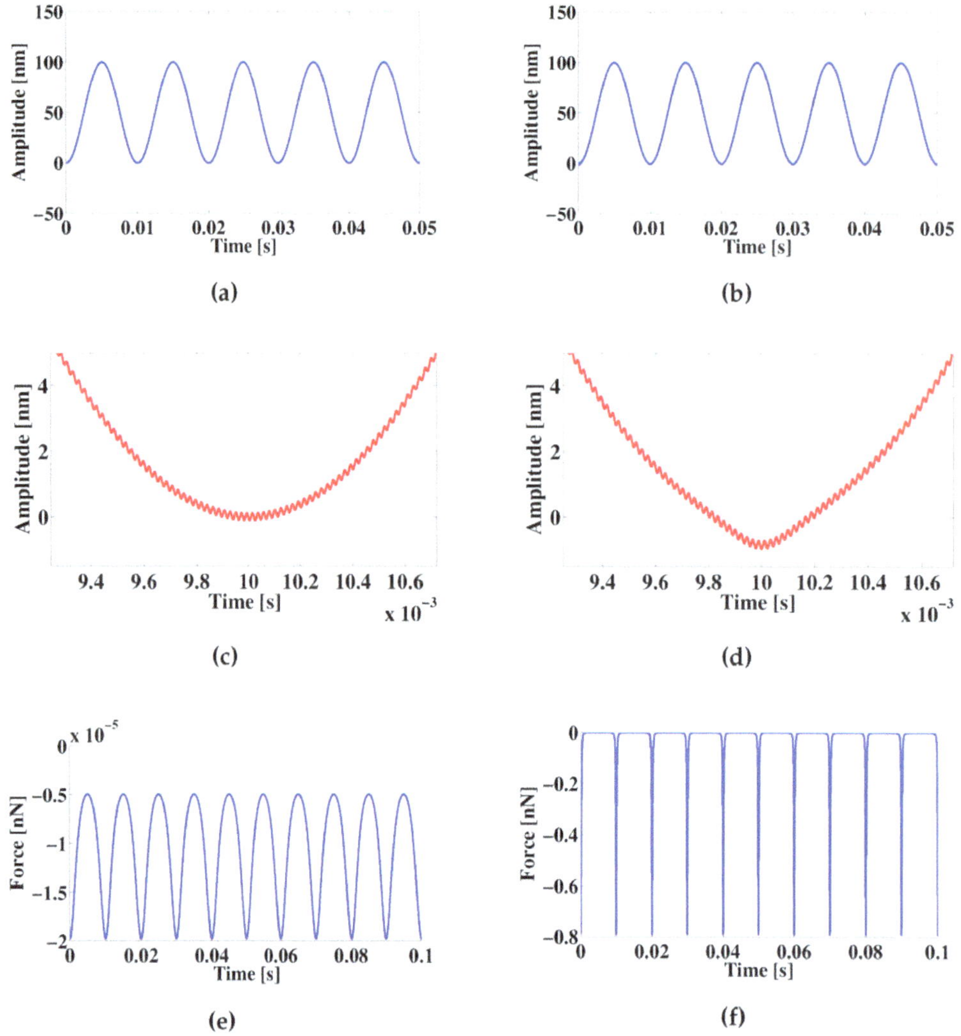

Figure 7. Example of a simulated cantilever tip behavior due to a harmonic motion. (**a**,**b**) Amplitude representation of a harmonic displacement with the tip placed for dz1and dz2 respectively; (**c**,**d**) Zoom of the displacement amplitude around the region where the tip is closer to the sample; a deformation of the harmonic motion increases as the minimum tip to sample distance is reduced; (**e**,**f**) Corresponding force effect for dz1 and dz2.

The resulting OFI signal contains the information of the reference velocity, as well as of the target displacement, as is displayed in Figure 8b by the frequency peaks induced over the spectra due to the Doppler effect. It should also be noted that the normalized amplitude of the frequency peaks caused by the tip motion is dependent on the force resulting in a value of 0.4383 AU for d_{z1} and 0.4389 AU for d_{z2}. For the presented simulation, the frequency resolution is limited to 10 Hz. In the case of non-multiple frequencies, the peaks caused by the cantilever vibration become wider, and as in the previous case, their amplitude also changes as a function of their proximity to the simulated sample.

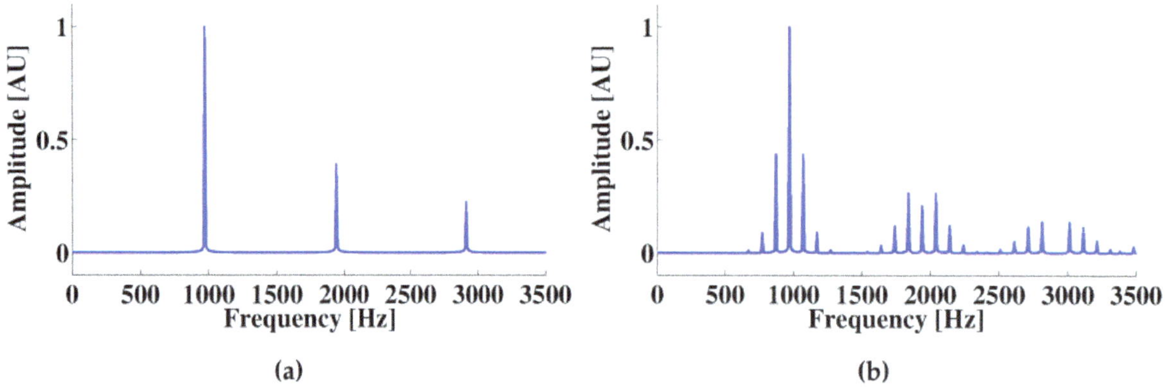

Figure 8. (**a**) Frequency spectra of a simulated mechanically-modulated OFI sensor signal without target motion; (**b**) The effect of the cantilever motion over the frequency spectra; different peaks are observed around the main OFI frequency and its corresponding harmonics; the frequency difference between the main OFI peak and the highest amplitude peak around it is equal to the target vibration frequency.

After analyzing the spectral effects, the simulated mechanically-modulated OFI signal is subjected to the algorithm discussed in Section 2. In the case of the 100-nm sample distance, the reconstruction follows the simulated cantilever displacement with an RMS error of 7.66 nm and a standard deviation of 7.70 nm, as illustrated in Figure 9. In the case of a distance of 0.50 nm, no significant differences are observed in the reconstruction; however, the RMS error and standard deviation increase respectively in 0.002 nm. The errors in the reconstruction can be related to non-linearities induced by large displacements. This is best observed if we double the velocity of the ramp reference displacement, which results in 3.84 nm and a standard deviation of 3.86 nm, respectively. This is also observed when a 1-nm amplitude displacement is applied as a cantilever motion in the simulation. The resulting RMS error in the reconstruction is reduced to a change in path length of 0.161 nm and a standard deviation of 0.1615 nm.

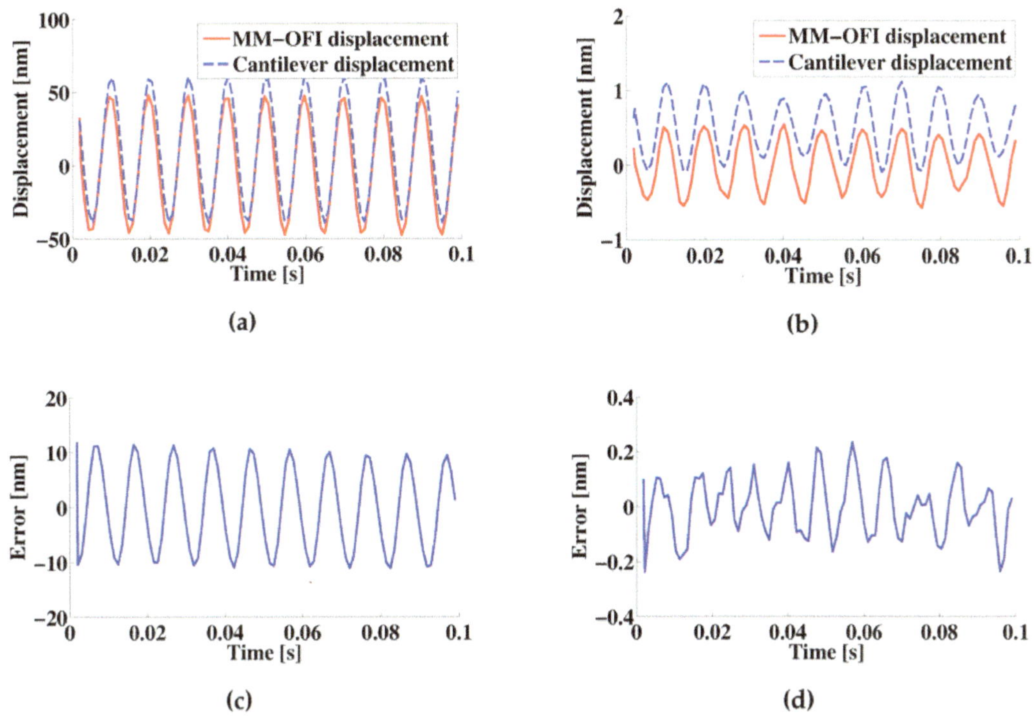

Figure 9. (**a**,**b**) Reconstructed displacements and (**c**,**d**) Reconstruction errors for a 100-nm and 1-nm amplitude displacements. In (**a**), a 10-nm offset is included on the 100 nm displacement for visibility purposes. Similarly, in (**b**), a 1-nm offset is included on the 1-nm displacement for visibility purposes. The legend MM-OFI is used to refer to the mechanically-modulated OFI signal.

Experimental Results

On the first test, a set of mechanically-modulated OFI measurements was obtained in regular lab working conditions at a room temperature of 25 °C. A 100-nm peak to peak oscillation was induced on the cantilever, and the signals were retrieved following the steps described in Section 2, resulting in the waveform shown in Figure 10. Reconstruction errors can be attributed to three major sources: (1) the linearity of the reconstruction algorithm; (2) mechanical fixation issues; and (3) mechanical noise caused by acoustic and mechanic vibration of the building itself. Low SNR signals may also induce errors because of the increased difficulty to detect the position of the OFI signal fringes. This effect, however, is limited for the measurements presented in this work, as the sensor is calibrated to work close to C=1.

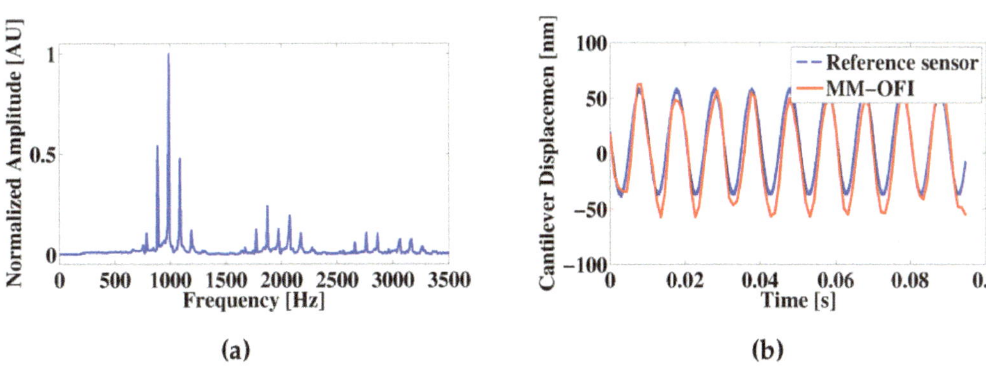

Figure 10. (**a**) Frequency analysis of a mechanically-modulated OFI signal during an experimental cantilever vibrating at 100 Hz with a peak to peak amplitude of 100 nm; (**b**) Comparison between the internal capacitive sensor response of the PI-LISA PZT (dashed line) and the mechanically-modulated OFI sensor (solid line).

For the case presented, the difference between the PZT sensor response and the mechanically-modulated OFI measurement can be explained by mechanical fixation issues between the PI-LISA PZT and the cantilever holder and by possible linearity issues in the reconstruction method. Another possible explanation to the difference might be the detection of surface forces over-imposed on the reference displacement as they are within the 20 to 30-nm range, which would represent forces in the range of 0.3 nN to 18 nN. It should be noted that the frequency spectra displayed in Figure 10a show two peaks at 886 Hz and 1088 Hz from the expected 987-Hz frequency. The difference between the central frequency and any of these peaks corresponds to the main driving frequency of the sinusoidal modulation (100 Hz). These peaks also appear at the second and third harmonic of the OFI spectra (1975 and 2962 Hz) with smaller decay than those corresponding to the OFI signal.

In order to remove some of the issues observed in the previous measurement, the holder grip mechanism was modified increasing the rigidity of the mechanical structure. After a new calibration process, a new set of measurements was acquired under normal laboratory conditions, which include the presence of random mechanical noise and acoustic vibrations. An example of signals acquired for a 50-μm displacement in the X-axis with steps of 0.5 μm is presented in Figure 11. As is observed, there exist different punctual events, which cause errors in the reconstructed measurement, in some cases with amplitudes circa λ, as shown in Figure 11b. In most cases, the measurement holds the expected 100-nm amplitude value, as observed in Figure 11c. This stability in the detected amplitude and frequency is also confirmed in the representation of the Fourier spectra for each of the OFI acquired measurements (Figure 11d). The spectra do not show any discernible variations in the amplitude of the signal (which can be obtained by multiplying by $\lambda/4$ the normalized amplitude) showing a mean value of 92 nm with a standard deviation of 8 nm for 100 measurements. This, again, has good agreement with the expected capacitive sensor response.

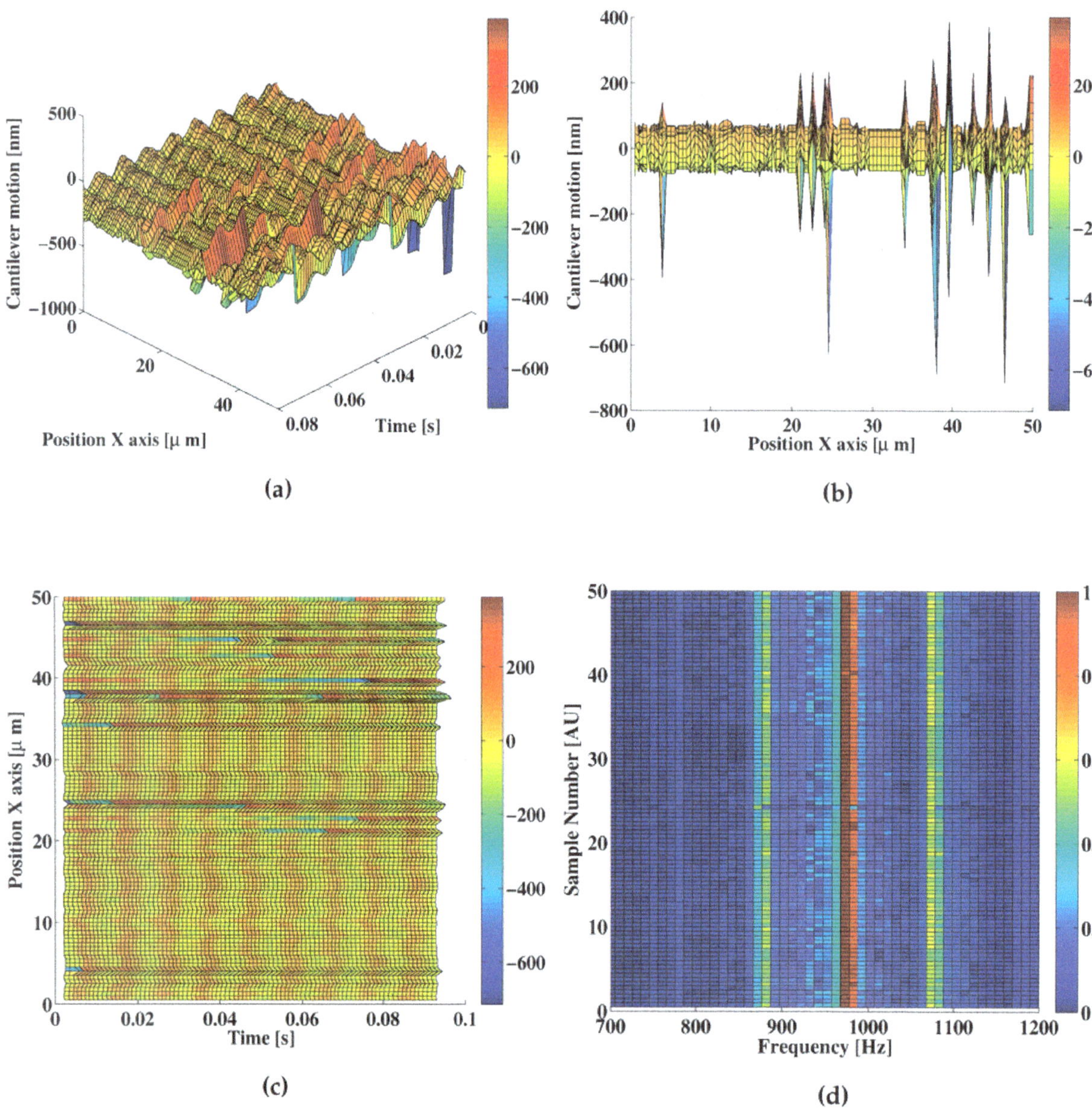

Figure 11. Example of an experimental A=100-nm cantilever displacement reconstruction under uncontrolled acoustic and mechanical noise conditions. The displacement is shown in different projections: (**a**) 3D; (**b**) top and (**c**) lateral view; (**d**) The top view of the Fourier spectra around the main harmonic of the OFI measurement signal for each of the points acquired in the X-axis.

A second set of measurements was acquired under more controlled conditions. In this case the system was automatized to perform the measurements automatically while there was no one on the room. In principle, this reduced possible sources of acoustic noise during the measurement, and also, sources of mechanical noise induced by footsteps in the room. Again, signals were acquired for a 50 μm displacement in the X axis with steps of 0.5 μm as presented in Figure 12. As it is shown in Figure 12a,c, no spikes of amplitude larger than 100 nm are visible. A random variation on the detected amplitude is observed in Figure 12b, leading us to believe that the proximity to the sample surface may be having an effect on the measurement. These variations, nonetheless, cannot be quantified at the moment. In most cases, however, the measurement holds the expected 100 nm amplitude value as observed in Figure 12c. The stability in the detected amplitude and frequency is also confirmed in the representation of the Fourier spectra for each of the OFI acquired measurements (Figure 12d). The Fourier spectra of the signals shows a mean value of 92 nm with a standard deviation of 8 nm for 100 measurements which again is in good agreement with the results of the capacitive sensor.

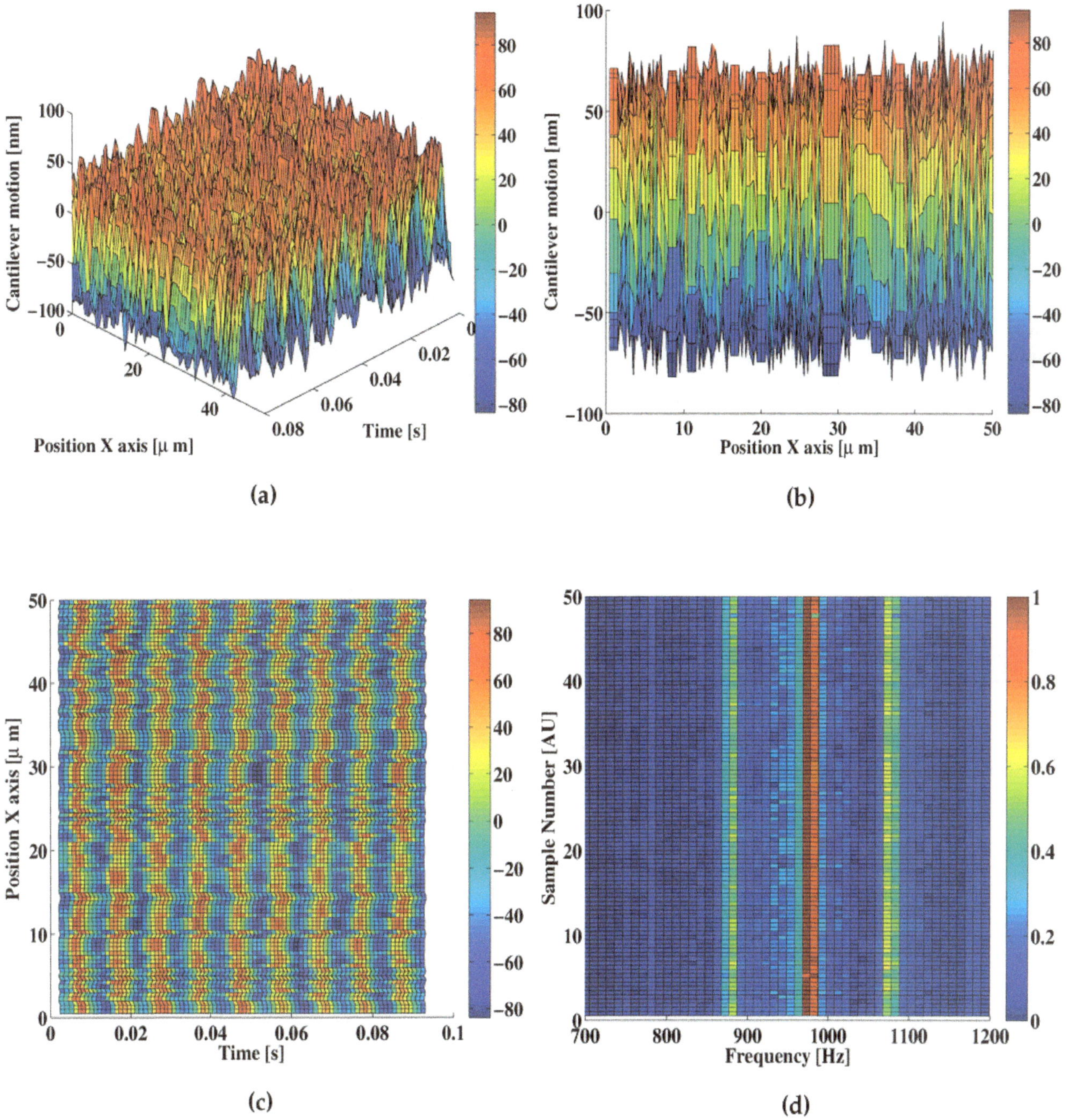

Figure 12. Example of an experimental A=100-nm cantilever displacement reconstruction under low acoustic noise conditions. The displacement is shown in different projections: (**a**) 3D; (**b**) Top and (**c**) Lateral view; (**d**) Top view of the Fourier spectra around the main harmonic of the OFI measurement signal for each of the points acquired in the X-axis.

The off-line computing time for each signal containing takes in average 0.695 s, which shows feasibility for an on-line signal processing using the mechanically modulated OFI method.

DISCUSSION

The results obtained in the previous section show the feasibility of a mechanically-modulated OFI sensor for detecting a cantilever sinusoidal motion with peak to peak amplitudes going from 100 to 200 nm. Previous results with the technique [25] show that a practical resolution of 3 nm can be reached, thus allowing the capture of 10-nm amplitudes with sufficient detail.

By means of simulation, it has also been shown that it is possible to reproduce the displacement of lower figures, such as is the case of the 1-nm displacement shown in Figure 9, since the linearity of the sensor is increased for smaller changes. Similar results are also shown in [23], where it is observed that the OFI readings behave linearly for small changes. Nevertheless, attaining such figures seems difficult to reach in practice due to the several sources of noise that might affect either the target or the sensor itself in its present configuration. In particular, it has been observed that the system is extremely sensitive to mechanical and acoustic fluctuations in the room. Therefore, it is required to increase the isolation of the system with the room acoustic conditions, which might be achieved by casing the system in a sound-proof box. External vibrations can be reduced by setting the system on an absorbent material (e.g., marble), which would reduce the amount of vibrations transmitted from the floor into the instrument. Another possibility would be to work on vibration amplitudes with frequencies larger than 200 Hz to reduce possible vibrations inherit to the building structure. In terms of mechanical noise reduction, it is also possible to change the measurement direction towards the X- or Y-axis. This, however, may introduce an issue with sample handling in the case of liquid-based media. In terms of electrical noise, it has been observed that the system is capable of obtaining low errors if the SNR of the signal is larger that 10 dB; for lower values, the signal tends to degrade, reducing the fringe visibility and, thus, affecting the fringe occurrence time detection. For SNRs above 10 dB, detected RMS errors on the signal typically stay below $\lambda/100$. The signals presented in this work presented a minimum SNR in the order of 20 dB.

As mentioned before, the normalized Fourier spectra of the mechanically-modulated OFI signal also contain practical information of the displacement, as it can be used to detect the information of the mean velocity used to generate the reference signal and also to compute the main frequency of the sample displacement in the case of the harmonic motion. It should also be noted that for the case where the cantilever was subjected to forces in an order lower than 1 nN, a change on the peak amplitude at the corresponding Doppler frequency shift with a value of 0.0006 units was detected. In principle, this variation is negligible; however, it should be expected that as the sample frequency is increased and reaches values closer to the cantilever resonance frequency, the amplitude will be amplified.

It is also to be noticed that this work focuses on measurements at low frequencies compared to typical measurements in the area that typically work over the kHz range. It is stressed that the work described in this paper is to prove the feasibility of the mechanically-modulated OFI method on the detection of microcantilever displacement in the nanometric scale. The measurements were performed taking into account the limitations of the mechanical systems used to provide both the reference modulation and the cantilever modulation. While working within the selected frequency range can be partially the cause of the noise induced over the measurement, the obtained results show that, nonetheless, it is possible to retrieve a cantilever displacement with nanometric amplitudes.

CONCLUSIONS

In this paper, we have shown the feasibility of using mechanically-modulated OFI-based sensors for the measurement of microcantilever motion. To the extent of our knowledge, this is the first time that this method has been applied to the measurement of microcantilever motions. The technique, as it is described, is able to recover the motion of a commercial microcantilever usually employed in AFM systems, suggesting that the sensor should be able to detect the motions of micrometric cantilevers, as long as the displacement is within the resolution of the device. While experimental conditions and available materials did not allow testing the method for amplitudes below 100 nm, larger amplitude motions show good agreement with the theory and simulations presented. Furthermore, the frequency analysis shows good correspondence with the theory showing the appearance of secondary peaks in the Fourier spectra with the correct amplitude relative to the normalized amplitude of the principal OFI harmonic and with a shift in frequency corresponding to the main frequency of the cantilever motion.

There exist several factors that require further work before creating a practical mechanically-modulated OFI sensor for cantilever motion measuring. First of all, a better mechanical noise isolation system is required since, as observed in Figure 11, some noise sources may drastically hamper the results on the measurements. Some possible solutions to achieve this isolation are described in Section 4. Furthermore, the change of some conditions on the tip holder and the tip itself may benefit the measurement. These changes include reducing the angle between the tip and the horizontal plane and removing the aluminum coating on the cantilever. In the first case, a reduction on the possible fringe detection errors during the signal processing caused by the angle of the cantilever is to be expected. In the second case, the aluminum coating of the cantilever can cause issues due to the bimetallic effect, which introduces strain on the cantilever, causing some degree of bending due to the difference of thermal expansion coefficients between the reflective and substrate materials of the cantilever. This has also been remarked by other authors, e.g., in [18,23], as an issue with classic cantilevers. Even more problematic for the OFI case is the large reflectivity of aluminum, which may change the OFI working point to a strong feedback regime. In the measurements proposed above, a small amount of de-focusing has allowed removing the issue at the cost of having a larger spot size and lower SNR, which also may limit the accuracy of the method.

Towards a profile reconstruction algorithm, further work is required to perform a complete interpretation of the acquired data. For these purposes, a tip force characterization system needs to be included in the processing. The method may largely benefit from the use of other types of OFI setups, such as the one discussed in [39], as they work closer to the bandwidth frequency of more classical commercial instruments. It is important to remark that in its current setup, the sensor is limited by the mechanical modulation imposed by a PZT, thus limiting the reference working frequency as a function of the displacement amplitude. In principle, the sensor described in this paper should be able to measure displacements with frequencies in the order of 1 to 2 kHz with high definition and up to 10 kHz with a lower definition. Further steps will also include the reconstruction of well-controlled sample surfaces in order to clarify the limits on the vertical and lateral resolution of the system.

ACKNOWLEDGMENTS

The authors would like to thank the Spanish Ministry of Science and Innovation for the funding provided through Plan Nacional I+D+I in project DPI2014-56881-R. Francisco J. Azcona would like to thank AGAUR for the funding received with pre-doctoral Grant 2012FI_BI 00240. Ajit Jha would like to thank the European Union funding via the Erasmus Mundus program for the fellowship. Special thanks to María del Mar Álvarez and Carlos Domínguez from the National Center of Nanoscience and Nanotechnology Research (ICN2) and Crina Crojocaru and Jose Francisco Trull from the Nonlinear Dynamics, Nonlinear Optics and Lasers (DONLL) at UPC for helping us with the characterization of equipment and for giving us access to their facilities.

AUTHOR CONTRIBUTIONS

The research presented in this manuscript was performed with the aid of all the authors. Francisco J. Azcona collaborated in the conception of the technique, experimental setup and signal processing. He has also collaborated in the writing process of the manuscript. Ajit Jha collaborated with the signal processing and provided useful insights along the drafting of the manuscript. Carlos Yañez helped during the signal acquisition and helped in the revision of the manuscript. Reza Atashkhooei collaborated with the experiment design, conception of the technique and discussion. Santiago Royo co-wrote the manuscript and provided helpful insight on the method.

CONFLICTS OF INTEREST

The authors declare no conflict of interest.

REFERENCES

1. Binning, G.; Quate, C.F.; Gerber, C. Atomic Force Microscope. *Phys. Rev. Lett.* **1986**, *56*, 930–934.
2. Boudaoud, M.; Haddab, Y.; Gorrec, Y.L.; Lutz, P. Noise characterization in millimeter sized micromanipulation systems. *J. Mech.* **2011**, *21*, 1087–1097.
3. Pai, P.; Chen, L.; Tabib-Azar, M. Fiber Optic Magnetometer with Sub-Pico Tesla Sensitivity for Magneto-Encephalography. In Proceedings of the 2014 IEEE SENSORS, Valencia, Spain, 2–5 November 2014; pp. 722–725.
4. Cherry, A.; Abadie, J.; Piat, E. Microforce sensor for microbiological applications based on a floating magnetic principle. In Proceedings of the 2007 IEEE International Conference on Robotics and Automation, Rome, Italy, 10–14 April 2007; pp. 1504–1509.
5. Giessib, F.J. Advances in atomic force microscopy. *Rev. Mod. Phys.* **2003**, *75*, 949–983.
6. Lee, J.; Ju, S.; Kim, I.T.; Jung, S.H.; Min, S.J.; Kim, C.; Sim, S.J.; Kim, S.K. Evaluation of Chemical Interactions between Small Molecules in the Gas Phase Using Chemical Force Microscopy. *Sensors* **2015**, *15*, 30683–30692.
7. Kasas, S.; Thomson, N.H.; Smith, B.L.; Hansma, P.K.; Miklossy, J.; Hansma, H.G. Biological Applications of the AFM: From Single Molecules to Organs. *Int. J. Imaging Syst. Technol.* **1997**, *8*, 151–161.
8. Martinez, V.; Behr, P.; Drechsler, U.; Polesel-Maris, J.; Potthoff, E.; Böros, J.; Zambelli, T. SU-8 hollow cantilevers for AFM cell adhesion studies. *J. Micromech. Microeng.* **2016**, *26*, 1–10.
9. Zhu, X.; Siamantouras, E.; Liu, K.K.; Liu, X. Determination of work of adhesion of biological cell under AFM bead indentation. *J. Mech. Behav. Biomed.* **2016**, *56*, 77–86.
10. Turner, J.A.; Wiehn, J.S. Sensitivity of flexural and torsional vibration modes of atomic force miscroscope cantilevers to surface stiffness variations. *Nanotechnology* **2001**, *12*, 322–330.
11. Hsu, J.C.; Lee, H.L.; Chang, W.J. Flexural vibration frequency of atomic force microscope cantilevers using the Timoshenko beam model. *Nanotechnoly* **2007**, *18*, 285503.
12. Sader, J.E.; Chon, J.W.M.; Mulvaney, P. Calibration of rectangular atomic force microscope cantilevers. *Rev. Sci. Instrum.* **1999**, *70*, 3967–3969.
13. Burnham, N.A.; Chen, X.; Hodges, C.S.; Matei, G.A.; Thoreson, E.J.; Roberts, C.J.; Davies, M.C.; Tendler, S.J.B. Comparison of calibration methods for atomic-force microscopy cantilevers. *Nanotechnology* **2003**, *14*, 1–6.
14. Green, C.P.; Lioe, H.; Cleveland, J.P.; Proksch, R.; Mulvaney, P.; Sader, J.E. Normal and torsional spring constants of atomic force microscope cantilevers. *Rev. Sci. Instrum.* **2004**, *75*, 1988–1996.
15. Raiteri, R.; Grattarola, M.; Butt, H.J.; Skládal, P. Micromechanical cantilever-based biosensors. *Sens. Actuators B Chem.***2001**, *79*, 115–126.
16. Putman, C.A.J.; Grooth, B.G.D.; Hulst, N.F.V.; Greve, J. A detailed analysis of the optical beam deflection technique for use in atomic force microscopy. *J. Appl. Phys.* **1992**, *72*, 6–12.
17. Thundat, T.; Warmack, R.J.; Chen, G.Y.; Allison, D.P. Thermal and ambient-induced deflections of scanning force microscope cantilevers. *Appl. Phys. Lett.* **1994**, *64*, 2894.
18. Ramos, D.; Mertens, J.; Calleja, M.; Tamayo, J. Study of the Origin of Bending Induced by Bimetallic Effect on Microcantilever. *J. Sens.* **2007**, *7*, 1757–1765.
19. Rugar, D.; Mamin, H.J.; Guethner, P. Improved fiber-optic interferometer for atomic force microscopy. *Appl. Phys. Lett.***1989**, *55*, 2588–2590.

20. Wang, W.M.; Grattan, K.T.V.; Palmer, A.W.; Boyle, W.J.O. Self-mixing interference inside a single-mode diode laser for optical sensing applications. *J. Lightwave Technol.* **1994**, *12*, 1577–1587.
21. Bosch, T.; Servagent, N.; Donati, S. Optical feedback interferometry for sensing application. *Opt. Eng.* **2001**, *40*, 20–27.
22. Taimre, T.; Nicolić, M.; Bertling, K.; Lim, Y.L.; Bosch, T.; Rakić, A.D. Laser feedback interferometry: A tutorial on the self-mixing effect for coherent sensing. *Adv. Opt. Photonics* **2015**, *7*, 570–631.
23. Ovryn, B.; Andrews, J.H. Measurement of changes in optical path length and reflectivity with phase-shifting laser feedback interferometry. *Appl. Opt.* **1999**, *38*, 1959–1967.
24. Larsson, D.; Greve, A.; Hvam, J.M.; Boisen, A.; Yvind, K. Self-mixing interferometry in vertical-cavity surface-emitting lasers for nanomechanical cantilever sensing. *Appl. Phys. Lett.* **2009**, *94*, 091103-1–091103-3.
25. Azcona, F.J.; Atashkhooei, R.; Royo, S.; Mendez-Astudillo, J. A nanometric displacement measurement system using differential optical feedback interferometry. *IEEE Photonics Technol. Lett.* **2013**, *25*, 2074–2077.
26. Zabit, U.; Bony, F.; Bosch, T.; Racik, A.D. A Self-Mixing Displacement Sensor With Fringe-Loss Compensation for Harmonic Vibrations. *IEEE Photonics Technol. Lett.* **2010**, *22*, 410–412.
27. Norgia, M.; Giuliani, G.; Donati, S. Absolute Distance Measurement with Improved Accuracy Using Laser Diode Self-Mixing Interferometry in a Closed Loop. *IEEE Trans. Instrum. Meas.* **2007**, *56*, 1894–1900.
28. Scalise, L.; Yu, Y.; Giuliani, G.; Plantier, G.; Bosch, T. Self-Mixing Laser Diode Velocimetry: Application to Vibration and Velocity Measurement. *IEEE Trans. Instrum. Meas.* **2004**, *53*, 223–232.
29. Arasanz, A.; Azcona, F.J.; Royo, S.; Jha, A.; Pladellorens, J. A new method for the acquisition of arterial pulse wave using self-mixing interferometry. *Opt. Laser Technol.* **2014**, *63*, 98–104.
30. Donati, S. Responsivity and Noise of Self-Mixing Photodetection Schemes. *IEEE J. Quantum Electron.* **2011**, *47*, 1428–1433.
31. Lang, R.; Kobayashi, K. External optical feedback effects on semiconductor injection laser properties. *IEEE J. Quantum. Electron.* **1980**, *16*, 347–355.
32. Acket, A.; Leenstra, D.; Boef, A.J.D.; Verbeek, B.H. The influence of feedback intensity on longitudinal mode properties and optical noise in index guided semiconductor lasers. *IEEE J. Quantum. Electron.* **1984**, *20*, 1163–1169.
33. Kliese, R.; Taimre, T.; Bakar, A.A.A.; Lim, Y.L.; Bertling, K.; Nikolić, M.; Perchoux, J.; Bosch, T.; Rakić, A.D. Solving self-mixing equations for arbitrary feedback levels: A concise algorithm. *Appl. Opt.* **2014**, *53*, 3723–3736.
34. Wang, M. Fourier transform method for self-mixing interference signal analysis. *Opt. Laser Technol.* **2001**, *33*, 409–416.
35. Bernal, O.D.; Zabit, U.; Bosch, T. Study of a laser feedback phase under self-mixing leading to improved phase unwrapping for vibration sensing. *IEEE Sens. J.* **2013**, *13*, 4962–4971.
36. Dean, P.; Valavnis, A.; Keeley, J.; Bertling, K.; Lim, Y.L.; Alhathool, R.; Chowdry, S.; Taimre, T.; Li, L.H.; Indjin, D.; et al. Coherent three-dimensional terahertz imaging through self-mixing in a quantum cascade laser. *Appl. Phys. Lett.* **2013**, *103*, 181112-1–181112-4.
37. Giuliani, G.; Bozzi-Pietra, S.; Donati, S. Self-mixing laser diode vibrometer. *Meas. Sci. Technol.* **2003**, *14*, 24–32.
38. Atashkhooei, R.; Royo, S.; Azcona, F.J. Adaptive autofocus technique for speckle control in optical feedback interferometry. *Sens. Actuators A Phys.* **2014**, *216*, 243–248.
39. Jha, A.; Azcona, F.J.; Royo, S. Frequency modulated optical feedback interferometry for nanometric scale vibrometry. *IEEE Photonics Technol. Lett.* **2016**, *28*, 1217–1219.

CHAPTER 17

DIAGNOSTIC MEASUREMENTS FOR POWER TRANSFORMERS

Stefan Tenbohlen[1], Sebastian Coenen[2], Mohammad Djamali[1], Andreas Müller[3], Mohammad Hamed Samimi[1] and Martin Siegel [1]

[1]Institute of Power Transmission and High Voltage Technology, University of StutFtgart, Stuttgart 70569, Germany

[2]General Electric Grid Solutions; Moenchengladbach 41065, Germany

[3]TÜV Süd Energietechnik, Filderstadt 70794, Germany

ABSTRACT

With the increasing age of the primary equipment of the electrical grids there exists also an increasing need to know its internal condition. For this purpose, off- and online diagnostic methods and systems for power transformers have been developed in recent years. Online monitoring is used continuously during operation and offers possibilities to record the relevant stresses which can affect the lifetime. The evaluation of these data offers the possibility of detecting oncoming faults early. In comparison to this, offline methods require disconnecting the transformer from the electrical grid and are used during planned inspections or when the transformer is already failure suspicious. This contribution presents the status and current trends of different diagnostic techniques of power transformers. It provides significant tutorial elements, backed up by case studies, results and some analysis. The broadness and improvements of the presented diagnostic techniques show that the power transformer is not anymore a black box that does not allow a view into its internal condition. Reliable and accurate condition assessment is possible leading to more efficient maintenance strategies.

INTRODUCTION

The reliability of the electrical power system depends on the performance and availability of its components such as power transformers. Due to the increasing age of the transformer population condition assessment and thus onsite diagnostics are important issues to secure a reliable operation of the electrical power system. During the last decades, major works are being carried out for development of reliable and accurate condition assessment techniques [1]. Offline methods require disconnecting the transformer from the power network and are mainly used during scheduled inspections or when the transformer is already failure suspicious. In comparison to this, online methods are used during the operation and offer a possibility to record the condition under realistic operating conditions (e.g., electric field, load, and temperature). Monitoring is the continuous application of online measurement techniques which allows trending and the early detection of an oncoming fault by automatic evaluation of these data. The applicability of the different condition assessment techniques are depicted in Table 1.

Citation: Tenbohlen, S.; Coenen, S.; Djamali, M.; Müller, A.; Samimi, M.H.; Siegel, M. Diagnostic Measurements for Power Transformers. Energies 2016, 9, 347. DOI: 10.3390/en9050347

Table 1. Status of different condition assessment techniques. DGA: dissolved gas analysis; PD: partial discharge, IEC: International Electrotechnical Commission, UHF: ultra-high frequency, FRA: frequency response analysis, PDC: polarization and depolarization currents, FDS: frequency domain spectroscopy.

Method	Offline	Online	Monitoring	Offsite
Ageing of oil (e.g., color, moisture, and tan δ)	xxx	xXx	x [1]	xxx
Furan in oil analysis	xX	XX	-	XX
Gas-in-oil analysis (DGA)	xxx	xXx	xXx	xxx
PD (IEC 60270)	xxx	XX	X	XXX
Unconventional PD-measurement (e.g., UHF PD measurement)	xx	XX	X	XX
Transfer function (FRA)	xxX	X	-	XXX
Dielectric diagnostic (PDC and FDS)	xx	-	-	XX
Thermal monitoring	-	-	XX	-
Degree of polymerization (DP-value)	-	-	-	XXX

XXX: generally accepted or standardized; XX: accepted by different users; X: under investigation or consideration; [1] moisture measurement; and -: not applicable.

Localized internal insulation failures can lead to catastrophic breakdown. To reduce such risk, power transformers have to pass a range of factory tests including one for partial discharge (PD) activity before acceptance and commissioning. Onsite PD measurements are often restricted by high site interference. Hence the electrical PD-measurement set-up according to International Electrotechnical Commission (IEC) 60270 [2] usually has sensitivity limitations for onsite/online measurements. In contrast, the sensitivity of unconventional ultra-high frequency (UHF) PD measurements is sufficient and normally not affected by external disturbances. Especially in noisy surroundings, it is a helpful method to support other PD diagnostic techniques for example acoustic localization of PD.

Frequency response analysis (FRA) compares measured transfer functions of power transformers. Deviations between frequency response curves indicate electrical or mechanical damages of windings. Interpretation of the deviations between transfer functions is not a standardized process and promises to be more reliable and meaningful if skilled knowledge about the transformer design is incorporated.

Especially, the established dissolved gas analysis (DGA) can provide relevant information about internal transformer faults. Natural aging, thermal and electrical failures generate typical fault gases, which dissolve in the insulating oil. By analyzing these fault gas profiles, evaluation of kind and severity of the fault can be estimated. Online monitoring of the content of dissolved gases is a powerful tool, enabling the control of fault gas development continuously and thus guaranteeing an optimal energy transmission.

Another factor is the assessment of the transformer's ageing condition. Ageing depends directly on operating temperature and time. The limiting factor of lifetime is the degradation of paper insulation. This goes hand in hand with moisture production from the ageing process. Thus, by knowing the moisture content in the solid insulation, it is possible to get an idea of the ageing condition and decide further actions. Moisture assessment can be done in various ways. For onsite offline measurements, dielectric spectroscopy is a viable tool. Moisture has an effect on the dielectric response of the insulating system. Among different methods frequency domain spectroscopy (FDS) will be presented in detail.

The lifetime performance of a power transformer strongly depends on the temperatures its materials have been exposed to. Because of their significant effect, a profound knowledge about these temperatures is of great interest. Due to transient load conditions, changing ambient conditions, adjustable cooling systems and their design and operation principles, transformers represent a rather complex thermal system. In addition, the particular design of a certain transformer, with its applied materials and installed components, complicates precise simulations tremendously. However, knowledge about the exact temperature distribution inside a transformer may not be necessary within the scope of every desired application. For instance, the determination of a constant, non-critical level of overload might already be achieved by means of a simplified thermal model which additionally can offer the possibility of monitoring of the operational state of the cooling unit.

FAILURE STATISTIC OF POWER TRANSFORMERS

Accurate information about service experience of high voltage (HV) equipment is of great importance for both manufacturers and utilities of such equipment. It helps the manufacturers to improve their products, and provides important inputs for the utilities when buying equipment, organizing maintenance and benchmarking their performance. Statistical analysis of past failure data can show useful features with respect to future failure behavior. Equipment reliability data are also required when assessing the overall reliability of an electric power system. Furthermore, international standards applicable to HV equipment are being improved based on service experience and reliability data.

International council of large electric systems (CIGRE) working group A2.37 therefore developed a questionnaire to collect utility failure statistics for power transformers in a standardized way [3]. Besides information about the population under investigation, failure data were collected for various groups of transformers in terms of the failure locations, failure causes, failure modes, actions, external effects and others parameters.

A major failure was defined as any situation that required the transformer to be removed from service for a period longer than seven days for investigation, remedial work or replacement. The necessary repairs should have involved major remedial work, usually requiring the transformer to be removed from its installation site and returned to the factory. A major failure would require at least the opening of the transformer or the tap changer tank, or an exchange of the bushings. A reliable indication that the transformer condition prevents its safe operation is considered a major failure, if remedial work (longer than seven days) was required for restoring it to the initial service capability. In some cases, failures were also assigned as major, if remedial work was shorter than seven days and extensive work with oil processing had to be done (e.g., exchange of bushings).

The working group collected 964 major failures which occurred in the period 1996–2010, within a total population of 167,459 transformer-years, contributed by 58 utilities from 21 countries [3]. The year of manufacture of the units span from the 1950s up to 2009 and the reference periods range from 3 years to 11 years. Because the number of operational transformers was only provided for one year, the total number of transformer-years (population per utility) was calculated under the assumption that the number of transformers in operation was constant during the reference period. The number of transformers was multiplied with the length of the reference period in years to obtain an estimate of the total number of transformer-years. Failure rate was calculated according to Equation (1):

$$\lambda = \frac{n_1 + n_2 + \ldots + n_i}{(N_1 + N_2 + \ldots + N_i) \times T} \times 100\% \quad (1)$$

where n_i = number of failures in i-th year; N_i = number of transformers operating in the i-th year; and T = reference period (normally one year)

The overall failure rate of substation and generator step-up units were all within 1% (Table 2). The failure data of the full population were analyzed as a function of the primary location (component) in the transformer where the failure was initiated. Figure 1 shows the failure location analysis for substation transformers with voltages 100 kV and above, respectively.

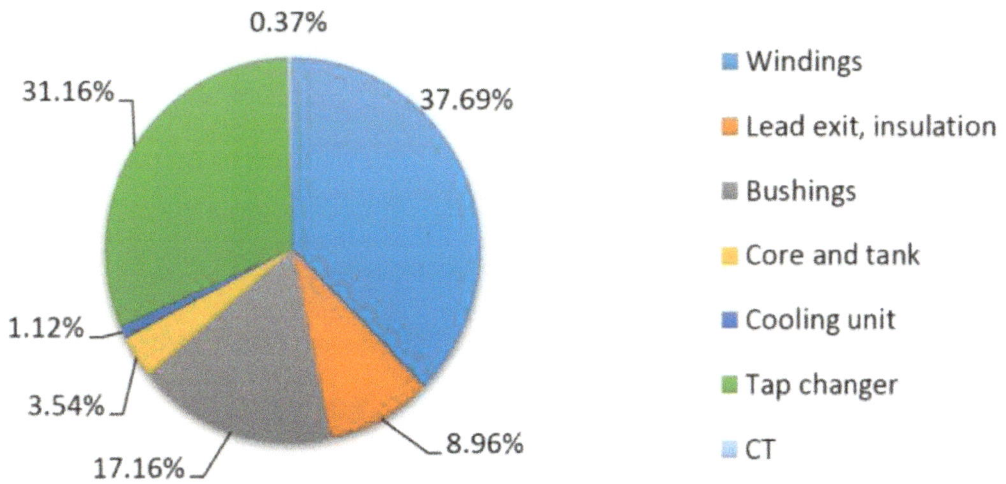

Figure 1. Failure locations of substation transformers (>100 kV) (based on 536 major failures) [4].

Table 2. Investigated population and failure rates of substation transformers [4].

Population information	Highest system voltage (kV)						
	69 ⩽ kV < 100	100 ⩽ kV < 200	200 ⩽ kV < 300	300 ⩽ kV < 500	500 ⩽ kV < 700	kV ⩾ 700	All
Number of transformers	2962	10,932	4272	3233	434	348	22,181
Transformer years	15,267	64,718	37,017	25,305	4774	2991	150,072
Major failures	144	280	186	152	27	10	799
Failure rate	0.94%	0.43%	0.50%	0.60%	0.57%	0.33%	0.53%

Winding related failures were the largest contributor in both transformer applications. GSU transformers had a higher contribution of winding (48%) and lead exit failures (13%) than substation transformers (38% and 6%, respectively). Substation transformers, on the other hand, had a higher contribution of tap changer related failures (31%) than GSU transformers (12%). The contributions of bushing related failures were similar in both transformer applications.

The high percentage of failures related to windings and lead exits indicates the need of condition assessment especially for these components. Thus, the current state of the art of the most important techniques to monitor these parts will be described in the next sections.

PARTIAL DISCHARGE DIAGNOSTICS

Measuring Technique

Electrical PD-measurement according to IEC 60270 [2] usually has sensitivity limitations for onsite/online measurements because of high noise levels in field. The "UHF PD measuring method" is based on the fact, that PD under oil are very fast electrical processes which radiate electromagnetic waves with frequencies in the ultrahigh range (UHF: 300–3000 MHz). Due to the moderately attenuated propagation of UHF signals inside the transformer, the sensitivity to electromagnetic wave detection is good [5,6,7,8]. UHF sensors, as shown in Figure 2, can be installed at the transformer during full operation through the oil filling valve. As a result of electromagnetical shielding characteristics of the grounded transformer tank against external disturbances, a clear decision can normally be made about the internal PD activity of the transformer under test.

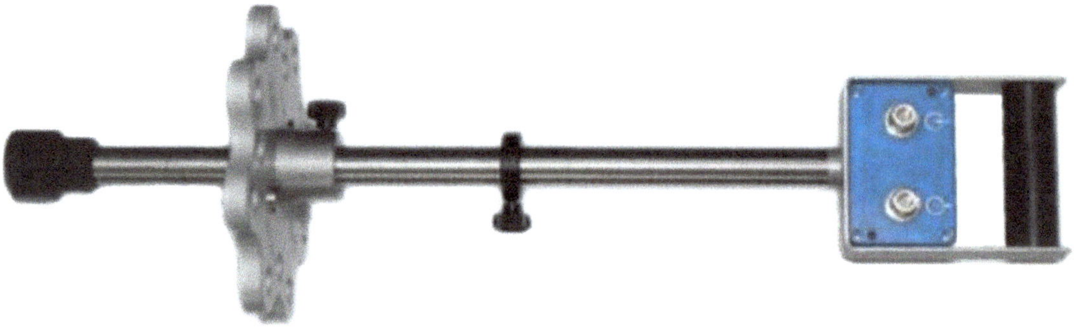

Figure 2. UHF PD probe for standard oil filling valve.

When electrical or UHF PD measurements record PD activity, a three dimensional localization of PD is the next step for risk evaluation. This can be performed based on arrival time measurement of acoustical PD signals [9] using piezo-electric sensors mounted on transformer tank walls [10]. However, acoustic PD signals are affected by distortion within the tank from the windings, core and supporting structures in the transit path. The solution is to use a combination of the two methods: Sensitive UHF PD signals to provide triggering for the averaging of acoustic PD signals for de-noising. By using averaged signals, the acoustic PD pulses remain constructively overlapped, whereas the background noise is averaged to nearly zero [9].

Calibration of Partial Discharge Measurements

The fundamental difference between the electrical PD measurement according to IEC 60270 [2] and the unconventional UHF PD measurement is their physical values. The apparent charge level in terms of picocoulomb (pC) of the electrical measurement is determined by integration of the recharging current. The electromagnetic radiation of the PD is measured in terms of millivolt (mV) using UHF sensors [11]. There is always a partly unknown propagation path because both methods cannot measure directly at the defect. Thus, the actual level of PD (pC or mV) remains unknown within a power transformer. Nevertheless, a common acceptance level for the apparent charge in terms of pC has become widely accepted, especially in factory acceptance tests (FAT). Due to disturbances by ambient noise and existing corona discharge activities, there are certain drawbacks for the use in the field or for online monitoring purposes. In theory, both measured variables contain the same information. A linear physical relation between measured UHF antenna voltage (in mV) and apparent charge (in pC) of the electrical measurement can be demonstrated in a constant laboratory setup with simple geometry [8]. However, it is not given for complex structures like power transformers. UHF antennas measure electromagnetic emissions of PD directly in oil inside a transformer tank, which acts as a "Faraday cage". It becomes apparent that usually UHF measurement is advantageous regarding suppression of external disturbances. This makes it suitable for both, offsite measurement at routine tests under laboratory conditions with low ambient noise and especially onsite with usually high noise levels, e.g., after transportation and installation of transformers as a site acceptance test (SAT). These considerations make the UHF method interesting as supplement for transformer routine tests, when its sensitivity can be defined and demonstrated case by case. Both, the measureable electric and the electromagnetic PD levels, are influenced by the:

- type and actual level of the PD source;
- signal attenuation in the coupling path;
- sensor sensitivity (the UHF antenna or the coupling capacitor and the quadrupole);
- attenuation of the measurement cable and the sensitivity of the measurement device.

The influence of the electric setup (coupling capacitor and quadrupole) and the measurement device can be corrected using calibration for the electric measurement according to [2]. In order to determine the sensor sensitivity of the UHF sensor, its antenna factor (AF) must be known, which can be determined by means of an oil-filled gigahertz-transversal-electro-magnetic setup (GTEM) cell [12].

The sensitivity of an antenna depends on its design in relation to the electromagnetic wavelength. Antennas are usually described by different characteristic values, e.g., by the antenna gain or the antenna aperture. For antennas which are not defined by a physical area, such as monopoles or dipoles, the effective length l_{eff} or the AF can be used which is the following:

$$AF(f) = \frac{E(f)}{U(f)} \qquad (2)$$

where $U(f)$ is the voltage at the antenna terminals and $E(f)$ is the electric field strength at the antenna.

An appropriate special designed oil-filled GTEM cell [12] is used for the evaluation of the antenna sensitivity. A GTEM cell is an expanded coaxial conductor where a defined electromagnetic field can be applied to the equipment under test (EUT) without interference from the ambient electromagnetic environment. In the cell, a test volume is defined in which the EUT is situated. In the test volume, the cell provides a homogeneous electric field distribution E_{hom} and an orthogonal magnetic field of the TEM wave, ideally. In addition, the electric field strength E_{hom} in the test volume has to be known for AF calculation of the EUT.

The AF of a UHF PD sensor can be measured using the transmission factor (S_{21}) (Figure 3). The measurement setup consists of the oil-filled GTEM cell with installed UHF sensor and the vector network analyzer (VNA). The sensor insertion depth should be the same like in the later field measurements; here an insertion depth of 50 mm from tank wall is used.

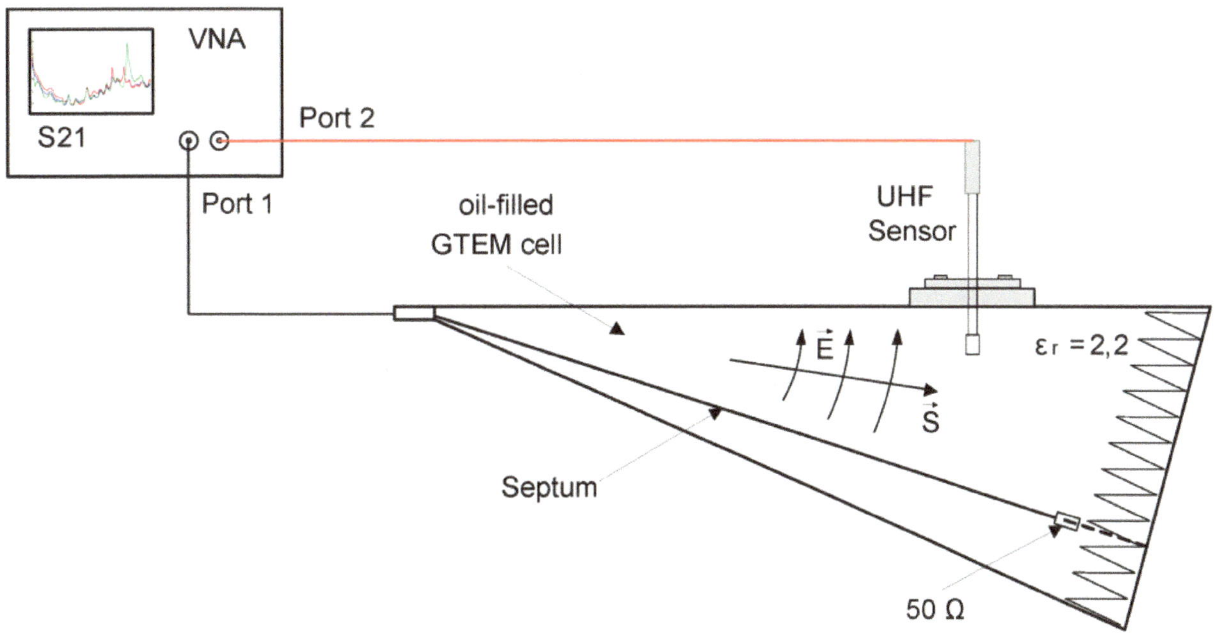

Figure 3. Transmission measurement (S_{21}) for antenna factor (AF) determination [13].

In this setup, the input port of the GTEM cell is excited with a sinusoidal frequency sweep from 300 kHz to 3 GHz generated by the VNA. The second port of the VNA simultaneously measures the resulting voltage at the output of the UHF sensor. The resulting transmission factor S_{21} can be converted into the AF of the UHF sensor if the electric field strength in the test volume is taken into account. Two different AF of a UHF drain valve sensor are presented in Figure 4. Compared to a real transformer, the GTEM cell characterization measurement only considers the influence of the sensor and can therefore be regarded as first step reaching a calibration procedure.

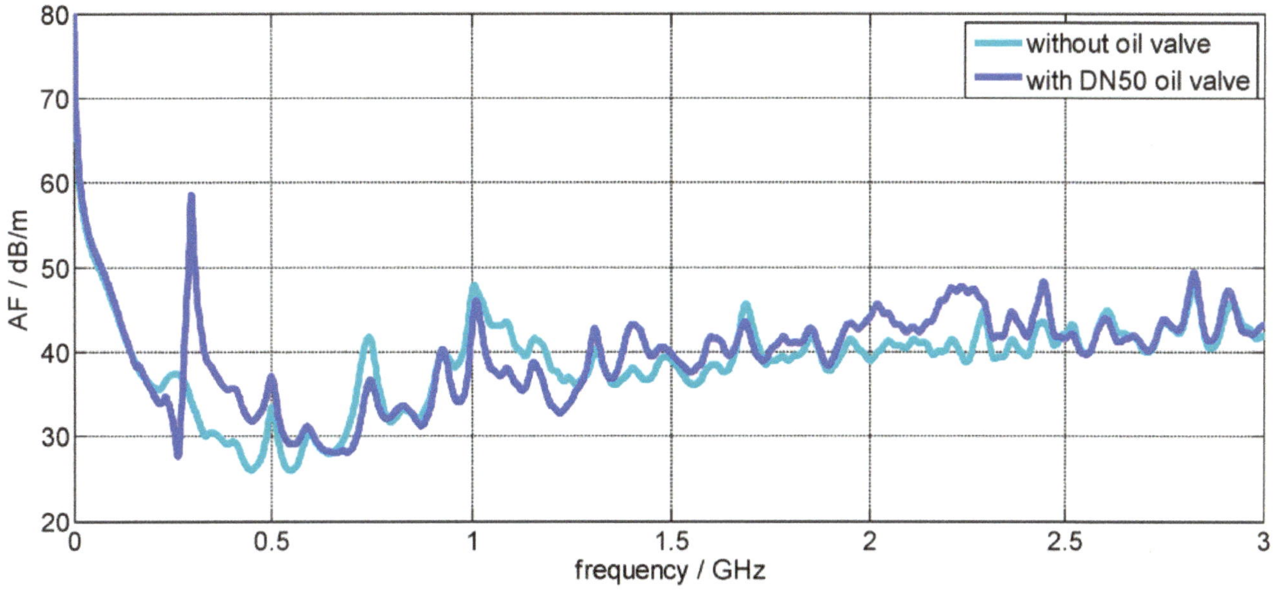

Figure 4. AF of UHF sensor measured in gigahertz-transversal-electro-magnetic (GTEM) cell [13].

To achieve a comparable method, UHF measurement systems require calibration including a validation of the UHF antenna sensitivity installed at the transformer. In a first step, a known UHF calibration impulse is injected into the measurement setup without antenna in order to calibrate the cable and the measurement device itself (Figure 5) [13].

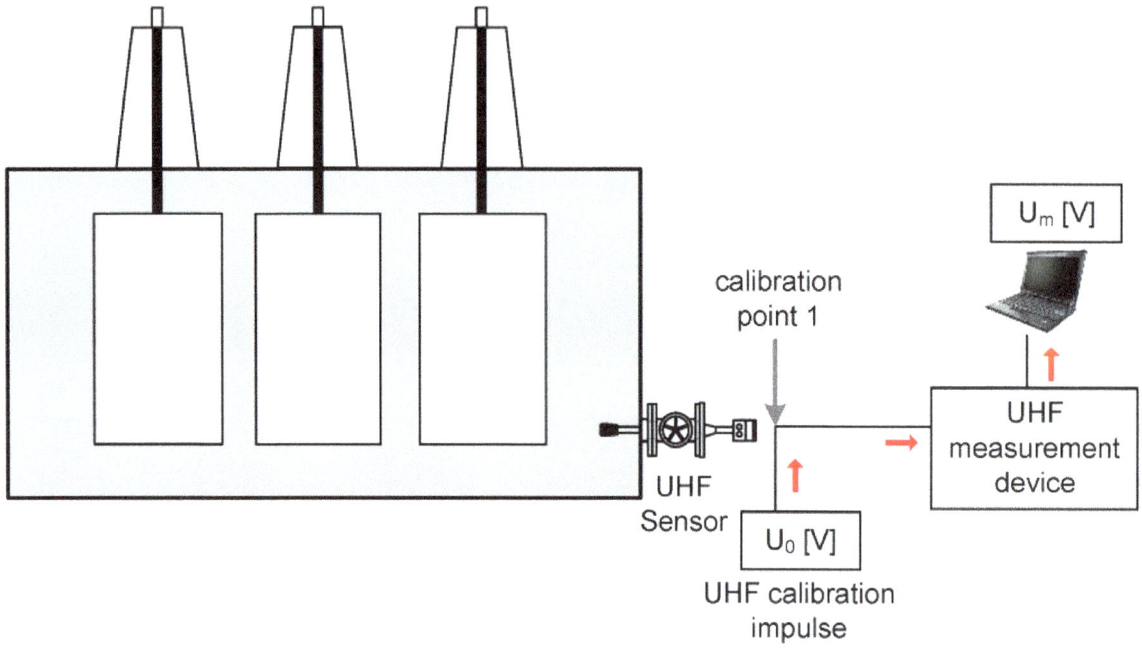

Figure 5. Calibration of measurement device and cables [13].

From this calibration measurement, the calibration factor K_1 can be calculated:

$$K_1 = \frac{U_0}{U_m} \tag{3}$$

In a second step, the sensor's characteristic is included into the calibrated path using its $AF(f)$. The known transfer function provided by the AF allows the shifting of the calibration point from the injection point of the calibrator to the UHF antenna inside the transformer. In order to simplify the calibration procedure, the frequency dependent $AF(f)$ can be reduced to a scalar correction factor AF_s which represents most common occurring UHF PD frequencies with sufficient accuracy. It is proposed to use the mean $AF(f)$ from 300 MHz to 1 GHz as scalar.

$$AF_s = \underset{300\text{MHz} \leqslant f \leqslant 1\text{GHz}}{\text{mean}} AF(f) \tag{4}$$

The resulting AF_s can be used in its delogarithmized form K_2 to correct time domain signals.

$$K_2 = 10^{\frac{AF_S}{20}} \tag{5}$$

Figure 6 shows an example of simplifying the $AF(f)$ to a mean AF_s and the new calibration point, which is shifted inside the transformer to the UHF antenna.

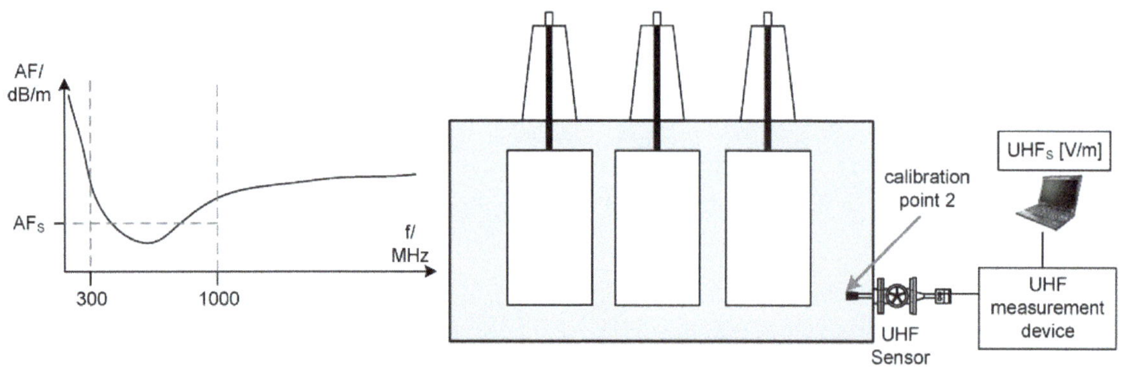

Figure 6. Calibration of the UHF sensor.

The idea of using a mean value of the frequency dependent $AF(f)$ is only valid for broadband UHF measurement systems. When using a narrowband measurement system, the actual AF at the used center frequency should be used for K_2.

The complete UHF calibration factor K_{UHF} is calculated:

$$K_{UHF} = K_1 \times K_2 \tag{6}$$

An impulse U_m measured with the UHF measurement system can now be corrected and a value correlated to the electrical field emitted of a PD is displayed. This value can be called "apparent UHF signal" (UHF_S) similar to "apparent charge (q_S)" of the electrical PD measurement. It is called apparent as it is not directly related to the actual PD value itself but allows comparison of different measurement systems (including UHF sensors, cables and measuring devices) due to its calibration.

$$UHF_S = K_{UHF} \times U_m \quad (7)$$

Online Monitoring of Partial Discharges

Online monitoring of power transformers, which supports established diagnosis methods, is steadily gaining importance. Continuous measurement and trend analysis allows to detect and trace undesirable changes at an early state. For PD, UHF monitoring represents an advantageous technique, because the measurement is done inside the tank and is thus much less sensitive to external noise. Additionally, it is applicable to transformers in service. The considerable amount of data generated requires appropriate evaluation; partially automated analysis is inevitable. A 50-year-old unit generator transformer with a rated voltage of 110/10 kV and a rated power of 120 MVA was monitored [14]. An online UHF PD measurement system recorded the UHF PD signals with 35 dB amplification and 9 MHz bandwidth at 505 MHz center frequency. Phase L1 was used for phase correlation. Due to the fact the generating unit is only in operation on demand, the transformer is not continuously in service. Measurements are available for approximately 65 days from 2009 to 2012. This case study presents an approach using phase-resolved PD pattern analysis. Typical patterns from known PD sources are reduced to an abstracted shape which uniquely characterizes the form of the PD source. This template is compared to measured phase resolved partial discharge pattern (PRPD) patterns gained from the monitoring data. Comparison between pattern and template is calculated by two-dimensional normalized cross-correlation algorithm. Source tracking over time is evaluated using continuous correlation. By introducing a set of templates for correlation, the progress of individual PD sources is determined.

Cross-correlation is an algorithm for recognition of patterns within images. The higher the similarity between two images, the higher is their correlation factor. In this contribution, the normalized cross-correlation is used providing values of correlation coefficients between −1 and +1 for each matrix element. Thus, cross-correlations of different images become comparable. A correlation coefficient of 1 indicates an exact match of the template (never occurs in practical pattern analysis), while −1 represents an area where image intersection and template are opposed (negative image, never occurs in practical pattern analysis). Figure 7 shows three typical PD patterns of this transformer.

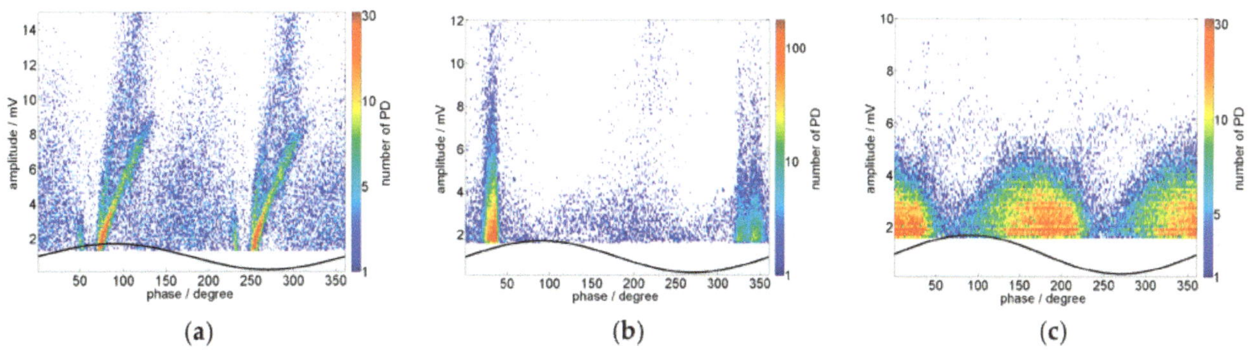

Figure 7. UHF phase resolved partial discharge patterns (PRPDs): (**a**) Pattern 1; (**b**) Pattern 2; and (**c**) Pattern 3 [14].

Patterns should be traceable over time. Therefore, the constant PD data stream is divided into segments with constant duration. For each segment, a PRPD pattern is generated and then cross-correlated with the templates. Determining an adequate duration period depends on the behavior of the source over time. Pattern 1 from Figure 7 shows high volatility. Therefore, duration is set to one minute. The maximum value of the correlation matrix represents the correlation coefficient for the time segment. An example is plotted in Figure 8. The correlation coefficient is calculated by using a template which resembles a pattern as shown in Figure 7 (Pattern 1). The correlation coefficient is shown in red, aligned to the left axis. For comparison, the number of PD/min is plotted (black, right axis).

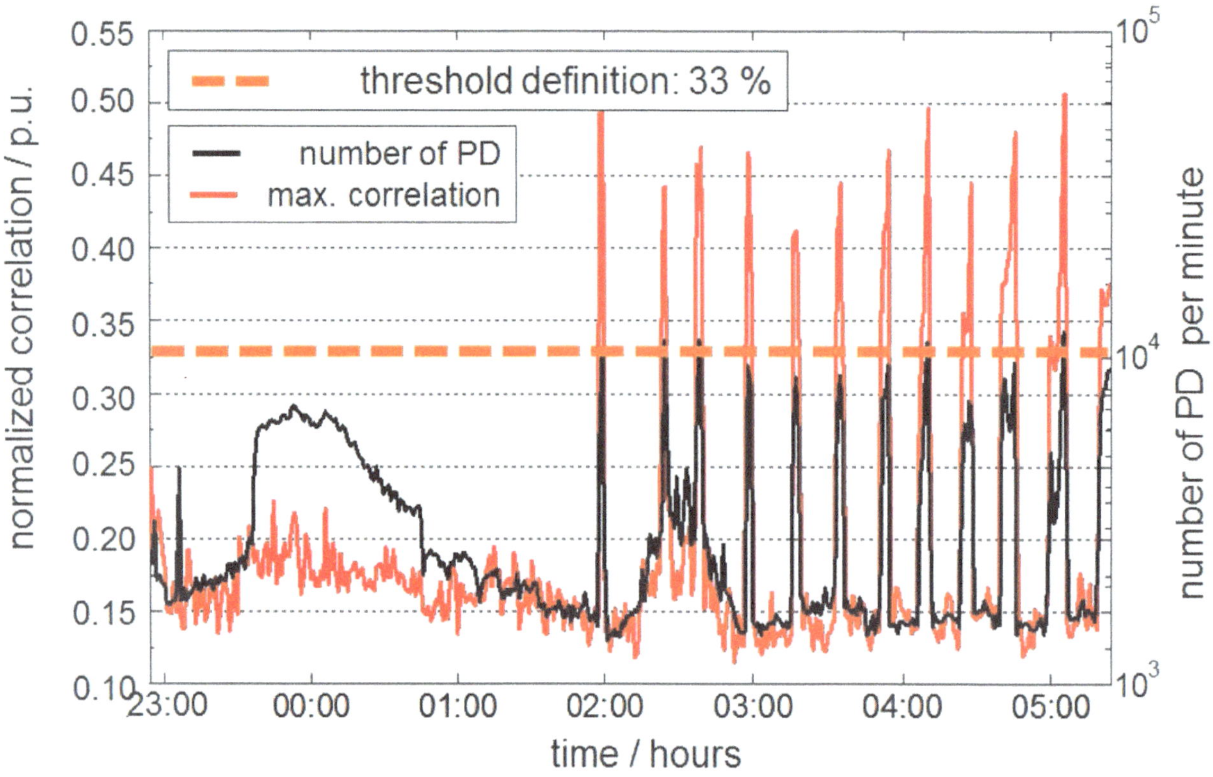

Figure 8. Number of PD per minute and correlation coefficient [14].

Figure 9 shows a bar graph for each pattern. Pattern 1 is only present 15% of the time showing intermitted behavior. Pattern 2 has a higher rate of appearance and can be detected 40% of the measurement time. Pattern 3 is the dominating source which can be detected 60% of the time over the entire period.

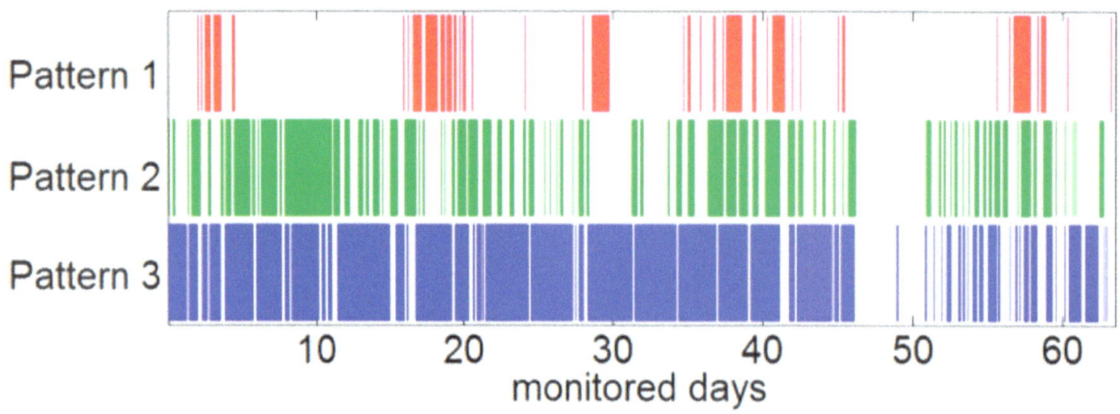

Figure 9. Results of recognition algorithm of determined Patterns 1–3 [14].

The essential benefit of the presented method is its application on very large PD datasets (e.g. from monitoring systems). In the case presented, 65 days of monitoring data were evaluated. Therefore, three PRPD patterns being typical for this specific transformer are evaluated. Using cross-correlation, it is possible to track patterns over the monitoring period in terms of their appearance and their phase positions.

Localization of Partial Discharges

Because of increasing DGA values, a 333 MVA, 400/220kV single-phase autotransformer was tested onsite and online for PD. The high noise level in field strongly disturbed the electrical PD measurements made according to [2] at frequencies lower than 1 MHz. Source of noise was in that case the 400 kV bus bar above the transformer producing audible corona discharge [15].

Consequently, UHF PD measurements for PD detection were performed in combination with acoustic PD measurements for localization of the PD source in order to get reliable results. In this case, the transformer was equipped with three oil valves and therefore three identical UHF Sensors were installed. Figure 10 shows the positions of the UHF PD sensors (UHF 1–UHF 3) and the acoustic PD sensors (A1–A6).

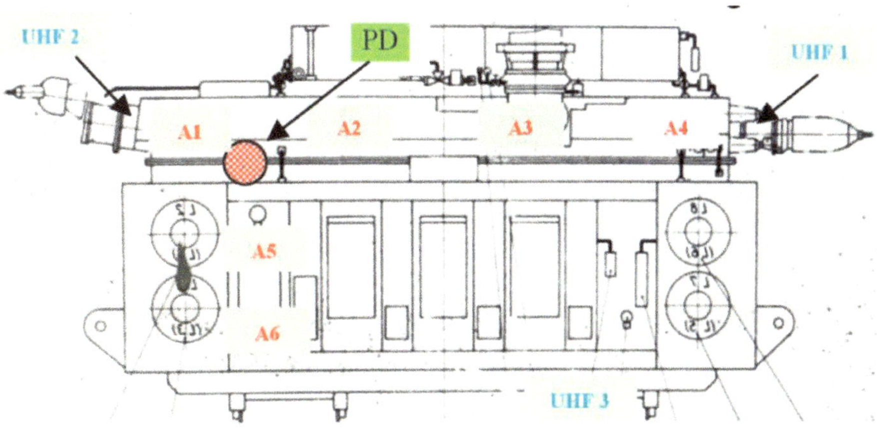

Figure 10. Positions of UHF and acoustic PD sensors and the localized PD source in a 333 MVA/400 kV single phase autotransformer [15].

At nominal voltage, UHF signals from internal PD sources were detectable with all three UHF sensors. The travelling speed of the electromagnetic waves is approximately 2/3 of speed of light inside the transformer. Thus, for localization, UHF signals are detectable almost the same time PDs occur. Conversely, the speed of acoustic waves is 1400 m/s, producing transit times within the range of milliseconds. With at least three acoustic sensors and corresponding time of flights, it is possible to calculate the intersection of the spheres and thus the PD location. It must be assumed that the acoustic waves travel directly in the line of sight from the PD source to the sensor without any reflections. Furthermore, the acoustical PD localization process also has to deal with acoustic waves travelling faster through the steel tank wall than through transformer oil. The time of flights of the acoustic signals can be computed accurately with the help of the Hinkley criterion [9], which is based on the signal energy of the measured signal [15].

As illustrated in Figure 10, the position of the located PD source is in the vicinity of the tap changer. Inaccuracy is thereby within the range of approx. 40 cm on all space axes. The time of flight differences were measured with six sensors, which are the three UHF sensors (UHF 1–UHF 3) and the three acoustic sensors (A2, A5, A6) placed near to the PD source [15].

After transportation of the transformer to the manufacturer the PD localization result was confirmed by an IEC triggered acoustic PD localization in a shielded test area and the transformer was detanked for repair. The visual inspection of the active parts at the tap changer confirmed the localization results. After repair procedure, the transformer passed the acceptance test without any indication of PD activity and was put back into service [15].

FREQUENCY RESPONSE ANALYSIS

Frequency response measurements are a widely applied technique for power transformer winding failure detection after lightning strike, short circuit or transport and are considered to be usually more sensitive than the conventional short-circuit voltage measurements [16]. Deviations between frequency responses indicate mechanical and/or electrical changes of the active part [17]. The basic principle of FRA is the investigation of differences between frequency responses. The electrical transfer behavior in the frequency range up to 2 MHz, in other words the frequency response, is identified by numerous characteristic resonances, which depend on the stray capacitances and inductances that are determined by the geometry and arrangement of active part and transformer tank [18]. The transfer function *TF*(*f*) as a state variable displays the geometric structural condition like a finger print. Known damage types, e.g., deformations, have an effect on the transfer functions like shifting, creating or disappearing of resonances [19]. However, interpretation of particular differences between two transfer function curves is the missing link between failure identification, measurement and assessment of the transformer. Figure 11shows the frequency ranges with their corresponding winding parts being sensitive towards mechanical changes as identified by CIGRE WG A2.26 [20]. For smaller power transformers, the frequency ranges of interest tend to be shifted towards higher frequency.

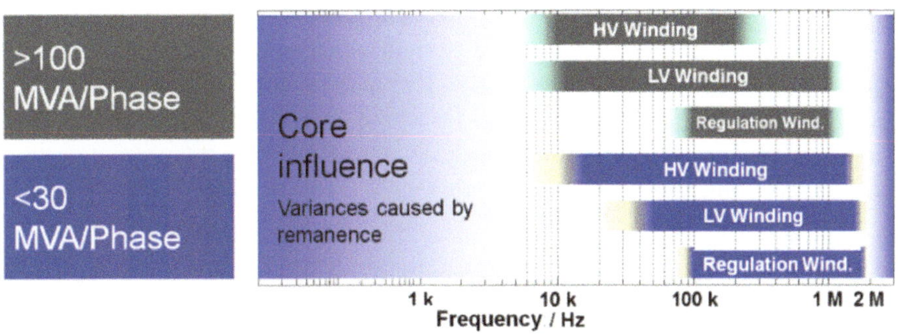

Figure 11. Frequency ranges with known sensitivity towards changes of the winding geometry [20].

Measuring Technique

The two most commonly used test types for transfer function measurement of power transformers are the so-called end-to-end transfer function measurement $\underline{TF}_{EE}(f)$ and the capacitive inter-winding (CI) measurement $\underline{TF}_{CI}(f)$ [20]. Figure 12shows the associated connection diagrams. The obtained transfer function of a measured phase then is:

$$\underline{TF}_{EE}(f) = \frac{\underline{U}_{2,EE}}{\underline{U}_1} \tag{8}$$

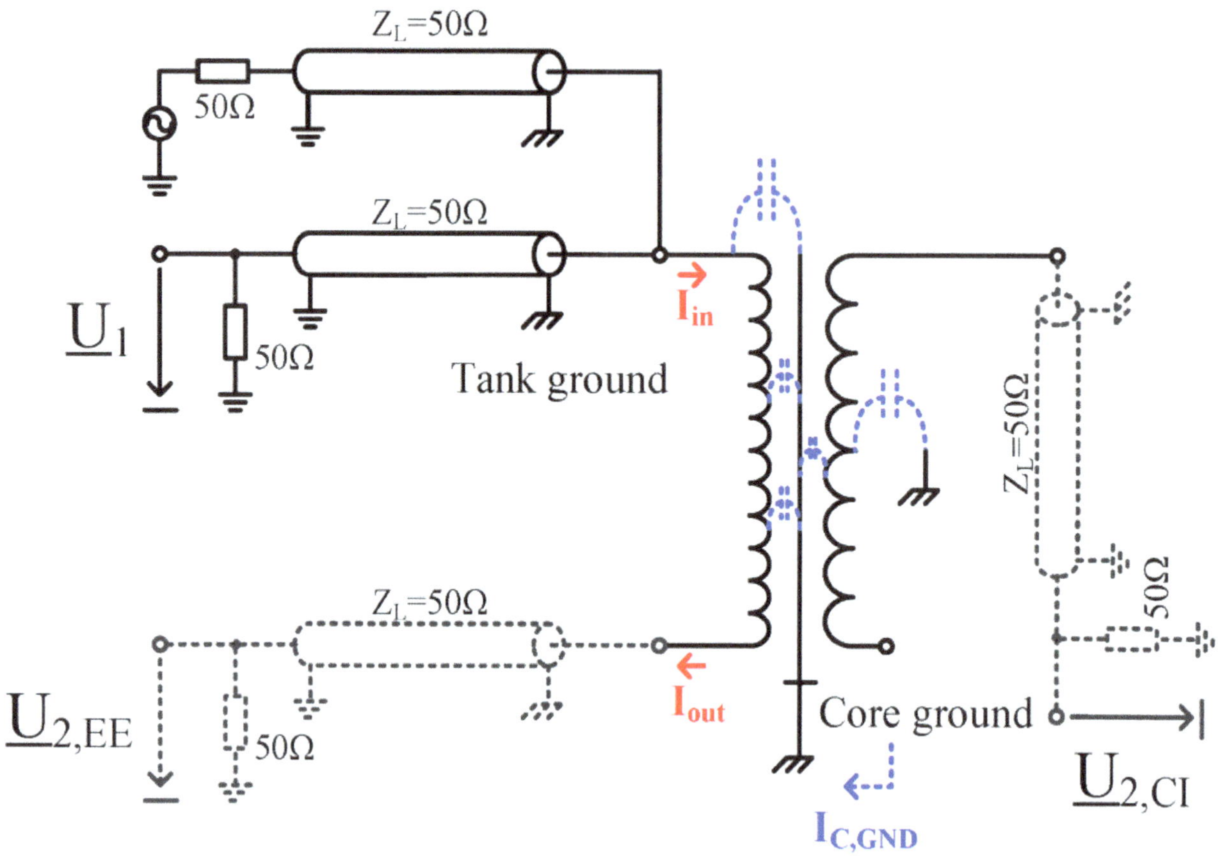

Figure 12. Connection schemes for measurements of end-to-end (EE) and capacitive inter-winding (CI) frequency response.

For the second measurement type, the transfer function is:

$$\underline{TF}_{CI}(f) = \frac{\underline{U}_{2,CI}}{\underline{U}_1} \tag{9}$$

Each connection can be used for measuring the frequency response. However, some studies [21,22] showed that the inter-winding connections are more sensitive to common mechanical changes.

Influencing Factors

In addition to mechanical changes of the windings, other factors may also affect the frequency response of a transformer [23]. This may result in a false interpretation, *i.e.*, an external factor influences the frequency response, but it is conceived to be a mechanical change. These factors are mainly the temperature, grounding practices, moisture contents, and shunt resistors [23,24]. The best situation occurs when two measurements are carried out in similar conditions, which requires that all of these factors be mentioned in the test sheet.

In order to explain misinterpretation, a case study is presented here. Figure 13 shows two traces of the middle phase of a 30 kVA/10 kV transformer measured at two different temperatures. The measurements were carried out on the same day and in a healthy condition. As can be seen, there are differences between these two traces. The amount of the difference extracted from these traces by numerical indices is approximately equal to that for 3 mm of axial displacement. As a result, this situation may be interpreted as a mechanical fault which is not the case. Other factors mentioned above may also change the frequency response and, therefore, special attentions should be paid to them during the measurement and the interpretation.

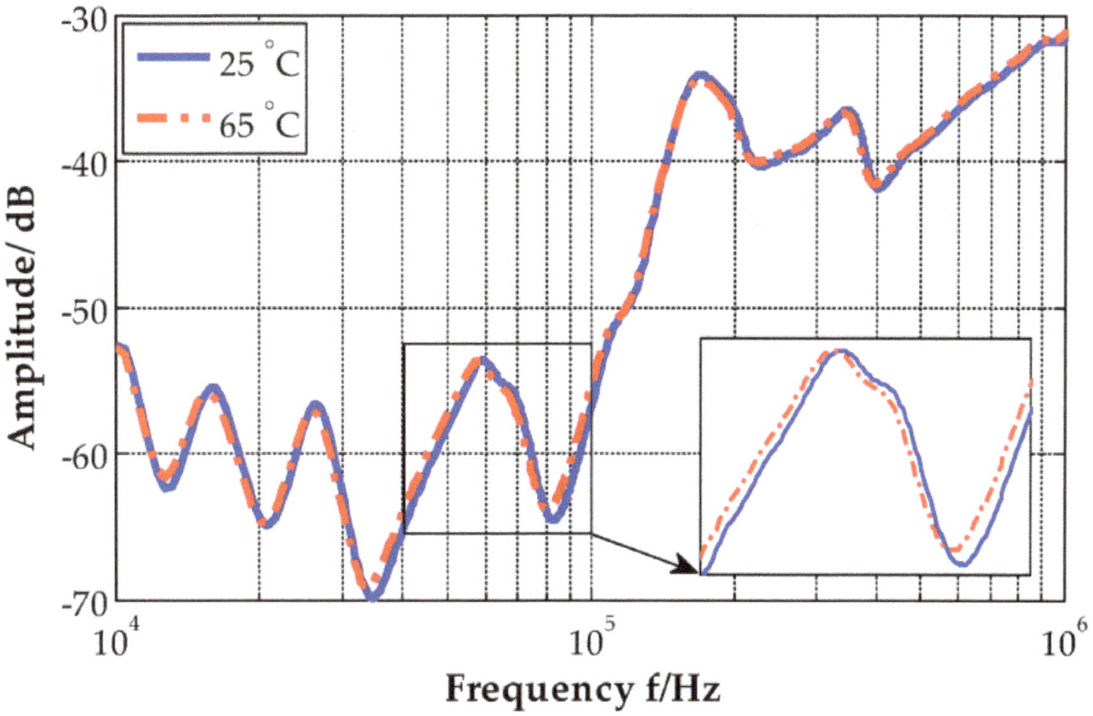

Figure 13. The middle phase frequency response of a three-phase oil transformer for two different temperatures. The traces correspond to the CI connection.

Interpretation of Frequency Response Measurement

The interpretation algorithms in the literature can be categorized into two main groups. In the first group, the transformer is modeled using circuit elements, and mechanical deformations are simulated by changing the values of these circuit elements [25,26,27,28]. After simulating different types and extents of mechanical deformations, comparison between the FRA traces extracted from the model with the one recorded from a real transformer can reveal the existence of a mechanical deformation in the transformer. In the second group, a numerical index is extracted from the FRA trace measured newly and the fingerprint. The interpretation, then, is carried out based on the values of indices [29,30,31,32,33,34].

Different amendments can be carried out regarding the interpretation; a case is mentioned here as an instance. Most of the numerical indices are based on only the magnitude response of the transfer function. However, the transfer function has a magnitude and a phase value in each frequency sample, *i.e.* a vector in the complex plane for each frequency sample, Figure 14. The index, Euclidean distance (ED) [24], calculates the magnitude differences of each frequency and, then, computes the root sum squared of them over the frequency range, Figure 14a. A new index, complex distance (CD), can be defined to include the phase information; it calculates the distance between two samples in the complex plane, Figure 14b,c shows the amounts of both indices for different steps of axial displacements in a winding. As can be seen, including the phase response does not impair the linearity of the index, but increases the sensitivity of the index to the mechanical change significantly,*i.e.*, it makes the detection of the mismatches between two FRA traces simpler. Plenty of such works still need to be carried out to provide a reliable assessment algorithm for the FRA interpretation.

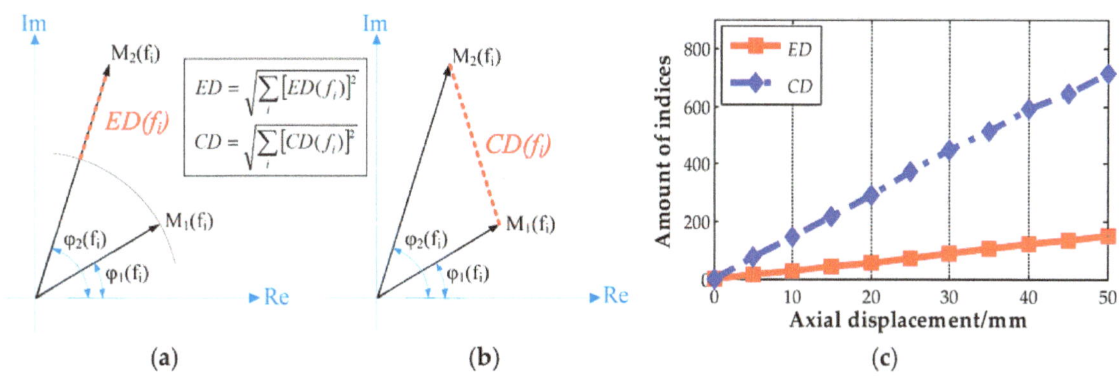

Figure 14. (**a**, **b**) The *i*th samples of the new and the fingerprint frequency response in the complex plane. *M* and ϕ corresponds to the magnitude and phase values of a sample; and (**c**) comparison of the two indices for different amounts of axial displacements [35].

Figure 15 shows a representative case of transfer functions of all phases measured at a 200 MVA 110 kV/65 kV three phase power transformer. There are characteristic frequency bands that reflect different electrical properties of the transformer [33]. The lower frequency bands FB1 and FB2 are dominated by the core magnetizing inductance and winding series capacitance, while FB3 and FB4 are related to the leakage inductances of the windings and the capacitances of adjacent windings [36,37]. The high frequency part of FB4 up to 2 MHz represents smaller details of the active part geometry together with the winding leads. Additionally, the bushing capacitances influence the high frequency part of the transfer functions [38].

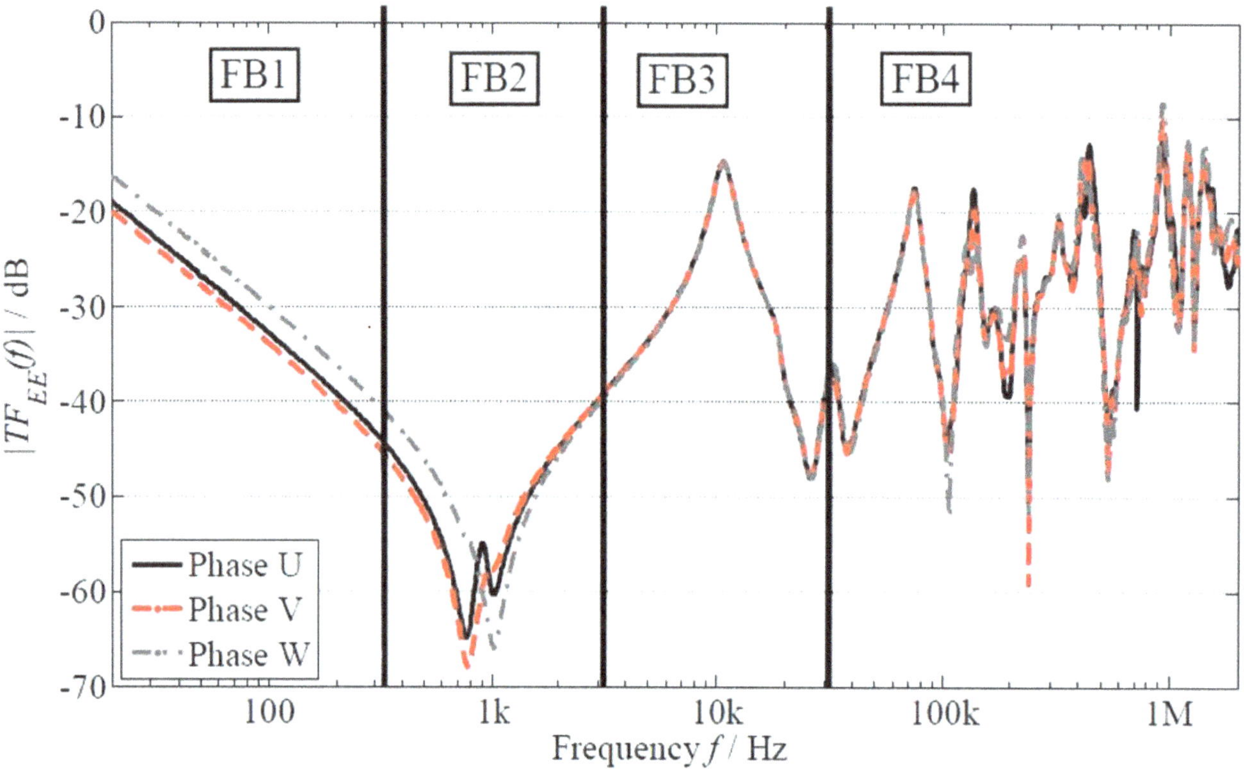

Figure 15. Characteristic frequency bands of measured end-to-end frequency responses of a three phase power transformer [33].

Differences between transfer functions of different phases are greater in FB1, FB2 and in the upper part of FB4. For FB1 and FB2, the differences can be explained by the unbalance of the core (magnetic circuit) and remanence of the core. Mismatching between phases in the upper part of FB4 can be explained by unbalances of the ground capacitances and the practical realization of the winding lead outs [39].

FB3 and the lower part of FB4 reflect the most important part of the frequency response. Deviations occurring in this frequency range indicate mechanical changes such as bulk winding movement or buckling of windings [33].

DISSOLVED GAS ANALYSIS

The established DGA provides relevant information about internal transformer faults. Natural aging, thermal and electrical failures generate typical fault gases, which dissolve in the insulating oil. By analyzing these fault gas profiles evaluation of kind and severity of the fault can be estimated [40,41]. Usually, DGA assesses the severity of emerging faults using annually taken oil samples and evaluates both, ratios of different gases and long term gas generation rates [42]. The applicability of generation rates is enforced by the raising number of installed DGA monitoring systems, which provide continuous trend analysis [43]. The existing DGA standards are based on relatively complicated procedures, which can be well controlled under laboratory conditions. The field application with online monitoring systems, however, requires robustness and high long time precision for reliable trend development, which is the main focus of online monitoring.

Dynamics between Oil and Gas Phase

A feasible gas based diagnosis requires the consideration of both, gas generation and gas losses. Gases escaping through the conservator tank may lead to misleading concentration levels and underrated gas generation rates especially for free breathing transformers [44]. In addition, the other way round is possible: gases from the ambient air, mainly nitrogen and oxygen, are slowly dissolving in the conservator tank oil and are afterwards mixed with the main tank oil, too [45]. An adequate model can help to assume the actual fault gas generation rate of a transformer. Thus, the severity of an upcoming fault could be determined more accurately for free breathing transformers. The loss of dissolved fault gases mainly depends on the conservator tank design. The oil is in direct contact to ambient air at free breathing conservator tanks. Fault gases can evaporate into the atmosphere and air gases can dissolve in the transformer oil. Free breathing conservator tanks for power transformers are common in several European and other countries [46]. Gas exchange does not occur at hermetically sealed transformers, transformers with a gas blanket or transformers with a membrane installed in the conservator tank.

The oil transports dissolved fault gases from the main into the conservator tank. There, the different fault gas concentrations between oil and air are equalized by diffusion, which results in a loss of fault gases and an increase of ambient air gases. Breathing depends on the exchange rate between main tank and conservator. The exchange rate depends on several factors: the entire oil volume of the transformer, the loading behavior of the transformer, the ambient temperature, the design of the conservator tank and its connection to the main tank. The oil flow is driven by two effects. The temperature dependent oil density forces a volume change of the oil and hence a flow from the main tank into the conservator tank at rising temperatures because oil expands with app. 0.076%/K. Additionally, a convection effect occurs (natural circulation), which is caused by the temperature gradient between main tank top oil temperature and the oil temperature in the conservator tank which approximately equals ambient temperature.

The flow rate for 26 different power transformers is calculated from thermal monitoring data in order to evaluate the effects of breathing [47]. The calculation only considers the oil flow due to volume change. The transformers' nominal power varies between 31.5 MVA and 850 MVA including generator step-up units, grid couplers and wind park transformers. Therefore, the loading and the cooling types are diverse. The resulting flow rates cover a wide range from 4 L/day to 36 L/day per 1000 L oil volume for grid coupling and wind park transformers, both with volatile load factors. Generation step-up units provide a flow rate of 1.8–6.2 L/day per 1000 L oil volume, with stable load factors. Oil directed water forced (ODWF) units provide even lower flow rates because of a temperature controlled cooling system.

Gas Evaporation via Conservator

The exchange of gases at the boundary surface between oil and ambient air in the conservator tank is simulated in a laboratory setup. Barrels are used as substitutes for a conservator tank. These barrels have a total volume of about 217 L. To vary the surface area between oil and ambient air inside, one experiment is carried out with an upright standing barrel with a boundary surface of about 0.26 m^2 and one with a horizontal barrel and a boundary surface of about 0.48 m^2. The oil volume is kept constant at 100 (±5) L. Thus, there is a gas volume with about 117 L over the oil, which is open to the atmosphere. All experiments use mineral oil. Before each experiment, the oil is degassed, dried, and filtered at 60 °C with a vacuum oil treatment system. After degassing a fault gas mix is dissolved in the oil using a porous Polytetrafluoroethylene (PTFE) plastic.

The influence of the boundary surface between oil and ambient air on the loss of fault gases is examined by the two surface areas 0.26 m^2 and 0.48 m^2. The experiments are performed at room temperature. Throughout the experiment the oil is not moved. Figure 16a shows the concentration profile of hydrogen for the two different boundary surfaces. The curves are exponential partial regressions to the corresponding measured values, with $C(t) = C_0 \exp(-\lambda t)$. The gas loss factor λ has the unit $1/h$ and depends on the boundary surface area.

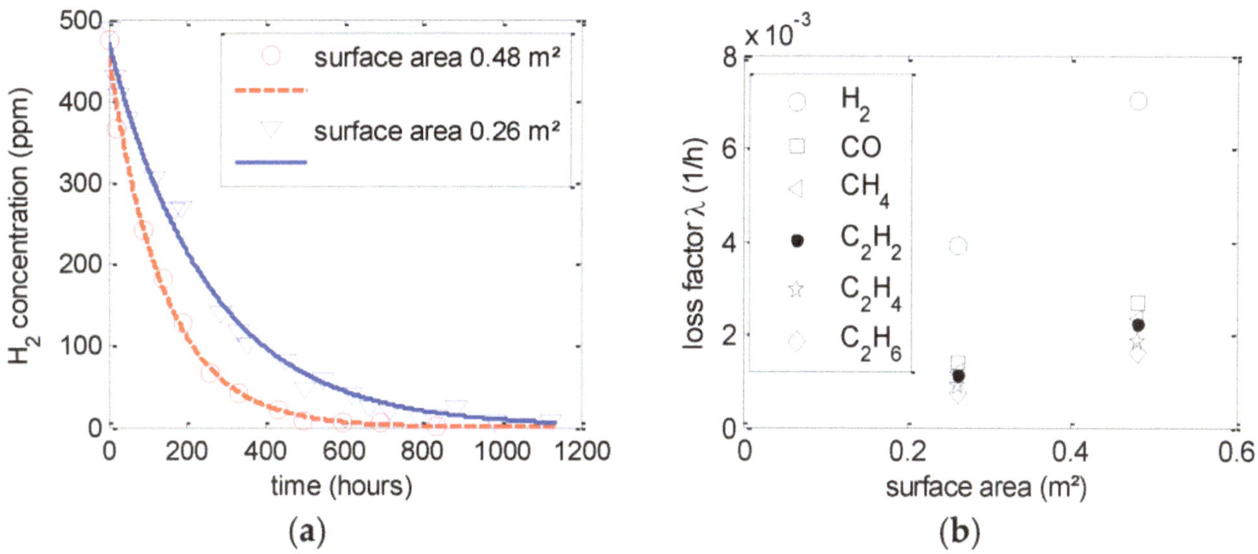

Figure 16. (**a**) Concentration trend of hydrogen for different boundary surface areas; and (**b**) gas loss factor over surface area for different fault gases.

The decrease of hydrogen concentration is faster for the larger surface area. Besides hydrogen, also other fault gases are analyzed with the same experimental setup. Figure 16b shows the gas loss factors λ of all gases for both surface areas. With a linear increasing surface area the gas loss factor also increases linearly. Thus, more gases evaporate. Hydrogen evaporates at the fastest rate, followed by CO and CH_4. C_2H_6 evaporates the slowest.

As the evaporation is also dependent on the influence of temperature, four different temperatures are investigated: room temperature (about 22 °C), 35 °C, 50 °C and 65 °C. All tests are performed using the same oil volume (100 L) and with a constant surface area between oil and ambient air (0.26 m^2). Figure 17a shows the fitted gas loss factors λ for all gases at different temperatures. The gas loss factors over temperature show an exponential increase. An exemplary exponential gradient over temperature is plotted for CO. Again, H_2 has a higher gas loss factor and is more volatile than the other presented gases.

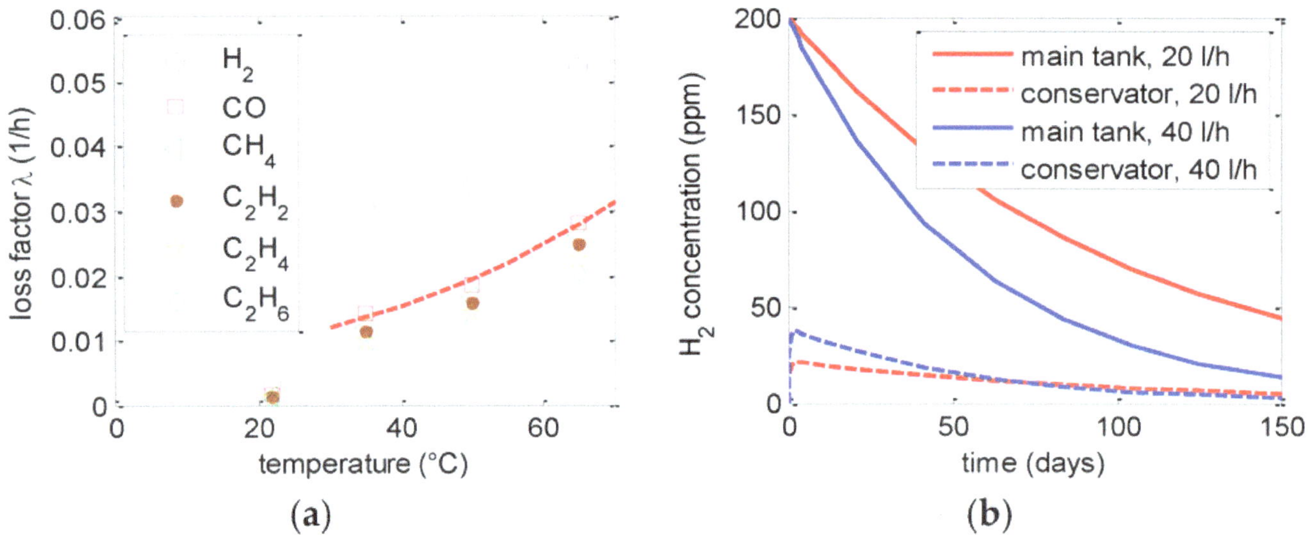

Figure 17. (**a**) Gas loss factors of different fault gases over temperature. Dashed line: exemplary exponential gradient for CO; and (**b**) concentration trend for H_2 in main tank (straight line) and conservator (dashed line) at different oil exchange rates.

The large deviation between the gas loss factor at 22 °C and 35 °C may be explained by the experimental procedure. No heating of the test barrel is necessary at room temperature (22 °C). At higher temperatures (35–65 °C), a ring-shaped barrel heater is used. Due to the radial heat input a convection effect occurs in the oil. This leads to an increased mixing of the oil and thus to a higher gas loss factor.

Calculation of Fault Gas Losses

The conducted experiments show the possibility to calculate the gas losses of a free breathing conservator tank by taking the oil temperature and the boundary surface area into account. Changes of the gas concentration in the main tank influence the concentration in the conservator tank and *vice versa*. Therefore, the derivative of each concentration is affected by the concentrations of both, main and conservator tank.

A system of two differential equations is used for modeling the gas transport (Equations (10) and (11)).

$$\frac{\partial K(t)}{\partial t} = \mathrm{m} \times K(t) + \mathrm{n} \times A(t) \tag{10}$$

$$\frac{\partial A(t)}{\partial t} = \mathrm{p} \times A(t) + \mathrm{q} \times K(t) \tag{11}$$

where $K(t)$ and $A(t)$ are the time-dependent gas concentrations of the main tank oil K and of the conservator tank oil A, respectively; and m, n, p and q are constants that depend on the volume, the gas loss factor and the oil exchange rate.

The exemplary transformer has an oil volume of 42,000 L for the simulation. The free breathing conservator tank of this transformer has an oil filled volume of 1700 L. The main tank contains a concentration of 200 ppm dissolved hydrogen as starting condition. At the beginning, there is no dissolved hydrogen in the conservator tank. The exchange between the main and conservator tank is assumed constant at 20 L/h (40 L/h) corresponding to the results shown in Section 5.2. Every hour, 20 L oil flow into the conservator and back into the main tank. Temperatures are also considered constant. The gas loss factor in the conservator tank is set to $\lambda_{H2} = 0.1$, which corresponds to a surface area of 1.5 m^2 and an average temperature of 15 °C. The λ_{H2} value is extrapolated from the values measured in the experiment presented above.

Figure 17b shows the concentration trend for hydrogen with the previously made assumptions. The oil concentration of hydrogen in the main tank decreases exponentially. The concentration in the conservator first increases and then decreases exponentially. At a starting concentration for H_2 of 200 ppm the loss corresponds to 0.086 ppm/h or 2.06 ppm/d for an oil exchange of 20 L/h. At higher oil exchange rates, the losses also rise. The comparison between the calculations for 20 L/h and 40 L/h shows, that the fault gas loss is strongly influenced by the exchange of oil between main and conservator tank. For an improved model further laboratory tests as well as the evaluation of existing measurements from real transformers have to be performed.

Online Monitoring of Fault Gases

A 600 MVA/380 kV generator step-up unit monitored by an IR-based multi gas analyzer showed in the beginning of July a strong and continuous increase of dissolved combustible gases (Figure 18). Because of the high TDCG rate (total dissolved combustible gases) of more than 100 ppm/day, in parallel to the DGA monitoring system oil samples were analyzed in the laboratory daily. This offered the possibility to check the deviation between monitoring and laboratory results. The dissolved gas was extracted by headspace and vacuum technology. Hydrocarbons showed very good agreement between monitoring and laboratory values (Figure 18a). The highest deviations occurred with hydrogen, which could be attributed to its high diffusivity. However, even though the oil samples analyzed in the laboratory revealed slightly different absolute gas values, the failure cause and trend corresponded fairly well with the monitoring results.

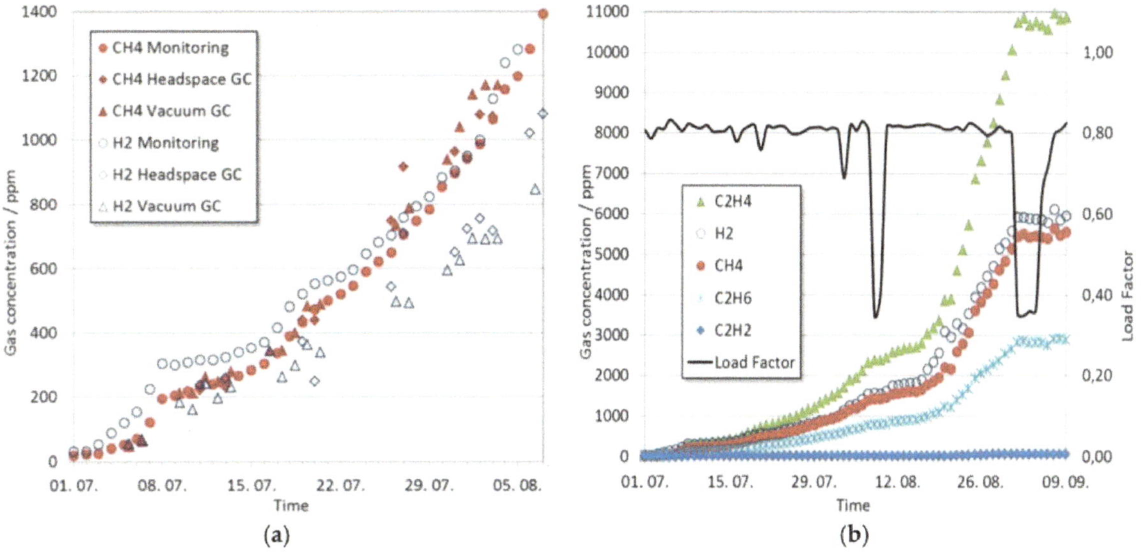

Figure 18. Development of gas concentration of a 600 MVA/380 kV GSU transformer: (**a**) first 5 weeks; and (**b**) 10 weeks.

Most notably hydrogen, methane, ethane and ethylene were generated in high amounts. The Duval triangle indicated a thermal fault with temperatures above 700 °C. The concentration of CO and CO_2 stayed almost constant, which indicated no involvement of paper in the fault. Because the amount of acetylene increased during the first 50 days quiet slowly just by approximately 15 ppm melting of the core laminations was also excluded as failure cause.

In the beginning of August, the total amount of dissolved combustible gases (approximately 5000 ppm) and respective generation rate exceeded the values given as "Condition 4" according to IEEE Std C57.104 [42], which suggests to remove the transformer from service under these circumstances. Despite of the high dissolved combustible gas amounts (Figure 18b) no undissolved gases could be detected in the Buchholz relay. Additionally, the very low level of acetylene led to the decision to keep the transformer in service until a spare transformer would be in place. If the rise of acetylene had exceeded 5 ppm/day in an interval of two days or an abrupt rise of 5 ppm/h occurred, the transformer would have been switched off immediately. Due to the continuous monitoring of the different fault gases, this strategy of supervised extension of operation was possible in order to prevent switching off the power station with its high financial impact.

The reduction of load factor showed a stagnation of the gas generation. Based on the DGA results as potential failure causes problems with current leads (e.g., increased resistance of selector contacts or crimped connections) or circulating currents were taken into account. Unfortunately the repetitive changing of tap positions did not change the gassing behavior. After a further increase of the gassing rate the loading of the transformer was reduced by 50%. Due to this load reduction, the gas production stopped immediately, which at least could be interpreted as a temporary intermission of the failure. The following load increase also showed no resumption of the failure.

On 10 September 2012, the transformer was switched off as scheduled and exchanged by a spare unit provided at the time. A visual inspection of the tank revealed that due to mechanical problems the tank got in contact with the magnetic core sheets, which resulted in circulating currents via the normal core grounding system. This circulating current is induced by the load dependent stray flux and can reach levels of some hundred amperes. At the contact point of tank and core a hot spot with oil carbon was visible indicating the decomposition of the oil. Additionally, the copper cable for the normal grounding of the core outside on top of the tank was melted, which led to the intermission of the failure.

With this example of a large generator step-up transformer it is shown how the careful attention of the dynamic fault gas development offers the possibility to keep a faulty transformer longer in service. This enables the utility to keep the power station longer in operation in order to provide a spare transformer.

MOISTURE MEASUREMENT

Motivation

Moisture is a critical parameter of the insulating system for several reasons. One is that the dielectric strength of insulating liquid is strongly affected by moisture. The risk of PDs and the bubble effect (evaporation of water vapor under high load conditions) is higher. Water is also hazardous to the cellulose based insulation as it accelerates further ageing. Thus, by knowing the water content of a transformer's solid insulation it is possible to get an idea of its ageing condition and thus plan further actions.

The first source of water in the lifetime of a transformer can be from improper drying in the factory. Another can be moisture ingress from the atmosphere. This can happen for free-breathing transformers if the air-drying system is poorly maintained. Another source is during repair in the field where there are no possibilities of dry storage. The last, if not the most important one is moisture originating from ageing of paper and pressboard. This means that even a perfectly serviced transformer will become wet with increasing age. There are studies which show that the ageing accelerates with higher moisture content and at the same time water is produced due to ageing (Figure 19) [48]. One can conclude that lifetime estimations of cellulosic insulating systems do not only depend on operating temperature but also on moisture content of solid insulation. This is the reason why water assessment in combination with other diagnostic techniques can help finding the health condition of a transformer.

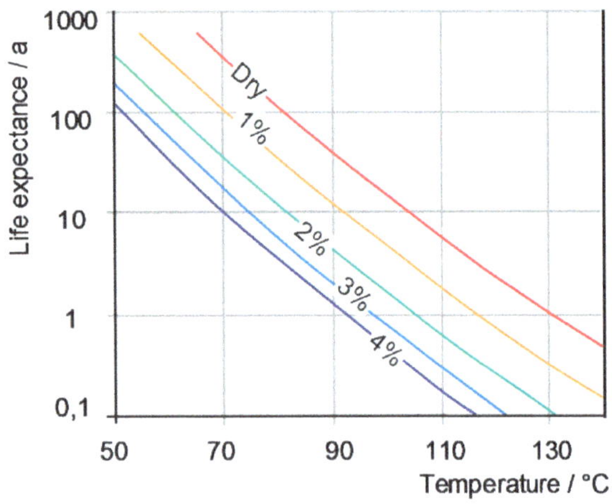

Figure 19. Life expectance of oil impregnated paper depending on temperature and water content [48].

Methods of Moisture Assessment

The aim is to know the water content of the critical solid insulation which is the thin paper. It is nearly impossible to measure this directly as the paper cannot be sampled for measurement without destroying it. Here only indirect methods can help. Basically there are two groups of techniques: first are the moisture equilibrium charts and second are the dielectric response measurements. The equilibrium diagrams use the fact that the moisture in the paper-oil system reaches an equilibrium and will be distributed between oil and paper. Thus, by knowing the temperature and water content of the oil, one can estimate the water content in the solid. If the thermal and kinetic equilibrium cannot be guaranteed, these charts can give erroneous estimations additionally to sources of error during oil sampling and measurement in the laboratory [49].

Dielectric Response Methods

Dielectric diagnostic methods assess moisture in paper or pressboard from dielectric properties like return voltage, polarization and depolarization currents (PDC) and dissipation factor. A time domain current measurement records the charging and discharging currents of the insulation. They are usually called PDC. FDS are derived from dissipation factor measurements, yet with a frequency range particularly enhanced for low frequencies. Figure 20 shows the typical s-shaped curve of the dissipation factor via frequency and the scientifically agreed interpretation scheme for a power transformer.

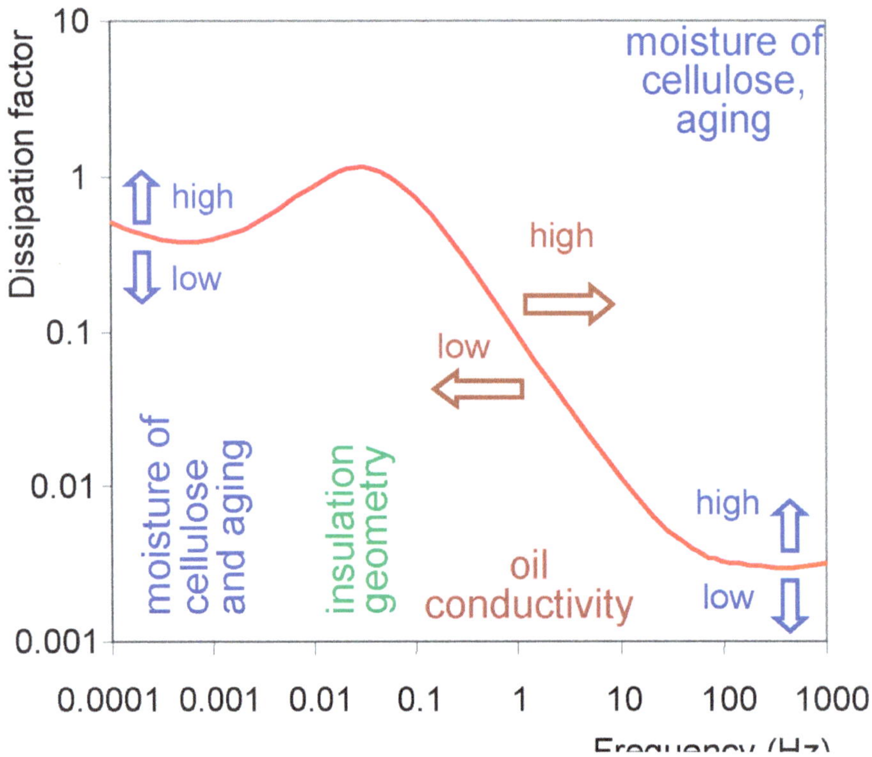

Figure 20. Interpretation of the dissipation factor *vs.* frequency of an oil-paper-insulation system [50].

The transformer under investigation has to be offline. The dielectric response of the oil-paper system should be measured between HV and low voltage (LV) winding. On one side a variable frequency voltage source is connected and on the other winding the resulting current is recorded. With this information the loss factor can be calculated. These currents are very low and can easily be disturbed in the field. One protective means onsite is guarding, which avoids surface currents over the bushings from LV to HV. These fault currents would impede the measurement of the correct dielectric response and thus make the water assessment difficult.

THERMAL MONITORING FOR LIFE ESTIMATION AND DYNAMIC RATING

Advanced Thermal Model

As dynamic thermal models are used in most online monitoring system applications, their empirical verification, statistical evaluation and fundamental development are of key importance [51,52]. Thermal models are based on the modeling of heat transfer phenomena from top of the tank to the ambient with simple equations. However, their accuracy is limited in some cases because they consider the thermal resistance of a transformer as a constant value. Therefore, there is a need to improve the accuracy of these models with a more accurate determination of the thermal resistance.

An advanced thermal model is used to calculate the top-oil temperature which is based on IEC 60076-7 model [53]. The thermal-electrical analogy of the model is shown in Figure 21. The thermal resistances of this model are temperature dependent values and affected by the bottom-oil temperature and the temperature of the oil in the cooler [54].

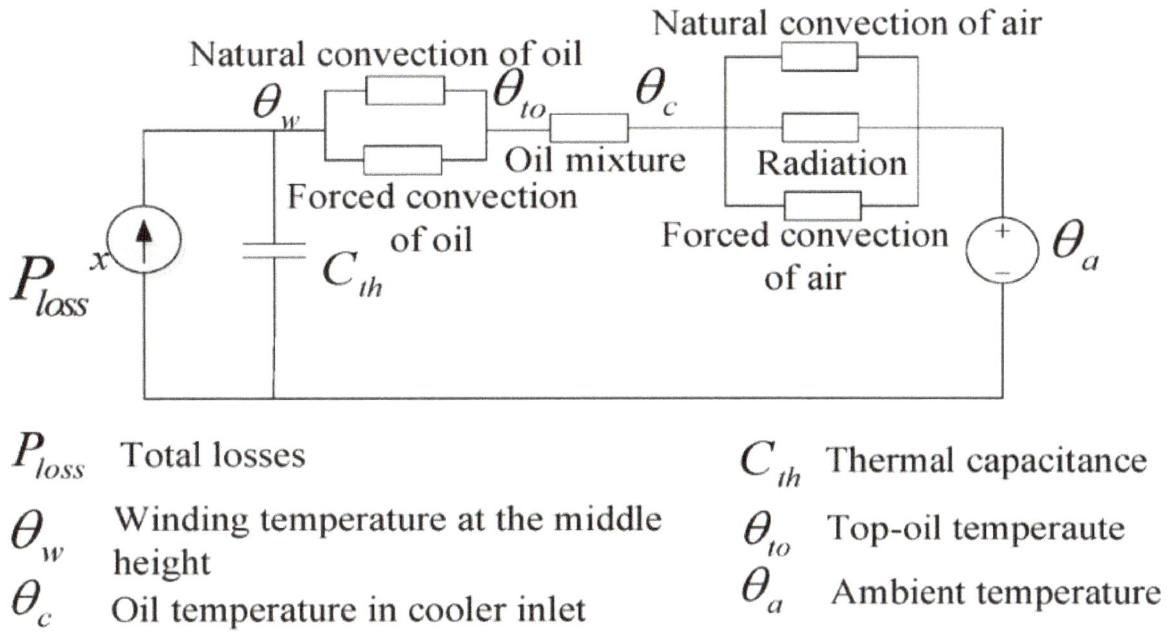

Figure 21. Thermal-electrical analogy of the advanced thermal model [54].

In order to use a physical thermal model in an online monitoring system, the performance and accuracy of the model should be firstly validated. The model is applied on an oil directed air forced (ODAF) transformer whose main characteristics are listed in Table 3. For this transformer, the model has been parameterized using a dataset measured in June 2001, from beginning to the end of the month, during normal operation of the transformer. However, for validation of the model, the dataset measured in July 2001, from beginning to the end of the month, is used. The measured and calculated top-oil temperature using the advanced thermal model, load factor, ambient temperature, and number of running pumps and fans in July, from beginning to the end of the month, is illustrated in Figure 22a.

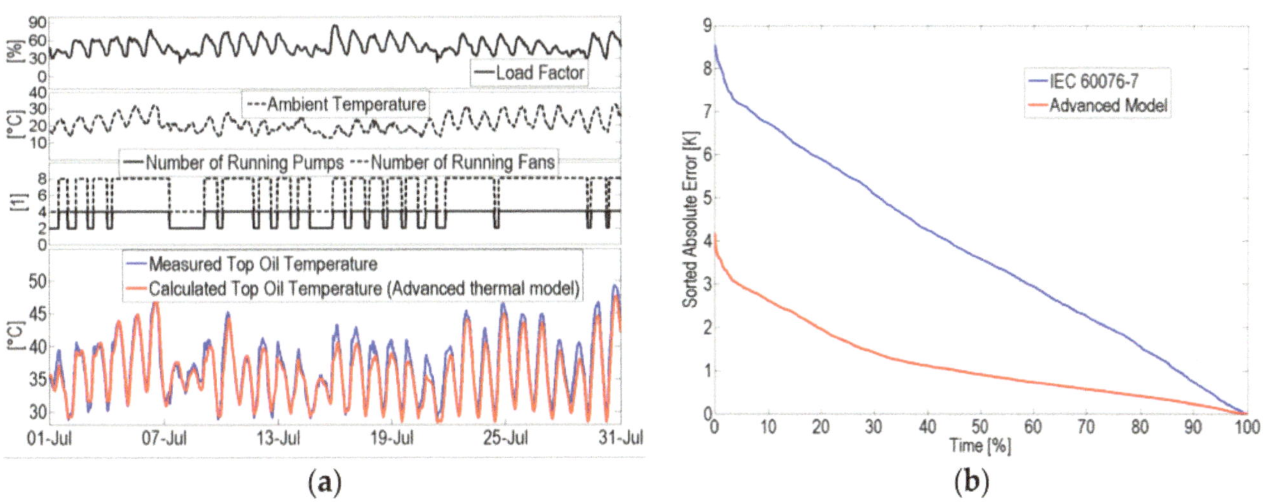

Figure 22. (**a**) The calculated and the measured top-oil temperature during the normal operation of the transformer in July; and (**b**) the error duration curves of IEC 60076-7 and the advanced thermal model.

Table 3. Characteristics of the transformer under study. ODAF: oil directed air forced.

Characteristic	Value	Characteristic	Value	Characteristic	Value
Power (MVA)	333	Mass of active parts and tank (to)	200	Number of Fans	8
Short circuit losses (kW)	510	Mass of oil (to)	50	Number of pumps	4
No load losses (kW)	47	Thermal capacitance (W·s/K)	199,050	Cooling system	ODAF

In order to compare the performance of the advanced thermal model with the model proposed in IEC 60076-7 [53], the error duration curve of each model is used which is shown in Figure 22b. The error duration curve shows the sorted absolute error dependent on the per cent time duration inside the observation time-window. As can be seen, the advanced thermal model represents better performance compared with IEC 60076-7 model for the calculation of the top-oil temperature. Due to the accuracy and simplicity of the advanced model, it can be used in an online monitoring system for monitoring of the cooling system and calculation of the overload capability of transformers.

Monitoring of Cooling Unit

The transformer as described in Table 3 experienced a malfunction in a part of its cooling system due to a failure of the power supply which was not in accordance with the information supplied by the control system. The failure occurred on 21 December 2001 at 6:00 a.m. when the control system aimed to switch on two additional fans in order to reduce the oil temperature. However, due to the failure in the power supply of the fans, they did not start to run. The site check revealed later that two out of eight fans failed to operate while the pump related to the failed fans kept in operation.

The faulty operation of the cooling system can be detected based on a proper online algorithm for analysis of the deviation of the calculated top-oil temperature from the measured top-oil temperature [53,55]. Figure 23 shows the alarm signal of the monitoring system, the number of running fans provided by the control system, and the standardized error between the calculated and the measured top-oil temperature for the described malfunction of the fans. As can be seen, after failure of the two fans, the calculated temperature starts to deviate from the measured temperature indicating that a failure occurred in the cooling system. The standardized error exceeds the maximum allowable value leading to an alarm on 21 December 2001 at 9:00 a.m. In other words, the failure in the cooling system is detected 3 h (three steps) after the occurrence. Therefore, online thermal monitoring provides the ability to detect a malfunction of the cooling system with a short response time.

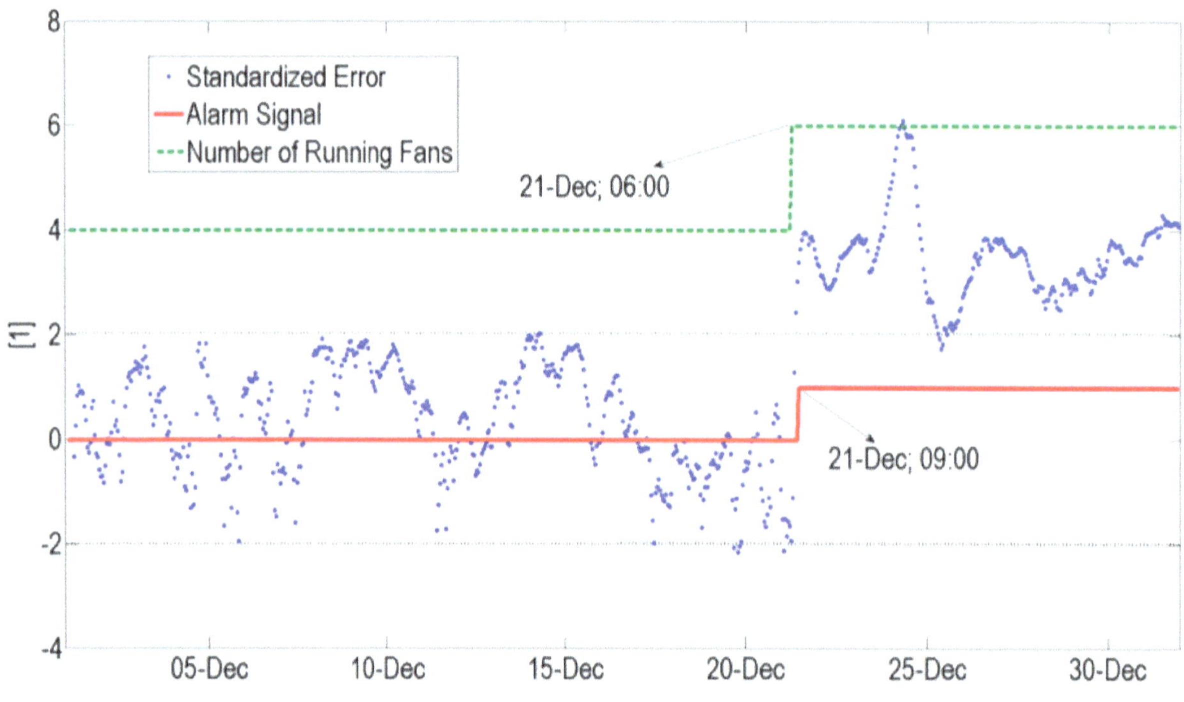

Figure 23. Standardized error between measured and calculated oil temperature for faulty operation of fans.

Assessment of Overload Capability

According to [53], a higher-than-rated load may be applied, but regarding the relative thermal ageing rate this loading should be equivalent to the rated load at normal ambient temperature (40 °C). This is achieved by taking advantage of low ambient temperatures during the rest of the load cycle. In other words, long-term overloading can be achieved as long as the hot spot temperature of the transformer does not exceed the allowed hot spot temperature corresponding to unity relative aging rate whose value for non-thermally upgraded paper is 98 °C.

Thus the hot spot temperature is the key word for the calculation of the overload capability of the transformer.

The hot spot temperature Θ_{hs} is calculated using the top-oil temperature Θ_{to} as follows [53]:

$$\Theta_{hs} = \Theta_{to} + H \times g_r \times k^y \tag{12}$$

where H is the hot spot factor. For medium and large power transformer, the value of $H = 1.3$ can be considered [53]. By solving this equation for the load factor k the long term overload capability of transformer is calculated considering that the hot spot temperature should not exceed 98 °C. It is assumed that the full cooling power of the transformer is in operation. Therefore, the overload capability OV depends only upon the ambient temperature Θ_a. This dependency can be written as a linear equation [56]:

$$OV = a \times \Theta_a + b \tag{13}$$

Figure 24a illustrates the load factor, ambient temperature as well as the overload capability of the transformer according to Table 3. Of course, the loading capability of the transformer is increased at low ambient temperature. The dependency of the overload capability on the ambient temperature is depicted in Figure 24b. It should be noted that the cooling unit is an over-dimensioned by 33%, because two out of eight fans of this transformer are spare fans, which cause a very high overload capability at low ambient temperature and even at the nominal ambient temperature of 40 °C. However, limitations concerning other parts of a transformer, e.g., bushings and On-load tap changer (OLTC) during overloading periods, should also be kept in mind.

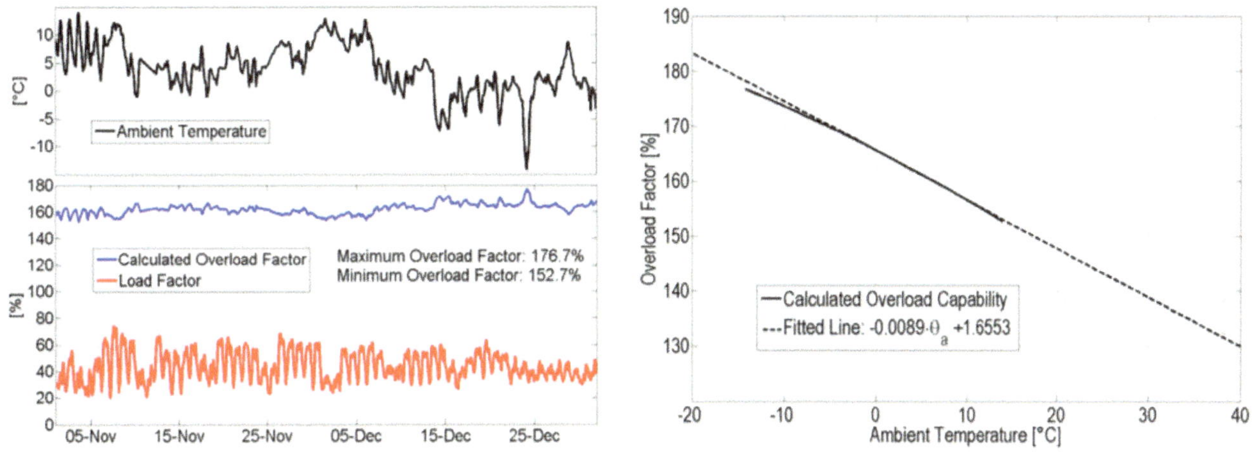

(a) (b)

Figure 24. (**a**) Ambient temperature, long-term overload factor and load factor; and (**b**) dependency of overload capability on ambient temperature [56].

CONCLUSIONS

PD-measurement set-up according to [2] often has sensitivity limitations in case of onsite/online measurements because of the high noise level in field. UHF PD measurements can easily support PD measurements even online and are mostly sensitive enough against PD. As a result of shielding characteristics of the transformer tank against external electromagnetic waves, a clear decision can normally be made concerning the PD activity of the test object. A calibration procedure for the UHF method is proposed as it is necessary to ensure reproducibility and comparability of UHF measurements. Acoustic PD measurements for localization purposes should not be used exclusively and benefit from a sensitive trigger method as, e.g., UHF for double checking and noise suppression.

FRA is a helpful diagnostic tool when clearing up transformer failures after short circuit events or drop out after lightning strokes. Time-based comparison of frequency responses of a certain transformer unit promises most reliable assessments, as design-based or type-based comparisons are always critical due to natural unbalances between phases in design and inaccuracies of the production process.

The established DGA provides relevant information about internal transformer faults. A feasible gas based diagnosis for free breathing transformers requires the consideration of both gas generation and gas losses. An adequate model can help to assume the actual fault gas generation rate of a transformer. Thus, the severity of an upcoming fault could be determined more accurately for free breathing transformers.

Dielectric response methods deliver reliable results about the moisture condition of a transformer. They can be applied onsite even under conditions where the equilibrium state cannot be guaranteed. Disturbances can be reduced by proper connections and guarding.

An improved thermal model which takes the temperature dependency of thermal resistances into account allows thermal monitoring with a high accuracy. Thus, overloading capabilities dependent on ambient temperatures can be determined. During overloading periods, the cooling system of a power transformer plays a decisive role in transferring the heat generated in the windings to the surrounding ambient. Faulty operation of the cooling system can be detected by the proposed algorithm with a short response time.

The broadness and improvements of the presented diagnostic techniques show that the power transformer is no longer a black box that does not allow a view into its internal condition. Reliable and accurate condition assessment is possible leading to more efficient maintenance strategies.

ACKNOWLEDGMENTS

The authors wish to thank the editors of Energies, and Issouf Fofana for their kind invitation to present this feature article.

AUTHOR CONTRIBUTIONS

Sebastian Coenen and Martin Siegel contributed to the chapter 3. Mohammad Hamed Samimi contributed to chapter 4. Andreas Müller contributed to chapter 5. Mohammad Djamali contributed to chapter 7.

Conflicts of Interest

The authors declare no conflict of interest.

Abbreviations

AF	Antenna factor
CI	Capacitive inter-winding
CIGRE	International council of large electric systems
DGA	Dissolved gas analysis
EUT	Equipment under test
FB	Frequency band
FRA	Frequency response analysis
GSU	Generator step-up unit
GTEM	Gigahertz-transversal-electro-magnetic
IEC	International Electrotechnical Commission
ODAF	Oil directed air forced
ODWF	Oil directed water forced
OLTC	On-load tap changer
PD	Partial discharge
PRPD	Phase resolved partial discharge pattern
PTFE	Polytetrafluoroethylene
TDCG	Total dissolved combustible gases
TF	Transfer function
UHF	Ultra high frequency

REFERENCES

1. Chakravorti, S.; Dey, D.; Chatterjee, B. *Recent Trends in the Condition Monitoring of Transformers*; Power Systems Springer-Verlag: London, UK, 2013.
2. IEC 60270. *High-Voltage Test Techniques–Partial Discharge Measurements*; International Electrotechnical Commission: Geneva, Switzerland, 2000.
3. Tenbohlen, S.; Jagers, J.; Vahidi, F.; Bastos, G.; Desai, B.; Diggin, B.; Fuhr, J.; Gebauer, J.; Krüger, M.; Lapworth, J.; *et al.Transformer Reliability Survey*; Technical Brochure 642 CIGRE: Paris, France, 2015.
4. Tenbohlen, S.; Jagers, J.; Bastos, G.; Desai, B.; Diggin, B.; Fuhr, J.; Gebauer, J.; Krüger, M.; Lapworth, J.; Manski, P.; *et al.* Development and Results of a Worldwide Transformer Reliability Survey. In Proceedings of the CIGRE SC A2 Colloquium, Shanghai, China, 20–25 September 2015.
5. Gulski, E.; Strehl, T.; Muhr, M.; Tenbohlen, S.; Meijer, S.; Judd, M.D.; Bodega, R.; Lemke, E.; Jongen, R.A.; Coenen, S.;*et al. Guidelines for Unconventional Partial Discharge Measurements*; Technical Brochure 444; CIGRE: Paris, France, 2010.
6. Judd, M.D. Experience with UHF Partial Discharge Detection and Location in Power Transformers. In Proceedings of the Electrical Insulation Conference, Annapolis, MD, USA, 5–8 June 2011.
7. Fuhr, J. Benefits and Limits of Advanced Methods Used for Transformer Diagnostics. In Proceedings of the IEEE Electrical Insulation Conference, Montreal, QC, Canada, 31 May–3 June 2009.
8. Coenen, S.; Tenbohlen, S.; Markalous, S.M.; Strehl, T. Sensitivity of UHF PD measurements in power transformers.*IEEE Trans. Dielectr. Electr. Insul.* **2008**, *15*, 1553–1558.
9. Markalous, S.M.; Tenbohlen, S.; Feser, K. Detection and location of partial discharges in power transformers using acoustic and electromagnetic signals. *IEEE Trans. Dielectr. Electr. Insul.* **2008**, *15*, 1576–1583.
10. Sikorski, W.; Siodla, K.; Moranda, H. Location of partial discharges sources in power transformers based on advanced auscultatory technique. *IEEE Trans. Dielectr. Electr. Insul.* **2012**, *19*, 1948–1956.
11. Venkatesan, S.; Stewart, B.G.; Judd, M.D.; Reid, A.J.; Fouracre, R.A. Analysis of PD in Voids Using Correlation Plots of RF Radiated Energy and Apparent Charge. In Proceedings of the International Conference on Solid Dielectrics, Winchester, UK, 8–13 July 2007.
12. Siegel, S.; Tenbohlen, S. Design of an Oil-filled GTEM Cell for the Characterization of UHF PD Sensors. In Proceedings of the International Conference on Condition Monitoring and Diagnosis (CMD), Jeju, Korea, 21–25 September 2014.
13. Coenen, S.; Siegel, M.; Luna, G.; Tenbohlen, S. Parameters influencing Partial Discharge Measurements and their Impact on Diagnosis. In Proceedings of the Monitoring and Acceptance Tests of Power Transformers CIGRE Session, Paris, France, 22–26 August 2016.
14. Müller, A.; Beltle, M.; Siegel, M.; Tenbohlen, S. Assessment of UHF PD Monitoring Data by Means of Pattern Recognition. In Proceedings of the 18th International Symposium on High Voltage Engineering, Seoul, Korea, 25–30 August 2013.
15. Coenen, S.; Tenbohlen, S. Location of PD sources in power transformers by UHF and acoustic measurements. *IEEE Trans. Dielectr. Electr. Insul.* **2012**, *19*, 1934–1940.
16. Gomez-Luna, E.; Aponte Mayor, G.; Gonzalez-Garcia, C.; Pleite Guerra, J. Current status and future trends in frequency-response analysis with a transformer in service. *IEEE Trans. Power Deliv.* **2013**, *28*, 1024–1031.
17. Tang, W.H.; Wu, Q. *Condition Monitoring and Assessment of Power Transformers Using Computational Intelligence*; Power Systems Springer-Verlag: London, UK, 2011.
18. Wang, Z.D.; Li, J.; Sofian, D.M. Interpretation of transformer FRA responses; Part I: Influence of winding structure.*IEEE Trans. Power Deliv.* **2009**, *24*, 703–710.
19. Rahimpour, E.; Christian, J.; Feser, K.; Mohseni, H.

Transfer function method to diagnose axial displacement and radial deformation of transformer windings. *IEEE Trans. Power Deliv.* **2003**, *18*, 493–505.

20. Cigre, W.G. *Mechanical Condition Assessment of Transformer Windings using Frequency Response Analysis (FRA)*; Brochure 342; CIGRE: Paris, France, 2008.
21. Jayasinghe, J.A.S.B.; Wang, Z.D.; Jarman, P.N.; Darwin, A.W. Winding movement in power transformers: A comparison of FRA measurement connection methods. *IEEE Trans. Dielectr. Insul.* **2006**, *13*, 1342–1349.
22. Samimi, M.H.; Tenbohlen, S.; Shayegani Akmal, A.A.; Mohseni, H. Effect of different connection schemes, terminating resistors and measurement impedances on the sensitivity of the FRA method. *IEEE Trans. Dielectr. Insul.* **2016**. unpublished work.
23. Reykherdt, A.A.; Davydov, V. Case studies of factors influencing frequency response analysis measurements and power transformer diagnostics. *IEEE Electr. Insul. Mag.* **2011**, *27*, 22–30.
24. Samimi, M.H.; Tenbohlen, S.; Shayegani Akmal, A.A.; Mohseni, H. Effect of Terminating and Shunt Resistors on the FRA Method Sensitivity. In Proceedings of the International Power System Conference, Tehran, Iran, 23–25 November 2015.
25. Bagheri, M.; Phung, B.T.; Blackburn, T. Influence of Temperature and Moisture Content on Frequency Response Analysis of Transformer Winding. *IEEE Trans. Dielectr. Electr. Insul.* **2014**, *21*, 1393–1404.
26. Mitchell, S.D.; Welsh, J.S. Modeling power transformers to support the interpretation of frequency-response analysis.*IEEE Trans. Power Del.* **2011**, *26*, 2705–2717.
27. Darwin, A.W.; Sofian, D.; Wang, Z.D.; Jarman, P.N. Interpretation of Frequency Response Analysis (FRA) Results for Diagnosing Transformer Winding Deformation. In Proceedings of the CIGRE 2009 6th Southern Africa Regional Conference, Cape Town, South Africa, August 2009.
28. Rahimpour, E.; Tenbohlen, S. Experimental and theoretical investigation of disc space variation in real high-voltage windings using transfer function method. *IET Electric Power Appl.* **2010**, *4*, 451–461.
29. Ghanizade, A.J.; Gharehpetian, G.B. ANN and cross-correlation based features for discrimination between electrical and mechanical defects and their localization in transformer winding. *IEEE Trans. Dielectr. Electr. Insul.* **2014**, *21*, 2374–2382.
30. Karimifard, P.; Gharehpetian, G.B.; Tenbohlen, S. Determination of axial displacement extent based on transformer winding transfer function estimation using vector-fitting method. *Eur. Trans. Electr. Power* **2008**, *18*, 423–436.
31. Heindl, M.; Tenbohlen, S.; Kraetge, A.; Krüger, M.; Velásquez, J.L. Algorithmic Determination of Pole-Zero Representations of Power Transformers' Transfer Function for Interpretation of FRA Data. In Proceedings of the 16th International Symposium of High Voltage Engineering, Cape Town, South Africa, 24–28 August 2009.
32. Rahimpour, E.; Jabbari, M.; Tenbohlen, S. Mathematical comparison methods to assess transfer functions of transformers to detect different types of mechanical faults. *IEEE Trans. Power Del.* **2010**, *25*, 2544–2555.
33. Heindl, M.; Tenbohlen, S.; Velásquez, J.; Kraetge, A.; Wimmer, R. Transformer Modeling Based on Frequency Response Measurements for Winding Failure Detection. In Proceedings of the 2010 International Conference on Condition Monitoring and Diagnosis, Tokyo, Japan, 6–11 September 2010; pp. 201–204.
34. Nirgude, P.; Ashokraju, D.; Rajkumar, A.; Singh, B. Application of numerical evaluation techniques for interpreting frequency response measurements in power transformers. *IET Sci. Meas. Technol.* **2008**, *2*, 275–285.
35. Samimi, M.H.; Tenbohlen, S.; Shayegani Akmal, A.A.; Mohseni, H. Using the Complex Values of the Frequency Response to Improve Power Transformer Diagnostics. In Proceedings of the 24th Iranian Conference on Electrical Engineering (ICEE), Shiraz, Iran, 10–12 May 2016.
36. Abeywickrama, N.; Serdyuk, Y.V.; Gubanski, S.M. Effect of core magnetization on frequency response analysis (FRA) of power transformers. *IEEE Trans. Power Deliv.* **2008**, *23*, 1432–1438.
37. Abu-Siada, A.; Hashemnia, N.; Islam, S.; Masoum, M.A.S. Understanding power transformer frequency response analysis signatures. *IEEE Electr. Insul. Mag.* **2013**, *29*, 48–56.
38. Wang, M.; Vandermaar, A.J.; Srivastava, K.D. Transformer winding movement monitoring in service—Key factors affecting FRA measurements. *IEEE Electr. Insul. Mag.* **2004**, *20*, 5–12.
39. Shintemirov, A.; Tang, W.; Wu, Q. Transformer winding condition assessment using frequency response analysis and evidential reasoning. *IET Electr. Power Appl.* **2010**, *4*, 198–212.
40. Bakar, N.A.; Abu-Siada, A.; Islam, S. A review of dissolved gas analysis measurement and interpretation techniques.*IEEE Electr. Insul. Mag.* **2014**, *30*, 39–49.
41. Möllmann, A.; Pahlavanpour, B. New guidelines for the interpretation of dissolved gas analysis in oil-filled transformers. *Electra* **1999**, *186*, 31–51.
42. *IEEE Guide for the Interpretation of Gases Generated in Oil Immersed Transformers*; IEEE Std. C57.104; IEEE Power & Energy Society: New York, NY, USA, 2008.
43. *Report on Gas Monitors for Oil-filled Electrical Equipment*; Cigré Technical Brochure 409; Cigre: Paris, France, 2010.
44. Anderson, R.; Roderick, U.R.; Jaakkola, V.; Östman, N. *The Transfer of Fault Gases in Transformers and Its Effect upon the Interpretation of Gas Analysis Data*; Cigré Report 12-02; Cigre: Paris, France, 1976.
45. Bräsel, E.; Bräsel, O.; Sasum, U. *Gashaushalt bei Transformatoren der offenen Bauart—Neue Erkenntnisse*; EW-Medien und Kongresse: Frankfurt, Germany, ew, Jg. 109, Heft 14–15; pp. 56–59. 2010. (In German)

46. *Recent Developments in DGA Interpretation*; Cigré Technical Brochure 296; Cigre: Paris, France, 2006.
47. Müller, A.; Tenbohlen, S. Analysis of Fault Gas Losses through the Conservator Tank of Free-Breathing Power Transformers. In Proceedings of the 18th International Symposium on High Voltage Engineering, Seoul, Korea, 25–30 August 2013.
48. Lundgaard, L.E.; Hansen, W.; Linhjell, D.; Painter, T.J. Aging of oil-impregnated paper in power transformers. *IEEE Trans. Power Deliv.* **2004**, *19*, 230–239.
49. Koch, M.; Tenbohlen, S.; Stirl, T. Diagnostic application of moisture equilibrium for power transformers. *IEEE Trans. Power Deliv.* **2010**, *25*, 2574–2581.
50. Koch, M.; Krüger, M.; Tenbohlen, S. Comparing Various Moisture Determination Methods for Power Transformers. In Proceedings of the CIGRE South Africa Regional Conference, Cape Town, South Africa, 17–21 August 2009.
51. Susa, D.; Lehtonen, M.; Nordman, H. Dynamic thermal modelling of power transformers. *IEEE Trans. Power Deliv.***2005**, *20*, 197–204.
52. Lesieutre, B.C.; Hagman, W.H.; Kirtley, J.L., Jr. An improved transformer top oil temperature model for use in an online monitoring and diagnostic system. *IEEE Trans. Power Deliv.* **1997**, *12*, 249–256.
53. IEC 60076-7. *Power Transformers—Part 7: Loading Guide for Oil Immersed Power Transformers*; International Electrotechnical Commission: Geneva, Switzerland, 2005.
54. Djamali, M.; Tenbohlen, S. A Dynamic Top Oil Temperature Model for Power Transformers with Consideration of the Tap Changer Position. In Proceedings of the 19th International Symposium on High Voltage Engineering, Pilsen, Czech Republic, 23–28 August 2015.
55. Djamali, M.; Tenbohlen, S. Malfunction detection of the cooling system in air-forced power transformers using online thermal monitoring. *IEEE Trans. Power Deliv.* **2016**. unpublished work.
56. Tenbohlen, S.; Djamali, M. A Dynamic Top-Oil Temperature Model for Online Assessment of Overload Capability of Power Transformers. In Proceedings of the CIGRE SC A2 Colloquium, Shanghai, China, 20–25 September 2015.

CHAPTER 18

GROUND THERMAL DIFFUSIVITY CALCULATION BY DIRECT SOIL TEMPERATURE MEASUREMENT. APPLICATION TO VERY LOW ENTHALPY GEOTHERMAL ENERGY SYSTEMS

José Manuel Andújar Márquez, Miguel Ángel Martínez Bohórquez and Sergio Gómez Melgar

Escuela Técnica Superior de Ingeniería, Universidad de Huelva, Ctra. Palos de la Ftra.-Huelva s/n, Huelva 21819, Spain

ABSTRACT

This paper presents a methodology and instrumentation system for the indirect measurement of the thermal diffusivity of a soil at a given depth from measuring its temperature at that depth. The development has been carried out considering its application to the design and sizing of very low enthalpy geothermal energy (VLEGE) systems, but it can has many other applications, for example in construction, agriculture or biology. The methodology is simple and inexpensive because it can take advantage of the prescriptive geotechnical drilling prior to the construction of a house or building, to take at the same time temperature measurements that will allow get the actual temperature and ground thermal diffusivity to the depth of interest. The methodology and developed system have been tested and used in the design of a VLEGE facility for a chalet with basement at the outskirts of Huelva (a city in the southwest of Spain). Experimental results validate the proposed approach.

INTRODUCTION

Geothermal exchange technology takes advantage of the thermal energy stored in the surface area of the Earth (first 100 m). Up to 10–15 m deep approximately, ground heat is supplied by the sun and rain. From there the underground temperature increases about 3 °C per 100 m depth, due to the internal thermal energy of the Earth. On average, the underground temperature at 10 m depth remains constant throughout the year and substantially equal to the average temperature of the place [1,2,3,4].

Initially geothermal exchange systems were developed for heating in cold climates, hence its development in northern European countries, USA and Canada, but they are also suitable for cooling, increasing their profitability and interest in countries of southern latitudes. In cooling mode, heat is extracted from the building and this is transferred to the ground. In heating mode, heat is extracted from the Earth and it is transferred to the building (Figure 1). Of course the underground temperature remains unchanged (or it does so very slightly depending on depth), but the room temperature is what changes.

Citation: Andújar Márquez, J.M.; Martínez Bohórquez, M.Á.; Gómez Melgar, S. Ground Thermal Diffusivity Calculation by Direct Soil Temperature Measurement. Application to very Low Enthalpy Geothermal Energy Systems. Sensors 2016, 16, 306. DOI: 10.3390/s16030306

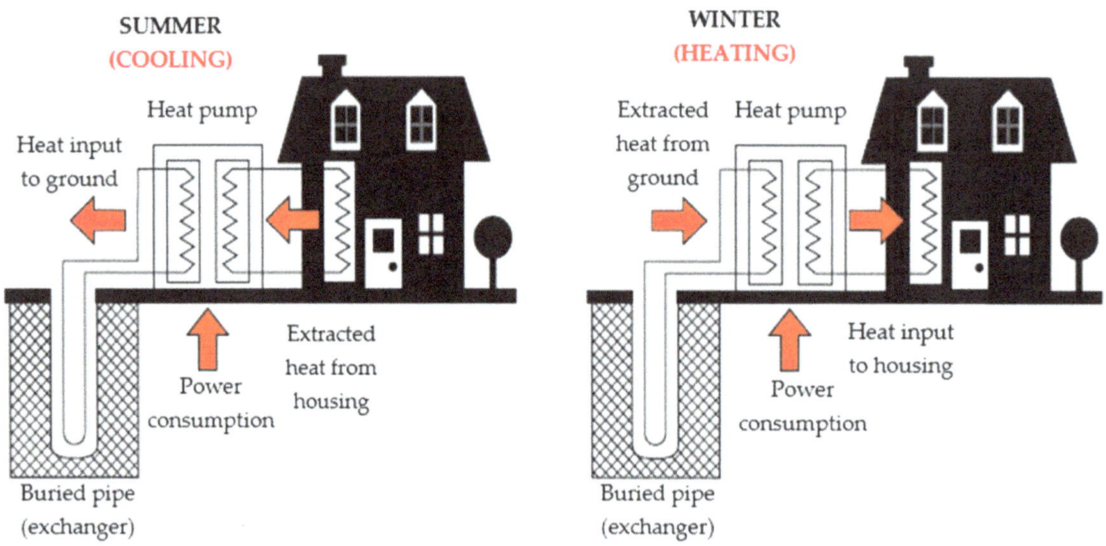

Figure 1. Typical operation of a geothermal exchange system with heat pump.

Geothermal energy is called very low enthalpy (very low enthalpy geothermal energy, VLEGE) when the heat transfer from the ground occurs at low depth and low temperature (usually less than 40 °C) [1]. To exploit effectively the heat capacity of the ground, a heat-exchanger system has to be constructed [5]. In practice, heat exchange is performed by internal fluid flowing through a collector (buried pipe), which is usually water, glycol water or even air. The geothermal exchange can be performed through a heat pump [6], whereupon the internal and external circuits of the building become independent (see Figure 1), or directly by circulating the geothermal fluid through the network building's HVAC (heating, ventilating, and air conditioning) by floor heating, radiators or air outlets for example.

As it is well known the coefficient of performance (COP) of a heat pump is inversely proportional to the temperature difference of the heat sources [2,4,7]; in the case of the VLEGE it is not big and substantially independent of seasonal variations. Even so the COP of the heat pump of a VLEGE system is around 4, which means that the system captures an average of 3 kWh of thermal energy from the ground per kWh consumed from the mains supply to power the heat pump. For this reason VLEGE is a very competitive alternative compared with conventional alternatives for heating and cooling. Furthermore from the environmental point of view the impact of a VLEGE system is much lower than a conventional air conditioning system.

Since VLEGE uses the heat accumulated in the more superficial layers of the Earth, collectors are usually placed horizontally adopting different shapes (Figure 2) and trying always to have the largest surface area in contact with the ground [8]. The sizing of these collectors is carried out according to the thermal performance of the soil, which depends on its composition, density, water content and of course on the depth. The two key soil parameters for designing a VLEGE system are the *soil temperature* at the collectors depth $T(z)$ and the *ground thermal diffusivity* at that depth (αz). By definition *thermal diffusivity*, α (m^2/s), is the thermal conductivity divided by density and specific heat capacity at constant pressure, and it represents the ability of a material to conduct thermal energy relative to its ability to store thermal energy. Since ground thermal diffusivity depends on soil type, density and water content, the range of values that can take is very broad. For instance in [9] ground thermal diffusivity values ranging from 1.72×10−6 m2/s to 3×10−6 m2/s, depending on the soil characteristics. Another example: for a sandy soil with a density of 1.46×103 Kg/m3, [10] provides values ranging between 0.3×10−6 m2/s to 1.1×10−6 m2/s, depending on its water content. If we continue researching on the specialized literature, we can find very different value ranges. This is because even in the same soils (same composition), if their density and/or moisture degrees are different their α values as well. This means that for a given soil and depth, α can varies throughout the year, depending on rainfall. Therefore it makes sense to work with average values.

Figure 2. Collectors typical arrangement in a VLEGE facility.

Assuming that the soil behavior is uniform (for stricter mathematical study beyond the scope of this paper can be found [11] among other references), which as we know is not true in practice, ground thermal diffusivity values are tabulated according to the soil composition, so ground thermal diffusivity can be estimated by tables. In a more realistic way, ground thermal diffusivity can be measured in the laboratory from a test tube of the ground or performing a thermal response test (TRT) [12,13,14].

A TRT is an *in-situ* measurement method which is currently the most exact way to determine the ground effective thermal properties; it is the most reliable and the only recommended for large or medium VLEGE facilities, however it is a complex and expensive method. Therefore, the TRT is impractical to carry out in most VLEGE facilities (which are small and domestic) so that designers resort to using tables. This implies an approach to the composition, water content and compaction of the ground at the depth where the collectors are buried; consequently the probability of making mistakes can be high. Incorrect design of the VLEGE facility (diameter and collector length, type of material, type of fluid and heat exchanger) cause malfunction and poor performance of the heat pump. This is a serious problem: dig up collectors and change them is expensive and sometimes, depending on the characteristics of the facility, impossible. Consequently, this may result in the need to use external heating/cooling, which call into serious question the energy efficiency of the facility.

This paper develops a methodology and instrumentation system for the indirect measurement of the thermal diffusivity of a ground at a given depth from measuring its temperature at that depth. The methodology and developed system have been tested and used in the design of a VLEGE facility for a chalet with basement at the outskirts of Huelva (a city in the southwest of Spain).

The paper is laid out as follows: Section 2 explains the theoretical foundation that develops the methodology for calculating the ground thermal diffusivity at a given depth from knowledge of the ground temperature at that depth.Section 3 explains the developed hardware/software instrumentation system for measuring ground temperature at a given depth accurately and at low cost. Section 4 is devoted to experimentation in a real scenario and obtained results; these demonstrate the usefulness of the proposed solution. Finally, Section 5 provides some conclusions about the work done and its results.

THEORETICAL FOUNDATION

The energy from solar radiation and other atmospheric agents is continuously transferred to the surface of the Earth, which causes effects on its temperature in depths close to the surface. The observed soil temperature at various depths has been studied decades ago [15,16,17,18]. Since then, no new theoretical approaches in this field are found in literature, but obviously what has evolved with technological advancement are measuring systems and data processing.

It is well known that the typical annual cycles of monthly average soil temperatures at the surface and at depths close to the surface, follow a pattern of easy fit to a simple harmonic function. Then, the temperature in the soil surface can be represented by the following function:

$$T_{soil\ surface}(t) = T_m - T_p\cos(\omega t - \varphi) \qquad (1)$$

where:

T_m= Annual average temperature of soil in the stable layer; it is commonly set to the average temperature of air, practically the average temperature in the place (°C).

T_p= Amplitude (°C); the peak deviation of the function from zero. In this case the annual amplitude of the monthly average temperature cycle in the place.

t= Time coordinate (s). To set a starting time (t = 0), it begins to run from 1 January at 0 s. This way of measuring the time obviously result in a phase shift since the beginning of the sinusoid in general will not coincide with t = 0.

ω= Angular frequency (rad/s). The rate of change of the function argument in units of radians per second; $\omega=2\pi/T$, where T is the period of the sinusoid; in this case the annual temperature cycle, *i.e.*, $T=365.242189\times24\times3600=3.1557\times10^7$ s.

φ= Phase (rad). When φ is non-zero, the entire waveform appears to be shifted in time by the amount φ/ω seconds. A negative value represents a delay, and a positive value represents an advance.

The soil temperature oscillation near the surface for the stand point of heat conduction theory has been discussed since decades in various texts of heat transfer [19,20,21]. The subsurface is treated as a semi-infinite domain, $\Omega=\{x=(x,y,z)^T: -\infty<x,y<\infty, 0\le z<\infty\}$. The subsurface temperature, T(x,t), at any point x and time t is provided by the heat conduction equation:

$$\rho C\frac{\partial T(\mathbf{x},t)}{\partial t} = \nabla\cdot[\mathbf{k}\nabla T(\mathbf{x},t)] \qquad (2)$$

where ρ is the average soil density (Kg/m^3), C the soil specific heat capacity (J/Kg·K) and k the soil thermal conductivity (W/m·K). The three soil parameters ρ, C and k, vary, to different degrees, in space and time due to soil heterogeneity and changing water content. Soil anisotropy gives rise to the thermal conductivity tensor k, whose principle components are aligned with the coordinate system, such that the off-diagonal components of this tensor are $K_{ij}=0$ for i≠j. Equation (2) is subject to an initial condition:

$$T(x,0)=T_m \qquad (3)$$

This temperature varies in response to atmospheric fluctuations at the ground surface (z=0), which manifest themselves through the boundary condition (1): T(x,y,0,t). By solving the heat equation in the transient state for a semi-infinite medium whose surface temperature is imposed by Equation (1), the soil temperature *versus* depth only can be obtained:

$$T_{soil}(z,t) = T_m - T_p e^{-z\sqrt{\frac{\omega}{2\alpha}}}\cos\left(\omega t - \varphi - z\sqrt{\frac{\omega}{2\alpha}}\right) \qquad (4)$$

where z is the depth (m) and α is the ground thermal diffusivity (m^2/s) given by:

$$\alpha = \frac{k}{\rho C} \qquad (5)$$

Equation (4) allows us to show some important features of the evolution of soil temperature with depth:

- As the depth increases the second term of Equation (4) tends to zero, which means that the soil temperature tends to the annual average in the place (T_m).
- The amplitude of the temperature variation decreases exponentially with depth.
- The phase shift increases with depth.

As an illustrative example, let's consider a place where the annual average temperature is 11 °C (Tm in Equation (4)), the annual peak to peak amplitude of the monthly average temperature cycle (T_{pp}) is 25 °C (then T_p in Equation (4) is 25/2=12.5°C), and whose physical soil characteristics, constant regardless of the depth, are: ρ = 2000 Kg/m³, k = 2 W/(m·K) and C = 99 J/(Kg·K). The application of Equation (4) to this place allows obtaining the temperature behavior of the soil at different depths. This is shown in Figure 3, where the behavior of the soil temperature is drawn from the surface to a depth of 4 m in steps of 0.5 m.

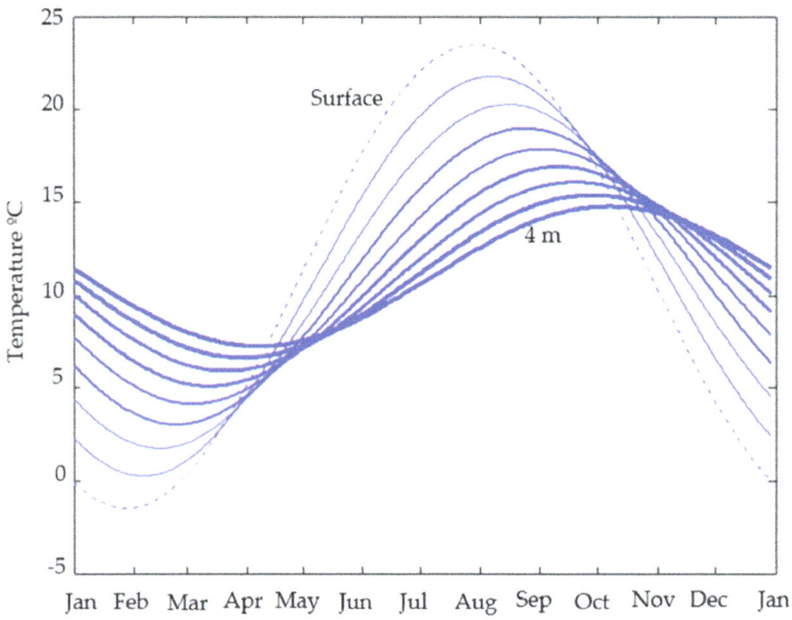

Figure 3. Typical temperature behavior of the soil close to the surface at different depths. The curves show the evolution of the temperature from the surface (0 m) to 4 m depth in steps of 0.5 m.

Note in Figure 3 that, as the depth increases, the amplitude of thermal fluctuations decreases, and their maximum and minimum are going phased out (due to the thermal inertia of the soil itself). At 4 m deep, the phase shift reached two months. As the depth increases annual fluctuations in soil temperature are diminishing and it is getting closer to the annual average of the place. In the real world, and for a practical case, in depths between the surface and about 10 m, the temperature of the soil varies depending on the depth and anything that disturbs the type, compactness and uniformity of the ground and of course moisture or groundwater (anything that changes the ground thermal diffusivity). This means that in general the ground thermal diffusivity will not be constant, whereby the distribution of the actual curves may not be as uniform as in Figure 3. This actual behavior goes unnoticed when considering the ground thermal diffusivity of a soil at a given depth from tables, since it is considered uniform.

From Equation (4) and considering a ground thermal diffusivity obtained by tables for example, we can calculate the evolution of the soil temperature at any depth between the surface and the deep where the soil temperature tends to the annual average in the place (T_m), *i.e.*, around 10 m. Nevertheless we know that ground thermal diffusivity is not constant, which prevents knowing the degree of accuracy in the calculation of the temperature at the considered depth. But what can we do if we have precise ground temperature measurements at that depth?

From Equation (4), the minimum (T_L) and maximum (T_H) temperatures of the soil that take place during the annual cycle for any depth (z) can be calculated. This is:

$$T_L = T_m - T_p e^{-z\sqrt{\frac{\omega}{2\alpha}}} \quad (6)$$

$$T_H = T_m + T_p e^{-z\sqrt{\frac{\omega}{2\alpha}}} \quad (7)$$

From any of the above two equations the ground thermal diffusivity at any depth z can be calculated. For example from Equation (7):

$$\alpha_z = \frac{\omega}{2}\left(\frac{z}{\ln \frac{T_p}{T_H - T_m}}\right)^2 \quad (8)$$

T_m and T_p values are easy to know from meteorological stations or by measuring at place, so if you can measure accurately T_H over annual cycle for a z depth, Equation (8) allows obtaining the ground thermal diffusivity without need to make estimates by tables or calculating it by the costly and complex TRT. Similar reasoning could be made with respect to T_L. Certainly α_z can vary along a seasonal cycle (to find instantaneous values you can use Equation (4)), but in the absence of extreme situations such as flooding for example, at depths far of the surface (the interest for VLEGE) there is less influence of surface weather, therefore under normal conditions the ground thermal diffusivity will have an average value close to that obtained by Equation (8).

DEVELOPED INSTRUMENTATION SYSTEM FOR MEASURING GROUND TEMPERATURE

Figure 4 shows the block diagram of the developed instrumentation system for measuring ground temperature. The system is monitored and controlled remotely by two microcontroller cards via a virtual instrument (VI) from anywhere with an Internet connection. Now we will be explaining each of the blocks.

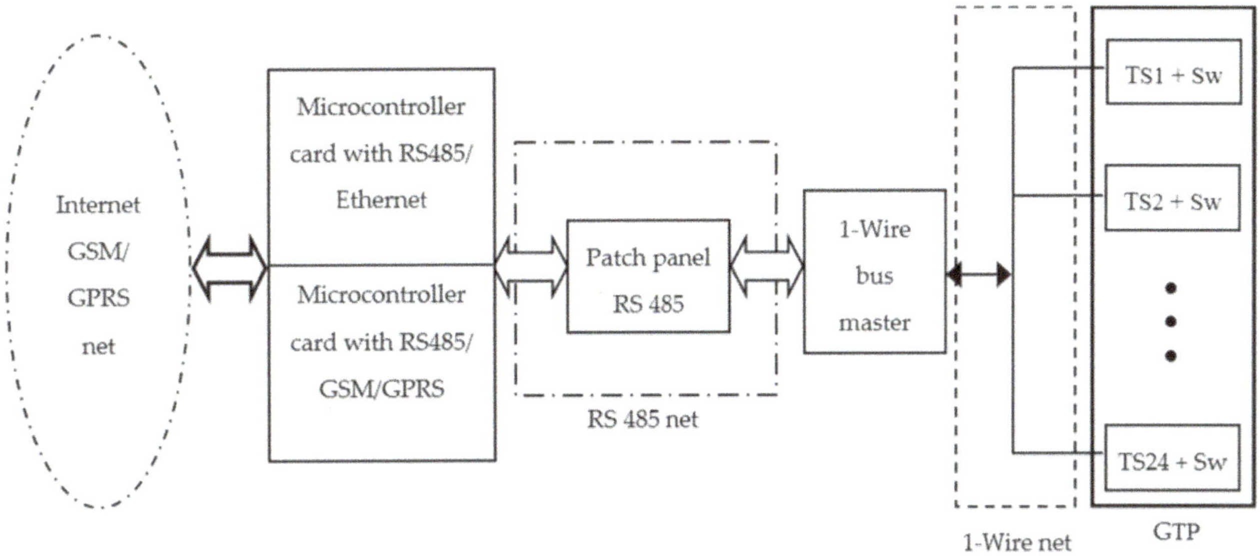

Figure 4. Block diagram of the developed instrumentation system for measuring ground temperature.

Ground Temperature Probe

For continuous measurement of the temperature of the ground at different depths, a ground temperature probe (GTP) was designed and built using a PVC pipe of 5 m length and 100 mm diameter. In it, 24 temperature sensors (TS1–TS24) with digital output and a 1-wire bus connection (patented by the authors, [22]) are placed. Each TS can be connected/disconnected to the 1-wire bus by an addressable electronic switch (Sw) which function will be explain later in this section. TS are spaced 20 cm each, and they are glued to a metal ring to favor the thermal transmittance between ground and sensor (see Figure 5a). Metal rings are glued to the pipe and securely fastened by a polyethylene clamp (see Figure 5b). All wiring flows inside the pipe and it is filled with polyurethane foam to TS protect. 24 TS are located from level 0.2 m to 4.80 m deep. The first 20 cm of GTP must be above the surface to facilitate connections and recovery, and at the tip of the GTP it is not advisable to place TS as it is the part that suffers most in the introduction to the field. The GTP can be disassembled into two sections of 2.5 m for easy transport (see Figure 5b). Of course we could have another GTP length; however it has done so for constructive interest of the house where the experiment was conducted. That is, when a chalet or house with basement is built, an excavation of approximately 3 m around the perimeter of the house is required. From this depth it is easier and cheaper to dig about 2 m further by a backhoe loader. Therefore the maximum depth to build at low cost a VLEGE facility in a chalet or house with basement is around 5 m. Of course you can build the VLEGE facility deeper; however this requires a work infrastructure which makes much more expensive the construction. In any case, obviating the construction work details that are not the subject of this paper, developed system experimentation (see Section 4) was carried out with ground temperature measurements between 0 and 4.80 m depth.

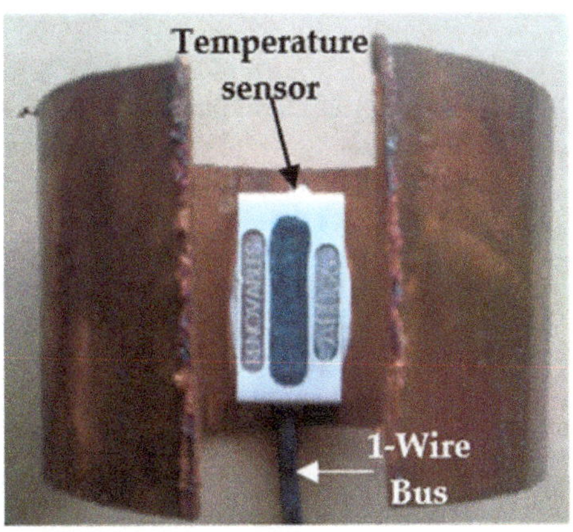

(a)

(b)

Figure 5. (**a**) Temperature sensor attached to the metal ring; (**b**) GTP disassembled into two sections and arrangement of the sensors on the probe.

The 24 TS are connected via 1-wire bus (see Figure 4 and Figure 5a). This is a low-cost bus for digital communication over twisted-pair cable with 1-wire components. As each TS has a unique 64-bit serial code, it is possible to connect multiple TS to the same 1-wire bus quickly, simply and very cheaply. Using the code from each TS, its temperature measurement can be addressed from a PC or microcontroller board. The 1-wire bus is not a bus designed to operate in industrial environments (where noise can be an important factor to consider) because it is not differential. In [23] it is shown as providing robustness to the 1-wire bus and how to connect the TS to the bus.

Regarding the connection of sensors in the GTP and its electronic control, a robust solution it was implemented as required for our system. With respect to TS communications via the 1-wire bus, if one of the connected TS fails, the rest of the bus can become unusable. When this occurs in a facility of easy access, the damaged TS is changed and the problem is solved. However, for our application is a major problem because to solve it would have to dig up the GTP. In order to prevent and resolve this potential problem, each TS can be connected/disconnected to the 1-wire bus by an addressable electronic switch (Sw). In Figure 6 detail of TS + Sw connection of Figure 4 is shown. The switch has been implemented by the DS2413 integrated circuit and allows physically disconnecting the TS of the 1-wire bus by the GND line.

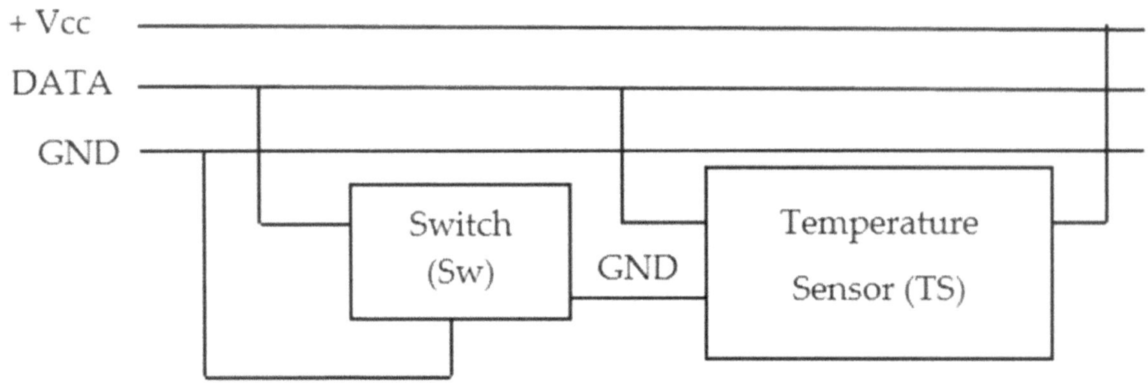

Figure 6. 1-Wire bus connection of the temperature sensor + switch set.

The scheme of Figure 6, also serves another important function: when a TS branch is connected to a 1-wire bus there is a mismatch impedance at the branch point which can cause reflections to the bus. To avoid it, a 100 Ω series resistor between DATA and GND lines on each TS should be placed. The internal resistance of the switch fulfills this task, avoiding the need to place an additional external resistance [24].

The 1-Wire Bus Master

A detailed explanation of the 1-wire bus master (see Figure 4) can be found in [23]; it is designed around PIC 18F252 and is able to addressing simultaneously two 1-wire networks of 50 TS each. For this application we are only using 24 sensors, so that the same 1-wire bus master also applies for deeper studies where more sensors would be required.

The RS 485 Patch Panel

The RS 485 patch panel (see Figure 4) has the function to allow connection to multiple sensors (room temperature, irradiance, speed and direction wind, humidity, *etc.*) or other devices to the system. In this way a single communications channel as that of Figure 4 can monitor multiple variables, although the measurement points are spaced hundreds of meters.

Choosing the RS 485 bus is based on its robust features in terms of noise performance (it is a differential bus) and its ability to carry data over long distances.

Microcontroller Cards

The two microcontroller cards (see Figure 4) are based on Arduino, an open-source prototyping platform based on easy-to-use hardware and software. These cards control communications with devices connected to patch-panel and store measured data in a SD memory. In addition they manage online external communications either via Ethernet or GSM/GPRS. Although data are sent in real time, they are also stored in one of the microcontroller cards for safety, so if for any reason communications fail, always be possible recover the data. The system is designed so that in those locations that do not have Internet connection, you can use the mobile communications GSM/GPRS. In the worst case, if not even there is mobile phone coverage we can always use a radio modem link. This last case is not included in Figure 4.

Monitoring and Control Software

For full monitoring and control of the developed instrumentation system, we have developed a VI in LabVIEW™, which is located on a computer at the headquarters of the research group to which the authors belong, at the Higher Technical School of Engineering of the University of Huelva. In this way the whole facility operates as a remote lab [25].

The VI consists of different sub VI with screens that display real time information. It also enables data processing for several kinds of operations and also provides multiple graphs types. The data can be processed by the VI itself or exported in various formats like spreadsheet for example. This will be discussed in more detail in Section 4 of this manuscript.

Figure 7 shows the VI screen with real-time measurement of the 24 TS of the GTP. Figure 8 shows the VI screen to detect the failure of a sensor. Failure detection is performed by comparison between adjacent sensors. The temperature gap is adjustable for each application and when it is exceeded an alarm is triggered. When a sensor malfunction it can be disconnected remotely from the VI thanks to developed electronics (see Figure 6).

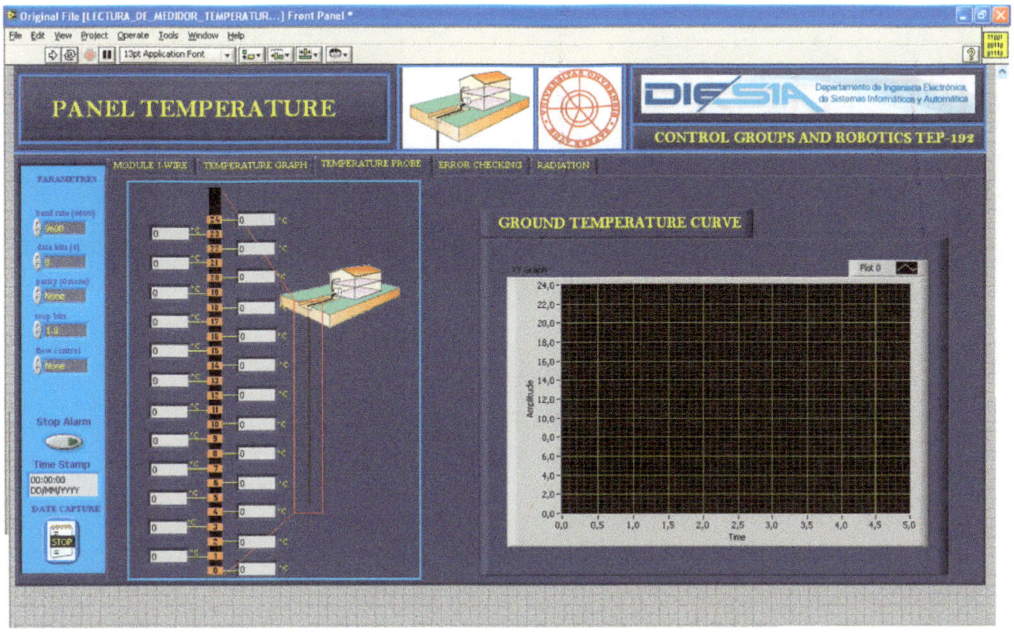

Figure 7. Real time temperature measurement of the 24 TS of the GTP.

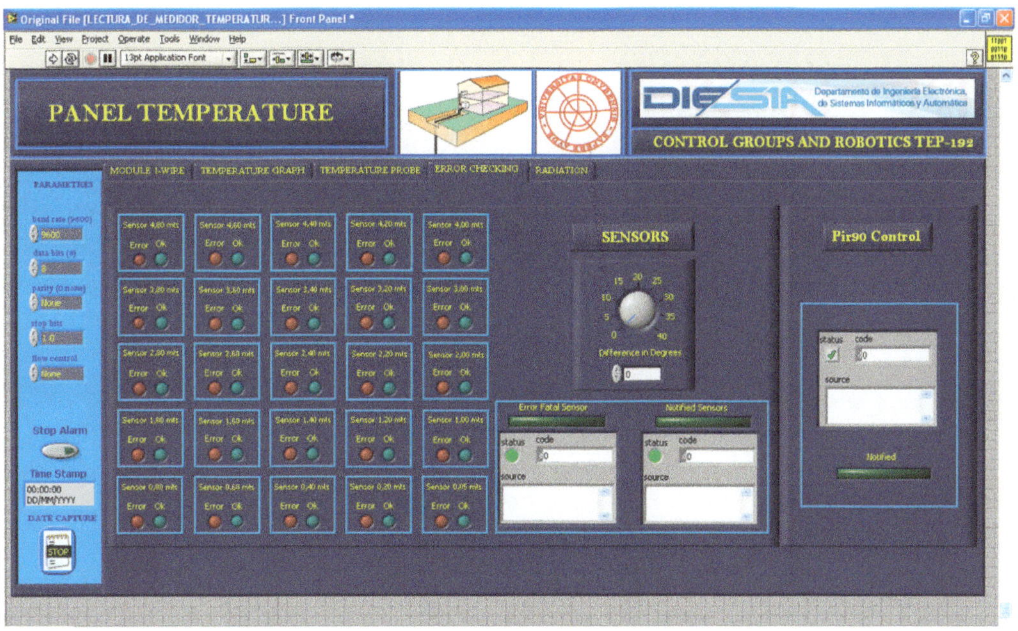

Figure 8. Screen that show faults in any TS of the GTP and allows adjustment of all parameters.

EXPERIMENTATION AND RESULTS

In order to put into practice and test the developed method and instrumentation system, we have had the opportunity to participate in the construction of a chalet with basement located at the southwest of Spain. Here we have taken soil temperature measurements with the developed GTP for about a year. At the building site local climate is warm tempered by the influence of the Atlantic Ocean with an annual average temperature around 19 °C. The coldest month is usually January (some years December or February), with an average temperature around 12 °C, and the warmest July (some years August), with an average temperature around 30 °C. The annual rainfall is around 500 mm and it has over 3000 h of sunshine and only about 60 cloudy days per year.

The chalet construction work has served as a real laboratory for about a year. In Figure 9a geotechnical drilling machine that was also exploited to make the drill through which the probe was introduced into a steel pipe (Figure 9b) is shown. Then, once the probe was placed in its final position, the steel pipe was removed. Later the holes between the probe and the ground were filled and compacted carefully. Finally the perimeter was watered abundantly, in order to favor the GTP contact with the ground along its whole length. The measurement campaign began when this artificial humidity was dissipated, several days later.

(a) (b)

Figure 9. (**a**) Drilling machine; (**b**) Steel pipe for introducing the probe.

The initial end of the probe, once buried, remains outside (see Figure 10a). The 1-wire bus that picks up signals from the 24 TS (see Figure 4 and Figure 5) goes from the GTP to a waterproof box (protection category IP 65) which houses the 1-wire bus master. From here the RS 485 bus goes to another larger size waterproof box which houses the remaining elements ofFigure 4 (see Figure 10b).

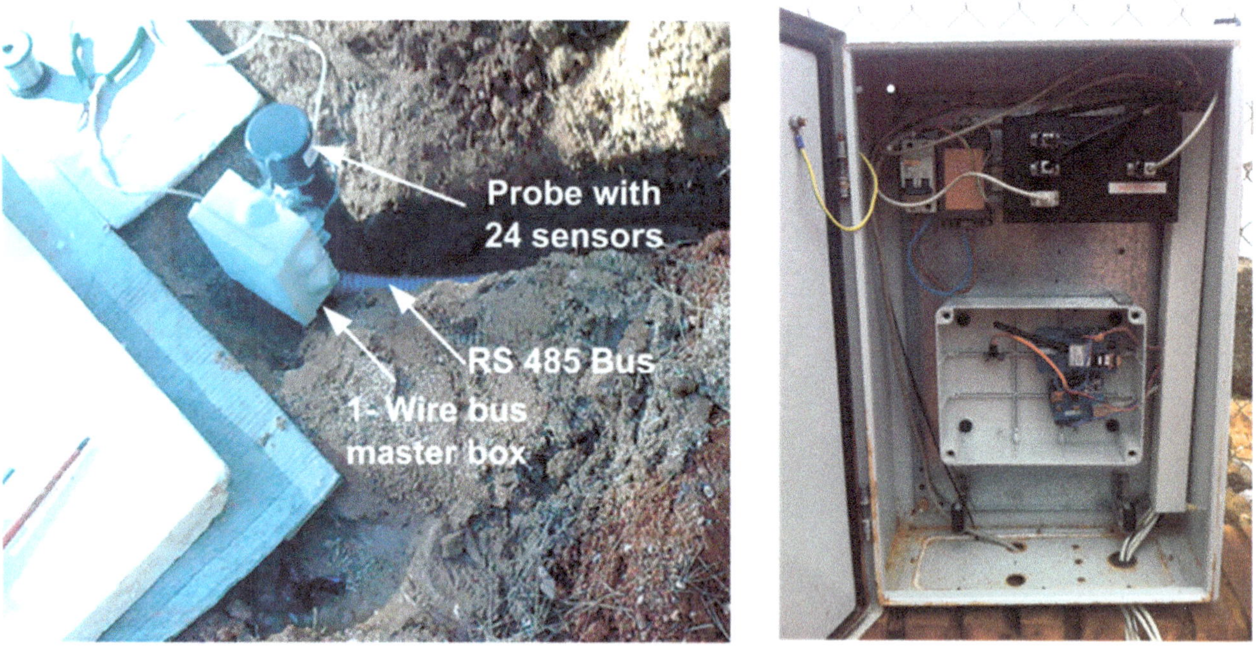

(a) (b)

Figure 10. (**a**) Details of the buried GTP tip, box which houses the 1-wire bus master and RS 485 bus; (**b**) Waterproof box housing remainder elements of the measurement system.

The buried GTP has been acquiring data for about a year, continuously (six measurements per minute) from 12 months (February 2103 to January 2014). Figure 11 shows the average monthly temperatures regarding depth measured by the GTP. Note clearly that as the depth increases the soil temperature tends to the annual average temperature of the place (Tm in Equation (4) and 19.9 °C measured in the considered time period). In fact a temperature annual gap around 20 °C in the surface becomes only around 5 °C at 4.80 m depth.

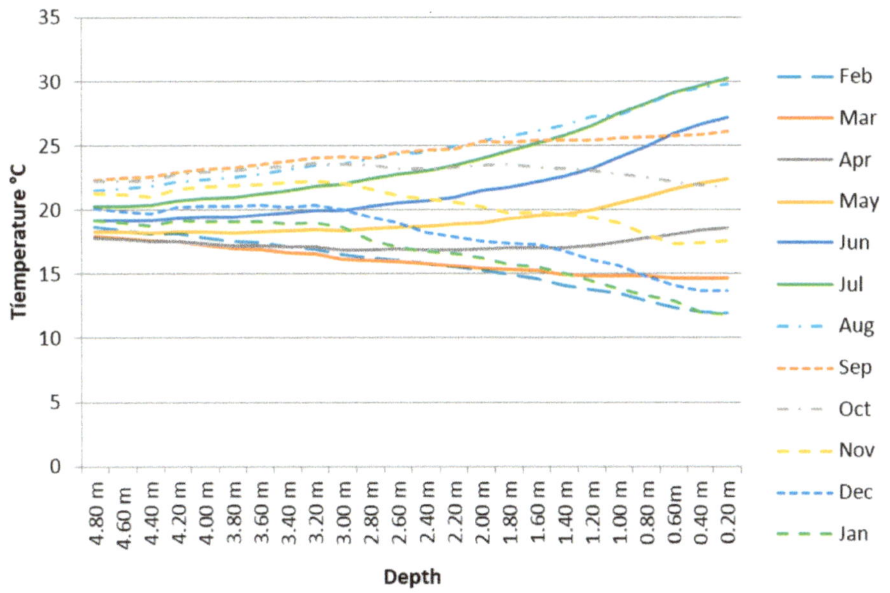

Figure 11. Behavior of the average monthly temperature regarding soil depth.

Figure 12 shows for the same time period the average daily temperatures at different depths (from 1 m to 4.8 m). Notice the sinusoidal profile as we have explained in Section 2 of this paper and how just 4.80 m deep, soil temperature is close all year to the annual average temperature in the place (19.9 °C). Note also the timing difference (phase shift) between sinusoidal functions which is due to the soil thermal inertia. This is also applicable to Figure 11 in the sense that the behavior of a surface average monthly temperature does not correspond with that expected for this month as the depth increases.

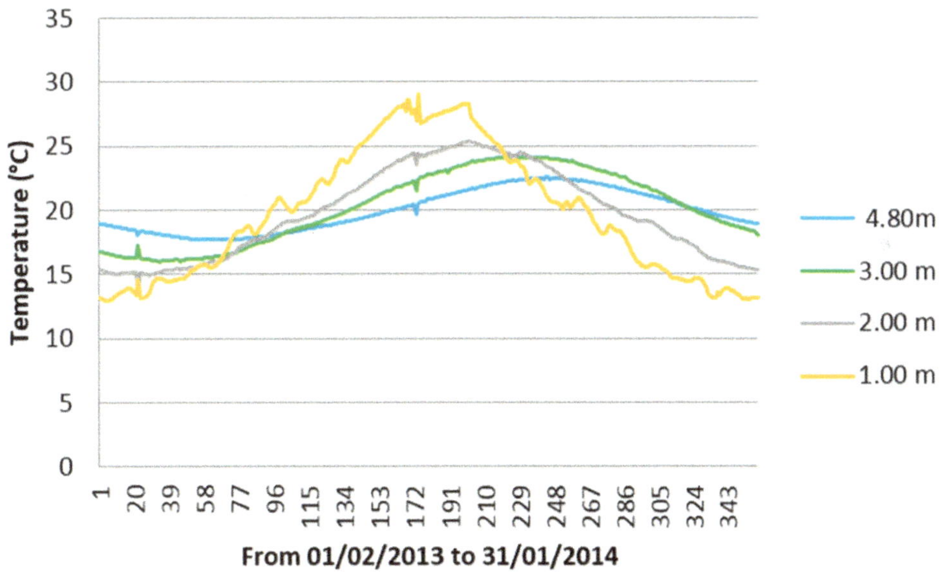

Figure 12. Average daily temperatures at different depths from 1 m to 4.8 m.

As we can observe in Figure 12 the sinusoidal profile regarding depth is each time better, which indicates that the influence of the surface weather decreases with depth, so the temperature profile is each time softer and closer to a sinusoid. This is also applicable to Figure 11.

Calculation of the Actual Ground Thermal Diffusivity

In order to probe the power and usefulness of the developed methodology and system we have entrusted to a specialized company preparation of a geological report of the soil in the experimentation place regarding depth. This report asserts that from surface to a 4.80 depth there are five different kind of soil. In particular at 4.80 m depth the soil is a mix of sand and gravel. Moreover, the report asserts that the ground water level is 5.20 m depth so at 4.80 m the soil is wet.

Table 1 (our own elaboration from [10]) shows thermal conductivity, volumetric heat capacity and thermal diffusivity for different kinds of soil. With the help of Table 1 and the data obtained in the geological report we must decide the ground thermal diffusivity coefficient to use. However, depending on the amount of humidity and the mix of gravel and sand the range varies from 0.19×10^{-6} m^2/s to 1.72×10^{-6} m^2/s. It is clearly too high a range to decide without any further criteria.

Table 1. Thermal conductivity, volumetric heat capacity and thermal diffusivity for different kinds of soil.

Rock Type	Thermal Conductivity (W/mK)			Volumetric Heat Capacity (MJ/m³K)	Thermal Diffusivity (10^6 m²/s)		
	Min	Typ	Max		Min	Typ	Max
Basalt	1.3	1.7	2.3	2.6	0.5	0.65	0.88
Greenstone	2	2.6	2.9	2.9	0.69	0.90	1
Gabbro	1.7	1.9	2.5	2.6	0.65	0.73	0.96
Granite	2.1	3.4	4.1	3	0.7	1.13	1.37
Peridotite	3.8	4	5.3	2.7	1.41	1.48	1.96
Gneiss	1.9	2.9	4	2.4	0.79	1.21	1.67
Marble	1.3	2.1	3.1	2	0.65	1.05	1.55
Mica schist	1.5	2	3.1	2.2	0.68	0.91	1.41
Shale sedimentary	1.5	2.1	2.1	2.5	0.6	0.84	0.84
Limestone	2.5	2.8	4	2.4	1.04	1.17	1.67
Loam	1.5	2.1	3.5	2.3	0.65	0.91	1.52
Quartzite	3.6	6	6.6	2.2	1.64	2.73	3
Salt	5.3	5.4	6.4	1.2	4.42	4.5	5.33
Sandstone	1.3	2.3	5.1	2.8	0.46	0.82	1.82
Siltstones and argillites	1.1	2.2	3.5	2.4	0.46	0.92	1.46
Dry gravel	0.4	0.4	0.5	1.6	0.25	0.25	0.31
Water saturated gravel	1.8	1.8	1.8	2.4	0.75	0.75	0.75
Dry sand	0.3	0.4	0.55	1.6	0.19	0.25	0.34
Water saturated sand	1.7	2.4	5	2.9	0.59	0.83	1.72
Dry clay/silt	0.4	0.5	1	1.6	0.25	0.31	0.62
Water saturated clay/silt	0.9	1.7	2.3	3.4	0.26	0.5	0.68
Peat	0.2	0.4	0.7	3.8	0.05	0.10	0.18

At this stage it is where we see the power and usefulness of the developed methodology and system (even more if we had not geological report). Thus, from Equation (8), with Tm (19.9 °C) and Tp (12 °C) measured in the place and with T_H (22.5 °C) measured by GTP, we can calculate the real ground thermal diffusivity value to 4.80 m depth. This is:

$$\alpha_z = \frac{\pi}{T}\left(\frac{z}{\ln\frac{T_p}{T_H - T_m}}\right)^2 = \frac{\pi}{3.1557 \times 10^7}\left(\frac{4.8}{ln\frac{12}{22.5-19.9}}\right)^2 = 0.98 \times 10^{-6}\frac{\mathrm{m}^2}{\mathrm{s}} \tag{9}$$

The result corresponds, as we can expect, with a mixed sand and gravel soil and with a determined moisture grade. This composition and soil state is irrelevant to us (for this application), because what really matters is the value of αz to 4.80 m depth (α4.80).

CONCLUSIONS

This paper tries to reflect a couple of year of work of studying, analyzing and experimenting with soil temperature measures in order to find a practical and low cost way to dispose of the soil principal variable measurements (temperature and ground thermal diffusivity) to design VLEGE systems in a proper way. The studies conducted and conclusions reached are not only applicable in VLEGE systems, but also in construction, agriculture, biology, *etc.*

Focusing on VLEGE systems, the fluid/ground heat exchange lies in the fact that the temperature difference between the incoming fluid (e.g., air) and the ground varies during the year and depends on the depth to which the collector (pipe) is buried. The amplitude of the ground temperature variation decreases exponentially with depth and it tends to stabilize to the average temperature of the place, therefore at greater depth higher heating/cooling gain.

The methodology and instrumentation system developed in this work can be a useful tool for architects, geologists and engineers, since it gets the real values of the soil variables contrary to exclusively theoretical or empirical methodologies. Incorrect sizing of the VLEGE facility cause malfunction and poor performance of the heat pump. This is a serious problem: dig up collectors and change them is expensive and sometimes, depending on the characteristics of the facility, impossible. Consequently, this may result in the need to use external heating/cooling, which call into serious question the energy efficiency of the facility.

It seems that in an actual construction work is not always possible to perform a one year long ground temperature measurement campaign. This is true, however by observing Figure 3 and Figure 12 is easily verified that is not necessary. The sinusoidal behavior of the soil temperature implies that one needs only a part of the function to rebuild it fully. Really one only needs to get data around T_H or T_L and it is easy because you know (in basis on the timing difference or phase shift between the maximum/minimum soil temperature at chosen depth and on the surface) the days of the year when that will happen.

ACKNOWLEDGMENTS

This work is a contribution of the DPI2013-43870 Project supported by the Spanish Ministry of Economy and Competitiveness. This provides funds for covering the costs to publish in open access.

AUTHOR CONTRIBUTIONS

Sergio Gómez Melgar conceived the experiments. Miguel Ángel Martínez Bohórquez designed and performed the experiments. José Manuel Andújar Márquez wrote the paper.

CONFLICTS OF INTEREST

The authors declare no conflict of interest.

REFERENCES

1. Buzăianu, A.; Csáki, I.; Moțoiu, P.; Popescu, G.; Thorbjornsson, I.; Ragnarsodottir, K.R.; Guðlaugsson, S.; Goubmunson, D. Recent advances of the basic concepts in geothermal turbines of low and high enthalpy. Adv. Mater. Res. 2015, 1114, 233–238.

2. Carmo, C.; Elmegaard, B.; Nielsen, M.P.; Detlefsen, N. Empirical platform data analysis to investigate how heat pumps operate in real-life conditions. In Proceedings of the 24th Iir International Congress of Refrigeration (ICR2015), Yokohama, Japan, 16–22 August 2015.
3. Droulia, F.; Lykoudis, S.; Tsiros, I.; Alvertos, N.; Akylas, E.; Garofalakis, I. Ground temperature estimations using simplified analytical and semi-empirical approaches. Solar Energy 2009, 83, 211–219.
4. Graf, S.; Lanzerath, F.; Sapienza, A.; Frazzica, A.; Freni, A.; Bardow, A. Prediction of scp and cop for adsorption heat pumps and chillers by combining the large-temperature-jump method and dynamic modeling. Appl. Therm. Eng. 2016,98, 900–909.
5. Florides, G.; Kalogirou, S. Ground heat exchangers—A review of systems, models and applications. Renew. Energy2006, 32, 2461–2478.
6. Kavanaugh, S.P.; Rafferty, K.D. Ground-Source Heat Pumps: Design of Geothermal Systems for Commercial and Institutional Buildings; ASHRAE: Atlanta, GA, USA, 1997.
7. Li, C.; Mao, J.; Xing, Z.; Zhou, J.; Li, Y. Analysis of geo-temperature restoration performance under intermittent operation of borehole heat exchanger fields. Sustainability 2016, 8.
8. Fujii, H.; Yamasaki, S.; Maehara, T.; Ishikami, T.; Chou, N. Numerical simulation and sensitivity study of double-layer Slinky-coil horizontal ground heat exchangers. Geothermics 2013, 47, 61–68.
9. Taylor, S.A.; Ashcroft, G.L. Physical Edaphology: The Physics of Irrigated and Nonirrigated Soils; W.H. Freeman and Co.: San Francisco, CA, USA, 1972.
10. Hillel, D. Introduction to Soil Physics; Academic press: New York, NY, USA, 1982.
11. Ciriello, V.; Bottarelli, M.; Di Federico, V.; Tartakovsky, D.M. Temperature fields induced by geothermal devices. Energy 2015, 93, 1896–1903.
12. Fujii, H.; Okubo, H.; Nishi, K.; Itoi, R.; Ohyama, K.; Shibata, K. An improved thermal response test for u-tube ground heat exchanger based on optical fiber thermometers. Geothermics 2009, 38, 399–406.
13. Gehlin, S. Thermal response test method development and evaluation. 2002.
14. Sanner, B.; Hellström, G.; Spitler, J.; Gehlin, S. Thermal response test–current status and world-wide application. In Proceedings of the World Geothermal Congress 2005, Antalya, Turkey, 24–29 April 2005; pp. 24–29.
15. Carson, J.E. Analysis of soil and air temperatures by fourier techniques. J. Geophys. Res. 1963, 68, 2217–2232.
16. Kusuda, T.; Achenbach, P.R. Earth Temperature and Thermal Diffusivity at Selected Stations in the United States; DTIC Document AD0472916; National Bureau of Standards: Gaithersburg, MD, USA, May 1965.
17. Penrod, E.; Elliott, J.; Brown, W. Soil temperature variation (1952–1956) at lexington, kentucky. Soil Sci. 1960, 90, 275–283.
18. Penrod, E.; Walton, W.; Terrell, D. A method to describe soil temperature variation. J. Soil Mech. Found. Div. 1958, 84, 1–21.
19. Carslaw, H.S.; Jaeger, J.C. Conduction of Heat in Solids, 2nd ed.; Oxford Clarendon Press: Oxford, UK, 1959.
20. Sepaskhah, A.; Boersma, L. Thermal conductivity of soils as a function of temperature and water content. Soil Sci. Soc. Am. J. 1979, 43, 439–444.
21. Yuan, Y.; Ji, H.; Du, Y.; Cheng, B. Semi-analytical solution for steady-periodic heat transfer of attached underground engineering envelope. Build. Environ. 2008, 43, 1147–1152.
22. Huelva, E.U.D. Dispositivo Para Captar la Temperatura Superficial de un Objeto. ES 2340013 B1, 22 February 2011.
23. Martínez, M.; Andújar, J.; Enrique, J. Temperature measurement in pv facilities on a per-panel scale. Sensors 2014, 14, 13308–13323.
24. Bohórquez, M.A.M.; Enrique Gómez, J.M.; Andújar Márquez, J.M. A new and inexpensive temperature-measuring system: Application to photovoltaic solar facilities. Solar Energy 2009, 83, 883–890.
25. Andujar Marquez, J.; Mateo Sanguino, T. Design of virtual and/or remote laboratories. A practical case. Revista Iberoamericana de Autom. e Inf. Ind. 2010, 7, 64–72.

CHAPTER 19

APPLICATION OF A HEAT FLUX SENSOR IN WIND POWER ELECTRONICS

Elvira Baygildina [1], Liudmila Smirnova [1], Kirill Murashko [1], Raimo Juntunen [1], Andrey Mityakov [1], Mikko Kuisma [1], Olli Pyrhönen [1], Pasi Peltoniemi [1], Katja Hynynen [1], Vladimir Mityakov [2] and Sergey Sapozhnikov [2]

[1]Department of Electrical Engineering, Lappeenranta University of Technology, P.O. Box 20, Lappeenranta FI-53851, Finland

[2]Department of Thermodynamics and Heat Transfer, Saint-Petersburg State Polytechnical University, Polytechnicheskaya 29, Saint-Petersburg 195251, Russia

ABSTRACT

This paper proposes and investigates the application of the gradient heat flux sensor (GHFS) for measuring the local heat flux in power electronics. Thanks to its thinness, the sensor can be placed between the semiconductor module and the heat sink. The GHFS has high sensitivity and yields direct measurements without an interruption to the normal power device operation, which makes it attractive for power electronics applications. The development of systems for monitoring thermal loading and methods for online detection of degradation and failure of power electronic devices is a topical and crucial task. However, online condition monitoring (CM) methods, which include heat flux sensors, have received little research attention so far. In the current research, an insulated-gate bipolar transistor (IGBT) module-based test setup with the GHFS implemented on the base plate of one of the IGBTs is introduced. The heat flux experiments and the IGBT power losses obtained by simulations show similar results. The findings give clear evidence that the GHFS can provide an attractive condition monitoring method for the thermal loading of power devices.

INTRODUCTION

Alongside the growing complexity and power density of the modern power systems, the thermal management of these systems becomes crucial. A considerable proportion of the power loss in the system is converted into heat, which leads to a temperature rise of the power components. An insulated-gate bipolar transistor (IGBT) power module as a key component of the wind power converter is particularly prone to a number of potential failure modes because of the thermal stress. As it experiences high voltages and currents, the IGBT module suffers from high switching and conduction losses. Moreover, such extreme conditions as wind gusts and high reactive power injection during low-voltage ride-through can significantly increase the losses [1,2].

Power cycles generated in the IGBT module induce local temperature excursions over the multilayered structure of the device. Besides the vertical temperature gradient due to the thermal resistance of the module at a steady state, the dynamic temperature excursions of different magnitudes take place between the layers due to the different thermal capacitances of

Citation: Baygildina, E.; Smirnova, L.; Murashko, K.; Juntunen, R.; Mityakov, A.; Kuisma, M.; Pyrhönen, O.; Peltoniemi, P.; Hynynen, K.; Mityakov, V.; Sapozhnikov, S. Application of a Heat Flux Sensor in Wind Power Electronics. Energies 2016, 9, 456. DOI: 10.3390/en9060456

the materials. The latter causes the cyclic thermo-mechanical stress in the module, which induces wear-out mechanisms. The research of the failure mechanisms caused by the reason of different thermal expansion properties is carried out in [3,4,5,6]. However, the structural deformation of the module and the change in thermal parameters of the device result in changed heat conduction properties and the heat flux distribution pattern in the IGBT module.

To ensure a high reliability performance of the component, efficient measurement tools and an appropriate design of the cooling system are needed. Currently, the IGBT junction temperature measurement methods, such as optical methods, temperature sensors implemented inside the module, and methods applying temperature sensitive electrical parameters (TSEP), are widely studied. The use of the optical methods (infrared (IR) sensors, optical fibers, IR microscope, and IR camera) gives the possibility of capturing a device temperature map. However, the optical systems cannot be used if the device is packaged. Thus, it is not possible to perform online temperature measurement during the device operation. Considering the physical contact methods, temperature sensors are usually soldered onto the chip or mounted on the direct copper bonded (DCB) substrate. Depending on the type of physical contact, either the point or multiple, the sensors can provide the local or spatial temperature measurements. The disadvantage of the physical contact method is the need for a mechanical access to the chip [7,8]. The TSEP methods use the predefined relationship between the electrical parameters, such as the internal gate resistance R_{Gint}, the collector-emitter saturation voltage V_{CEsat}, the gate-emitter voltage V_{GE} or the saturation current I_{sat}, and the IGBT junction temperature [7,9,10]. However, the TSEP approach involves implementation of additional measurement circuitry and a need for calibration.

Considering conventional methods of estimating the IGBT loss, high-precision estimation of the IGBT on-state voltage and current is required to obtain accurate results [11,12]. The approach based on power loss calculation calls for knowledge of the thermal impedance to achieve the loss/temperature profiles, because the IGBT electrical characteristics are functions of junction temperature [5]. Therefore, the practical implications of this method in a multi-device system are highly complicated because of the excessive use of data and short sample time [13].

Thus, the need for real-time and direct monitoring of the thermal load cycles of IGBT devices has made the heat flux sensor (HFS) an attractive approach to be adopted in wind power converters. The HFS allows online monitoring of the dissipated power of the IGBT module without interrupting the operation of the converter. Moreover, the HFS provides an opportunity to optimize the thermal performance by regulating the heat transfer [14].

Various HFS technologies have been developed so far. However, there are only two possible principles of operation: measuring the flux based on spatial temperature distribution or temporal temperature variations.

Currently, one of the available HFSs is the thermopile heat flux sensor [15,16]. Its operation is based on a spatial temperature gradient, where the temperature difference is measured by thermocouples bonded on both the upper and lower surfaces of the sensor. Thermocouples can be connected in series, thereby obtaining several thermoelectric junctions in order to amplify the voltage output.

A heat measuring method based on temperature transients is adopted by thin film resistance temperature devices (RTD), coaxial thermocouples, and calorimeters [17,18]. In order to measure the heat flux as a function of time, it is necessary to manage the temperature history by the methods discussed in [19]. The basic principle of the methods in [19] is based on the proportionality of the heat flux to the temperature gradient. However, it is difficult to obtain reliable measurements using these methods, since they assume a one-dimensional heat transfer and uniform temperature distribution as it changes only with time.

The present technologies associated with the HFS and heat measuring methods are implemented in a wide variety of industrial and research applications. Usually, the method of implementation involves mounting the sensor into or onto the surface of a solid material. Depending on the area of application, heat flux sensors are selected based on individual characteristics such as response time, sensitivity, experienced levels of temperatures, and heat flux.

The aforementioned heat measurement techniques are widely used in the industry. However, their disadvantage is the indirect information of the heat flux. Either the temperature measured in the transient or the spatial temperature distribution require additional interpretation of the data. This study aims to adopt the heat flux measurements in a power device using a gradient heat flux sensor (GHFS). The GHFS allows direct measurements of the heat flux based on the Seebeck effect of the tilted layered materials. The structure of the sensor is provided by anisotropic thermoelements so that the thermoelectric electromotive Force (EMF) along the length of the sensor is achieved.

In this paper, the HFS is used to monitor the thermal stress of an IGBT module. Sine-wave-modulated current is applied to induce thermal load dynamics similar to the operation of the module in the wind power converter. Several steady state conditions are tested to compare the measured heat flux and the estimated IGBT power loss.

Section 2 describes the technology and basic characteristics of the GHFS. In Section 3, the focus is on the modeling of the spatial heat flux distribution within the module and possible heat power dissipated through the GHFS attached to the base plate of the IGBT. The application of the heat flux sensor in the test setup and an analysis of the test results are described inSection 4 and Section 5, respectively. Interpretation and practical application of the obtained results are discussed in Section 6.

GRADIENT HEAT FLUX SENSOR TECHNOLOGY

The thermoelectric effect induced in the material with a temperature gradient is called the Seebeck effect. Heat flux sensors based on the Seebeck effect are classified into longitudinal and transverse ones according to the direction of the induced thermo-EMF (Figure 1) [20,21]. In the longitudinal HFS (Figure 1a), the heat flow vector $\vec{q}$ and the vector of the thermo-EMF $\vec{E}$ || are collinear. Thus, to increase the volt-watt sensitivity, the thickness of the sensor should be large enough (e.g., 1–2 mm). A relatively large thickness leads to two main drawbacks: a high thermal resistivity, which can reach the order of 10^{-3} m^2·K/W, and a slow time response (e.g., 10–100 s) [21].

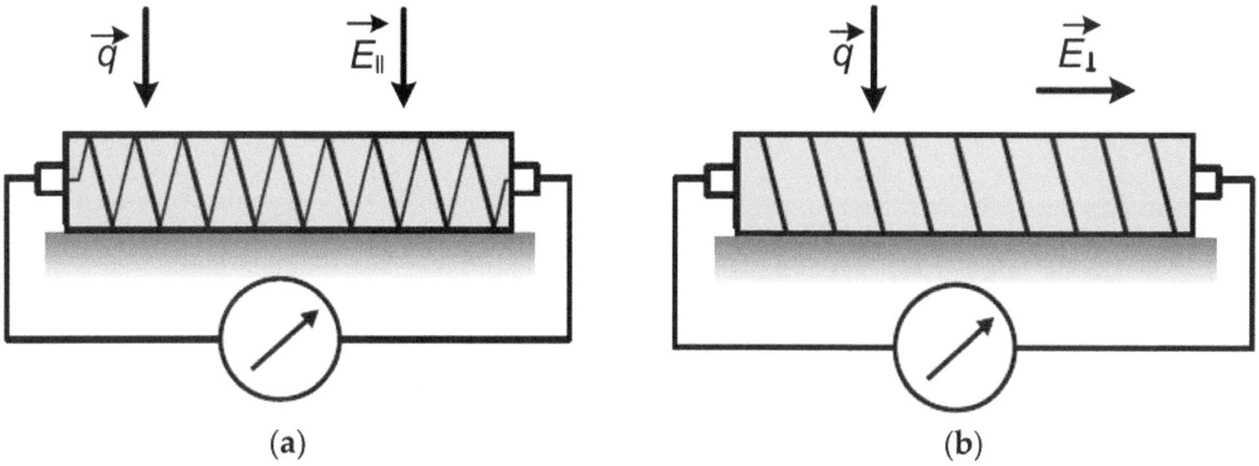

Figure 1. Heat flux sensors: (**a**) longitudinal; (**b**) and transverse [21]. Copyright Springer 2012.

The operation principle of the GHFS is based on the transverse Seebeck effect (Figure 1b). The sensitivity of the GHFS depends on the length of the sensor as the thermo-EMF $\vec{E}\perp$ is induced in the direction normal to the heat flow $\vec{q}$. This effect is achieved by highly anisotropic thermoelements, whose crystallographic axis does not coincide with the geometrical axis of the sensor plate.

The current study focuses on the GHFS made of single-crystal bismuth. The artificial tilted layered material of the sensor yields the anisotropic properties of thermoelectric conductivity and the EMF induced in the material. The volt-watt sensitivity of such a sensor can be expressed as:

$$S_0 = \frac{E}{qA_{sensor}}, \tag{1}$$

where E is the thermoelectric EMF, q is the heat flux density, and A_{sensor} is the area of the sensor plate. The sensitivity of the GHFS has been tested under different heat flux values and environmental conditions. The obtained sensitivity was constant, and the maximum calibration error was 0.64%. The details of calibration process are described in [21].

The sensor is constructed of series-connected bismuth strips, where the width of a single strip is 0.1–0.2 mm (Figure 2) [22]. The thermo-EMF induced in a single bismuth plate is relatively low, and thus, the sensitivity can be increased by series connection.

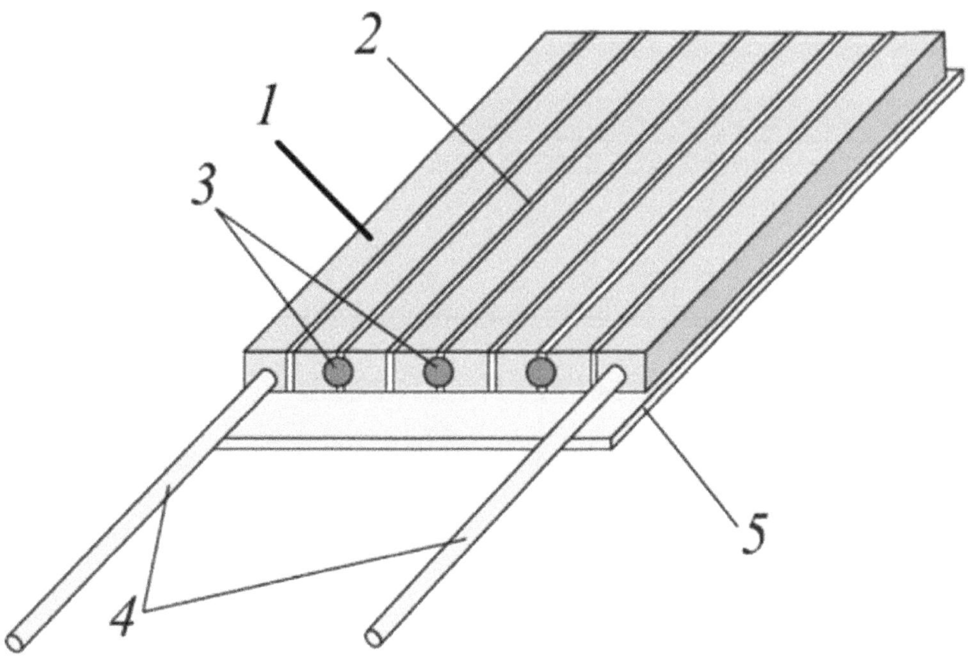

Figure 2. Bismuth-based gradient heat flux sensor, (1) single plate of bismuth; (2) insulation; (3) solder; (4) connecting terminals; (5) mica base [21]. Copyright Springer 2012.

The thermo-EMF of such a sensor (Figure 2):

$$E_{tot} = qES_0n, \quad (2)$$

where n is the number of bismuth plates.

With the thermal conductivity of the bismuth material 7.45 W/(m·K) and the thickness of the plate 0.1–0.2 mm, the thermal resistivity is in the order of 10^{-5} m²·K/W. The maximum working temperature is limited by the melting point of bismuth, which is 544 K.

In the tests performed in the current study, the GHFS has the thickness of 0.2 mm, the volt-watt sensitivity of 15.9 mV/W, and the area of 0.85 cm². The response time of the sensor is in the range of 10^{-9}–10^{-8} s [20,22]. However, considering the IGBT module as a system with a specific thermal impedance, the bandwidth of the measured heat flux dynamics is limited by the thermal transients associated with the heat flow from the chip to the base plate.

HEAT FLUX SENSOR APPLICATION IN AN IGBT MODULE FOR A WIND POWER CONVERTER

The thermal load fluctuations of IGBTs in the modern large-scale wind power converters can be classified into several time domains with a specific time constant range and loading behavior [23,24]. In Figure 3a, the wind profile and the corresponding IGBT power loss dissipation in an Infineon FZ400R17KE4 IGBT module of a 3L ANPC converter prototype are shown. The results were obtained based on the 100 kW wind turbine modelled in Matlab Simulink. The thermal performance of the IGBT was analyzed using PLECS Blockset [25]. The dissipated heat power is averaged by the fundamental frequency of 50 Hz. In Figure 3b, the instantaneous IGBT power losses in the steady state at the converter nominal power and the switching frequency of 2 kHz are shown. Obviously, the IGBT thermal stress during the wind turbine operation is significant even for the dynamics associated with the stress drivers in different time domains.

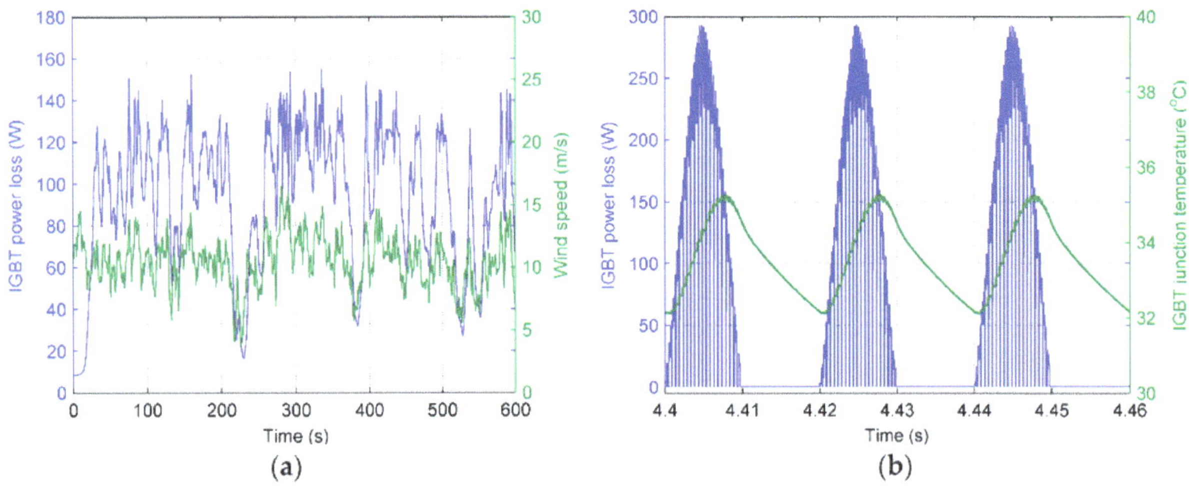

Figure 3. Wind speed and insulated-gate bipolar transistor (IGBT power losses: (**a**) averaged by the fundamental frequency *vs.* time; (**b**) instantaneous IGBT power losses and junction temperature in the steady state.

In order to investigate the temperature and flux distribution inside the switch, the Finite Element Method (FEM) model of the module placed on an aluminum heat sink was generated in Comsol Multiphysics. The mesh of the FEM model has 920407 tetrahedral elements and the run time is 120 s. The cross-section of the model is shown in Figure 4, and the characteristics of the materials used in the simulation are listed in Table 1.

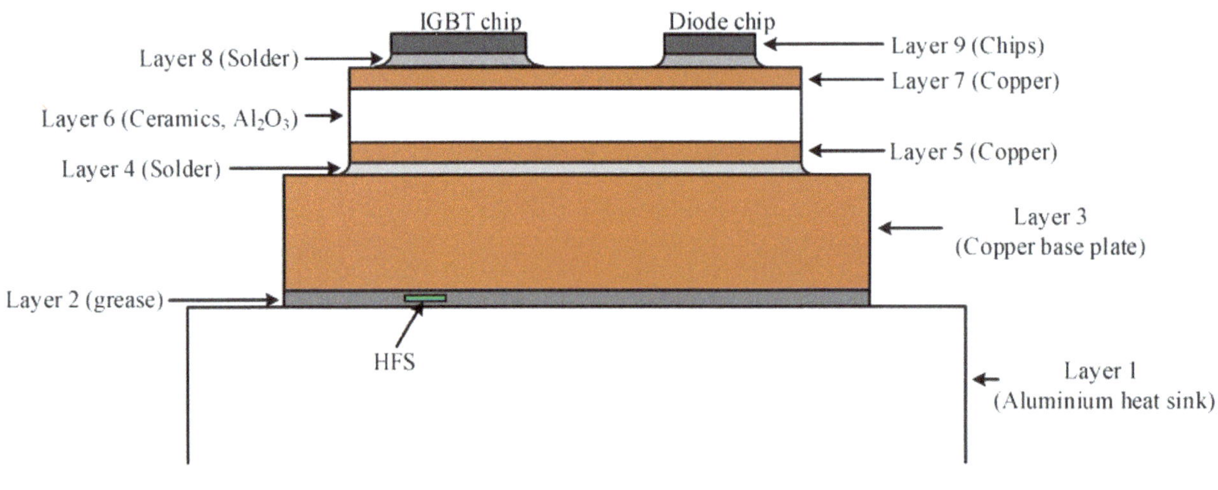

Figure 4. Cross-section of the model.

Table 1. Properties of the materials in the model.

Layer	Material	Thickness, mm	Thermal Conductivity, W/(m·K)
1	aluminum	120	205
2	thermal grease	1	1
3	copper	3	400
4	solder	0.08	50
5	copper	0.15	400
6	Al_2O_3	0.42	27
7	copper	0.15	400
8	solder	0.08	50
9	silicon	0.22	130

In the model, the HFS is located in layer 2 and attached to the base plate. The thickness of the sensor is 0.2 mm, and the thermal conductivity is 0.8 W/(m·K). Thermal grease is applied on both sides of the sensor so that it will be possible to attach the sensor to the surface of the base plate and the heat sink. Such a thick thermal grease layer was obtained because the practical task is also to attach the thermocouple of rather thick diameter, as discussed in Section 4.

It is pointed out that the heat flow modeled here assumes that all heat from the chip is dissipated through the heat sink. In practical applications, as a result of the conduction heat transfer within the IGBT and the smaller thermal resistance in the direction from the chip to the heat sink, the majority of heat dissipates through the bottom side of the base plate. Furthermore, considering the path in the direction to the edge of the base plate, which has a smaller area than the bottom surface of the base plate, the heat transfer is determined by free (or natural) convection, and thus, the thermal resistance is much higher in this direction. The FEM modelling of the heat flux has shown that only 2% of the heat dissipates through the sides of the IGBT module, thus the free convection at the rim of the baseplate can be omitted. These facts makes it reasonable to attach the HFS to the bottom surface of the base plate.

In the FEM modelling, the ambient temperature is assumed to be 25 °C. The heat generated in the IGBT (13.2 W) is dissipated from all four chips by forced convection from the fins of the heat sink, and the convective coefficient is 5 W/(m^2·K). The surface temperature distribution is shown in Figure 5. The heat flux density through the thermal grease is shown in Figure 6, where the nonuniformity of the heat flux is observed. Thus, 13.2 W of the dissipated power in the IGBT corresponds to 33.2 °C of the chip temperature and 1870 W/m^2 of the heat density dissipated through the HFS. The average heat flux density over the base plate surface is 2025 W/m^2. Because the thermal conductivities of the thermal grease and the sensor are not the same, the heat flux through the HFS has an 8% lower density than the average density over the base plate. Obviously, the thermal conduction under the chip where the HPS is located is disturbed. In order to mitigate the disturbance of the heat conduction, a possible solution would be to place the HFS between the IGBT chips. However, in this case the influence of the other chips on the HFS should be considered. Another solution could be integrating the sensor close to the chip, but relatively low thermal conductivity and inappropriately high thickness of the sensor makes this solution unacceptable.

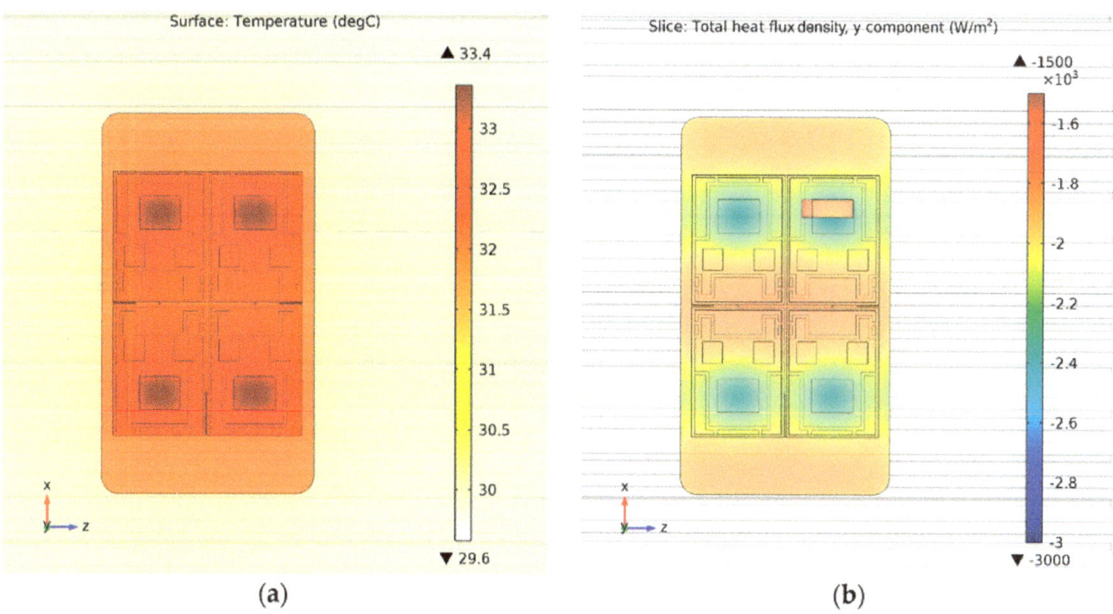

Figure 5. Top view of the model: (**a**) temperature distribution; (**b**) heat flux density distribution.

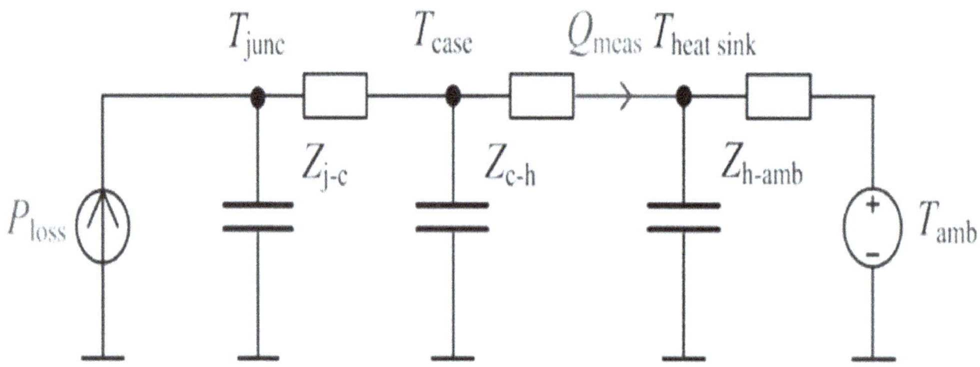

Figure 6. Thermal model of IGBT module.

The thermal model of the IGBT is presented in Figure 6. It includes the thermal impedances of the junction to the case Z_{j-c}, the case to the heat sink Z_{c-h}, and the heat sink to the ambient Z_{h-a}. Because the HFS is attached on the surface of the base plate, the location of the heat flux measurement point is as indicated by Q_{meas} in the thermal model. The thermal impedances of the FZ400R17KE4 IGBT module are presented in Table 2.

Table 2. Thermal impedances of the FZ400R17KE4 IGBT module.

Thermal Impedance	Z_{j-c} Layer 1	Z_{j-c} Layer 2	Z_{j-c} Layer 3	Z_{j-c} Layer 4	Z_{c-h}
R_{iIGBT}, K/kW	2	9.2	42.6	6.3	18
τ_{iIGTB}, ms	0.8	13	50	600	-

HEAT FLUX SENSOR IN THE TEST SETUP

The HFS has been tested to monitor the dissipated power of the Infineon FZ400R17KE4 IGBT module. Two IGBT modules are connected in parallel to the direct current (DC) voltage source (Figure 7). The operating switching algorithm of the IGBT is based on sinusoidal pulse-width modulation (PWM) with a 1 kHz carrier and the fundamental frequencies of 0.1 and 7 Hz. The gate driver provides inverse operation of two IGBTs. Numerical modeling of the scheme in Figure 7 is carried out using PLECS blockset in Matlab Simulink (Figure 8).

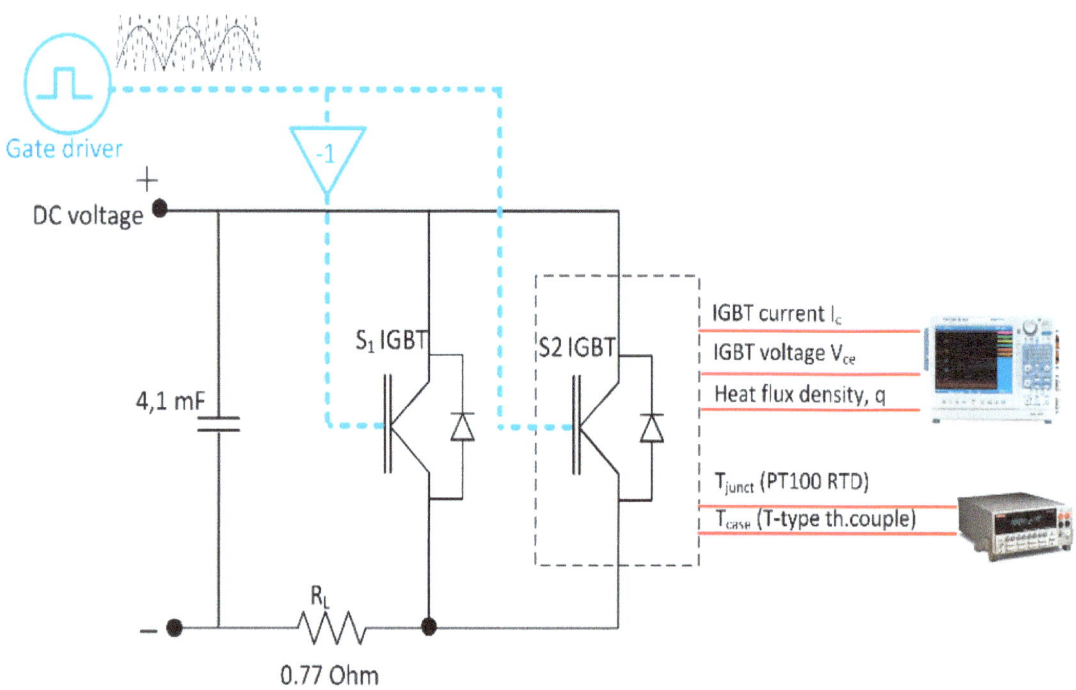

Figure 7. Test setup for the IGBT heat flux measurements.

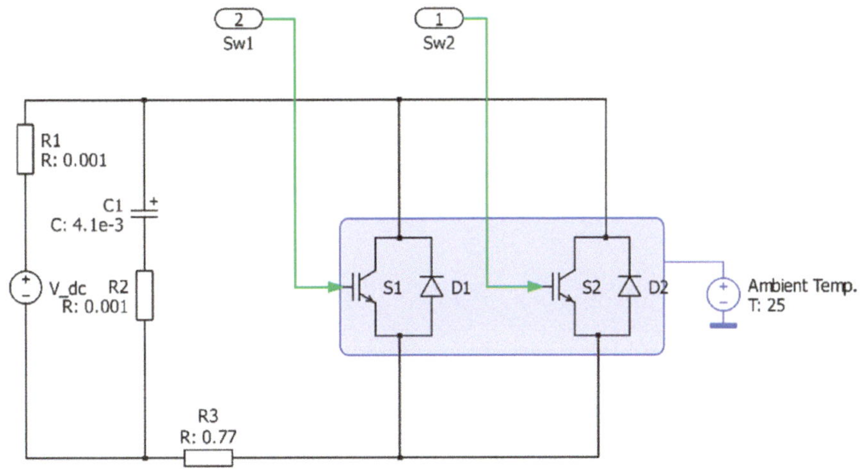

Figure 8. PLECS model for the IGBT heat flux measurements.

The DC voltage, the corresponding DC current, and the fundamental frequency of the IGBT control algorithm are the input parameters of the circuit, which were varied to achieve several steady state operating points. The fundamental frequency signal of the IGBT control algorithm is a rectified sine wave. Thus, the power loss in the IGBT S2 averaged by switching frequency is formed as half-wave sinusoids with the fundamental frequencies of 0.1 and 7 Hz. Since, the IGBT S1 operates in an inverse mode, the corresponding power loss in S1 is formed as the inversed half-wave sinusoids.

The measurements obtained in the test were the collector-emitter voltage V_{ce}, the collector current I_c of the S_2 IGBT module, and the heat flux density q measured by the GHFS. These signals were observed by a DL850 Scope Corder. In addition, the IGBT case temperature was measured using a T-type thermocouple, which was attached to the base plate using the thermal grease. The temperatures were observed in the LabView interface by a Keithley 2700 data logger.

The collector-emitter voltage V_{ce} and the collector current I_c of the IGBT were measured to obtain the instantaneous IGBT power loss P_{loss} and to make a comparison with the heat flux Q.

Two IGBT modules were placed on the air-cooled heat sink. The GHFS was attached on the surface of the base plate of the S_2 IGBT module using thermal grease (Figure 9). The position of the sensor was predefined based on the location of the four IGBT chips inside the module so that the sensor was placed under one of the IGBT chip positions.

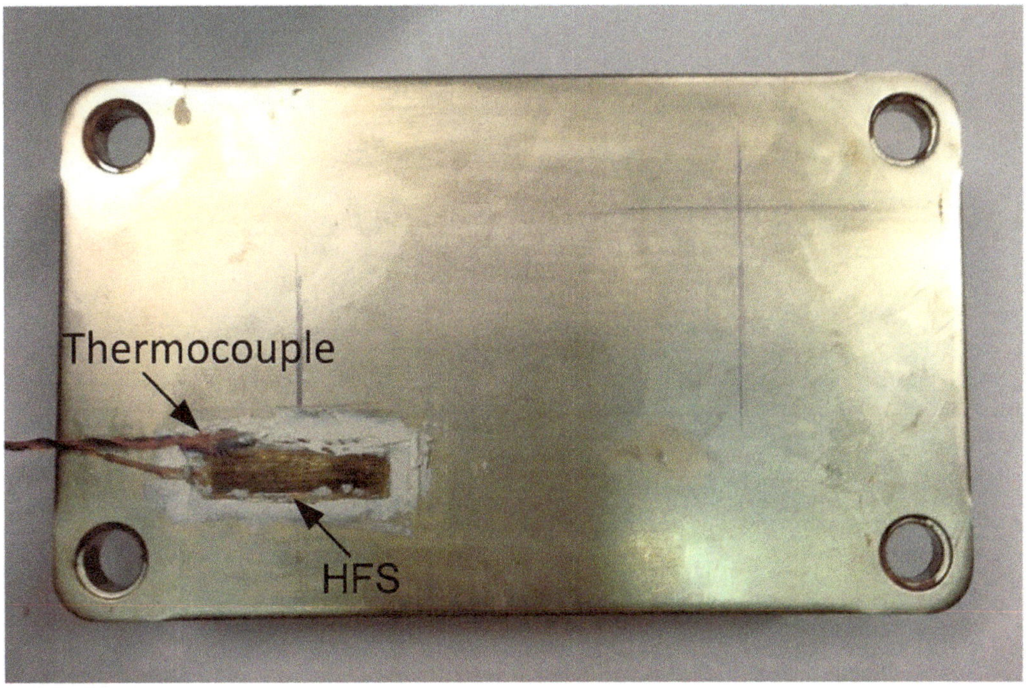

Figure 9. Attachment of the sensor to the base plate.

The induced EMF of the GHFS is in the range of millivolts. Thus, in order to detect the results by the DL850 Scope Corder, an amplifier connected to the GHFS terminals was used. The EMF was also measured by the Keithley 2700 data logger to record and visualize the heat flux density q in the LabView interface.

Therefore, the induced thermoelectric EMF of the GHFS should be interpreted to the heat flux Q determined from Equations (1) and (2) as follows:

$$Q = qA_{baseplate} \quad (3)$$

$$q = \frac{E_{tot}}{S_0 A_{sensor}} \quad (4)$$

where Q is the heat flux (W) and $A_{baseplate}$ is the area of the IGBT base plate (m^2). The extrapolation of the heat flux density in Equation (3) could cause an error in case of nonuniform heat flux distribution over the baseplate. The comparison of the heat flux test results with the numerical calculations are presented in Section 5.

The parameters used in the test are presented in Table 3.

Table 3. Heat flux sensor and IGBT parameters used for the heat flux calculation.

Parameter	Value
$A_{baseplate}$, cm^2	65.3
A_{sensor}, cm^2	0.85
S_0, V/W	15.9×10^{-3}

ANALYSIS OF THE TEST RESULTS

The IGBT heat flux dissipation is observed in three steady state conditions, that is, with load currents of 15, 30, and 60 A, which respectively correspond to 6.9, 13.7, and 29.5 W of the average dissipated IGBT power. Unfortunately, due to the voltage limit of the power source used, one cannot apply higher currents in the test setup. The heat flux density and the IGBT case temperature performance are visualized in a LabView interface (Figure 10). The case temperature and the heat flux have transient dynamics, due to the total thermal capacity of the system. It includes the convection properties of the cooler air and the thermal impedance from the case to the heat sink, and from the heat sink to the ambient environment.

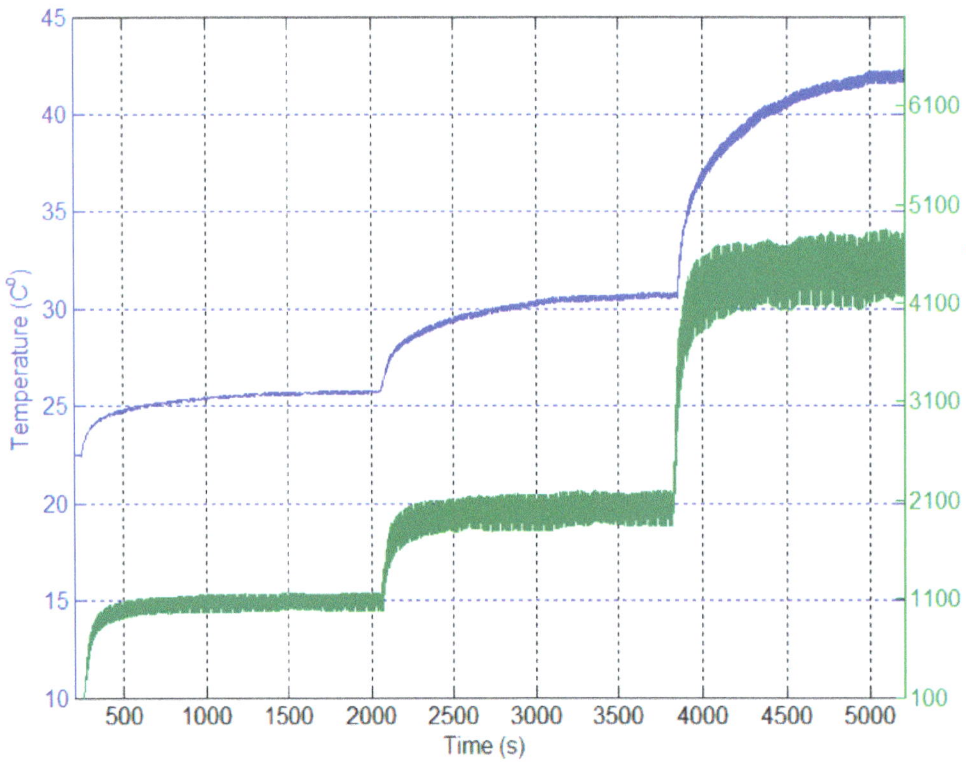

Figure 10. Heat flux density and case temperature in three steady state conditions measured at 15, 30, and 60 A switching current with fundamental and switching frequency of 0.1 Hz and 1 kHz, correspondingly.

In Figure 11, the measured heat flux densities at 0.1 and 7 Hz fundamental frequencies are presented. The heat flux density has a form of half-wave sinusoids with the periods of 10 and 0.14 s, respectively. The heat flux density at 0.1 Hz was measured by the Keithley 2700 data logger with an average sampling time of 0.13 s, and the heat flux density at 7 Hz was recorded by a DL850 Scope Corder with a 0.2 ms sampling time. Thus, significant noise content is visible at 7 Hz.

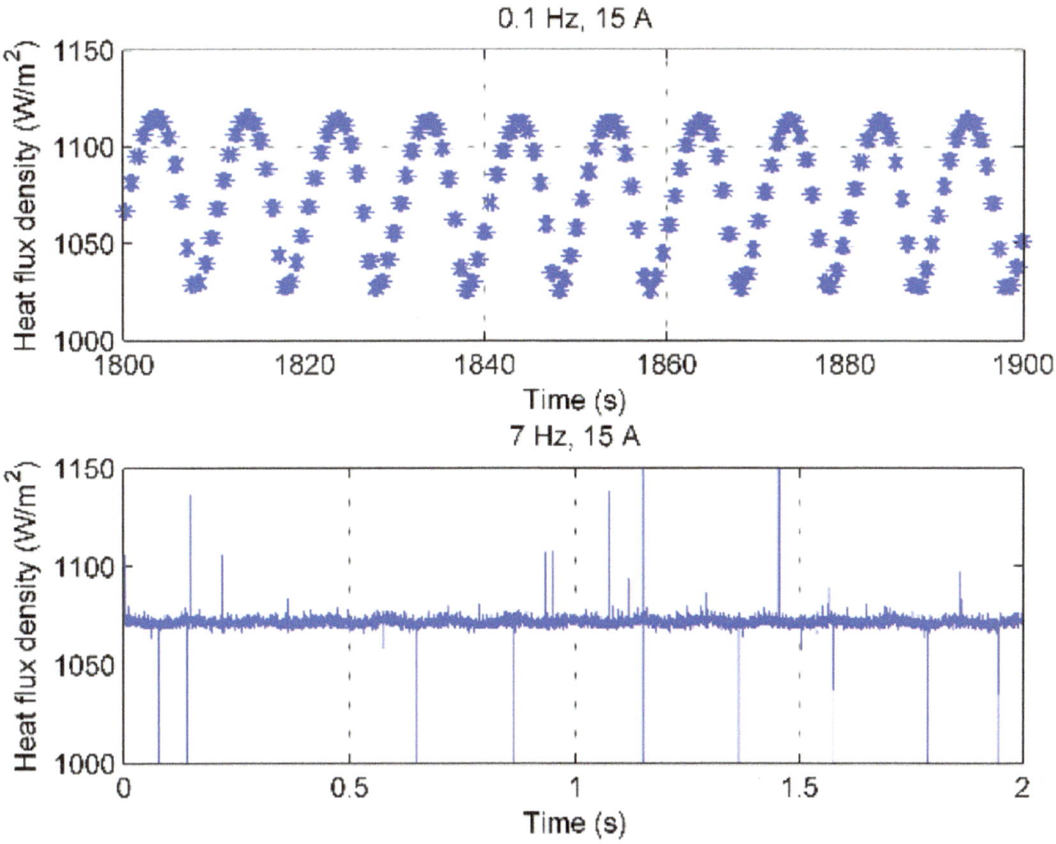

Figure 11. Heat flux at 0.1 Hz and 7 Hz with 15 A source current.

The IGBT collector-emitter voltage V_{ce} and the collector current I_c, measured within a 5 ms time to cover the IGBT switching are shown in Figure 12. The corresponding instantaneous and average power losses calculated based on V_{ce} and I_c and the measured heat loss Q are presented and shown to be equal. In Figure 13, the waveforms of the IGBT power losses modeled in PLECS (Figure 9), averaged by the switching frequency at 15, 30, and 60 A load current, are presented. Here, the power losses are formed as half-waves of sinusoids. The average values at 15, 30, and 60 A are 6.9, 13.7, and 29.5 W, respectively.

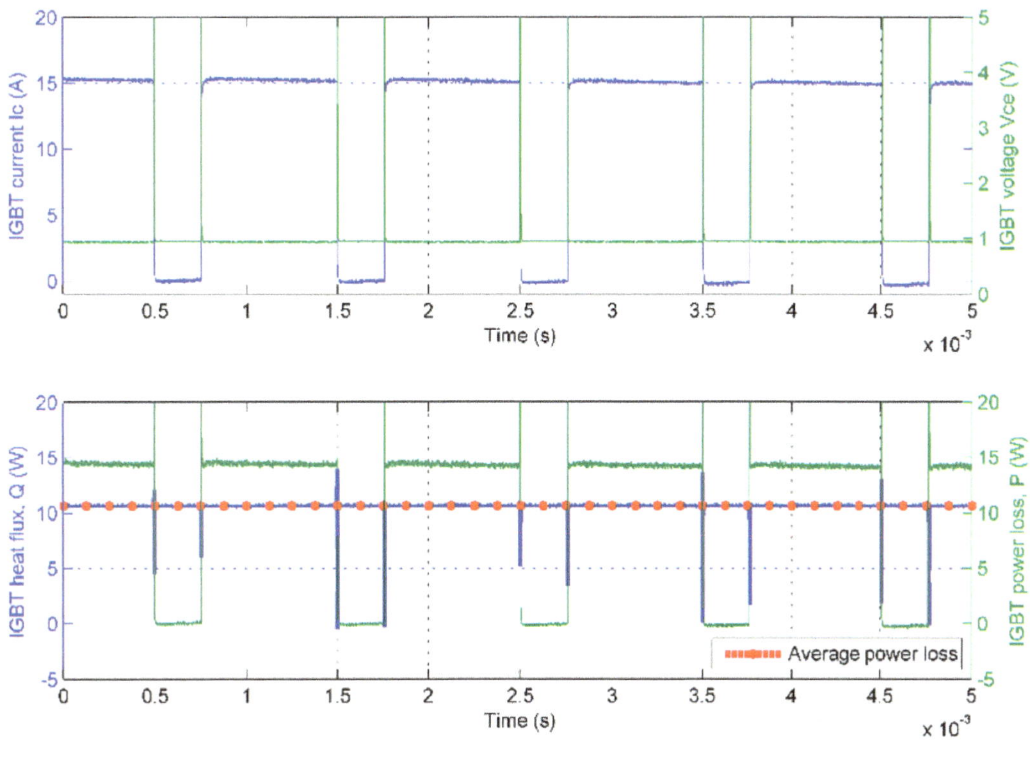

Figure 12. Collector-emitter voltage V_{ce}, collector current I_c, instantaneous and average IGBT power loss P, and heat flux Q within 5 ms period.

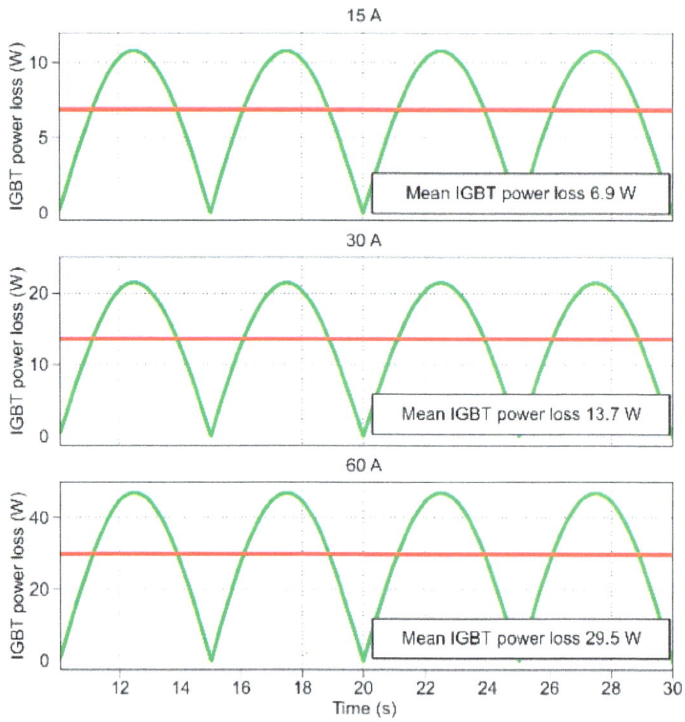

Figure 13. IGBT power losses averaged by the switching frequency at 15, 30, and 60 A.

No high switching frequency variation can be observed in the measured heat power because of the thermal impedance $Z_{j\text{-}c}$ of the IGBT from the junction to the case (Figure 6). Therefore, to compare the performance of the heat flux in steady state conditions, it is advisable to define both the heat flux Q and the power loss P_{loss} averaged by the fundamental frequency.

The calculated IGBT power losses P_{loss} and the measured heat flux Q, both averaged by the fundamental frequency of 0.1 Hz, are presented in Table 4. Thus, in the 15, 30, and 60 A steady states, the differences between P_{loss} and Q are 1.1%, 3.5%, and 3.7%, respectively.

Table 4. Comparison of the heat flux test results with the power loss defined by numerical modeling in Matlab PLECS.

Compared Signals	Steady State Conditions, DC Current		
	15 A	30 A	60 A
Power loss P_{loss} averaged by fundamental frequency, W (modeled by Matlab PLECS)	6.90	13.68	29.50
Measured heat flux Q averaged by fundamental frequency, W	6.98	13.20	28.40

The measured heat flux density q is compared with the results of the FEM modeling in Table 4. The modeled results indicate that the heat flux through the HFS has a 6.6%–8% lower density compared with the density averaged over the whole base plate (Figure 6). As mentioned in Section 3, this finding can be explained by the difference in the thermal conductivities between the HFS (0.8 W/(m·K)) and the thermal grease (1 W/(m·K)), which results in the distortion of the heat flux distribution over the base plate. However, the measured heat flux density results are similar to the modeled averaged density (Table 5).

Table 5. Comparison of the heat flux density test results with the numerical modeling.

Compared Signals		Steady State Conditions, DC Current		
		15 A	30 A	60 A
Heat flux density (FEM model), q_{FEM} W/m^2	Through HFS	992	1870	4060
	Average over base plate	1075	2025	4332
Measured heat flux density q, W/m^2		1074	2025	4347

Notes: HFS refers to heat flux sensor.

It should be noted that the small variations in the model parameters and the test setup, such as the thickness of the grease layer and the thermal conductivities of the material, could produce significant deviation to the chip temperature and the heat flux density distribution over the base plate.

INTERPRETATION AND PRACTICAL APPLICATION OF THE RESULTS

According to the experimental results, it can be concluded that the heat flux variations have a limited bandwidth that depends on the IGBT thermal impedance and the module package properties. The results observed at 7 Hz support the latter statement because of the extremely small amplitude of the sine-wave heat flux density. Therefore, the IGBT power loss oscillations corresponding to the IGBT switching and the variations in the 50 Hz load current cannot be tracked by the HFS. However, modern variable-speed wind turbines involve the operation of the generator-side inverters at lower frequencies. In doubly-fed drives, in particular, the rotational speed and the corresponding fundamental frequency of the load current are limited to around ±30% of their synchronous values [26]. Thus, at low operating frequencies, the IGBT thermal impedance has a lower attenuation effect on thermal cycling than at the 50 Hz operating frequency.

The HFS can be applied to indicate an IGBT failure associated with degradation. As mentioned above, the typical package-related failure mechanism, such as solder degradation, induces localized heating as a result of a rise in the thermal resistance [27]. The voiding effect, both under the silicon chip or the ceramic substrate, causes an uneven heat distribution over the chip or base plate surface, which may lead to a higher localized heat flux density. Measured heat flux density can be compared with the expected power losses of an appropriately functioning IGBT in order to observe a possible deviation. However, in order to provide online calculation of the expected power loss, knowledge of the current operating point is required. For this purpose, the electrical variables from the wind turbine data acquisition system, such as load current, the DC link voltage, the power factor, and the fundamental and switching frequencies, are needed. The measurement of all electrical parameters can be avoided if the correlation between the load current and the heat flux can be determined in advance. The deviation from the correlations would mean the degradation of the device. Because of the difference in the time constants of the IGBT power losses and the heat flux, it is reasonable to compare the signals averaged by the fundamental frequency. If a deviation is detected, one can conclude that degradation has taken place. This deviation can also be a result of another failure mechanism observed in actual systems [28,29], which is thermal grease degradation. The mechanisms such as the grease pump-out and dry-out also induce an increase in the grease thermal resistance [30] and the heat flux through the HFS.

As mentioned in Section 5, another attractive application of the HFS is the option to optimize the IGBT thermal performance. In particular, heat flux measurements can be used as a feedback signal for a cooling control system. In [14], the power losses of the device measured by the HFS were used to control the fluid flow rate in the liquid cooling systems. The purpose of the study [14] was to smooth the heat sink temperature during the variable load and thereby prolong the lifetime.

CONCLUSIONS

In this paper, an application of the HFS based on the transverse Seebeck effect in wind power electronics was described. The operation of the sensor was tested in an IGBT-based test setup. The control algorithm for the IGBT switches was programmed to generate output IGBT power loss dynamics resembling the operation in a wind turbine converter. In order to implement direct measurements of the heat flux dissipation of the IGBT, an HFS was attached onto the surface of the IGBT base plate using thermal grease. The HFS was placed under one of the four chips on the base plate. The heat flux performance was tested in steady state conditions. DC currents of 15, 30, and 60 A induced sine-wave IGBT heat dissipation of 6.98, 13.2, and 28.40 W, respectively, averaged by fundamental frequency. The measured average IGBT power loss deviates from the thermal model calculated in Matlab PLECS by 3.7% at the maximum.

The measured and FEM simulation results were compared in terms of heat flux density based on the model analyzed in Comsol Multiphysics, where the heat flux dissipation from the chip to the heat sink was assumed. The measured heat flux density was 8% higher than the density of the modeled heat flux through the sensor. However, the averaged heat flux density over the base plate was equal to the measured one. Here, it is pointed out that the distribution of the heat flux density over the base plate is highly dependent on the thermal conductivities, and the thickness of the thermal grease and the HFS.

The paper shows good results of direct measurements of the IGBT heat dissipation using the HFS, which can be an attractive and innovative approach to the thermal stress condition monitoring (CM). Heat flux measurements provide an opportunity to improve the reliability performance of the IGBT and provide online detection of the device aging or failure during wind turbine operation. By heat flux measurements, the thermal performance can be optimized and the lifetime of the device can be extended.

ACKNOWLEDGMENTS

This work was funded by Electrical Engineering Department, Lappeenranta University of Technology.

AUTHOR CONTRIBUTIONS

Kirill Murashko designed the test setup; Raimo Juntunen programmed the IGBT module switching algorithm; Elvira Baygildina and Liudmila Smirnova performed the experiments and analyzed the data; Andrey Mityakov, Vladimir Mityakov, and Sergey Sapozhnikov provided the gradient heat flux sensor; Mikko Kuisma provided the amplifier; Olli Pyrhönen, Pasi Peltoniemi, and Katja Hynynen gave useful suggestions and revised the manuscript; Elvira Baygildina wrote the paper.

CONFLICTS OF INTEREST

The authors declare no conflict of interest.

Abbreviations

The following abbreviations are used in this manuscript:

CM	Condition Monitoring
CTE	Coefficients of Thermal Expansion
DC	Direct Current
DCB	Direct Copper Bonded
EMF	Electromotive Force
FEM	Finite Element Method
GHFS	Gradient Heat Flux Sensor
HFS	Heat Flux Sensor
IGBT	Insulated-gate Bipolar Transistor
IR	Infrared
PWM	Pulse-Width Modulation
RTD	Resistance Temperature Device
TSEP	Temperature Sensitive Electrical Parameter

REFERENCES

1. Ma, K.; Blaabjerg, F.; Liserre, M. Thermal Analysis of Multilevel Grid-Side Converters for 10-MW Wind Turbines under Low-Voltage Ride Through. *IEEE Trans. Ind. Appl.* **2013**, *49*, 909–921.
2. Ma, K.; Blaabjerg, F.; Liserre, M. Reactive Power Influence on the Thermal Cycling of Multi-MW Wind Power Inverter.*IEEE Trans. Ind. Appl.* **2013**, *49*, 922–930.
3. Ikonen, M. Power Cycling Lifetime Estimation of IGBT Power Modules Based on Chip Temperature Modeling. Ph.D. Thesis, Lappeenranta University of Technology, Lappeenranta, Finland, 2012.
4. Ciappa, M. Selected failure mechanisms of modern power modules. *Microelectron. Reliab.* **2002**, *42*, 653–667.
5. Ma, K.; Bahman, A.S.; Beczkowski, S.; Blaabjerg, F. Complete Loss and Thermal Model of Power Semiconductors Including Device Rating Information. *IEEE Trans. Power Electron.* **2015**, *30*, 2556–2569.
6. Kostandyan, E.E.; Sørensen, J.D. Reliability Assessment of solder joints in power electronics modules by crack damage model for wind turbine applications. *Energies* **2014**, *4*, 2236–2248.
7. Avenas, Y.; Dupont, L.; Khatir, Z. Temperature Measurement of Power Semiconductor Devices by Thermo-Sensitive Electrical Parameters—A Review. *IEEE Trans. Power Electron.* **2012**, *27*, 3081–3092.
8. Blackburn, D.-L. Temperature measurements of semiconductor devices–A review. In Proceedings of the 20th Annual Semiconductor Thermal Measurement and Management Symposium, San Jose, CA, USA, 9 March–11 March 2004; pp. 70–80.
9. Baker, N.; Munk-Nielsen, S.; Liserre, M.; Iannuzzo, F. Online junction temperature measurement via internal gate resistance during turn-on. In Proceedings of the European Conference on Power Electronics and Applications (EPE 2014), Lappeenranta, Finland, 26–28 August 2014; pp. 1–10.
10. Ammous, A.; Allard, B.; Morel, H. Transient Temperature Measurements and Modeling of IGBT's Under Short Circuit. *IEEE Trans. Power Electron.* **1998**, *13*, 12–25.
11. Ghimire, P.; de Vega, A.R.; Beczkowski, S.; Munk-Nielsen, S.; Rannested, B.; Thogersen, P.B. Online Vce measurement and temperature estimation method for high power IGBT module in normal PWM operation. In Proceedings of the International Power Electronics Conference (IPEC-Hiroshima 2014—ECCE-ASIA), Hiroshima, Japan, 18–21 May 2014; pp. 2850–2855.
12. Automated Power Loss Characterization of High Power IGBT Modules. Available online: http://www.corpe.et.aau.dk/laboratory-facilities/ Automated+Power+Loss+Characterization+of+High+Power+IGBT+Modules/ (accessed on 29 January 2016).
13. Zhou, Z.; Knanniche, M.S.; Igic, P.; Kong, S.T.; Towers, M.; Mawby, P.A. A fast power loss calculation method for long real time thermal simulation of IGBT modules for a three/phase inverter system. In Proceedings of the European Conference on Power Electronics and Applications (EPE 2005), Dresden, Germany, 11–14 September 2005; pp. 1–10.
14. Smirnova, L. Electromagnetic and Thermal Design of a Multilevel Converter with High Power Density and Reliability. Ph.D. Thesis, Lappeenranta University of Technology, Lappeenranta, Finland, 2015.
15. Olivier, H. *Thin Film Gauges and Coaxial Thermocou ples for Measuring Transient Temperatures*; Shock Wave Laboratory: Aachen, Germany, 2003.
16. RdF Heat Flux Sensors. Available online: http://www.rdfcorp.com/products/hflux/hfs-c_01.shtml (accessed on 29 January 2016).
17. Hogendoorn, C.J.; de Lange, H.C.; van Steenhoven, A.A. Design optimization for fast heat-transfer gauges. *Meas. Sci. Technol.* **1998**, *9*, 428–434.
18. Kidd, C.T.; John, C.; Adams, J. Fast response heat-flux sensor for measurement commonality in hypersonic wind tunnels. *J. Spacecr. Rockets* **2001**, *38*, 719–729.
19. Webster, J.G. *Handbook for Measurement, Instrumentation and Sensors*; CRC Press LLC: Boca Raton, FL, USA, 1999; pp. 1130–1139.
20. Mityakov, A.V.; Sapozhnikov, S.Z.; Mityakov, V.Y.; Snarskii, A.A.; Zhenirovsky, M.I.; Pyrhönen, J.J. Gradient heat flux sensors for high temperature environments. *Sens. Actuators* **2012**, *176*, 1–9.
21. Sapozhnikov, S.Z.; Mityakov, V.Y.; Mityakov, A.V.; Pokhodun, A.I.; Sokolov, N.A.; Matveev, M.S. The calibration of gradient heat flux sensors. *Meas. Tech.* **2012**, *54*, 1155–1159.
22. Sapozhnikov, S.Z.; Snarskii, A.A.; Mityakov, V.Y.; Zhenirovsky, M.I.; Mityakov, A.V. Gradient heat

flux sensors based on the artificial anisotropic thermoelements. *Thermoelectricity* **2010**, *4*, 38–53.

23. Ma, K.; Liserre, M.; Blaabjerg, F.; Kerekes, T. Thermal Loading and Lifetime Estimation for Power Device Considering Mission Profiles in Wind Power Converter. *IEEE Trans. Power Electron.* **2015**, *30*, 590–602.
24. Baygildina, E.; Smirnova, L.; Ma, K.; Peltoniemi, P.; Pyrhönen, O. Power semiconductor lifetime estimation considering dynamics of wind turbine. In Proceedings of the Power Electronics and Machines for Wind and Water Applications IEEE Symposium (PEMWA), Milwaukee, WI, USA, 24–26 July 2014; pp. 1–6.
25. *User Manual of PLECS Blockset*; version 3.7; Plexim GmbH: Zürich, Switzerland, 2015; Available online: http://www.plexim.com/files/plecsmanual.pdf (accessed on 28 April 2016).
26. Akhmatov, V. Variable-speed wind turbines with doubly-fed induction generators. Part I: Modelling in dynamic simulation tools. *J. Wind Eng.* **2002**, *26*, 85–108.
27. Yang, S.; Xiang, D.; Bryant, A.; Mawby, P.; Ran, L.; Tavner, P. Condition Monitoring for device reliability in power electronic converters: A review. *IEEE Trans. Power Electron.* **2010**, *25*, 2734–2752.
28. Perpina, X.; Navarro, L.; Jorda, X.; Vellvehi, M.; Serviere, J.-F.; Mermet-Guyennet, M. Reliability and Lifetime Prediction for IGBT Modules in Railway Traction Chains. In *Reliability and Safety in Railway*; Perpinya, X., Ed.; InTech: Rijeka, Croatia, 2012; pp. 193–222.
29. Fischer, K.; Stalin, T.; Ramberg, H.; Thiringer, T.; Wenske, J.; Karlsson, R. *Investigation of Converter Failure in Wind Turbines*; Elforsk AB: Stockholm, Sweden, 2012.
30. Skuriat, R.; Li, J.F.; Agyakwa, P.A.; Mattey, N.; Evans, P.; Johnson, C.M. Degradation of thermal interface materials for high-temperature power electronics applications. *Microelectron. Reliab.* **2013**, *53*, 1933–1942.

CHAPTER 20

A NEW VOID FRACTION MEASUREMENT METHOD FOR GAS-LIQUID TWO-PHASE FLOW IN SMALL CHANNELS

Huajun Li [1], Haifeng Ji [1], Zhiyao Huang [1], Baoliang Wang [1], Haiqing Li [1] and Guohua Wu [2]

[1]State Key Laboratory of Industrial Control Technology, College of Control Science and Engineering, Zhejiang University, Hangzhou 310027, China

[2]Key Laboratory of Complex Systems Modeling and Simulation of Ministry of Education, Hangzhou Dianzi University, Hangzhou 310018, China

ABSTRACT

Based on a laser diode, a 12 × 6 photodiode array sensor, and machine learning techniques, a new void fraction measurement method for gas-liquid two-phase flow in small channels is proposed. To overcome the influence of flow pattern on the void fraction measurement, the flow pattern of the two-phase flow is firstly identified by Fisher Discriminant Analysis (FDA). Then, according to the identification result, a relevant void fraction measurement model which is developed by Support Vector Machine (SVM) is selected to implement the void fraction measurement. A void fraction measurement system for the two-phase flow is developed and experiments are carried out in four different small channels. Four typical flow patterns (including bubble flow, slug flow, stratified flow and annular flow) are investigated. The experimental results show that the development of the measurement system is successful. The proposed void fraction measurement method is effective and the void fraction measurement accuracy is satisfactory. Compared with the conventional laser measurement systems using standard laser sources, the developed measurement system has the advantages of low cost and simple structure. Compared with the conventional void fraction measurement methods, the proposed method overcomes the influence of flow pattern on the void fraction measurement. This work also provides a good example of using low-cost laser diode as a competent replacement of the expensive standard laser source and hence implementing the parameter measurement of gas-liquid two-phase flow. The research results can be a useful reference for other researchers' works.

INTRODUCTION

In the past decades, the studies and industrial applications of gas-liquid two-phase flow in small-channel systems, such as chemical reactors, heat exchangers, refrigeration processes and micro-evaporators *etc.*, have become a significant area [1,2,3]. Void fraction is an important parameter of the two-phase flow. Its on-line measurement is of great importance for the heat and mass transfer efficiency, the estimation of pressure drop, the process control and the measurement of other parameters of the two-phase flow [4,5,6,7,8,9]. However, with the decrease of the channel dimension, the measurement of the void fraction becomes more and more difficult. The conventional measurement methods cannot fulfill the growing requirements

Citation: Li, H.; Ji, H.; Huang, Z.; Wang, B.; Li, H.; Wu, G. A New Void Fraction Measurement Method for Gas-Liquid Two-Phase Flow in Small Channels. Sensors 2016, 16, 159. DOI: 10.3390/s16020159

of the industrial applications and the mechanism studies of the two-phase flow [1,2,3,4,5,6,7,8,9].

Due to its advantages of high spatial and temporal resolution, the optical measurement technique is a very attractive and useful approach to implement the parameter measurement of gas-liquid two-phase flow in small channels [10,11,12]. The conventional optical measurement methods can be mainly divided into three categories: optical probe method, visualization method, and laser-based method [10,11,12]. Because the optical probe is directly in contact with the detected fluid, the optical probe method will more or less disturb the practical flow of the fluid. In addition, the probes may be contaminated and unpredictable measurement error will arise [13,14,15]. The visualization method includes high-speed cameras, digital cameras, and optical tomography *etc.* [16,17,18]. Although the measurement performance of the visualization method is satisfactory, the visualization method has the disadvantages of high cost, complicated construction and higher requirement of the measurement environment [16,17,18]. Many laser-based methods have been proposed and widely studied, including laser Doppler velocimetry, laser induced fluorescence and particle image velocimetry, *etc.* However, the conventional laser based methods need an expensive laser source (e.g., Nd–Yag laser source or He-Ne laser source.) and a complicated measurement system (including seeding particle, fluorescent particle and objective lens, *etc.*) [19,20,21]. These methods also have the disadvantages of high cost and higher requirements of the measurement environment. Therefore, although significant technical achievements and progresses have been obtained, due to the above mentioned disadvantages, more research works should be undertaken to seek a more effective optical method to implement the parameter measurement of gas-liquid two-phase flow in small channels with the advantages of lower cost, simpler construction and better capability of the environment [10,11,12].

Currently, the techniques of information science and micro-electronics have been rapidly developed. As a new photo-electric device, the performance of photodiode sensor has been significantly improved. The dimension of photodiode sensing element has been greatly decreased and a photodiode array sensor can be successfully developed with much lower cost/price [22,23]. As a new kind of laser source, in some cases, the laser diode can be used as a low-cost alternative of the conventional expensive laser source [24,25]. These technical progresses have laid a solid foundation of developing a low-cost optical measurement system. Meanwhile, as a newly emerging signal processing technology, machine learning, which aims to implement data mining, pattern recognition and modeling, *etc.*, has been widely applied and studied in many research fields. Machine learning technology can provide useful approaches to make full use of the measurement information and hence to implement the parameter measurement successfully [26,27,28,29]. However, up to date, our knowledge and experience on the applications of the above new devices and machine learning technology to the parameter measurement of gas-liquid two-phase flow in small channels are very limited [4,5,6,7,8,9].

Based on a photodiode array sensor and a laser diode, this work aims to develop a low-cost void fraction measurement system and hence to propose a new optical measurement method for the void fraction measurement of gas-liquid two-phase flow in small channels by using machine learning technology.

VOID FRACTION MEASUREMENT SYSTEM AND MEASUREMENT SCHEME

Figure 1a shows the structure of the new void fraction optical measurement system, including a laser diode, an extender lens, a slit, a photodiode array sensor, a data acquisition unit and a microcomputer. The laser diode which is used to produce a beam of laser is YuanDa laser L63510P5 with a wavelength of 635 nm (the wavelength of the laser diode is chosen according to our experience and previous experimental results) and an output power of 5 mW. The extender lens and the slit are used to change the laser into a parallel laser sheet (The extender is used to extend the beam diameter and decrease the laser's divergence. In this work, the resulting laser through the extender lens has a diameter of 40 mm). The laser sheet passes through the transparent channel perpendicularly, and is absorbed, reflected or deflected by the gas-liquid two-phase flow inside the small channel. The exit laser, which contains the characteristic information of the two-phase flow, is recorded by the array sensor. The output signals of the array sensor are then transmitted to the microcomputer by the data acquisition unit. Figure 1b illustrates the layout of the photodiode array sensor. According to the required sensing area, the cost and the size of the sensing element, the array sensor consists of 72 (12 × 6) sensing elements. The outputs of the 72 sensing elements will be sent into the micropucter simultaneously. The sensing element is Vishay Telefunken PIN photodiode BPW34, which has a sensing area of 3.0 × 3.0 mm^2 (in this work, the Signal to Noise Ratio of the sensing element is about 30 dB). Meanwhile, it is necessary to indicate that the number of the sensing elements is determined by our previous experimental results. It is not an optimal number. To look for an optimal element number of the array sensor will be our further research.

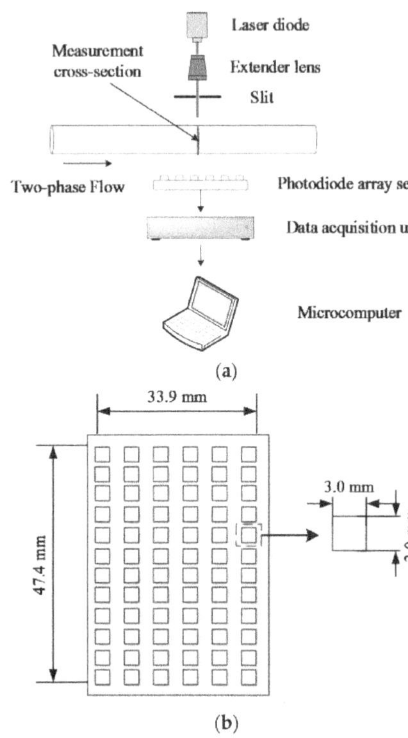

Figure 1. (**a**) The structure of the new void fraction optical measurement system; (**b**) The layout of the photodiode array sensor.

Research works have verified that the flow patterns of gas-liquid two-phase flow have significant influences on the measurement of the void fraction [1,2,3,4,5,6,30]. If we use one measurement model for the void fraction measurment, the measurement accuracies will not be satisfactory. An effective approach to solve this problem is to develop different measurement models for different flow patterns. Thus, the real time flow pattern identification result is necessary and the identification result is introduced to the void fraction measurment. In this work, the real-time flow pattern identification result is introduced into the void fraction measurement process. Meanwhile, for each typical flow pattern, a specific void fraction measurement model is developed. Figure 2 shows the scheme of the void fraction measurement. With the obtained measurement signals (a total of 72 groups of optical signals obtained by the array sensor), a feature vector is firstly extracted. According to the feature vector, the real-time flow pattern identification is then implemented. Finally, according to the flow pattern identification result, a relevant void fraction measurement model is selected to calculate the void fraction.

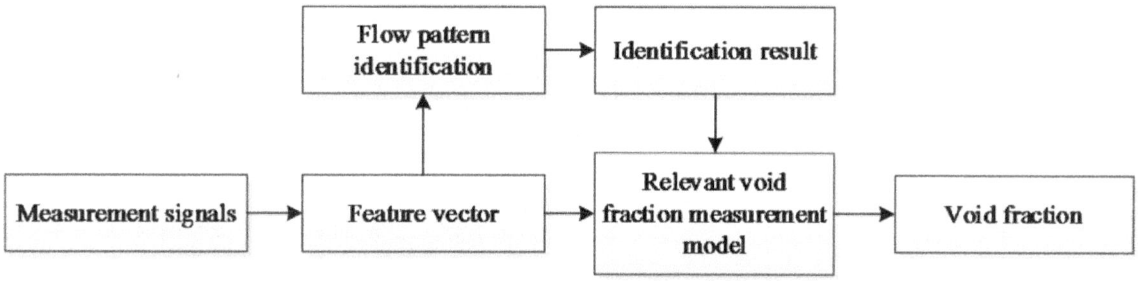

Figure 2. The scheme of the void fraction measurement.

To obtain the comprehensive information of the two-phase flow, the feature vector is extracted from the 72 measurement signals obtained by the photodiode array sensor (each measurement signal is obtained by one sensing element and contains a series of data points). The feature vector consists of two groups of statistical features which contain useful information of the gas-liquid two-phase flow, the mean values and the standard deviations of the 72 measurement signals.

The mean value represents the time-averaged characteristic of a measurement signal. The mean value of the measurement signal obtained by the k_{th} sensing element m_k is

$$m_k = \frac{1}{L}\sum_{i=1}^{L} u_k(i) \quad (1)$$

where L is the data length of the measurement signal, u_k is the measurement signal obtained by the k_{th} sensing element, and $k = 1,2,3,\ldots,72$.

The standard deviation represents the dispersion of a measurement signal. The standard deviation of the measurement signal obtained by the k_{th} sensing element d_k is

$$d_k = \sqrt{\frac{1}{L-1}\sum_{i=1}^{L}(u_k(i) - m_k)^2} \quad (2)$$

Thus, combining the two groups of statistical features, the feature vector x is obtained

$$x = [m_1,\ldots,m_{72},d_1,\ldots,d_{72}]^T \quad (3)$$

The flow pattern identification is a pattern recognition problem. The development of a void fraction measurement model is a modeling problem. As mentioned in Section 1, machine learning technology can provide many useful approaches to solve pattern recognition problems or modeling problems, such as k-nearest neighbor, linear discriminant analysis, and Bayes classifier, *etc.* can implement the pattern recognition [26,27,28,29], while linear regression, artificial neural network and Bayesian learning, *etc.* can implement the modeling [26,27,28,29].

Compared with the other pattern recognition techniques mentioned above, Fisher Discriminant Analysis (FDA) is a dimensionality reduction technique and can provide a linear transformation that maximizes the between-class scatter and minimizes the within-class scatter. FDA has been widely used in the pattern identification field, and its effectiveness has been verified [26,27,28,29,31,32]. Thus, in this work, FDA is adopted to implement the flow pattern identification. Compared with the other modeling techniques mentioned above, Support Vector Machine (SVM) is a valid machine-learning technique and is suitable for model developing in small sample conditions. SVM has good generalization ability and has been widely used in many fields for model development [26,27,28,29,33]. Therefore, in this work, SVM is selected to develop the void fraction measurement models.

FLOW PATTERN IDENTIFICATION

Four typical flow patterns of gas-liquid two-phase flow in small channels are investigated in this work, including bubble flow, slug flow, stratified flow, and annular flow. According to our experimental results, the measurement signals of the bubble flow and the slug flow have some similarities, while the measurement signals of the stratified flow and the annular flow have some similarities. Figure 3 shows typical groups of the measurement signals and the images of the four flow patterns. The measurement signals are obtained by a sensing element (*i.e.*, the sixth BPW34 diode at the fourth column of the photodiode array sensor). The images of the flow patterns are obtained by a high-speed camera (Intergrated Design Tools, Inc. (IDT) Redlake MotionXtra N-4).

As shown in Figure 3, in the bubble flow, when a gas bubble passes through the measurement cross-section, the measurement signal has a clear voltage decrease and stays steady when the channel is full of water. In the slug flow, the measurement signal also has clear voltage decrease when a gas slug approaches and leaves, while at the central part of the gas slug, the measurement signal remains invariable. The measurement signals of the bubble flow and the slug flow have some similarities, but the voltage decrease amplitudes of the signals are different. In the annular flow or the stratified flow, the measurement signals both display fluctuations. However, the amplitude of the measurement signal fluctuation in the annular flow is different from that in the stratified flow. Therefore, based on the above characteristics of the measurement signals in different flow patterns, in this work, the bubble flow and the slug flow are initially classified as one group (Group 1), and the stratified flow and the annular flow are initially classified as the other group (Group 2). Figure 4 shows the flowchart of the flow pattern identification. The process of the flow pattern identification has two key steps. The first step is to determine that the real-time flow pattern belongs to Group 1 or Group 2 by Classifier A. The second step is to determine the specific real-time flow pattern by Classifier B or C.

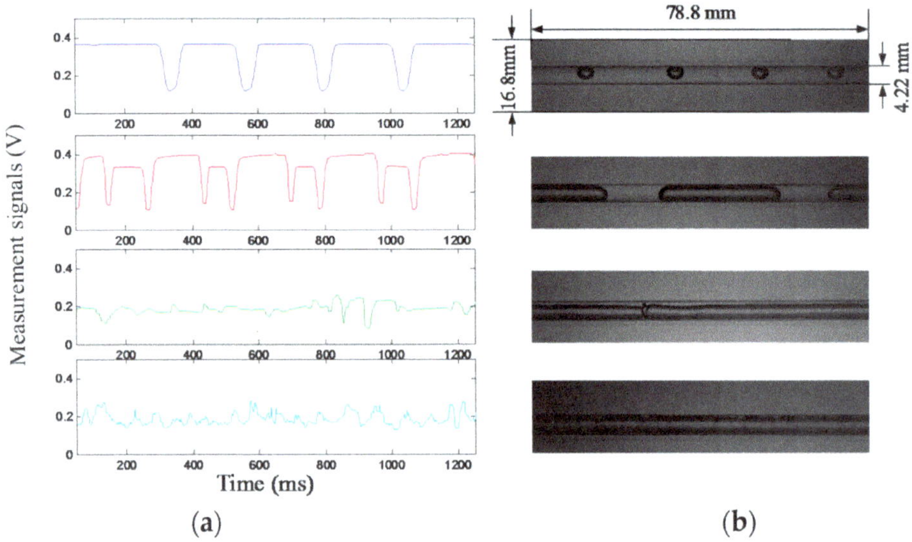

Figure 3. (**a**) Measurement signals of the four flow patterns obtained by one sensing element; (**b**) Images of the four flow patterns obtained by a high-speed camera.

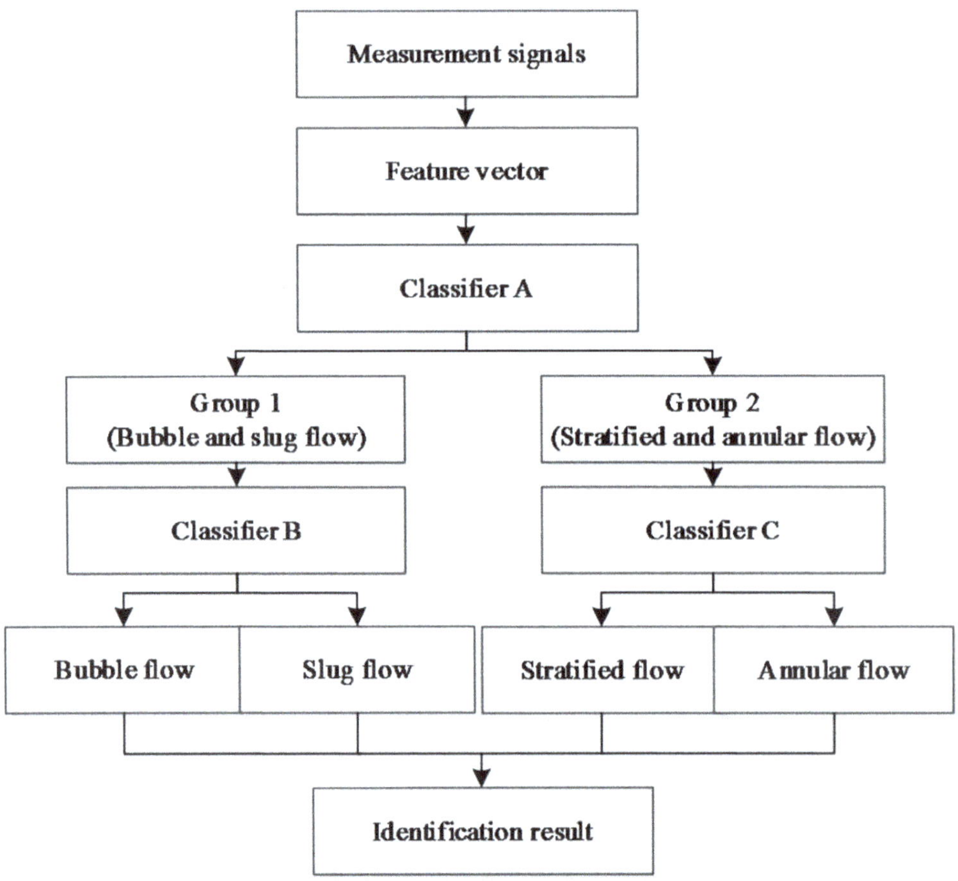

Figure 4. The flowchart of the flow pattern identification.

The three classifiers (Classifier A, B, and C) are developed by FDA and each classifier is aimed to solve a binary classification problem. The decision function of the binary classifier can be expressed as:

$$y(x) = \text{sign}\left[\omega^T x + \omega_0\right] \tag{4}$$

where y is the class label ($y = -1$ or 1), x is the feature vector, ω is the Fisher vector and ω_0 is the threshold. ω can be determined by solving the following problem:

$$J(\omega) = \max \frac{\omega^T S_b \omega}{\omega^T S_w \omega} \tag{5}$$

where S_w is the within-class-scatter matrix and S_b is the between-class-scatter matrix. Once the classifier is developed, by inputting the feature vector x into the decision function, the identification result can be obtained according to the class label$y$. The detailed description concerning FDA is available in [26,27,28,29].

VOID FRACTION MEASUREMENT MODEL DEVELOPMENT

Figure 5 shows the flowchart of the void fraction measurement model development. For each typical flow pattern, a specific void fraction measurement model is developed. According to experimental results, a training data set is constructed. The training data set includes the sample feature vectors extracted from the experimental measurement signals and the reference void fraction values. With the training data set, the void fraction measurement models of the four flow patterns (totally four models) are developed by SVM.

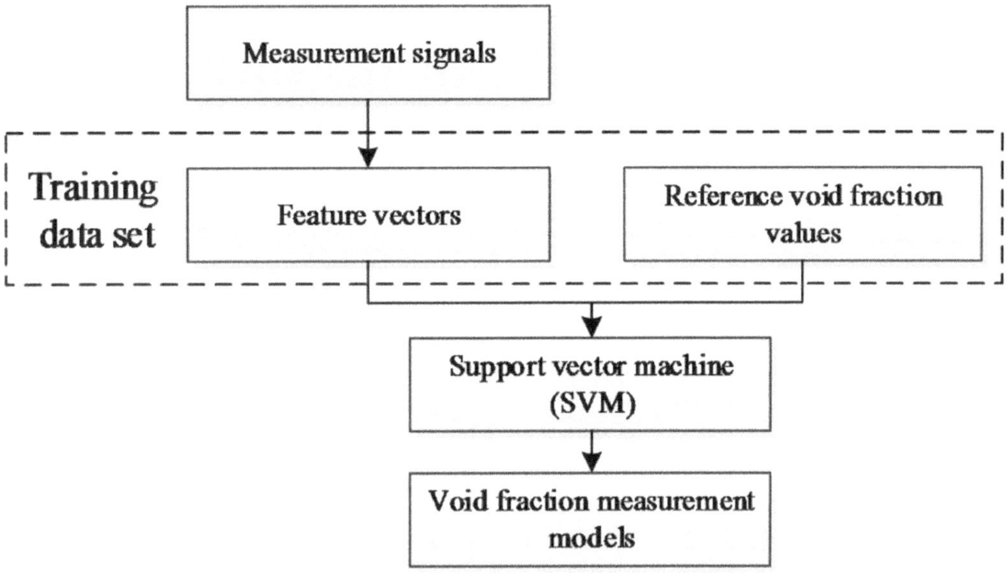

Figure 5. The flowchart of the void fraction measurement model development.

According to the principle of SVM, the void fraction measurement model can be expressed as:

$$\alpha(x) = \sum_{i=1}^{p} (\beta_i - \beta_i^*) K(x, x_i) + b \tag{6}$$

where α is the void fraction, x is the feature vector, and $\{x_i, \alpha_i\}_{i=1}^{p}$ is the training data set with p samples. $K(x, x_i)$ is kernel function. In this work, the radial basis function $K(x, x_i) = \exp(-|x - x_i|^2)/\sigma^2$ is selected as the kernel function, and $\sigma$is its diameter. b is a parameter. β_i and β_i* are the Lagrange multipliers, which are determined by solving the following optimization problem:

$$\begin{cases} min[\frac{1}{2}\sum_{i=1,j=1}^{p} (\beta_i - \beta_i^*)(\beta_j - \beta_j^*)K(x_i, x_j) - \sum_{i=1}^{p} (\beta_i - \beta_i^*)\alpha_i + \sum_{i=1}^{p} (\beta_i + \beta_i^*)\varepsilon] \\ \quad s.t. \begin{cases} 0 \leqslant \beta_i, \beta_i{}^* \leqslant C, i = 1, \ldots, p \\ \sum_{i=1}^{p} (\beta_i - \beta_i^*) = 0, i = 1, \ldots, p \end{cases} \end{cases} \tag{7}$$

where ε is the slack variable and C is the penalty factor. The detailed description concerning the SVM is available in [27,28,29].

PRACTICAL PROCESS OF THE VOID FRACTION MEASUREMENT

The practical process of the void fraction measurement is illustrated in Figure 6. Firstly, the 72 measurement signals of the gas-liquid two-phase flow are obtained by the array sensor, and the feature vector x of the signals is extracted. Secondly, the real-time flow pattern is identified. Then, according to the identification result, a relevant void fraction measurement model is selected. Finally, with the selected measurement model and the obtained feature vector x, the void fraction measurementα is obtained.

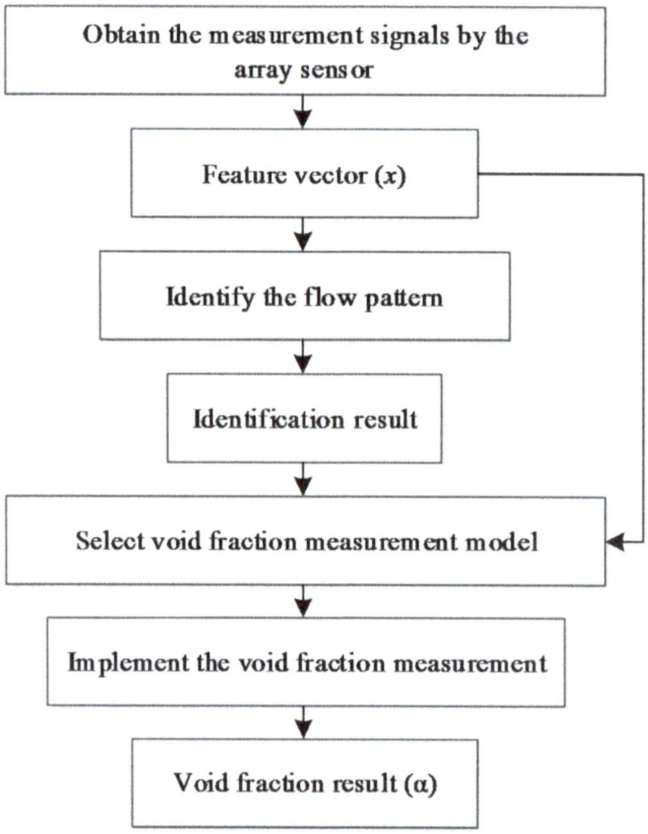

Figure 6. The practical process of the void fraction measurment.

EXPERIMENTAL SETUP

Figure 7 shows the experimental setup of the void fraction measurement system for gas-liquid two-phase flow in small channels. The gas and the liquid phase are driven into the small channel by syringe pumps or a nitrogen tank (if either the gas flowrate or the liquid flowrate is less than 3.6 L/h, the corresponding syringe pump is used; otherwise, the nitrogen tank is used). Nitrogen is used as the gas phase and its flowrate ranges from 0 to 1300 L/h. Tap water is used as the liquid phase, and its flowrate ranges from 0 to 20 L/h. The two phases mix at the mixer, and then the two-phase flow flows through a horizontal channel with a length of 0.95 m. The distance between the channel inlet and the photodiode array sensor is 0.25 m. The optical measurement signals of the two-phase flow are obtained by the photodiode array sensor and then sent to the microcomputer by the data acquisition unit. Meanwhile, an IDT Redlake MotionXtra N-4 high-speed camera (maximum fps (frames per second) @ full resolution: 3000 fps @ 1024 × 1024) is used to obtain the images of the flow patterns.

Four small channels with inner diameters (i.d.) of 4.22, 3.04, 2.16 and 1.08 mm, respectively, are used in the experiments. Four typical flow patterns including bubble flow, slug flow, stratified flow and annular flow are investigated. The sampling frequency of the photodiode array sensor is set to 1 kHz. The National Instruments cDAQ-9172 is selected as the data acquisition unit. The reference data of the void fraction is determined by the quick-closing valve method [4,5,6].

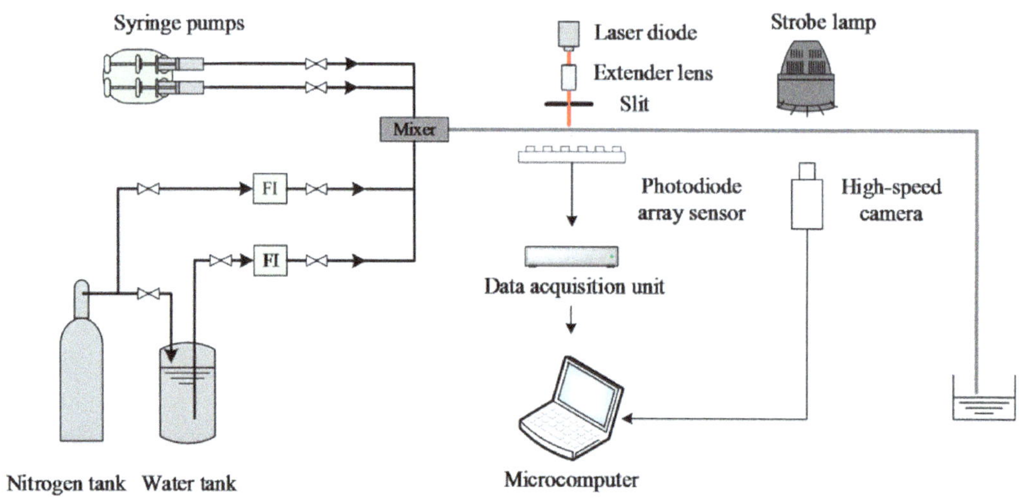

Figure 7. Experimental setup of the void fraction measurement system for gas-liquid two-phase flow in small channels.

EXPERIMENTAL RESULTS AND DISCUSSIONS

Experimental Results

Figure 8, Figure 9, Figure 10 and Figure 11 show the experimental results of the void fraction measurement in the four small channels. Compared with the reference void fractions obtained by the quick-closing valve method, the maximum absolute errors of the void fraction measurement in the four small channels are all less than 7%.

The experimental results indicate that the development of the void fraction measurement system is successful and the proposed void fraction measurement method is effective.

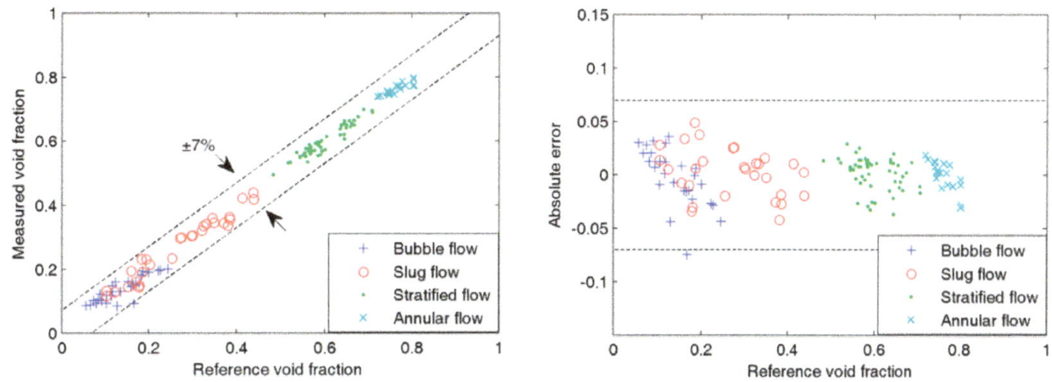

(a) (b)

Figure 8. Experimental results of the void fraction measurement in the 4.22-mm i.d. channel: (**a**) Comparison between measured and reference void fractions; (**b**) Absolute errors of the void fraction measurement.

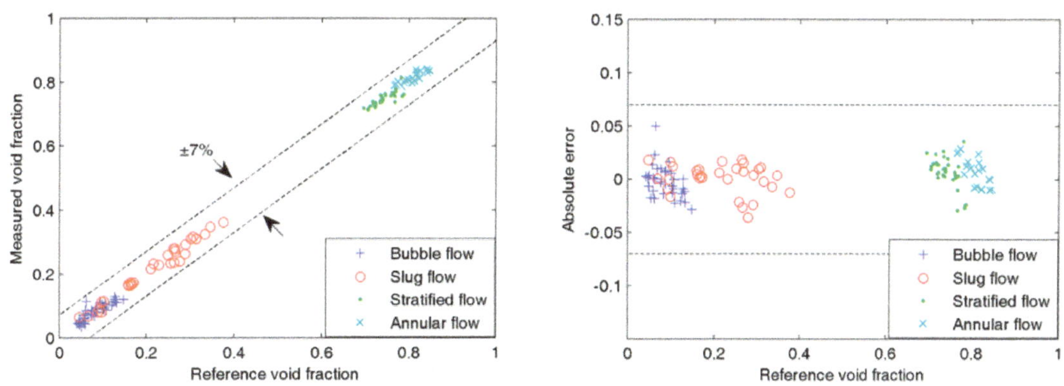

(a) (b)

Figure 9. Experimental results of the void fraction measurement in the 3.04-mm i.d. chanel: (**a**) Comparison between measured and reference void fractions; (**b**) Absolute errors of the void fraction measurement.

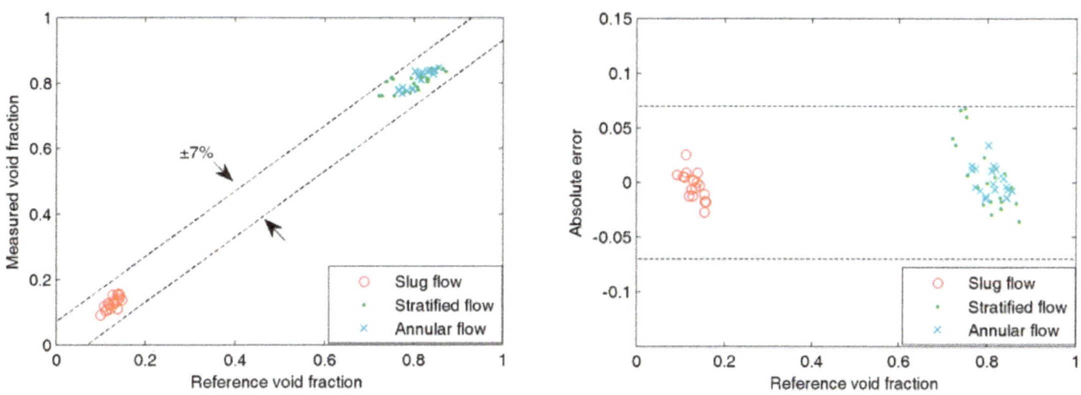

(a) (b)

Figure 10. Experimental results of the void fraction measurement in the 2.16-mm i.d. chanel: (**a**) Comparison between measured and reference void fractions; (**b**) Absolute errors of the void fraction measurement.

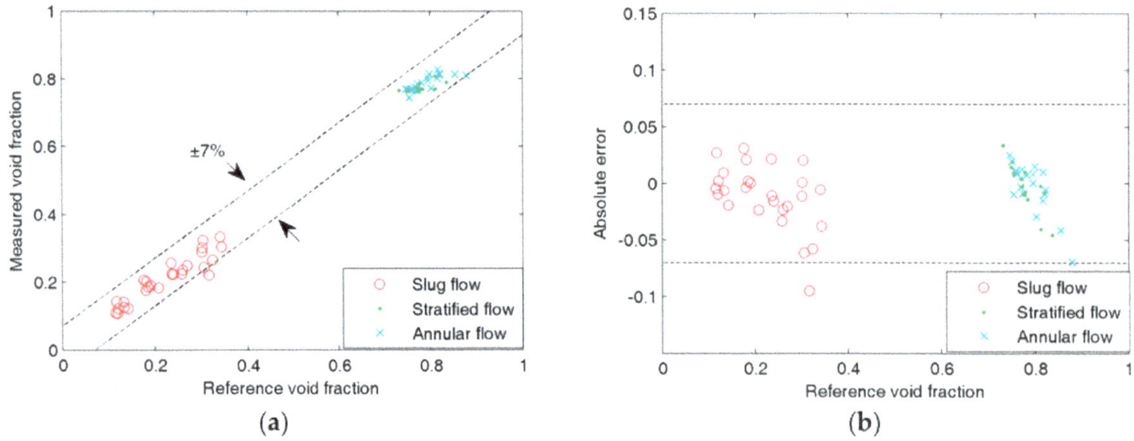

Figure 11. Experimental results of the void fraction measurement in the 1.08-mm i.d. chanel: (**a**) Comparison between measured and reference void fractions; (**b**) Absolute errors of the void fraction measurement.

DISCUSSION

In this work, a photodiode array sensor is used to obtain the signals of the exit laser. With 72 sensing elements, the array sensor has enough sensing area to obtain sufficient signals of the exit laser. Then, comprehensive information of the two-phase flow can be acquired from the obtained signals, and the void fraction measurement can be implemented. Meanwhile, in the proposed measurement method, a low-cost laser diode is used as the laser source. According to the current technique level, the performance indexes (such as output power, laser coherence and divergence, *etc.*) of the conventional standard laser sources (e.g., Nd: YAG laser source, He-Ne laser source, *etc.*) are comprehensively better than that of the laser diode. The only comparable performance index of the laser diode is the power stability (Power stability is the maximum drift with respect to mean power over eight hours. In this work, the laser diode has a power stability of 2%, while the standard laser sources such as THORLABS HNL050L He-Ne laser source have a power stability of 2.5% [34]). The proposed measurement method is mainly on the basis of the power changes and distribution of the exit laser. The power stability is the key performance index of the laser source. Thus, in this work, the advantage of the laser diode in power stability is fully utilized. Furthermore, with the supports of suitable machine learning techniques and the developed photodiode array sensor, the low-cost laser diode successfully acts as a competent replacement of the expensive standard laser source. Sufficient information concerning the characteristic of the two-phase flow is obtained and processed. Finally, the void fraction measurement is implemented.

To implement the void fraction measurement with a satisfactory accuracy, the influence of the flow pattern should be condersided. Compared with the normal scale channel, the flow characteristics of the two-phase flow in small channels have significant differences. As the dimension of the channel decreases, some flow patterns become common and the others are difficult to observe [2,3,4,5]. According to the experimental results, bubble flow is observed in the 4.22-mm and 3.04-mm i.d. channels but not in the 2.16-mm and 1.08-mm i.d. channels, while the slug flow, stratified flow and annular flow are all observed and investigated in the four small channels. These experimental results may provide useful reference for others' research work.

To overcome the influence of the flow pattern on the void fraction measurement, a void fraction measurement method is proposed. In this method, a specific void fraction measurement model is developed for each typical flow pattern. In practical measurement, the parameters of the models vary with the flow patterns, which means that the flow pattern indeed has significant influence on the void fraction measurment. To overcome the influence of the flow pattern, the flow pattern of the two-phase flow is identified at first, and then, according to the identification result, a relevant void fraction measurement model is selected to implement the void fraction measurement. The experimental results show that, with the introduction of the real-time flow pattern identification result, the influence of the flow pattern on the void fraction measurement has been significantly reduced.

CONCLUSIONS

In this work, based on a photodiode array sensor and a laser diode, a low-cost and simple-structure void fraction measurement system for gas-liquid two-phase flow in small channels is developed. A low-cost laser diode is adopted as the laser source and a 12 × 6 photodiode array sensor is used to obtain the information concerning the two-phase flow. Meanwhile, a new void fraction measurement method is proposed. The machine learning techniques (FDA and SVM) are adopted to implement the flow pattern identification and the development of the void fraction models. To overcome the influence of flow pattern on the void fraction measurement, the identification result is introduced to the void fraction measurement. Then, according to the identification result, a relevant void fraction measurement model is selected to implement the void fraction measurement.

Experiments are carried out in four small channels with different inner diameters of 4.22, 3.04, 2.16 and 1.08-mm, respectively. Four typical flow patterns including bubble flow, slug flow, stratified flow and annular flow are investigated. The maximum absolute error of the void fraction measurement is less than 7%. The experimental results show that the proposed void fraction measurement method is effective and the development of the measurement system is successful. The experimental results also show that the introduction of the flow pattern information can overcome the influence of the flow pattern on the void fraction measurement.

The research results also verify that the low-cost laser diode can act as a competent replacement of the expensive standard laser source if suitable signal processing techniques and information acquisition techniques are used. That can significantly reduce the cost of the laser based measurement system. This research work can provide a good reference for other researchers' works.

ACKNOWLEDGMENTS

The research is supported by the National Natural Science Foundation of China under Grant No. 61573312 and No. 11132008. This research is also supported by the Key Laboratory of Complex Systems Modeling and Simulation, Ministry of Education.

AUTHOR CONTRIBUTIONS

Huajun Li and Haifeng Ji conducted the experiments, analyzed the signals, and wrote the manuscript; Zhiyao Huang, Baoliang Wang, Haiqing Li, and Guohua Wu provided suggestions about the work and helped to modify the manuscript. All authors read and approved the final manuscript.

CONFLICTS OF INTEREST

The authors declare no conflict of interest.

REFERENCES

1. Crowe, C.T. *Multiphase Flow Handbook*; CRC Press: Boca Raton, FL, USA, 2005.
2. Kandlikar, S.; Garimella, S.; Li, D.; Colin, S.; King, M.R. *Heat Transfer and Fluid Flow in Minichannels and Microchannels*; Elsevier Press: Oxford, UK, 2005.
3. Cristiano, B.T.; Gherhardt, R. Flow boiling in micro-scale channels–Synthesized literature review. *Int. J. Refrig.* **2013**,*36*, 301–324.
4. Hetsroni, G. *Handbook of Multiphase Systems*; McGraw–Hill Press: New York, NY, USA, 1982.
5. Falcone, G.; Hewitt, G.F.; Alimonti, C. *Multiphase Flow Metering*; Elsevier Press: Oxford, UK, 2010.
6. Hewitt, G.F. *Measurement of Two Phase Flow Parameters*; Academic Press: New York, NY, USA, 1978.
7. Winkler, J.; Killion, J.; Garimella, S.; Fronk, B.M. Void fractions for condensing refrigerant flow in small channels: Part i literature review. *Int. J. Refrig.* **2012**, *35*, 219–245.
8. Winkler, J.; Killion, J.; Garimella, S.; Fronk, B.M. Void fractions for condensing refrigerant flow in small channels: Part ii: Void fraction measurement and modeling. *Int. J. Refrig.* **2012**, *35*, 246–262.
9. Triplett, K.A.; Ghiaasiaan, S.M.; Abdel-Khalik, S.I.; Lemouel, A.; Mccord, B.N. Gas–liquid two-phase flow in microchannels: Part ii: Void fraction and pressure drop. *Int. J. Multiphase Flow.* **1999**, *25*, 395–410.
10. Mayinger, F.; Feldmann, O. *Optical Measurements: Techniques and Applications*, 2nd ed.; Springer Press: Berlin, Germany, 2001.
11. Soid, S.N.; Zainal, Z.A. Spray and combustion characterization for internal combustion engines using optical measuring techniques—A review. *Energy* **2011**, *36*, 724–741.
12. Hodgkinson, J.; Tatam, R.P. Optical gas sensing: A review. *Meas. Sci. Technol.* **2012**, *24*, 111–123.
13. Cartellier, A. Simultaneous void fraction measurement, bubble velocity, and size estimate using a single optical probe in gas–liquid two–phase flows. *Rev. Sci. Instrum.* **1992**, *63*, 5442–5453.
14. Guet, S.; Fortunati, R.; Mudde, R.; Ooms, G. Bubble Velocity and Size Measurement with a Four–Point Optical Fiber Probe. *Part. Part. Syst. Char.* **2003**, *20*, 219–230.
15. Lim, H.J.; Chang, K.A.; Su, C.B.; Chen, C.Y. Bubble velocity, diameter, and void fraction Measurements in a multiphase flow using fiber optic reflectometer. *Rev. Sci. Instrum.* **2008**, *79*, 125105.
16. Chen, W.L.; Twu, M.C.; Pan, C. Gas-liquid two-phase flow in micro-channels. *Int. J. Multiphase Flow* **2002**, *28*, 1235–1247.
17. Charnay, R.; Revellin, R.; Bonjour, J. Flow pattern characterization for R–245fa in minichannels: Optical measurement technique and experimental results. *Int. J. Multiphase Flow* **2013**, *57*, 169–181.
18. Hanafizadeh, P.; Saidi, M.H.; Gheimasi, A.N.; Ghanbarzadeh, S. Experimental investigation of air–water, two-phase flow regimes in vertical mini pipe. *Sci. Iran.* **2011**, *18*, 923–929.
19. Vassallo, P.F.; Trabold, T.A.; Moore, W.E.; Kirouac, G.J. Measurement of velocities in gas-liquid two–phase flow using laser doppler velocimetry. *Exp. Fluids* **1993**, *15*, 227–230.
20. Waelchli, S.; Rohr, P.R.V. Two–phase flow characteristics in gas–liquid microreactors. *Int. J. Multiphase Flow* **2006**, *32*, 791–806.
21. Adrian, R.J. Particle-imaging techniques for experimental fluid-mechanics. *Annu. Rev. Fluid Mech.* **1991**, *23*, 261–304.
22. Revellin, R.; Dupont, V.; Ursenbacher, T.; Thome, J.R.; Zun, I. Characterization of diabatic two-phase flows in microchannels: Flow parameter results for R-134a in a 0.5 mm channel. *Int. J. Multiphase Flow* **2006**, *32*, 755–774.
23. Schleicher, E.; Silva, M.J.D.; Thiele, S.; Li, A.; Wollrab, E.; Hampel, U. Design of an optical tomograph for the investigation of single- and two-phase pipe flows. *Meas. Sci. Technol.* **2008**, *19*, 218–218.
24. Coldren, L.A.; Corzine, S.W. *Diode Lasers and Integrated Circuits*; John Wiley & Sons, Inc. Press: New York, NY, USA, 1995.
25. Nasim, H.; Jamil, Y. Diode lasers: From laboratory to industry. *Opt. Laser Technol.* **2014**, *56*, 211–222.
26. Duda, R.O.; Hart, P.E.; Stork, D.G. *Pattern Classification*, 2nd ed.; John Wiley & Sons, Inc. Press: New York, NY, USA, 2000.

27. Mitchell, T.M.; Carbonell, J.G.; Michalski, R.S. *Machine Learning*; Springer Press: Berlin Heidelberg, Germany, 2011.

28. Rogers, S.; Girolami, M. *A First Course in Machine Learning*; CRC Press: Boca Raton, FL, USA, 2012.

29. Alpaydin, E. *Introduction to Machine Learning*, 2nd ed.; MIT Press: Cambridge, MA, USA, 2004.

30. Cheng, L.; Ribatski, G.; Thome, J.R. Two-Phase Flow Patterns and Flow-Pattern Maps: Fundamentals and Applications. *Appl. Mech. Rev.* **2008**, *61*, 1239–1249.

31. Chiang, L.H.; Kotanchek, M.E.; Kordon, A.K. Fault diagnosis based on Fisher discriminant analysis and support vector machines. *Comput. Chem. Eng.* **2004**, *28*, 1389–1401.

32. Norazwan, M.N.; Mohd, A.H.; Che, R.C.H. Process Monitoring and Fault Detection in Non-Linear Chemical Process Based on Multi-Scale Kernel Fisher Discriminant Analysis. *Comput. Aided Chem. Eng.* **2015**, *37*, 1823–1828.

33. Cristianini, N.; Shawe-Taylor, J. *An Introduction to Support Vector Machines and Other Kernel–Based Learning Methods*; Ai Magazine Press: Menlo Park, CA, USA, 2000.

34. HNL050R Red HeNe Laser Syatem User Guide. Avaible online: http://www.thorlabschina.cn/thorproduct.cfm?partnumber=HNL050L (accessed on 11 January 2016).

INDEX

N

O

P

Q

R

S

T

U

V

W